Rational Function

$$f(x) = \frac{N(x)}{D(x)} = \frac{a_n x^n + \cdots + a_1 x + a_0}{b_m x^m + \cdots + b_1 x + b_0}$$

$N(x)$ and $D(x)$ are polynomials.

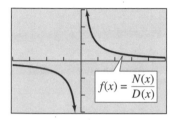

Domain: all real numbers, $D \neq 0$

Range: $(-\infty, 0), (0, \infty)$

y-intercept: $(0, f(x))$, if $f(0)$ exists

x-intercept(s): zeros of N

Vertical asymptotes: zeros of D

Horizontal asymptote:

$\quad y = 0$ if $n < m$

$\quad y = a_n/b_m$ if $n = m$

Exponential Function

$$f(x) = a^x, \; a > 0$$

$$f(x) = a^{-x}, \; a > 0$$

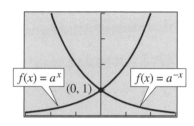

Domain: $(-\infty, \infty)$

Range: $(0, \infty)$

y-intercept: $(0, 1)$

Asymptote: $y = 0$

Logarithmic Function

$$f(x) = \log_a x, \; a > 1$$

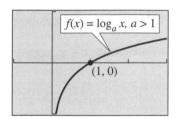

Domain: $(0, \infty)$

Range: $(-\infty, \infty)$

x-intercept: $(1, 0)$

Asymptote: $x = 0$

Sine Function

$$f(x) = \sin x$$

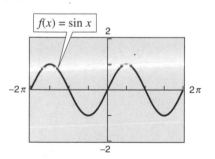

Domain: all real numbers

Range: $[-1, 1]$

Period: 2π

Cosine Function

$$f(x) = \cos x$$

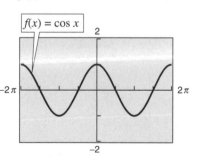

Domain: all real numbers

Range: $[-1, 1]$

Period: 2π

Tangent Function

$$f(x) = \tan x$$

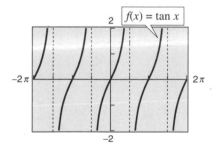

Domain: all $x \neq \dfrac{\pi}{2} + n\pi$

Range: $(-\infty, \infty)$

Period: π

Vertical asymptotes: $x = \dfrac{\pi}{2} + n\pi$

Cosecant Function

$f(x) = \csc x$

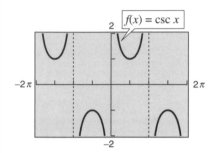

Domain: all $x \neq n\pi$

Range: $(-\infty, -1]$ and $[1, \infty)$
Period: 2π

Vertical asymptotes: $x = n\pi$

Secant Function

$f(x) = \sec x$

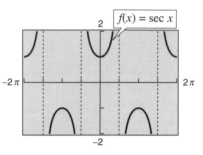

Domain: all $x \neq \dfrac{\pi}{2} + n\pi$

Range: $(-\infty, -1]$ and $[1, \infty)$
Period: 2π

Vertical asymptotes: $x = \dfrac{\pi}{2} + n\pi$

Cotangent Function

$f(x) = \cot x$

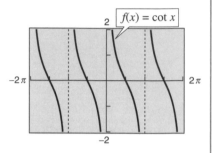

Domain: all $x \neq n\pi$

Range: $(-\infty, \infty)$
Period: π

Vertical asymptotes: $x = n\pi$

Inverse Sine Function

$f(x) = \arcsin x$

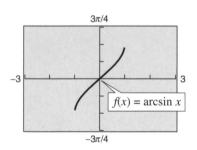

Domain: $[-1, 1]$

Range: $\left[-\dfrac{\pi}{2}, \dfrac{\pi}{2} \right]$

Inverse Cosine Function

$f(x) = \arccos x$

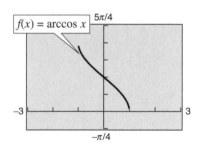

Domain: $[-1, 1]$

Range: $[0, \pi]$

Inverse Tangent Function

$f(x) = \arctan x$

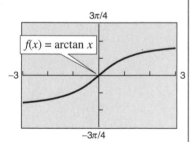

Domain: $(-\infty, \infty)$

Range: $\left(-\dfrac{\pi}{2}, \dfrac{\pi}{2} \right)$

Trigonometry

Trigonometry
A Graphing Approach

Third Edition

Ron Larson
Robert P. Hostetler
The Pennsylvania State University
The Behrend College

Bruce H. Edwards
University of Florida

With the assistance of David C. Falvo
The Pennsylvania State University
The Behrend College

Houghton Mifflin Company   Boston   New York

Editor-in-Chief: Jack Shira
Managing Editor: Cathy Cantin
Senior Associate Editor: Maureen Ross
Associate Editor: Laura Wheel
Assistant Editor: Carolyn Johnson
Supervising Editor: Karen Carter
Project Editor: Patty Bergin
Editorial Assistant: Kate Hartke
Art Supervisor: Gary Crespo
Senior Manufacturing Coordinator: Sally Culler
Marketing Manager: Michael Busnach
Freelance Development Editor: Michael Richards
Freelance Project Editor: Kathleen Deselle
Composition and Art: Meridian Creative Group
Cover Images: Meridian Creative Group
Cover Design: Gary Crespo

We have included examples and exercises that use real-life data as well as technology output from a variety of software. This would not have been possible without the help of many people and organizations. Our wholehearted thanks go to all for their time and effort.

Printed in the U.S.A.

Library of Congress Catalog Card Number: 00-133804

ISBN: 0-618-05293-3

23456789–DOW–04 03 02 01 00

Contents

A Word from the Authors

Welcome to *Trigonometry: A Graphing Approach*, Third Edition. In this revision we have focused on student success, accessibility, and flexibility.

Accessibility: Over the years we have taken care to write a text for the student. We have paid careful attention to the presentation, using precise mathematical language and clear writing to create an effective learning tool. We believe that every student can learn mathematics and we are committed to providing a text that makes the mathematics within it accessible to all students.

In the Third Edition, we have revised and improved upon many text features designed for this purpose. Our pedagogical approach includes presenting solutions to examples from multiple perspectives—algebraic, graphic, and numeric. The side-by-side format allows students to see that a problem can be solved in more than one way, and to compare the accuracy of the solution methods.

Technology has been fully integrated into the text presentation. Also, the *Exploration* and *Study Tip* features have been expanded. *Chapter Tests*, which give students an opportunity for self-assessment, now follow every chapter in the Third Edition. The exercise sets now contain both *Synthesis* exercises, which check students' conceptual understanding, and *Review* exercises, which reinforce skills learned in previous sections and chapters. Students also have access to several media resources—videotapes, *Interactive Trigonometry: A Graphing Approach* CD-ROM, and a *Trigonometry: A Graphing Approach* website— that provide additional text-specific support.

Student Success: During our past 30 years of teaching and writing, we have learned many things about the teaching and learning of mathematics. We have found that students are most successful when they know what they are expected to learn and why it is important to learn it. With that in mind, we have restructured the Third Edition to include a thematic study thread in every chapter.

Each chapter begins with a study guide called *How to Study This Chapter*, which includes a comprehensive overview of the chapter concepts (*The Big Picture*), a list of *Important Vocabulary* that is integral to learning *The Big Picture* concepts, a list of study resources, and a general study tip. The study guide allows students to get organized and prepare for the chapter.

An old pedagogical recipe goes something like this, "First I'm going to tell you what I'm going to teach you, then I will teach it to you, and finally I will go over what I taught you." Following this recipe, we have included a set of learning objectives in every section that outlines what students are expected to learn, followed by an interesting real-life application that illustrates why it is important to learn the concepts in that section. Finally, the chapter summary (*What did you learn?*), which reinforces the section objectives, and the chapter *Review Exercises*, which are correlated to the chapter summary, provide additional study support at the conclusion of each chapter.

Our new *Student Success Organizer* supplement takes this study thread one step further, providing a content-based study aid.

Flexibility: From the time we first began writing in the early 1970s, we have always viewed part of our authoring role as that of providing instructors with flexible teaching programs. The optional features within the text allow instructors with different pedagogical approaches to design their courses to meet both their instructional needs and the needs of their students. In addition, we provide several print and media resources to support instructors, including a new *Instructor Success Organizer*.

We hope you enjoy the Third Edition.

Ron Larson

Robert P. Hostetler

Bruce H. Edwards

Acknowledgments

We would like to thank the many people who have helped us prepare the text and the supplements package. Their encouragement, criticisms, and suggestions have been invaluable to us.

Third Edition Reviewers

Jamie Whitehead Ashby, Texarkana College; Teresa Barton, Western New England College; Diane Burleson, Central Piedmont Community College; Alexander Burstein, University of Rhode Island; Victor M. Cornell, Mesa Community College; Marcia Drost, Texas A & M University; Kenny Fister, Murray State University; Susan C. Fleming, Virginia Highlands Community College; Nicholas E. Geller, Collin County Community College; Betty Givan, Eastern Kentucky University; John Kendall, Shelby State Community College; Donna M. Krawczyk, University of Arizona; JoAnn Lewin, Edison Community College; David E. Meel, Bowling Green University; Beverly Michael, University of Pittsburgh; Jon Odell, Richland Community College; Laura Reger, Milwaukee Area Technical College; Craig M. Steenberg, Lewis-Clark State College; Mary Jane Sterling, Bradley University; Ellen Vilas, York Technical College. In addition, we would like to thank all the college algebra instructors who took the time to respond to our survey.

We would like to extend a special thanks to Ellen Vilas for her contributions to this revision.

We would like to thank the staff of Larson Texts, Inc. and the staff of Meridian Creative Group, who assisted in proofreading the manuscript, preparing and proofreading the art package, and typesetting the supplements.

On a personal level, we are grateful to our wives, Deanna Gilbert Larson, Eloise Hostetler, and Consuelo Edwards for their love, patience, and support. Also, a special thanks goes to R. Scott O'Neil.

If you have suggestions for improving this text, please feel free to write to us. Over the past two decades we have received many useful comments from both instructors and students, and we value these very much.

Ron Larson
Robert P. Hostetler
Bruce H. Edwards

Features Highlights

Student Success Tools

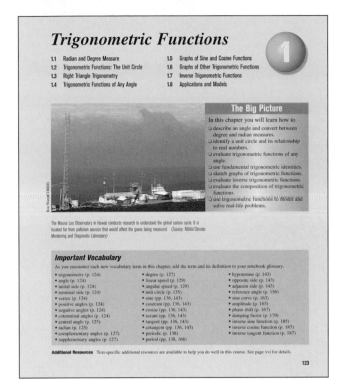

New Chapter Openers include:

▶ **The Big Picture**

An objective-based overview of the main concepts of the chapter.

▶ **Important Vocabulary**

Mathematical terms integral to learning *The Big Picture* concepts.

New Section Openers include:

▶ **"What you should learn"**

Objectives outline the main concepts and help keep students focused on *The Big Picture*.

▶ **"Why you should learn it"**

A real-life application or a reference to other branches of mathematics illustrates the relevance of the section's content.

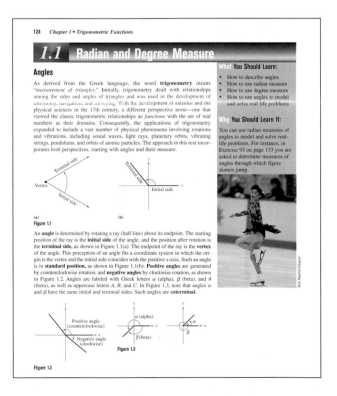

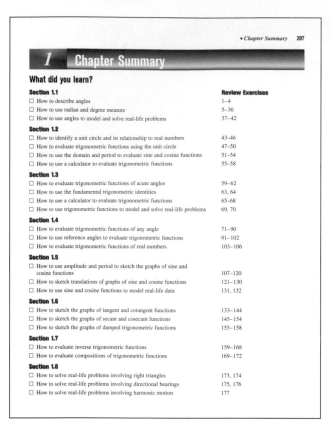

• Chapter Summary **207**

1 Chapter Summary

What did you learn?

Section 1.1	Review Exercises
☐ How to describe angles	1–4
☐ How to use radian and degree measure	5–36
☐ How to use angles to model and solve real-life problems	37–42

Section 1.2	
☐ How to identify a unit circle and its relationship to real numbers	43–46
☐ How to evaluate trigonometric functions using the unit circle	47–50
☐ How to use the domain and period to evaluate sine and cosine functions	51–54
☐ How to use a calculator to evaluate trigonometric functions	55–58

Section 1.3	
☐ How to evaluate trigonometric functions of acute angles	59–62
☐ How to use the fundamental trigonometric identities	63, 64
☐ How to use a calculator to evaluate trigonometric functions	65–68
☐ How to use trigonometric functions to model and solve real-life problems	69, 70

Section 1.4	
☐ How to evaluate trigonometric functions of any angle	71–90
☐ How to use reference angles to evaluate trigonometric functions	91–102
☐ How to evaluate trigonometric functions of real numbers	103–106

Section 1.5	
☐ How to use amplitude and period to sketch the graphs of sine and cosine functions	107–120
☐ How to sketch translations of graphs of sine and cosine functions	121–130
☐ How to use sine and cosine functions to model real-life data	131, 132

Section 1.6	
☐ How to sketch the graphs of tangent and cotangent functions	133–144
☐ How to sketch the graphs of secant and cosecant functions	145–154
☐ How to sketch the graphs of damped trigonometric functions	155–158

Section 1.7	
☐ How to evaluate inverse trigonometric functions	159–168
☐ How to evaluate compositions of trigonometric functions	169–172

Section 1.8	
☐ How to solve real-life problems involving right triangles	173, 174
☐ How to solve real-life problems involving directional bearings	175, 176
☐ How to solve real-life problems involving harmonic motion	177

▶ **"What did you learn?" Chapter Summary**

This chapter summary provides a concise, section-by-section review of the section objectives. These objectives are correlated to the chapter Review Exercises.

Flexibility and Accessibility

▶ **Algebraic, Graphical, and Numerical Approach**

- Many examples present solutions from multiple approaches—algebraic, graphical, and numerical.
- Solutions are displayed side-by-side.
- The multiple-approach format shows students different solution methods that can be used to reach the same answer.
- The solution format helps students expand their problem-solving abilities.

---

382 Chapter 5 • *Exponential and Logarithmic Functions*

Solving Exponential Equations

EXAMPLE 2 Solving Exponential Equations

Solve each equation.   **a.** $e^x = 72$ **b.** $3(2^t) = 42$

Algebraic Solution

a.
$$e^x = 72 \qquad \text{Write original equation.}$$
$$\ln e^x = \ln 72 \qquad \text{Take natural log of each side.}$$
$$x = \ln 72 \qquad \text{Inverse Property}$$
$$x \approx 4.277 \qquad \text{Use a calculator.}$$
The solution is $\ln 72 \approx 4.277$. Check this solution in the original equation.

b.
$$3(2^t) = 42 \qquad \text{Write original equation.}$$
$$2^t = 14 \qquad \text{Divide each side by 3.}$$
$$\log_2 2^t = \log_2 14 \qquad \text{Take log (base 2) of each side.}$$
$$x = \log_2 14 \qquad \text{Inverse Property}$$
$$x = \frac{\ln 14}{\ln 2} \qquad \text{Change-of-base formula}$$
$$x \approx 3.807 \qquad \text{Use a calculator.}$$
The solution is $\log_2 14 \approx 3.807$. Check this solution in the original equation.

Graphical Solution

a. Use a graphing utility to graph the left- and right-hand sides of the equation as $y_1 = e^x$ and $y_2 = 72$ in the same viewing window. Use the *intersect* feature or the *zoom* and *trace* features of the graphing utility to approximate the intersection point, as shown in Figure 5.24. So, the approximate solution is 4.277.

b. Use a graphing utility to graph $y_1 = 3(2^t)$ and $y_2 = 42$ in the same viewing window. Use the *intersect* feature or the *zoom* and *trace* features to approximate the intersection point, as shown in Figure 5.25. So, the approximate solution is 3.807.

Figure 5.24

Figure 5.25

EXAMPLE 3 Solving an Exponential Equation

Solve $4e^{2x} = 5$.

Algebraic Solution
$$4e^{2x} = 5 \qquad \text{Write original equation.}$$
$$e^{2x} = \frac{5}{4} \qquad \text{Divide each side by 4.}$$
$$\ln e^{2x} = \ln \frac{5}{4} \qquad \text{Take logarithm of each side.}$$
$$2x = \ln \frac{5}{4} \qquad \text{Inverse Property}$$
$$x = \frac{1}{2} \ln \frac{5}{4} \qquad \text{Solve for } x.$$
$$x \approx 0.112 \qquad \text{Use a calculator.}$$
The solution is $\frac{1}{2} \ln \frac{5}{4} \approx 0.112$. Check this solution in the original equation.

Graphical Solution

Rather than graph both sides of the equation you are solving as separate graphs, as you did in Example 2, another way to graphically solve the equation is to first rewrite the equation as $4e^{2x} - 5 = 0$, then use a graphing utility to graph $y = 4e^{2x} - 5$. Use the *zero* or *root* feature or the *zoom* and *trace* features of the graphing utility to approximate the value of x for which $y = 0$. From Figure 5.26, you can see that the zero occurs when $x \approx 0.112$. So, the approximate solution is 0.112.

Figure 5.26

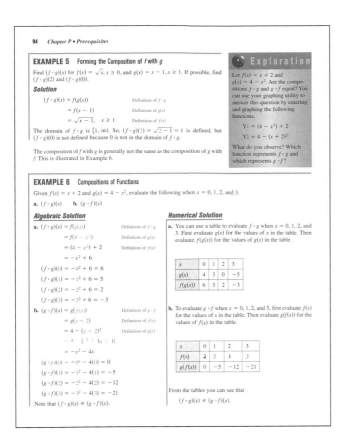

- Before introducing selected topics, *Exploration* engages students in active discovery of mathematical concepts and relationships, often through the power of technology.

- *Exploration* strengthens students' critical thinking skills and helps them develop an intuitive understanding of theoretical concepts.

- *Exploration* is an optional feature and can be omitted without loss of continuity in coverage.

▶ **Examples**

- Each example was carefully chosen to illustrate a particular mathematical concept or problem-solving skill.

- Every example contains step-by-step solutions, most with side-by-side explanations that lead students through the solution process.

- Many examples provide side-by-side solutions utilizing two separate approaches.

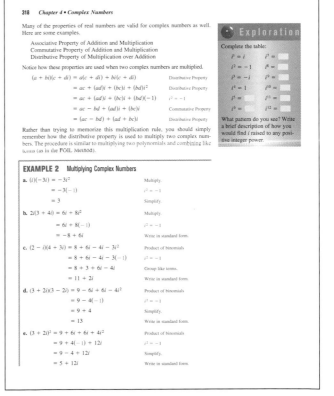

Supplements

Resources

Website (*college.hmco.com*)
Many additional text-specific study and interactive features for students and instructors can be found at the Houghton Mifflin website.

For the Student

Student Success Organizer

Study and Solutions Guide by Bruce H. Edwards (University of Florida)

Graphing Technology Guide by Benjamin N. Levy and Laurel Technical Services

Instructional Videotapes by Dana Mosely

Instructional Videotapes for Graphing Calculators by Dana Mosely

For the Instructor

Instructor's Annotated Edition

Instructor Success Organizer

Complete Solutions Guide by Bruce H. Edwards (University of Florida)

Test Item File

Problem Solving, Modeling, and Data Analysis Labs by Wendy Metzger (Palomar College)

Computerized Testing (Windows, Macintosh)

HMClassPrep Instructor's CD-ROM

An Introduction to Graphing Utilities

Graphing utilities such as graphing calculators and computers with graphing software are very valuable tools for visualizing mathematical principles, verifying solutions to equations, exploring mathematical ideas, and developing mathematical models. Although graphing utilities are extremely helpful in learning mathematics, their use does not mean that learning algebra is any less important. In fact, the combination of knowledge of mathematics and the use of graphing utilities allows you to explore mathematics more easily and to a greater depth. If you are using a graphing utility in this course, it is up to you to learn its capabilities and to practice using this tool to enhance your mathematical learning.

In this text there are many opportunities to use a graphing utility, some of which are described below.

Some Uses of a Graphing Utility

A graphing utility can be used to

- check or validate answers to problems obtained using algebraic methods.
- discover and explore algebraic properties, rules, and concepts.
- graph functions, and approximate solutions to equations involving functions.
- efficiently perform complicated mathematical procedures such as those found in many real-life applications.
- find mathematical models for sets of data.

In this introduction, the features of graphing utilities are discussed from a generic perspective. To learn how to use the features of a specific graphing utility, consult your user's manual or the website for this text found at *college.hmco.com.* Additionally, keystroke guides are available for most graphing utilities, and your college library may have a videotape on how to use your graphing utility.

The Equation Editor

Many graphing utilities are designed to act as "function graphers." In this course, you will study functions and their graphs in detail. You may recall from previous courses that a function can be thought of as a rule that describes the relationship between two variables. These rules are frequently written in terms of x and y. For example, the equation $y = 3x + 5$ represents y as a function of x.

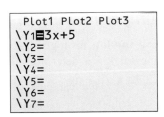

Figure 1

Many graphing utilities have an equation editor that requires an equation to be written in "$y =$" form in order to be entered, as shown in Figure 1. (You should note that your equation editor screen may not look like the screen shown in Figure 1.) To determine exactly how to enter an equation into your graphing utility, consult your user's manual.

The Table Feature

Most graphing utilities are capable of displaying a table of values with *x*-values and one or more corresponding *y*-values. These tables can be used to check solutions of an equation and to generate ordered pairs to assist in graphing an equation.

```
Plot1 Plot2 Plot3
\Y1◼3x/(X+2)■
\Y2=
\Y3=
\Y4=
\Y5=
\Y6=
\Y7=
```
Figure 2

To use the *table* feature, enter an equation into the equation editor in "*y* =" form. The table may have a setup screen, which allows you to select the starting *x*-value and the table step or *x*-increment. You may then have the option of automatically generating values for *x* and *y* or building your own table using the *ask* mode. In the *ask* mode, you enter a value for *x* and the graphing utility displays the *y*-value.

For example, enter the equation

$$y = \frac{3x}{x + 2}$$

into the equation editor, as shown in Figure 2. In the table setup screen, set the table to start at $x = -4$ and set the table step to 1. When you view the table, notice that the first *x*-value is -4 and each value after it increases by 1. Also notice that the Y₁ column gives the resulting *y*-value for each *x*-value, as shown in Figure 3. The table shows that the *y*-value when $x = -2$ is ERROR. This means that the variable *x* may not take on the value -2 in this equation.

X	Y1	
−4	6	
−3	9	
−2	ERROR	
−1	−3	
0	0	
1	1	
2	1.5	

X=−4

Figure 3

With the same equation in the equation editor, set the table to *ask* mode. In this mode you do not need to set the starting *x*-value or the table step, because you are entering any value you choose for *x*. You may enter any real value for *x*—integers, fractions, decimals, irrational numbers, and so forth. If you enter $x = 1 + \sqrt{3}$, the graphing utility may rewrite the number as a decimal approximation, as shown in Figure 4. You can continue to build your own table by entering additional *x*-values in order to generate *y*-values.

X	Y1	
2.7321	1.7321	

X=

Figure 4

If you have several equations in the equation editor, the table may generate *y*-values for each equation.

Creating a Viewing Window

A **viewing window** for a graph is a rectangular portion of the coordinate plane. A viewing window is determined by the following six values.

Xmin = the smallest value of *x*

Xmax = the largest value of *x*

Xscl = the number of units per tick mark on the *x*-axis

Ymin = the smallest value of *y*

Ymax = the largest value of *y*

Yscl = the number of units per tick mark on the *y*-axis

When you enter these six values into a graphing utility, you are setting the viewing window. Some graphing utilities have a standard viewing window, as shown in Figure 5.

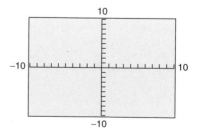

Figure 5

By choosing different viewing windows for a graph, it is possible to obtain very different impressions of the graph's shape. For instance, Figure 6 shows four different viewing windows for the graph of

$$y = 0.1x^4 - x^3 + 2x^2.$$

Of these, the view shown in part (a) is the most complete.

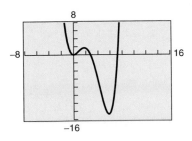

(a)

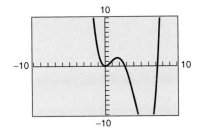

(b)

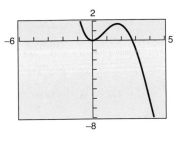

(c)

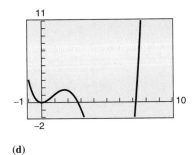

(d)

Figure 6

On most graphing utilities, the display screen is two-thirds as high as it is wide. On such screens, you can obtain a graph with a true geometric perspective by using **a square setting**—one in which

$$\frac{\text{Ymax} - \text{Ymin}}{\text{Xmax} - \text{Xmin}} = \frac{2}{3}.$$

One such setting is shown in Figure 7. Notice that the x and y tick marks are equally spaced on a square setting, but not on a standard setting.

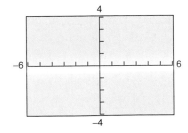

Figure 7

To see how the viewing window affects the geometric perspective, graph the semicircles $y_1 = \sqrt{9 - x^2}$ and $y_2 = -\sqrt{9 - x^2}$ in a standard viewing window. Then graph y_1 and y_2 in a square window. Note the difference in the shapes of the circles.

Zoom and Trace Features

When you graph an equation, you can move from point to point along its graph using the *trace* feature. As you trace the graph, the coordinates of each point are displayed, as shown in Figure 8. The *trace* feature combined with the *zoom* feature allows you to obtain better and better approximations of desired points on a graph. For instance, you can use the *zoom* feature of a graphing utility to approximate the x-intercept(s) of a graph [the point(s) where the graph crosses the x-axis]. Suppose you want to approximate the x-intercept(s) of the graph of $y = 2x^3 - 3x + 2$.

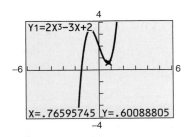

Figure 8

Begin by graphing the equation, as shown in Figure 9(a). From the viewing window shown, the graph appears to have only one *x*-intercept. This intercept lies between −2 and −1. By zooming in on the intercept, you can improve the approximation, as shown in Figure 9(b). To three decimal places, the solution is $x \approx -1.476$.

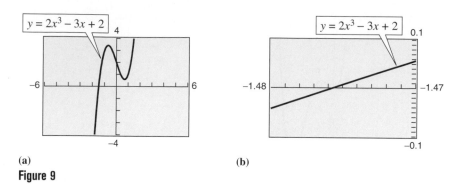

(a)   (b)

Figure 9

Here are some suggestions for using the *zoom* feature.

1. With each successive zoom-in, adjust the *x*-scale so that the viewing window shows at least one tick mark on each side of the *x*-intercept.

2. The error in your approximation will be less than the distance between two scale marks.

3. The *trace* feature can usually be used to add one more decimal place of accuracy without changing the viewing window.

Figure 10(a) shows the graph of $y = x^2 - 5x + 3$. Figures 10(b) and 10(c) show "zoom-in views" of the two *x*-intercepts. From these views, you can approximate the *x*-intercepts to be $x \approx 0.697$ and $x \approx 4.303$.

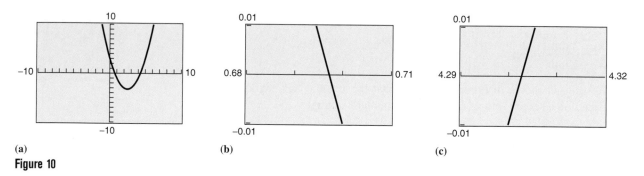

(a)   (b)   (c)

Figure 10

Zero or Root Feature

Using the *zero* or *root* feature, you can find the real zeros of functions of the various types studied in this text—polynomial, exponential, logarithmic, and trigonometric functions. To find the zeros of a function such as $f(x) = \frac{3}{4}x - 2$, first enter the function as $y_1 = \frac{3}{4}x - 2$. Then use the *zero* or *root* feature, which may require entering lower and upper bound estimates of the root, as shown in Figure 11.

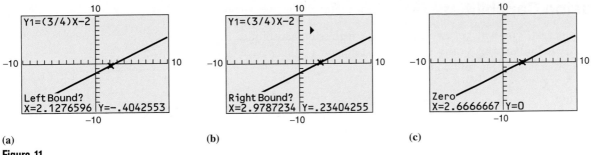

(a)   **(b)**   **(c)**

Figure 11

In Figure 11(c), you can see that the zero is $x = 2.6666667 \approx 2\frac{2}{3}$.

Intersect Feature

To find the points of intersection of two graphs, you can use the *intersect* feature. For instance, to find the points of intersection of the graphs of $y_1 = -x + 2$ and $y_2 = x + 4$, enter these two functions and use the *intersect* feature, as shown in Figure 12.

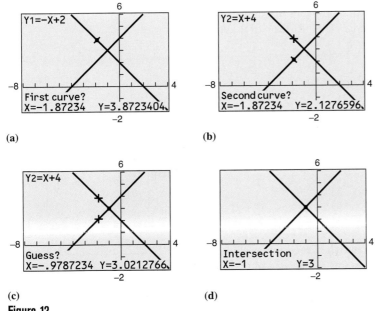

(a)   **(b)**

(c)   **(d)**

Figure 12

From Figure 12(d), you can see that the point of intersection is $(-1, 3)$.

Regression Capabilities

Throughout the text, you will be asked to use the regression capabilities of a graphing utility to find models for sets of data. Most graphing utilities have built-in regression programs for the following.

Regression	*Form of Model*
Linear	$y = ax + b$
Quadratic	$y = ax^2 + bx + c$
Cubic	$y = ax^3 + bx^2 + cx + d$
Quartic	$y = ax^4 + bx^3 + cx^2 + dx + e$
Logarithmic	$y = a + b \ln(x)$
Exponential	$y = ab^x$
Power	$y = ax^b$
Logistic	$y = \dfrac{c}{1 + ae^{-bx}}$
Sine	$y = a \sin(bx + c) + d$

For instance, you can find the linear regression model for the average hourly wages y (in dollars per hour) of production workers in manufacturing industries from 1987 through 1997 shown in the table. (Source: U.S. Bureau of Labor Statistics)

Year	1987	1988	1989	1990	1991	1992
y	9.91	10.19	10.48	10.83	11.18	11.46

Year	1993	1994	1995	1996	1997
y	11.74	12.06	12.37	12.78	13.17

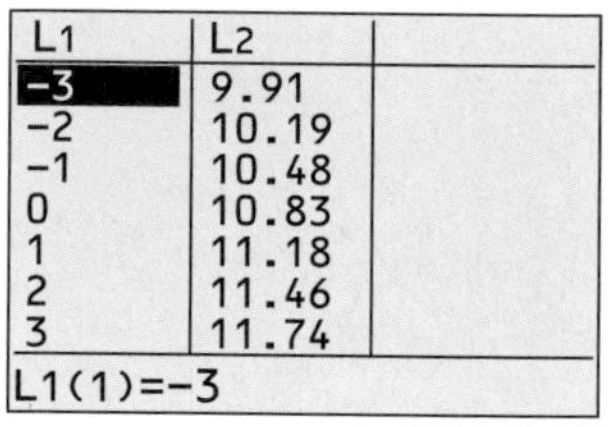

First, let $x = 0$ correspond to 1990 and enter the data into the list editor, as shown in Figure 13. Note that the list in the first column contains the years and the list in the second column contains the hourly wages that correspond to the years. Run your graphing utility's built-in linear regression program to obtain the coefficients a and b for the model $y = ax + b$, as shown in Figure 14. So, a linear model for the data is

$$y \approx 0.321x + 10.83.$$

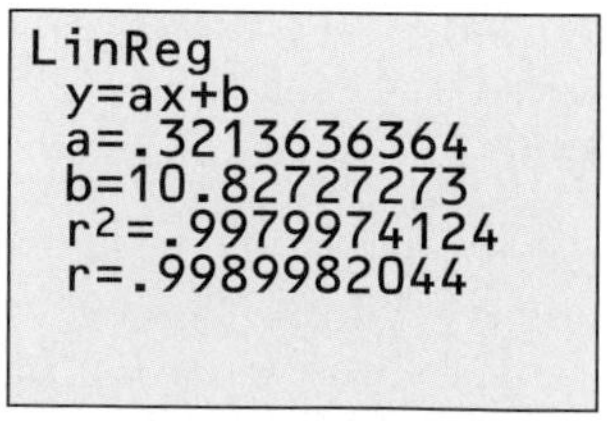

When you run some regression programs, you may obtain an "r-value," which gives a measure of how well the model fits the data. The closer the value of $|r|$ is to 1, the better the fit. For the data in the table above, $r \approx 0.999$, which implies that the model is a very good fit.

```
L1      | L2
-3      | 9.91
-2      | 10.19
-1      | 10.48
0       | 10.83
1       | 11.18
2       | 11.46
3       | 11.74
L1(1)=-3
```

Figure 13

```
LinReg
 y=ax+b
 a=.3213636364
 b=10.82727273
 r2=.9979974124
 r=.9989982044
```

Figure 14

Prerequisites

Superstock

The Big Picture

In this chapter you will learn how to

❑ evaluate algebraic expressions.

❑ plot points, use the Distance and Midpoint Formulas, and sketch graphs of equations.

❑ solve linear, quadratic, polynomial, radical, and absolute value equations.

❑ find and use the slopes of lines to write and graph linear equations.

❑ evaluate functions and find their domains.

❑ analyze graphs of functions and identify transformations, combinations, and compositions of functions.

❑ find inverses of functions.

One and one-half percent of all households own a horse as a pet. Some owners board their horses on farms. (Source: American Veterinary Medical Association)

Important Vocabulary

As you encounter each new vocabulary term in this chapter, add the term and its definition to your notebook glossary.

- real numbers (p. 2)
- real number line (p. 2)
- absolute value (p. 5)
- variables (p. 6)
- algebraic expression (p. 6)
- rectangular coordinate system (p. 12)
- ordered pair (p. 12)
- Distance Formula (p. 14)
- Midpoint Formula (p. 16)
- solution point (p. 17)
- graph of an equation (p. 17)

- intercepts (pp. 17, 28)
- equation (p. 27)
- solution (p. 27)
- extraneous (p. 28)
- point of intersection (p. 32)
- slope (p. 43)
- point-slope form (p. 45)
- function (p. 56)
- domain (p. 56)
- range (p. 56)
- implied domain (p. 60)

- graph of a function (p. 70)
- greatest integer function (p. 75)
- step function (p. 75)
- even function (p. 76)
- odd function (p. 76)
- rigid transformations (p. 87)
- nonrigid transformations (p. 87)
- arithmetic combination (p. 91)
- composition (p. 93)
- inverse function (p. 102)
- one-to-one (p. 106)

Additional Resources Text-specific additional resources are available to help you do well in this course. See page xvi for details.

P.1   Real Numbers

Real Numbers

Real numbers are used in everyday life to describe quantities such as age, miles per gallon, container size, and population. Real numbers are represented by symbols such as

$$-5, 9, 0, \tfrac{4}{3}, 0.666 \ldots, 28.21, \sqrt{2}, \pi, \text{ and } \sqrt[3]{-32}.$$

Here are some important subsets of the set of real numbers.

$$\{1, 2, 3, 4, \ldots\} \qquad \text{Set of natural numbers}$$

$$\{0, 1, 2, 3, 4, \ldots\} \qquad \text{Set of whole numbers}$$

$$\{\ldots -3, -2, -1, 0, 1, 2, 3, \ldots\} \qquad \text{Set of integers}$$

A real number is **rational** if it can be written as the ratio p/q of two integers, where $q \neq 0$. For instance, the numbers

$$\frac{1}{3} = 0.3333 \ldots, \quad \frac{1}{8} = 0.125, \quad \text{and} \quad \frac{125}{111} = 1.126126 \ldots = 1.\overline{126}$$

are rational. The decimal representation of a rational number either *repeats* (as in $\frac{173}{55} = 3.1\overline{45}$) or *terminates* (as in $\frac{1}{2} = 0.5$). A real number that cannot be written as the ratio of two integers is called **irrational.** Irrational numbers have infinite nonrepeating decimal representations. For instance, the numbers

$$\sqrt{2} \approx 1.4142136 \quad \text{and} \quad \pi \approx 3.1415927$$

are irrational. (The symbol $\approx$ means "is approximately equal to.")

Real numbers are represented graphically by a **real number line.** The point 0 on the real number line is the **origin.** Numbers to the right of 0 are positive, and numbers to the left of 0 are negative, as shown in Figure P.1. The term **nonnegative** describes a number that is either positive or zero.

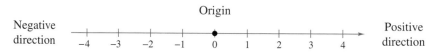

Figure P.1   *The Real Number Line*

There is a *one-to-one correspondence* between real numbers and points on the real number line. That is, every point on the real number line corresponds to exactly one real number called its **coordinate,** and every real number corresponds to exactly one point on the real number line, as shown in Figure P.2.

Every point on the real number line corresponds to exactly one real number.

Figure P.2   *One-to-One Correspondence*

Every real number corresponds to exactly one point on the real number line.

What You Should Learn:

- How to represent and classify real numbers
- How to order real numbers and use inequalities
- How to find the absolute values of real numbers and the distance between two real numbers
- How to evaluate algebraic expressions
- How to use the basic rules and properties of algebra

Why You Should Learn It:

Real numbers are used in every aspect of our daily lives, such as the variance of a budget in Exercises 81–84 on page 10.

Ordering Real Numbers

One important property of real numbers is that they are **ordered.**

Definition of Order on the Real Number Line

If a and b are real numbers, a is **less than** b if $b - a$ is positive. This order is denoted by the **inequality**

$a < b.$

This relationship can also be described by saying that b is **greater than** a and writing $b > a$. The inequality $a \leq b$ means that a is **less than or equal to** b, and the inequality $b \geq a$ means that b is **greater than or equal to** a. The symbols <, >, ≤, and ≥ are **inequality symbols.**

Geometrically, this definition implies that $a < b$ if and only if a lies to the *left* of b on the real number line, as shown in Figure P.3.

Figure P.3 *a < b if and only if a lies to the left of b.*

EXAMPLE 1 Interpreting Inequalities

Describe the subset of real numbers represented by each inequality.

a. $x \leq 2$ **b.** $x > -1$ **c.** $-2 \leq x \leq 3$

Solution

a. The inequality $x \leq 2$ denotes all real numbers less than or equal to 2, as shown in Figure P.4(a).

b. The inequality $x > -1$ denotes all real numbers greater than -1, as shown in Figure P.4(b).

c. The inequality $-2 \leq x < 3$ means that $x \geq -2$ *and* $x < 3$. The "double inequality" denotes all real numbers between -2 and 3, including -2 but not including 3, as shown in Figure P.4(c).

Inequalities can be used to describe subsets of real numbers called **intervals.** In the bounded intervals below, the real numbers a and b are the **endpoints** of each interval.

A computer animation of this example appears in the *Interactive* CD-ROM and *Internet* versions of this text.

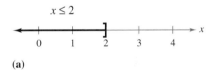

(a)

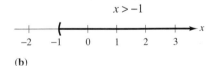

(b)

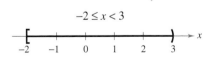

(c)

Figure P.4

Bounded Intervals on the Real Number Line

Notation	Interval Type	Inequality	Graph
$[a, b]$	Closed	$a \leq x \leq b$	
(a, b)	Open	$a < x < b$	
$[a, b)$		$a \leq x < b$	
$(a, b]$		$a < x \leq b$	

The symbols ∞, **positive infinity,** and $-\infty$, **negative infinity,** do not represent real numbers. They are simply convenient symbols used to describe the unbounded-edness of an interval such as $(1, \infty)$ or $(-\infty, 3]$.

Unbounded Intervals on the Real Number Line

Notation	Interval Type	Inequality	Graph
$[a, \infty)$		$x \geq a$	
(a, ∞)	Open	$x > a$	
$(-\infty, b]$		$x \leq b$	
$(-\infty, b)$	Open	$x < b$	
$(-\infty, \infty)$	Entire real line		

EXAMPLE 2   Using Inequalities to Represent Intervals

Use inequality notation to describe each of the following.

a. c is at most 2.

b. All x in the interval $(-3, 5]$

Solution

a. The statement "c is at most 2" can be represented by $c \leq 2$.

b. "All x in the interval $(-3, 5]$" can be represented by $-3 < x \leq 5$.

EXAMPLE 3   Interpreting Intervals

Give a verbal description of each interval.

a. $(-1, 0)$ **b.** $[2, \infty)$ **c.** $(-\infty, 0)$

Solution

a. This interval consists of all real numbers that are greater than -1 and less than 0.

b. This interval consists of all real numbers that are greater than or equal to 2.

c. This interval consists of all negative real numbers.

The **Law of Trichotomy** states that for any two real numbers a and b, precisely one of three relationships is possible:

$$a = b, \quad a < b, \quad \text{or} \quad a > b. \qquad \text{Law of Trichotomy}$$

Absolute Value and Distance

The **absolute value** of a real number is its distance from the origin on the real number line.

If a is a real number, the **absolute value** of a is

$$|a| = \begin{cases} a, & \text{if } a \geq 0 \\ -a, & \text{if } a < 0 \end{cases}.$$

Notice from this definition that the absolute value of a real number is never negative. For instance, if $a = -5$, then $|-5| = -(-5) = 5$. The absolute value of a real number is either positive or zero. Moreover, 0 is the only real number whose absolute value is 0. So, $|0| = 0$.

EXAMPLE 4 Evaluating the Absolute Value of a Number

Evaluate $\dfrac{|x|}{x}$ for (a) $x > 0$ and (b) $x < 0$.

Solution

a. If $x > 0$, then $|x| = x$ and $\dfrac{|x|}{x} = \dfrac{x}{x} = 1$.

b. If $x < 0$, then $|x| = -x$ and $\dfrac{|x|}{x} = \dfrac{-x}{x} = -1$.

Properties of Absolute Value

1. $|a| \geq 0$ **2.** $|-a| = |a|$

3. $|ab| = |a||b|$ **4.** $\left|\dfrac{a}{b}\right| = \dfrac{|a|}{|b|}, \quad b \neq 0$

Absolute value can be used to define the distance between two points on the real number line. For instance, the distance between -3 and 4 is

$$|-3 - 4| = |-7| = 7$$

as shown in Figure P.5.

Distance Between Two Points on the Real Line

Let a and b be real numbers. The **distance between a and b** is

$$d(a, b) = |b - a| = |a - b|.$$

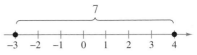

Figure P.5 *The distance between -3 and 4 is 7.*

Algebraic Expressions

One characteristic of algebra is the use of letters to represent numbers. The letters are **variables,** and combinations of letters and numbers are **algebraic expressions.** Here are a few examples of algebraic expressions.

$$5x, \qquad 2x - 3, \qquad \frac{4}{x^2 + 2}, \qquad 7x + y$$

Definition of Algebraic Expression

A collection of letters (**variables**) and real numbers (**constants**) combined using the operations of addition, subtraction, multiplication, division, and exponentiation is an **algebraic expression.**

The **terms** of an algebraic expression are those parts that are separated by *addition.* For example,

$$x^2 - 5x + 8 = x^2 + (-5x) + 8$$

has three terms: x^2 and $-5x$ are the **variable terms** and 8 is the **constant term.** The numerical factor of a variable term is the **coefficient** of the variable term. For instance, the coefficient of $-5x$ is -5, and the coefficient of x^2 is 1.

To **evaluate** an algebraic expression, substitute numerical values for each of the variables in the expression. Here are two examples.

A computer animation of this concept appears in the *Interactive* CD-ROM and *Internet* versions of this text.

Expression	*Value of* *Variable*	*Substitute*	*Value of* *Expression*
$-3x + 5$	$x = 3$	$-3(3) + 5$	$-9 + 5 = -4$
$3x^2 + 2x - 1$	$x = -1$	$3(-1)^2 + 2(-1) - 1$	$3 - 2 - 1 = 0$

When an algebraic expression is evaluated, the **Substitution Principle** is used. It states, "If $a = b$, then a can be replaced by b in any expression involving a." In the first evaluation shown above, for instance, 3 is *substituted* for x in the expression $-3x + 5$.

Basic Rules of Algebra

There are four arithmetic operations with real numbers: *addition, multiplication, subtraction,* and *division,* denoted by the symbols $+$, $\times$ or $\cdot$, $-$, and $\div$ or $/$. Of these, addition and multiplication are the two primary operations. Subtraction and division are the inverse operations of addition and multiplication, respectively.

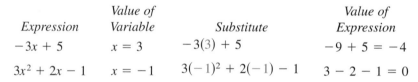

Subtraction: Add the opposite.   *Division: Multiply by the reciprocal.*

$$a - b = a + (-b) \qquad\qquad \text{If } b \neq 0, \text{ then } a/b = a\!\left(\frac{1}{b}\right) = \frac{a}{b}.$$

In these definitions, $-b$ is the **additive inverse** (or opposite) of b, and $1/b$ is the **multiplicative inverse** (or reciprocal) of b. In the fractional form a/b, a is the **numerator** of the fraction and b is the **denominator.**

Because the properties of real numbers below are true for variables and algebraic expressions, as well as for real numbers, they are often called the **Basic Rules of Algebra.** Try to formulate a verbal description of each property. For instance, the first property states that *the order in which two real numbers are added does not affect their sum.*

Basic Rules of Algebra

Let a, b, and c be real numbers, variables, or algebraic expressions.

Property		*Example*
Commutative Property of Addition:	$a + b = b + a$	$4x + x^2 = x^2 + 4x$
Commutative Property of Multiplication:	$ab = ba$	$(4 - x)x^2 = x^2(4 - x)$
Associative Property of Addition:	$(a + b) + c = a + (b + c)$	$(x + 5) + x^2 = x + (5 + x^2)$
Associative Property of Multiplication:	$(ab)c = a(bc)$	$(2x \cdot 3y)(8) = (2x)(3y \cdot 8)$
Distributive Properties:	$a(b + c) = ab + ac$	$3x(5 + 2x) = 3x \cdot 5 + 3x \cdot 2x$
	$(a + b)c = ac + bc$	$(y + 8)y = y \cdot y + 8 \cdot y$
Additive Identity Property:	$a + 0 = a$	$5y^2 + 0 = 5y^2$
Multiplicative Identity Property:	$a \cdot 1 = a$	$(4x^2)(1) = 4x^2$
Additive Inverse Property:	$a + (-a) = 0$	$5x^3 + (-5x^3) = 0$
Multiplicative Inverse Property:	$a \cdot \dfrac{1}{a} = 1, \quad a \neq 0$	$(x^2 + 4)\left(\dfrac{1}{x^2 + 4}\right) = 1$

Because subtraction is defined as "adding the opposite," the Distributive Properties are also true for subtraction. For instance, the "subtraction form" of $a(b + c) = ab + ac$ is

$$a(b - c) = ab - ac.$$

Properties of Negation and Equality

Let a, b, and c be real numbers, variables, or algebraic expressions.

Property	*Example*
1. $(-1)a = -a$	$(-1)7 = -7$
2. $-(-a) = a$	$-(-6) = 6$
3. $(-a)b = -(ab) = a(-b)$	$(-5)3 = -(5 \cdot 3) = 5(-3)$
4. $(-a)(-b) = ab$	$(-2)(-x) = 2x$
5. $-(a + b) = (-a) + (-b)$	$-(x + 8) = (-x) + (-8) = -x - 8$
6. If $a = b$, then $a + c = b + c$.	$\frac{1}{2} + 3 = 0.5 + 3$
7. If $a = b$, then $ac = bc$.	$4^2(2) = 16(2)$
8. If $a + c = b + c$, then $a = b$.	$1.4 - 1 = \frac{7}{5} - 1$
9. If $ac = bc$ and $c \neq 0$, then $a = b$.	$\dfrac{3}{4} = \dfrac{\sqrt{9}}{4}$

STUDY T!P

Be sure you see the difference between the opposite of a number and a negative number. If a is already negative, then its opposite, $-a$, is positive. For instance, if $a = -5$, then $-a = -(-5) = 5$.

Properties of Zero

Let a and b be real numbers, variables, or algebraic expressions.

1. $a + 0 = a$ and $a - 0 = a$ **2.** $a \cdot 0 = 0$

3. $\dfrac{0}{a} = 0, \quad a \neq 0$ **4.** $\dfrac{a}{0}$ is undefined.

5. Zero-Factor Property: If $ab = 0$, then $a = 0$ or $b = 0$.

Properties and Operations of Fractions

Let a, b, c, and d be real numbers, variables, or algebraic expressions such that $b \neq 0$ and $d \neq 0$.

1. Equivalent Fractions: $\dfrac{a}{b} = \dfrac{c}{d}$ if and only if $ad = bc$.

2. Rules of Signs: $-\dfrac{a}{b} = \dfrac{-a}{b} = \dfrac{a}{-b}$ and $\dfrac{-a}{-b} = \dfrac{a}{b}$

3. Generate Equivalent Fractions: $\dfrac{a}{b} = \dfrac{ac}{bc}, \quad c \neq 0$

4. Add or Subtract with Like Denominators: $\dfrac{a}{b} \pm \dfrac{c}{b} = \dfrac{a \pm c}{b}$

5. Add or Subtract with Unlike Denominators: $\dfrac{a}{b} \pm \dfrac{c}{d} = \dfrac{ad \pm bc}{bd}$

6. Multiply Fractions: $\dfrac{a}{b} \cdot \dfrac{c}{d} = \dfrac{ac}{bd}$

7. Divide Fractions: $\dfrac{a}{b} \div \dfrac{c}{d} = \dfrac{a}{b} \cdot \dfrac{d}{c} = \dfrac{ad}{bc}, \quad c \neq 0$

EXAMPLE 5   Properties and Operations of Fractions

a. $\dfrac{x}{5} = \dfrac{3 \cdot x}{3 \cdot 5} = \dfrac{3x}{15}$ Write equivalent fractions.

b. $\dfrac{x}{3} + \dfrac{2x}{5} = \dfrac{5 \cdot x + 3 \cdot 2x}{15} = \dfrac{11x}{15}$ Add fractions with unlike denominators.

c. $\dfrac{7}{x} \div \dfrac{3}{2} = \dfrac{7}{x} \cdot \dfrac{2}{3} = \dfrac{14}{3x}$ Divide fractions.

If a, b, and c are integers such that $ab = c$, then a and b are **factors** or **divisors** of c. A **prime number** is an integer that has exactly two positive factors: itself and 1. For example, 2, 3, 5, 7, and 11 are prime numbers. The numbers 4, 6, 8, 9, and 10 are **composite** because they can be written as the product of two or more prime numbers. The number 1 is neither prime nor composite. The **Fundamental Theorem of Arithmetic** states that every positive integer greater than 1 can be written as the product of prime numbers in precisely one way (disregarding order). For instance, the *prime factorization* of 24 is $24 = 2 \cdot 2 \cdot 2 \cdot 3$.

P.1 **E x e r c i s e s**

In Exercises 1–6, determine which numbers are (a) natural numbers, (b) integers, (c) rational numbers, and (d) irrational numbers.

1. $-9, -\frac{7}{2}, 5, \frac{2}{3}, \sqrt{2}, 0, 1, -4, -1$

2. $\sqrt{5}, -7, -\frac{7}{3}, 0, 3.12, \frac{5}{4}, -2, -8, 3$

3. $2.01, 0.666\ldots, -13, 0.010110111\ldots,$ $1, -10, 20$

4. $2.3030030003\ldots, 0.7575, -4.63, \sqrt{10}, -2,$ $0.03, -10$

5. $-\pi, -\frac{1}{3}, \frac{6}{3}, \frac{1}{2}\sqrt{2}, -7.5, -2, 3, -3$

6. $25, -17, -\frac{12}{5}, \sqrt{9}, 3.12, \frac{1}{2}\pi, 6, -4, 18$

In Exercises 7–10, use a calculator to find the decimal form of the rational number. If it is a nonterminating decimal, write the repeating pattern.

7. $\frac{5}{8}$ **8.** $\frac{1}{3}$

9. $\frac{41}{333}$ **10.** $\frac{6}{11}$

In Exercises 11–16, use a graphing utility to rewrite the rational number as the ratio of two integers.

11. 4.6 **12.** 12.3

13. 6.5 **14.** 3.81

15. -1.83 **16.** -2.490

In Exercises 17 and 18, approximate the numbers and place the correct inequality symbol (< or >) between them.

17.

18.

In Exercises 19–24, plot the two real numbers on the real number line. Then place the correct inequality symbol (< or >) between them.

19. $-4, -8$ **20.** $-3.5, 1$

21. $\frac{3}{2}, 7$ **22.** $1, \frac{16}{3}$

23. $\frac{5}{6}, \frac{2}{3}$ **24.** $-\frac{8}{7}, -\frac{3}{7}$

In Exercises 25–34, (a) verbally describe the subset of real numbers represented by the inequality, (b) sketch the subset on the real number line, and (c) state whether the interval is bounded or unbounded.

25. $x \le 5$ **26.** $x \ge -2$

27. $x < 0$ **28.** $x > 3$

29. $x \ge 4$ **30.** $x < 2$

31. $-2 < x < 2$ **32.** $0 \le x \le 5$

33. $-1 \le x < 0$ **34.** $0 < x \le 6$

In Exercises 35 and 36, use a calculator to order the numbers from least to greatest.

35. $\frac{7071}{5000}, \frac{584}{413}, \sqrt{2}, \frac{47}{33}, \frac{127}{90}$

36. $\frac{26}{15}, \sqrt{3}, 1.732, \frac{381}{220}, \frac{2103}{1214}$

In Exercises 37–46, use inequality and interval notation to describe the set.

37. x is negative.   **38.** z is at least 10.

39. y is nonnegative.   **40.** y is no more than 25.

41. c is at least 12 and at most 32.

42. p is less than 8 but no less than -1.

43. The dog's weight W is more than 45 pounds.

44. The annual rate of inflation r is expected to be at least 2.5%, but no more than 5%.

In Exercises 45 and 46, give a verbal description of the interval.

45. $(-6, \infty)$ **46.** $(-\infty, 4]$

In Exercises 47–56, evaluate the expression.

47. $|-10|$ **48.** $|0|$

49. $|3 - \pi|$ **50.** $|4 - \pi|$

51. $\frac{-5}{|-5|}$ **52.** $-3 - |-3|$

53. $-3|-3|$ **54.** $|-1| - |-2|$

55. $\frac{|x + 2|}{x + 2}$ **56.** $\frac{|x - 1|}{x - 1}$

The *Interactive* CD-ROM and *Internet* versions of this text contain step-by-step solutions to all odd-numbered Section and Review Exercises. They also provide Tutorial Exercises, which link to Guided Examples for additional help.

In Exercises 57–62, place the correct symbol ($<$, $>$, or $=$) between the pair of real numbers.

57. $|-3|$ ▨ $-|-3|$ **58.** $|-4|$ ▨ $|4|$

59. -5 ▨ $-|5|$ **60.** $-|-6|$ ▨ $|-6|$

61. $-|-2|$ ▨ $-|2|$ **62.** $-(-2)$ ▨ -2

In Exercises 63–70, find the distance between a and b.

63.

$a = -1$ .... $b = 3$

```
—●———+———+———+———+———●—→
 -1       0    1    2    3
```

64.

$a = -4$ .... $b = -\frac{3}{2}$

```
—●———+———+———●———+———→
 -4      -3   -2       -1
```

65. $a = 126, b = 75$ **66.** $a = -126, b = -75$

67. $a = -\frac{5}{2}, b = 0$ **68.** $a = \frac{1}{4}, b = \frac{11}{4}$

69. $a = \frac{16}{5}, b = \frac{112}{75}$ **70.** $a = 9.34, b = -5.65$

In Exercises 71 and 72, use the real numbers A, B, and C shown on the number line. Determine the sign of each expression.

```
    C  B            A
—●——●——+—————————●———→
       0
```

71. (a) $-A$ **72.** (a) $-C$

     (b) $B - A$ (b) $A - C$

In Exercises 73–80, use absolute value notation to describe the situation.

73. The distance between x and 5 is no more than 3.

74. The distance between x and -10 is at least 6.

75. y is at least six units from 0.

76. y is at most two units from a.

77. Distance   While traveling, you pass milepost 7, then milepost 18. How far do you travel during that time period?

78. Distance   While traveling, you pass milepost 103, then milepost 86. How far do you travel during that time period?

79. Temperature   The temperature was $50°$ at noon, then $27°$ at midnight. What was the change in temperature over the twelve-hour period?

80. Temperature   The temperature was $38°$ last night at midnight, then $78°$ at noon today. What was the change in temperature over the twelve-hour period?

Budget Variance   In Exercises 81–84, the accounting department of a company is checking to see whether the actual expenses of a department differ from the budgeted expenses by more than \$500 or by more than 5%. Fill in the missing parts of the table, and determine whether the actual expense passes the "budget variance test."

		Budgeted Expense, b	Actual Expense, a	$|a - b|$	$0.05b$
81.	Wages	\$112,700	\$113,356	▨	▨
82.	Utilities	\$9400	\$9772	▨	▨
83.	Taxes	\$37,640	\$37,335	▨	▨
84.	Insurance	\$2575	\$2613	▨	▨

Federal Deficit   In Exercises 85–90, use the bar graph, which shows the receipts of the federal government (in billions of dollars) for selected years from 1960 through 1998. In each exercise you are given the outlay of the federal government. Find the magnitude of the surplus or deficit for the year.   (Source: U.S. Office of Management and Budget)

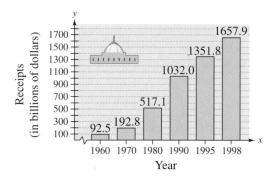

	Receipts, y	Outlay, x	$|y - x|$
85. 1960	▨	\$92.2 billion	▨
86. 1970	▨	\$195.6 billion	▨
87. 1980	▨	\$590.9 billion	▨
88. 1990	▨	\$1253.2 billion	▨
89. 1995	▨	\$1515.7 billion	▨
90. 1998	▨	\$1667.8 billion	▨

In Exercises 91–96, identify the terms. Then identify the coefficients of the variable terms of the expression.

91. $7x + 4$ **92.** $2x - 9$

93. $3x^2 - 8x - 11$ **94.** $7\sqrt{5}x^2 + 3$

95. $4x^3 + \dfrac{x}{2} - 5$ **96.** $3x^4 + 3x^3$

In Exercises 97–102, evaluate the expression for each value of *x*. (If not possible, state the reason.)

Expression		Values
97. $4x - 6$	(a) $x = -1$	(b) $x = 0$
98. $9 - 7x$	(a) $x = -3$	(b) $x = 3$
99. $x^2 - 3x + 4$	(a) $x = -2$	(b) $x = 2$
100. $-x^2 + 5x - 4$	(a) $x = -1$	(b) $x = 1$
101. $\dfrac{x+1}{x-1}$	(a) $x = 1$	(b) $x = -1$
102. $\dfrac{x}{x+2}$	(a) $x = 2$	(b) $x = -2$

In Exercises 103–112, identify the rule(s) of algebra illustrated by the equation.

103. $x + 9 = 9 + x$

104. $2\left(\frac{1}{2}\right) = 1$

105. $\dfrac{1}{h+6}(h + 6) = 1, \quad h \neq -6$

106. $(x + 3) - (x + 3) = 0$

107. $2(x + 3) = 2x + 6$

108. $(z - 2) + 0 = z - 2$

109. $1 \cdot (1 + x) = 1 + x$

110. $x + (y + 10) = (x + y) + 10$

111. $x(3y) = (x \cdot 3)y = (3x)y$

112. $\frac{1}{7}(7 \cdot 12) = \left(\frac{1}{7} \cdot 7\right)12 = 1 \cdot 12 = 12$

In Exercises 113–116, evaluate the expression. (If not possible, state the reason.)

113. $\dfrac{81 - (90 - 9)}{5}$

114. $10(23 - 30 + 7)$

115. $\dfrac{8 - 8}{-9 + (6 + 3)}$

116. $15 - \dfrac{3 - 3}{5}$

In Exercises 117–124, perform the operations. (Write fractional answers in simplest form.)

117. $\frac{3}{16} + \frac{5}{16}$

118. $\frac{6}{7} - \frac{4}{7}$

119. $\frac{5}{8} - \frac{5}{12} + \frac{1}{6}$

120. $\frac{10}{11} + \frac{6}{33} - \frac{13}{66}$

121. $\frac{x}{6} + \frac{3x}{4}$

122. $\frac{11}{x} \div \frac{3}{4}$

123. $12 \div \frac{1}{4}$

124. $\left(\frac{3}{5} \div 3\right) - \left(6 \cdot \frac{4}{8}\right)$

In Exercises 125–128, use a calculator to evaluate the expression. (Round your answer to two decimal places.)

125. $-3 + \frac{3}{7}$

126. $3\left(-\frac{5}{12} + \frac{3}{8}\right)$

127. $\dfrac{11.46 - 5.37}{3.91}$

128. $\dfrac{\frac{1}{5}(-8 - 9)}{-\frac{1}{3}}$

129. (a) Use a calculator to complete the table.

n	1	0.5	0.01	0.0001	0.000001
$5/n$					

(b) Use the result from part (a) to make a conjecture about the value of $\dfrac{5}{n}$ as n approaches 0.

130. (a) Use a calculator to complete the table.

n	1	10	100	10,000	100,000
$5/n$					

(b) Use the result from part (a) to make a conjecture about the value of $\dfrac{5}{n}$ as n increases without bound.

Synthesis

131. *Exploration*   Consider $|u + v|$ and $|u| + |v|$.

(a) Are the values of the expressions always equal? If not, under what conditions are they unequal?

(b) If the two expressions are not equal for certain values of u and v, is one of the expressions always greater than the other? Explain.

132. *Think About It*   Is there a difference between saying that a real number is positive and saying that a real number is nonnegative? Explain.

133. *Writing*   Describe the differences among the sets of natural numbers, integers, rational numbers, and irrational numbers.

True or False?   **In Exercises 134 and 135, determine whether the statement is true or false. Justify your answer.**

134. Let $a > b$, then $\dfrac{1}{a} > \dfrac{1}{b}$, where $a \neq 0$ and $b \neq 0$.

135. Because $\dfrac{a + b}{c} = \dfrac{a}{c} + \dfrac{b}{c}$, then $\dfrac{c}{a + b} = \dfrac{c}{a} + \dfrac{c}{b}$.

P.2   The Cartesian Plane and Graphs of Equations

The Cartesian Plane

Just as you can represent real numbers by points on a real number line, you can represent ordered pairs of real numbers by points in a plane called the **rectangular coordinate system,** or the **Cartesian plane,** after the French mathematician René Descartes (1596–1650).

The Cartesian plane is formed by using two real number lines intersecting at right angles, as shown in Figure P.6. The horizontal real number line is usually called the *x*-**axis,** and the vertical real number line is usually called the *y*-**axis.** The point of intersection of these two axes is the **origin,** and the two axes divide the plane into four parts called **quadrants.**

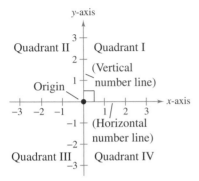

Figure P.6   *The Cartesian Plane*

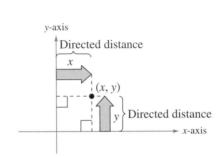

Figure P.7   *Ordered Pair (x, y)*

Each point in the plane corresponds to an **ordered pair** (x, y) of real numbers x and y, called **coordinates** of the point. The *x*-**coordinate** represents the directed distance from the *y*-axis to the point, and the *y*-**coordinate** represents the directed distance from the *x*-axis to the point, as shown in Figure P.7.

Directed distance (x, y) Directed distance
from *y*-axis $\searrow\nearrow$ $\nwarrow\swarrow$ from *x*-axis

The notation (x, y) denotes both a point in the plane and an open interval on the real number line. The context will tell you which meaning is intended.

EXAMPLE 1   Plotting Points in the Cartesian Plane

Plot the points $(-1, 2)$, $(3, 4)$, $(0, 0)$, $(3, 0)$, and $(-2, -3)$.

Solution

To plot the point $(-1, 2)$, imagine a vertical line through -1 on the *x*-axis and a horizontal line through 2 on the *y*-axis. The intersection of these two lines is the point $(-1, 2)$. This point is 1 unit to the left of the *y*-axis and 2 units up from the *x*-axis. The other four points can be plotted in a similar way (see Figure P.8).

What You Should Learn:

- How to plot points in the Cartesian plane
- How to represent data graphically using scatter plots, bar graphs, and line graphs
- How to use the Distance Formula to find the distance between two points
- How to use the Midpoint Formula to find the midpoint of a line segment
- How to sketch graphs of equations by point plotting
- How to sketch graphs of equations using a graphing utility
- How to find equations and sketch graphs of circles
- How to use graphs of equations in real-life problems

Why You Should Learn It:

The graph of an equation can help you see relationships between real-life quantities. For example, Exercise 119 on page 26 shows how a graph can be used to estimate the life expectancies of children born in the years 2002 and 2004.

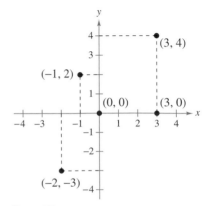

Figure P.8

Representing Data Graphically

The beauty of a rectangular coordinate system is that it allows you to see relationships between two variables. It is difficult to overestimate the importance of Descartes's introduction of coordinates to the plane. Today, his ideas are in common use in virtually every scientific and business-related field.

EXAMPLE 2   Sketching a Scatter Plot, Bar Graph, and Line Graph

From 1988 through 1997, the amount *A* (in millions of dollars) spent on archery equipment in the United States is given in the table, where *t* represents the year. (a) Sketch a scatter plot of the data. (b) Sketch a bar graph of the data. (c) Sketch a line graph of the data.   (Source: National Sporting Goods Association)

t	1988	1989	1990	1991	1992	1993	1994	1995	1996	1997
A	235	261	265	270	334	285	306	287	272	273

Solution

a. To sketch a *scatter plot* of the data given in the table, you simply represent each pair of values by an ordered pair (*t*, *A*) and plot the resulting points, as shown in Figure P.9. For instance, the first pair of values is represented by the ordered pair (1988, 235). Note that the break in the *t*-axis indicates that the numbers between 0 and 1988 have been omitted.

b. To create a *bar graph*, begin by drawing a vertical axis to represent the amount (in millions of dollars) and a horizontal axis to represent the year. Then for each value of *t* in the table, draw a vertical bar whose height corresponds to *A*. The bar graph is shown in Figure P.10.

c. To draw a *line graph*, begin by drawing a vertical axis to represent the amount (in millions of dollars). Then label the horizontal axis with years and plot the points given in the table. Finally, connect the points with line segments, as shown in Figure P.11.

STUDY T!P

You can use a graphing utility to sketch each of the graphs in Example 2. First, enter the data into the graphing utility. You can then use the *statistical plotting* feature to sketch a scatter plot, a bar graph, and a line graph.

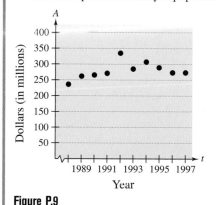

Amount Spent on Archery Equipment

Figure P.9

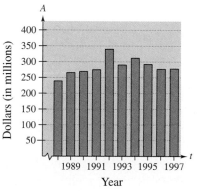

Amount Spent on Archery Equipment

Figure P.10

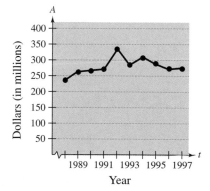

Amount Spent on Archery Equipment

Figure P.11

In Example 2, you could have let *t* = 1 represent the year 1988. In that case, the horizontal axis of each graph would not have been broken, and the tick marks would have been labeled 1 through 10 (instead of 1988 through 1997).

EXAMPLE 3   Interpreting a Population Model

The population (in millions) of North Carolina from 1990 to 1997 can be modeled by $y = 0.11x + 6.63$, where x is the time in years, with $x = 0$ corresponding to 1990. During which year did the population exceed 7 million?

Graphical Solution

You can use the *statistical plotting* feature of a graphing utility to create a bar graph of the population values obtained using the model. On the same viewing window, graph the line $y = 7$.

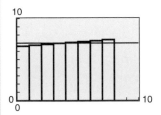

Figure P.12

From the graph in Figure P.12 you can see that the bars that correspond to $x = 0, 1, 2$, and 3 are less than 7. The bar that corresponds to $x = 4$ is the first bar greater than 7, so the population exceeded 7 million sometime during 1993.

Numerical Solution

You can use the *table* feature of a graphing utility to evaluate the model $y = 0.11x + 6.63$ for different years. Enter the model into the equation editor. Then create a table such as the one below.

X	Y1	
0	6.63	
1	6.74	
2	6.85	
3	6.96	
4	7.07	
5	7.18	
6	7.29	
X=3		

Figure P.13

From the table in Figure P.13, you can see that the value of y when $x = 3$ is 6.96, and the value of y when $x = 4$ is 7.07. So, the population exceeded 7 million when x is greater than 3 and less than 4, or sometime during 1993.

The Distance Formula

Recall from the Pythagorean Theorem that for a right triangle with hypotenuse of length c and sides of lengths a and b, you have $a^2 + b^2 = c^2$, as shown in Figure P.14. (The converse is also true. That is, if $a^2 + b^2 = c^2$, the triangle is a right triangle.)

Suppose you want to determine the distance d between two points (x_1, y_1) and (x_2, y_2) in the plane. With these two points, a right triangle can be formed, as shown in Figure P.15. The length of the vertical side of the triangle is $|y_2 - y_1|$, and the length of the horizontal side is $|x_2 - x_1|$. By the Pythagorean Theorem,

$$d^2 = |x_2 - x_1|^2 + |y_2 - y_1|^2$$
$$d = \sqrt{|x_2 - x_1|^2 + |y_2 - y_1|^2}$$
$$d = \sqrt{(x_2 - x_1)^2 + (y_2 - y_1)^2}.$$

This result is the **Distance Formula.**

Figure P.14

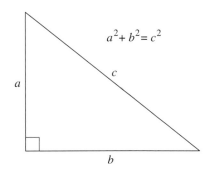

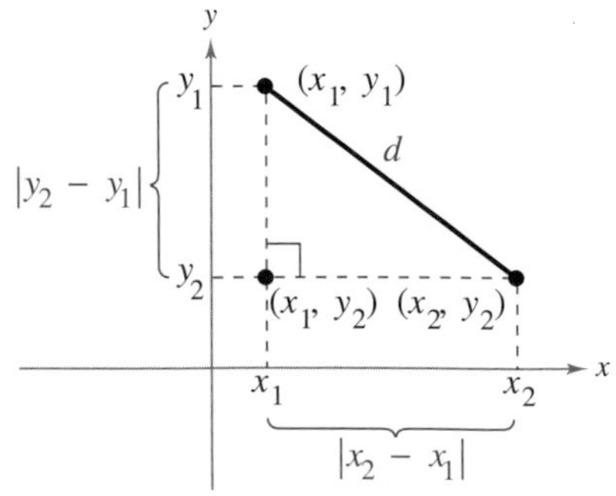

Figure P.15

The Distance Formula

The distance d between the points (x_1, y_1) and (x_2, y_2) in the plane is

$$d = \sqrt{(x_2 - x_1)^2 + (y_2 - y_1)^2}.$$

EXAMPLE 4   Finding a Distance

Find the distance between the points $(-2, 1)$ and $(3, 4)$.

Algebraic Solution

Let $(x_1, y_1) = (-2, 1)$ and $(x_2, y_2) = (3, 4)$. Then apply the Distance Formula as follows.

$$d = \sqrt{(x_2 - x_1)^2 + (y_2 - y_1)^2} \qquad \text{Distance Formula}$$

$$= \sqrt{[3 - (-2)]^2 + (4 - 1)^2} \qquad \begin{array}{l}\text{Substitute for}\\ x_1, y_1, x_2, \text{ and } y_2.\end{array}$$

$$= \sqrt{(5)^2 + (3)^2} \qquad \text{Simplify.}$$

$$= \sqrt{34}$$

$$\approx 5.83 \qquad \text{Use a calculator.}$$

So, the distance between the points is about 5.83 units.

You can use the Pythagorean Theorem to check that the distance is correct.

$$d^2 \stackrel{?}{=} 3^2 + 5^2 \qquad \text{Pythagorean Theorem}$$

$$\left(\sqrt{34}\right)^2 \stackrel{?}{=} 3^2 + 5^2 \qquad \text{Substitute for } d.$$

$$34 = 34 \qquad \text{Distance checks. } ✓$$

Graphical Solution

Use centimeter graph paper to plot the points A and B. Carefully sketch the line segment from A to B. Then use a centimeter ruler to measure the length of the segment.

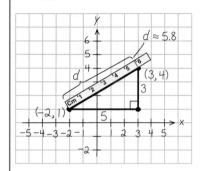

Figure P.16

The line segment measures about 5.8 centimeters, as shown in Figure P.16. So, the distance between the points is about 5.8 units.

EXAMPLE 5   Verifying a Right Triangle

Show that the points $(2, 1)$, $(4, 0)$, and $(5, 7)$ are vertices of a right triangle.

Solution

The three points are plotted in Figure P.17. Using the Distance Formula, you can find the lengths of the three sides as follows.

$$d_1 = \sqrt{(5 - 2)^2 + (7 - 1)^2} = \sqrt{9 + 36} = \sqrt{45}$$

$$d_2 = \sqrt{(4 - 2)^2 + (0 - 1)^2} = \sqrt{4 + 1} = \sqrt{5}$$

$$d_3 = \sqrt{(5 - 4)^2 + (7 - 0)^2} = \sqrt{1 + 49} = \sqrt{50}$$

Because

$$d_1{}^2 + d_2{}^2 = 45 + 5 = 50 = d_3{}^2$$

you can conclude that the triangle must be a right triangle.

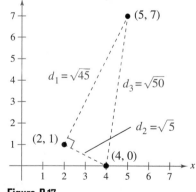

Figure P.17

The figure provided with Example 5 was not really essential to the solution. Nevertheless, it is strongly recommended that you develop the habit of including sketches with your solutions—even if they are not required.

The Midpoint Formula

To find the **midpoint** of the line segment that joins two points in a coordinate plane, you can simply find the average values of the respective coordinates of the two endpoints using the **Midpoint Formula.** (See Appendix A for a proof of the Midpoint Formula.)

> ### The Midpoint Formula
>
> The midpoint of the segment joining the points (x_1, y_1) and (x_2, y_2) is given by the Midpoint Formula
>
> $$\text{Midpoint} = \left(\frac{x_1 + x_2}{2}, \frac{y_1 + y_2}{2}\right).$$

EXAMPLE 6   Finding a Line Segment's Midpoint

Find the midpoint of the line segment joining the points $(-5, -3)$ and $(9, 3)$, as shown in Figure P.18.

Solution
Let $(x_1, y_1) = (-5, -3)$ and $(x_2, y_2) = (9, 3)$.

$$\text{Midpoint} = \left(\frac{x_1 + x_2}{2}, \frac{y_1 + y_2}{2}\right) \qquad \text{Midpoint Formula}$$

$$= \left(\frac{-5 + 9}{2}, \frac{-3 + 3}{2}\right) \qquad \text{Substitute for } x_1, y_1, x_2, \text{ and } y_2.$$

$$= (2, 0) \qquad \text{Simplify.}$$

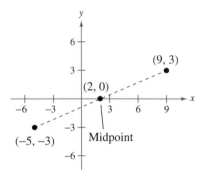

Figure P.18

EXAMPLE 7   Estimating Annual Sales

Winn-Dixie Stores had annual sales of $1.30 billion in 1996 and $1.36 billion in 1998. Without knowing any additional information, what would you estimate the 1997 sales to have been?   (Source: Winn-Dixie Stores, Inc.)

Solution
One solution to the problem is to assume that sales followed a linear pattern. With this assumption, you can estimate the 1997 sales by finding the midpoint of the segment connecting the points (1996, 1.30) and (1998, 1.36).

$$\text{Midpoint} = \left(\frac{1996 + 1998}{2}, \frac{1.30 + 1.36}{2}\right)$$

$$= (1997, 1.33)$$

So, you would estimate the 1997 sales to have been about $1.33 billion, as shown in Figure P.19. (The actual 1997 sales were $1.32 billion.)

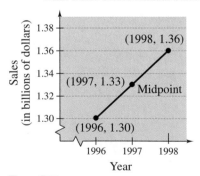

Figure P.19

The Graph of an Equation

News magazines often show graphs comparing the rate of inflation, the federal deficit, wholesale prices, or the unemployment rate to the time of year. Industrial firms and businesses use graphs to report their monthly production and sales statistics. Such graphs provide geometric pictures of the way one quantity changes with respect to another. Frequently, the relationship between two quantities is expressed as an equation. This section introduces the basic procedure for determining the geometric picture associated with an equation.

For an equation in variables x and y, a point (a, b) is a **solution point** if the substitution of $x = a$ and $y = b$ satisfies the equation. Most equations have *infinitely* many solution points. For example, the equation

$$3x + y = 5$$

has solution points $(0, 5)$, $(1, 2)$, $(2, -1)$, $(3, -4)$, and so on. The set of all solution points of an equation is the **graph of the equation.**

A computer animation of this example appears in the *Interactive* CD-ROM and *Internet* versions of this text.

EXAMPLE 8   Sketching a Graph by Point Plotting

Use point plotting and graph paper to sketch the graph of

$$y = x^2 - 2.$$

Solution

First, make a table of values by choosing several convenient values of x and calculating the corresponding values of y.

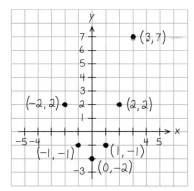

(a)

x	-2	-1	0	1	2	3
$y = x^2 - 2$	2	-1	-2	-1	2	7

Next, plot the corresponding solution points, as shown in Figure P.20(a). Finally, connect the points with a smooth curve, as shown in Figure P.20(b). This graph is called a *parabola*.

The points at which a graph touches or crosses an axis are the **intercepts** of the graph. For instance, in Example 8, the point $(0, -2)$ is the y-intercept of the graph because the graph crosses the y-axis at that point. The points $\left(0, -\sqrt{2}\right)$ and $\left(0, \sqrt{2}\right)$ are the x-intercepts of the graph because the graph crosses the x-axis at these points.

In this text, you will study two basic ways to create graphs: *by hand* and *using a graphing utility.* For instance, the graph in Figure P.20 was sketched by hand, and the graph in Figure P.22 was sketched using a graphing utility.

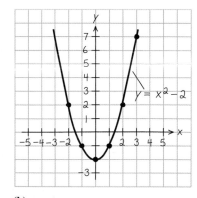

(b)
Figure P.20

Using a Graphing Utility

One of the disadvantages of the point-plotting method is that to get a good idea about the shape of a graph you need to plot *many* points. With only a few points, you could badly misrepresent the graph. For instance, consider the equation

$$y = \frac{1}{30}x(x^4 - 10x^2 + 39).$$

Suppose you plotted only five points: $(-3, -3), (-1, -1), (0, 0), (1, 1),$ and $(3, 3)$, as shown in Figure P.21(a). From these five points, you might assume that the graph of the equation is a straight line. That, however, is not correct. By plotting several more points, you can see that the actual graph is not straight at all, as shown in Figure P.21(b).

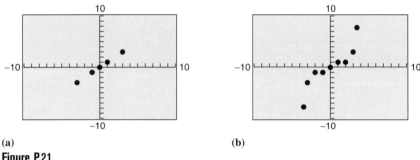

(a)   (b)

Figure P.21

From this, you can see that the point-plotting method leaves you with a dilemma. On the one hand, the method can be very inaccurate if only a few points are plotted. But on the other hand, it is very time-consuming to plot a dozen (or more) points. Technology can help solve this dilemma. Plotting several (even several hundred) points on a rectangular coordinate system is something that a computer or calculator can do easily.

The point-plotting method is the method used by *all* graphing utilities. Each computer or calculator screen is made up of a grid of hundreds or thousands of small areas called *pixels*. Screens that have many pixels per square inch are said to have a higher *resolution* than screens with fewer pixels.

Using a Graphing Utility to Graph an Equation

To graph an equation involving *x* and *y* on a graphing utility, use the following procedure.

1. Rewrite the equation so that *y* is isolated on the left side.
2. Enter the equation into a graphing utility.
3. Determine a *viewing window* that shows all important features of the graph.
4. Graph the equation.

EXAMPLE 9   Using a Graphing Utility to Graph an Equation

Use a graphing utility to graph $2y + x^3 = 4x$.

Solution

To begin, solve the equation for y in terms of x.

$$2y + x^3 = 4x \qquad \text{Write original equation.}$$

$$2y = -x^3 + 4x \qquad \text{Subtract } x^3 \text{ from each side.}$$

$$y = -\frac{1}{2}x^3 + 2x \qquad \text{Divide each side by 2.}$$

Now, by entering this equation into a graphing utility (using a standard viewing window), you can obtain the graph shown in Figure P.22.

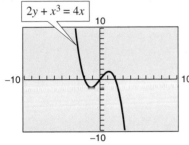

$2y + x^3 = 4x$

Figure P.22

By choosing different viewing windows for a graph, it is possible to obtain very different impressions of the graph's shape. For instance, Figure P.23 shows four different viewing windows for the graph of the equation in Example 9. None of these views shows *all* of the important features of the graph, as Figure P.22 does.

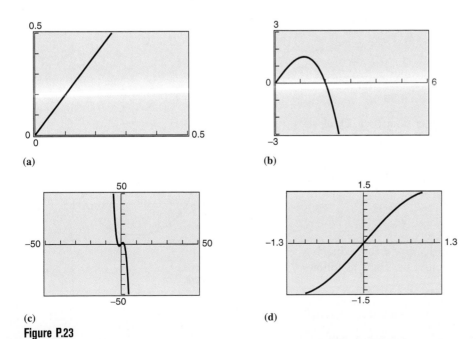

(a)

(b)

(c)

(d)

Figure P.23

STUDY T!P

Many graphing utilities are capable of creating a table, such as the following table, which shows some points of the graph in Figure P.22.

X	Y1	
-3	7.5	
-2	0	
-1	-1.5	
0	0	
1	1.5	
2	0	
3	-7.5	

X=-3

The Equation of a Circle

The Distance Formula provides a convenient way to define circles. A **circle of radius r** with center at the point (h, k) is shown in Figure P.24. The point (x, y) is on this circle if and only if its distance from the center (h, k) is r. This means that a **circle** in the plane consists of all points (x, y) that are a given positive distance r from a fixed point (h, k). Using the Distance Formula, you can express this relationship by saying that the point (x, y) lies on the circle if and only if

$$\sqrt{(x - h)^2 + (y - k)^2} = r.$$

By squaring both sides of this equation, you can obtain the **standard form of the equation of a circle.**

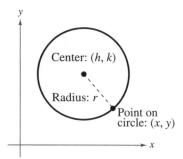

Figure P.24

Standard Form of the Equation of a Circle

The **standard form of the equation of a circle** is

$$(x - h)^2 + (y - k)^2 = r^2.$$

The point (h, k) is the **center** of the circle, and the positive number r is the **radius** of the circle. The standard form of the equation of a circle whose center is the origin is $x^2 + y^2 = r^2$.

EXAMPLE 10   Finding an Equation of a Circle

The point $(3, 4)$ lies on a circle whose center is at $(-1, 2)$, as shown in Figure P.25. Find an equation for the circle.

Solution
The radius r of the circle is the distance between $(-1, 2)$ and $(3, 4)$.

$$r = \sqrt{[3 - (-1)]^2 + (4 - 2)^2}$$
$$= \sqrt{16 + 4} = \sqrt{20}$$

So, the center of the circle is $(h, k) = (-1, 2)$, and the radius is $r = \sqrt{20}$. You can write the standard form of the equation of the circle as follows.

$$(x - h)^2 + (y - k)^2 = r^2 \qquad \text{Standard form}$$
$$[x - (-1)]^2 + (y - 2)^2 = \left(\sqrt{20}\right)^2 \qquad \text{Substitute for } h, k, \text{ and } r.$$
$$(x + 1)^2 + (y - 2)^2 = 20 \qquad \text{Equation of circle}$$

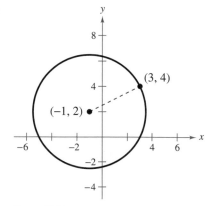

Figure P.25

The standard viewing window on many graphing utilities does not give a true geometric perspective. That is, perpendicular lines will not appear to be perpendicular and circles will not appear to be circular. To overcome this, you can use a *square setting*, as demonstrated in Example 11.

EXAMPLE 11   Sketching a Circle with a Graphing Utility

Use a graphing utility to graph

$$x^2 + y^2 = 9.$$

Solution

The graph of $x^2 + y^2 = 9$ is a circle whose center is the origin and whose radius is 3. To graph the equation, begin by solving the equation for y.

$x^2 + y^2 = 9$	Write original equation.
$y^2 = 9 - x^2$	Subtract x^2 from each side.
$y = \pm\sqrt{9 - x^2}$	Take square root of each side.

The graph of

$$y = \sqrt{9 - x^2} \qquad \text{Upper semicircle}$$

is the upper semicircle. The graph of

$$y = -\sqrt{9 - x^2} \qquad \text{Lower semicircle}$$

is the lower semicircle. Enter *both* equations in your graphing utility and generate the resulting graphs. In Figure P.26(a), note that if you use a standard viewing window, the two graphs do not appear to form a circle. You can overcome this problem by using a *square setting*, in which the horizontal and vertical tick marks have equal spacing, as shown in Figure P.26(b). On many graphing utilities, a square setting can be obtained by using a y to x ratio of 2 to 3. For instance, in Figure P.26(b), the y to x ratio is

$$\frac{Y_{max} - Y_{min}}{X_{max} - X_{min}} = \frac{4 - (-4)}{6 - (-6)} = \frac{8}{12} = \frac{2}{3}.$$

The *Interactive* CD-ROM and *Internet* versions of this text show every example with its solution; clicking on the *Try It!* button brings up similar problems. Guided Examples and Integrated Examples show step-by-step solutions to additional examples. Integrated Examples are related to several concepts in the section.

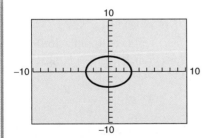

(a)

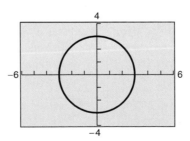

(b)

Figure P.26

Throughout this course, you will learn that there are many ways to approach a problem. Two of the three common approaches are illustrated in Example 12.

A Numerical Approach: Construct and use a table.

A Graphical Approach: Draw and use a graph.

An Algebraic Approach: Use the rules of algebra.

We recommend that you habitually use at least two approaches with every problem to help build your intuition and check that your answer is reasonable.

Application

You can develop mathematical models to represent real-world situations. Both a graphing utility and algebra can be used to understand and solve the problems posed.

> **STUDY T!P**
>
> In applications, it is convenient to use variable names that suggest real-life quantities: *d* for distance, *t* for time, and so on. Most graphing utilities, however, require the variable names to be *x* and *y*.

EXAMPLE 12   Running a Marathon

A runner runs at a constant rate of 4.9 miles per hour. The verbal model and algebraic equation relating distance run and elapsed time are as follows.

Verbal Model: Distance = Rate · Time   *Equation:* $d = 4.9t$

a. Determine how far the runner can run in 3.1 hours.

b. Determine how long it will take to run a 26.2-mile marathon.

Graphical Solution

a. To begin, use a graphing utility to graph the equation $d = 4.9t$. (Represent *d* by *y* and *t* by *x*.) Be sure to use a viewing window that shows the graph when $x = 3.1$. Then use the *value* feature or *zoom* and *trace* features of the graphing utility to estimate that when $x = 3.1$, the distance is $y \approx 15.2$ miles, as shown in Figure P.27(a).

b. Adjust the viewing window so that it shows the graph when $y = 26.2$. Use the *value* feature or *zoom* and *trace* features to estimate that when $y = 26.2$, the time is $x \approx 5.4$ hours, as shown in Figure P.27(b).

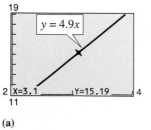

(a)   (b)

Figure P.27

Note that the viewing window on your graphing utility may differ slightly from those shown in Figure P.27.

Algebraic Solution

a. To find how far the runner can run in 3.1 hours, substitute 3.1 for *t* in the equation.

$$d = 4.9t \qquad \text{Write original equation.}$$
$$= 4.9(3.1) \qquad \text{Substitute 3.1 for } t.$$
$$\approx 15.2 \qquad \text{Use a calculator.}$$

So, a runner can run about 15.2 miles in 3.1 hours. Use estimation to check your answer. Because 4.9 is about 5 and 3.1 is about 3, the distance is about $5(3) = 15$. So, 15.2 is reasonable.

b. You can find how long it will take to run a 26.2-mile marathon as follows. (For help with solving linear equations see Appendix C.)

$$d = 4.9t \qquad \text{Write original equation.}$$
$$\frac{d}{4.9} = t \qquad \text{Divide each side by 4.9.}$$
$$\frac{26.2}{4.9} = t \qquad \text{Substitute 26.2 for } d.$$
$$5.3 \approx t \qquad \text{Use a calculator.}$$

So, it will take about 5.3 hours to run 26.2 miles.

P.2 E x e r c i s e s

In Exercises 1–4, sketch the polygon with the indicated vertices.

1. Triangle: $(-1, 1), (2, -1), (3, 4)$
2. Triangle: $(0, 3), (-1, -2), (4, 8)$
3. Square: $(2, 4), (5, 1), (2, -2), (-1, 1)$
4. Parallelogram: $(5, 2), (7, 0), (1, -2), (-1, 0)$

In Exercises 5 and 6, approximate the coordinates of the points.

5.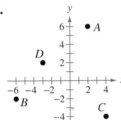
6.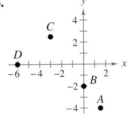

In Exercises 7–10, find the coordinates of the point.

7. The point is located three units to the left of the y-axis and four units above the x-axis.
8. The point is located eight units below the x-axis and four units to the right of the y-axis.
9. The point is located five units below the x-axis and the coordinates of the point are equal.
10. The point is on the x-axis and twelve units to the left of the y-axis.

In Exercises 11–20, determine the quadrant(s) in which (x, y) is located so that the condition(s) is (are) satisfied.

11. $x > 0$ and $y < 0$
12. $x < 0$ and $y < 0$
13. $x = -4$ and $y > 0$
14. $x > 2$ and $y = 3$
15. $y < -5$
16. $x > 4$
17. $(x, -y)$ is in the second quadrant.
18. $(-x, y)$ is in the fourth quadrant.
19. $xy > 0$
20. $xy < 0$

In Exercises 21 and 22, sketch a scatter plot of the data given in the table.

21. *Meteorology* The table shows the lowest temperature of record y (in degrees Fahrenheit) in Duluth, Minnesota, for each month x, where $x = 1$ represents January. (Source: NOAA)

x	1	2	3	4	5	6
y	-39	-33	-29	-5	17	27

x	7	8	9	10	11	12
y	35	32	22	8	-23	-34

22. *Business* The table shows the number y of Wal-Mart stores for each year x from 1992 through 1999. (Source: Wal-Mart Stores, Inc.)

x	1992	1993	1994	1995
y	2136	2440	2759	2943

x	1996	1997	1998	1999
y	3054	3406	3630	3815

23. *Fruit Crops* The table shows farmers' cash receipts (in millions of dollars) from fruit crops in 1996. Construct a bar graph for the data. (Source: U.S. Department of Agriculture)

Fruit	Receipts	Fruit	Receipts
Apples	1846	Oranges	1798
Cherries	264	Peaches	380
Cranberries	246	Pears	292
Grapes	2334	Plums and prunes	295
Lemons	228	Strawberries	770

24. *Oil Imports*   The table shows the amount of crude oil imported into the United States (in millions of barrels) for the years 1988 through 1997. Construct a line graph for the data and state what information the graph reveals.   (Source: Energy Information Administration)

Year	1988	1989	1990	1991	1992
Imports	1864	2133	2151	2110	2220

Year	1993	1994	1995	1996	1997
Imports	2477	2578	2643	2748	2918

In Exercises 25–30, find the distance between the points algebraically and verify graphically by using centimeter graph paper and a centimeter ruler.

25. $(6, -3), (6, 5)$ **26.** $(1, 4), (8, 4)$

27. $(-3, -1), (2, -1)$ **28.** $(-3, -4), (-3, 6)$

29. $(-2, 6), (3, -6)$ **30.** $(8, 5), (0, 20)$

In Exercises 31–42, (a) plot the points, (b) find the distance between the points, and (c) find the midpoint of the line segment joining the points.

31. $(1, 1), (9, 7)$ **32.** $(1, 12), (6, 0)$

33. $(-4, 10), (4, -5)$ **34.** $(-7, -4), (2, 8)$

35. $(-1, 2), (5, 4)$ **36.** $(2, 10), (10, 2)$

37. $\left(\frac{1}{2}, 1\right), \left(-\frac{5}{2}, \frac{4}{3}\right)$ **38.** $\left(-\frac{1}{3}, -\frac{1}{3}\right), \left(-\frac{1}{6}, -\frac{1}{2}\right)$

39. $(6.2, 5.4), (-3.7, 1.8)$

40. $(-16.8, 12.3), (5.6, 4.9)$

41. $(-36, -18), (48, -72)$

42. $(1.451, 3.051), (5.906, 11.360)$

In Exercises 43 and 44, (a) find the length of each side of a right triangle and (b) show that these lengths satisfy the Pythagorean Theorem.

43.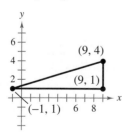

44.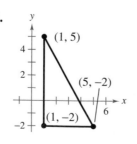

Business   **In Exercises 45 and 46, estimate the sales of a company in 1998, given the sales in 1996 and 2000. Assume the sales followed a linear pattern.**

45.

Year	1996	2000
Sales	$520,000	$740,000

46.

Year	1996	2000
Sales	$4,200,000	$5,650,000

In Exercises 47–54, determine whether the points lie on the graph of the equation.

	Equation		*Points*	
47.	$y = \sqrt{x + 4}$		(a) $(0, 2)$	(b) $(5, 3)$
48.	$y = x^2 - 3x + 2$		(a) $(2, 0)$	(b) $(-2, 8)$
49.	$y = 4 - |x - 2|$		(a) $(1, 5)$	(b) $(1.2, 3.2)$
50.	$y = \dfrac{1}{x^2 + 1}$		(a) $(0, 0)$	(b) $(3, 0.1)$
51.	$2x - y - 3 = 0$		(a) $(1, 2)$	(b) $(1, -1)$
52.	$x^2 + y^2 = 20$		(a) $(3, -2)$	(b) $(-4, 2)$
53.	$x^2y - x^2 + 4y = 0$		(a) $\left(1, \frac{1}{5}\right)$	(b) $\left(2, \frac{1}{2}\right)$
54.	$y = \frac{1}{3}x^3 - 2x^2$		(a) $\left(2, -\frac{16}{3}\right)$	(b) $(-3, 9)$

In Exercises 55–60, complete the table. Use the resulting solution points to sketch the graph of the equation. Use a graphing utility to verify the graph.

55. $y = -2x + 3$

x	-1	0	1	$\frac{3}{2}$	2
y					

56. $y = \frac{3}{2}x - 1$

x	-2	0	$\frac{2}{3}$	1	2
y					

57. $y = x^2 - 2x$

x	-1	0	1	2	3
y					

58. $y = 4 - x^2$

x	-2	-1	0	1	2
y					

59. $y = 3 - |x - 2|$

x	0	1	2	3	4
y					

60. $y = \sqrt{x - 1}$

x	1	2	5	10	17
y					

In Exercises 61–66, match the equation with its graph. [The graphs are labeled (a), (b), (c), (d), (e), and (f).]

(a)

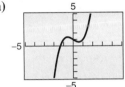

(b)

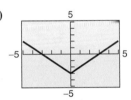

(c)

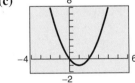

(d)

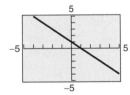

(e)

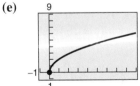

(f)

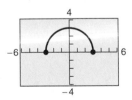

61. $y = 1 - x$ **62.** $y = x^2 - 2x$

63. $y = \sqrt{9 - x^2}$ **64.** $y = 2\sqrt{x}$

65. $y = x^3 - x + 1$ **66.** $y = |x| - 3$

In Exercises 67–80, sketch the graph of the equation.

67. $y = -3x + 2$ **68.** $y = 2x - 3$

69. $y = 1 - x^2$ **70.** $y = x^2 - 1$

71. $y = x^2 - 3x$ **72.** $y = -x^2 - 4x$

73. $y = x^3 + 2$ **74.** $y = x^3 - 1$

75. $y = \sqrt{x - 3}$ **76.** $y = \sqrt{1 - x}$

77. $y = |x - 2|$ **78.** $y = 4 - |x|$

79. $x = y^2 - 1$ **80.** $x = y^2 - 4$

In Exercises 81–94, use a graphing utility to graph the equation. Use a standard viewing window. Approximate any x- or y-intercepts of the graph.

81. $y = x - 5$ **82.** $y = (x + 1)(x - 3)$

83. $y = 3 - \frac{1}{2}x$ **84.** $y = \frac{2}{3}x - 1$

85. $y = x^2 - 4x + 3$ **86.** $y = \frac{1}{2}(x + 4)(x - 2)$

87. $y = x(x - 2)^2$ **88.** $y = \dfrac{4}{x^2 + 1}$

89. $y = \dfrac{2x}{x - 1}$ **90.** $y = \dfrac{4}{x}$

91. $y = x\sqrt{x + 6}$ **92.** $y = (6 - x)\sqrt{x}$

93. $y = \sqrt[3]{x}$ **94.** $y = \sqrt[3]{x + 1}$

In Exercises 95 and 96, use a graphing utility to sketch the graph of the equation. Begin by using a standard viewing window. Then graph the equation a second time using the specified viewing window. Which viewing window is better? Explain.

95. $y = \frac{5}{2}x + 5$ **96.** $y = -3x + 50$

Xmin = 0
Xmax = 6
Xscl = 1
Ymin = 0
Ymax = 10
Yscl = 1

Xmin = -1
Xmax = 4
Xscl = 1
Ymin = -5
Ymax = 60
Yscl = 5

In Exercises 97 and 98, describe the viewing window of the graph.

97. $y = 4x^2 - 25$ **98.** $y = x^3 - 3x^2 + 4$

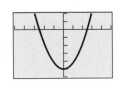   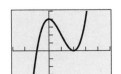

In Exercises 99–106, find the standard form of the equation of the specified circle.

99. Center: $(0, 0)$; radius: 3

100. Center: $(0, 0)$; radius: 5

101. Center: $(2, -1)$; radius: 4

102. Center: $\left(0, \frac{1}{3}\right)$; radius: $\frac{1}{3}$

103. Center: $(-1, 2)$; solution point: $(0, 0)$

104. Center: $(3, -2)$; solution point: $(-1, 1)$

105. Endpoints of a diameter: $(0, 0)$, $(6, 8)$

106. Endpoints of a diameter: $(-4, -1)$, $(4, 1)$

In Exercises 107–112, find the center and radius, and sketch the circle.

107. $x^2 + y^2 = 4$

108. $x^2 + y^2 = 16$

109. $(x - 1)^2 + (y + 3)^2 = 4$

110. $x^2 + (y - 1)^2 = 4$

111. $\left(x - \frac{1}{2}\right)^2 + \left(y - \frac{1}{2}\right)^2 = \frac{9}{4}$

112. $\left(x - \frac{2}{3}\right)^2 + \left(y + \frac{1}{4}\right)^2 = \frac{25}{9}$

In Exercises 113–116, solve for y and use a graphing utility to graph each of the resulting equations in the same viewing window. Adjust the viewing window so that a circle really does appear circular.

113. $x^2 + y^2 = 64$

114. $x^2 + y^2 = 49$

115. $(x - 1)^2 + (y - 2)^2 = 16$

116. $(x - 3)^2 + (y - 1)^2 = 25$

117. *Business* Starbucks Corporation had annual sales of \$696.5 million in 1996 and \$1308.7 million in 1998. Use the Midpoint Formula to estimate the 1997 sales. (Source: Starbucks Corporation)

118. *Finance* The dividends declared per share of Procter & Gamble Company from 1992 to 1998 can be approximated by the model

$$y = 0.464 + 0.091t, \quad 0 \le t \le 6$$

where y is the dividend (in dollars) and t is the time (in years), with $t = 0$ corresponding to 1992. (Source: The Procter & Gamble Company)

(a) Create a table showing the dividends y for the years 1992 through 1998. From your table, determine the year during which the dividend was \$0.65.

(b) Use your graphing utility to graph the model.

(c) Use the *zoom* and *trace* features of your graphing utility to verify your answer to part (a).

(d) Determine algebraically the value of y in 1997.

(e) Use the model to estimate the value of y in 2002.

119. *Population Statistics* The table gives the life expectancy of a child (at birth) in the United States for selected years from 1920 to 2000. (Source: U.S. National Center for Health Statistics)

Year	1920	1930	1940	1950
Life Expectancy	54.1	59.7	62.9	68.2

Year	1960	1970	1980	1990	2000
Life Expectancy	69.7	70.8	73.7	75.4	76.4

A model for the life expectancy during this period is

$$y = \frac{66.93 + t}{1 + 0.01t}$$

where y represents the life expectancy and t is the time in years, with $t = 0$ corresponding to 1950.

(a) Use the *zoom* and *trace* features of your graphing utility to find the life expectancy in 1925 and 1995. Verify your answers algebraically.

(b) Use your graphing utility to determine during which year the life expectancy was 72 years.

(c) Use the model to estimate the life expectancy in 2002 and 2004.

Synthesis

True or False? **In Exercises 120–122, determine whether the statement is true or false. Justify your answer.**

120. In order to divide a line segment into 16 equal parts, you would have to use the Midpoint Formula 16 times.

121. The points $(-8, 4)$, $(2, 11)$, and $(-5, 1)$ represent the vertices of an isosceles triangle.

122. The graph of a linear equation can have either no x-intercepts or only one x-intercept.

123. *Think About It* Find a and b if the x-intercepts of the graph of $y = (x - a)(x - b)$ are $(-2, 0)$ and $(5, 0)$.

124. *Writing* Explain how to find an appropriate viewing window for the graph of an equation.

P.3 Solving Equations Algebraically and Graphically

Equations and Solutions of Equations

An **equation** is a statement that two algebraic expressions are equal. To **solve** an equation in x means to find all values of x for which the equation is true. Such values are **solutions.** For instance, $x = 4$ is a solution of the equation $3x - 5 = 7$, because $3(4) - 5 = 7$ is a true statement.

The solutions of an equation depend upon the kinds of numbers being considered. For instance, in the set of rational numbers, $x^2 = 10$ has no solution because there is no rational number whose square is 10. However, in the set of real numbers the equation has the two solutions $\sqrt{10}$ and $-\sqrt{10}$.

An equation that is true for every real number in the domain of the variable is called an **identity.** For example, $x^2 - 9 = (x + 3)(x - 3)$ is an identity because it is a true statement for any real value of x.

An equation that is true for just *some* (or even none) of the real numbers in the domain of the variable is called a **conditional equation.** For example, the equation $x^2 - 9 = 0$ is conditional because $x = 3$ and $x = -3$ are the only values in the domain that satisfy the equation. The equation $2x + 1 = 2x - 3$ is also conditional because it is not true for any values of x. A **linear equation in one variable** x is an equation that can be written in the standard form $ax + b = 0$, where a and b are real numbers with $a \neq 0$. For a review of solving one- and two-step equations, see Appendix C.

To solve an equation involving fractional expressions, find the least common denominator (LCD) of all terms in the equation and multiply every term by this LCD. This procedure clears the equation of fractions.

Robert Holmes/CORBIS

EXAMPLE 1   Solving an Equation Involving Fractions

$$\frac{x}{3} + \frac{3x}{4} = 2 \qquad \text{Original equation}$$

$$(12)\frac{x}{3} + (12)\frac{3x}{4} = (12)2 \qquad \text{Multiply by the LCD.}$$

$$4x + 9x = 24 \qquad \text{Simplify.}$$

$$13x = 24 \qquad \text{Combine like terms.}$$

$$x = \frac{24}{13} \qquad \text{Divide each side by 13.}$$

Check

After solving an equation, check the solution in the original equation.

$$\frac{\frac{24}{13}}{3} + \frac{3\left(\frac{24}{13}\right)}{4} \stackrel{?}{=} 2 \qquad \text{Substitute } \tfrac{24}{13} \text{ for } x.$$

$$2 = 2 \qquad \text{Solution checks. } \checkmark$$

This close connection among x-intercepts and solutions described on page 29 is crucial to our study of algebra, and you can take advantage of this connection in two basic ways. You can use your algebraic "equation-solving skills" to find the x-intercepts of a graph, and you can use your "graphing skills" to approximate the solutions of an equation.

Finding Solutions Graphically

Polynomial equations of degree 1 or 2 can be solved in relatively straightforward ways. Polynomial equations of higher degrees can, however, be quite difficult to solve, especially if you rely only on algebraic techniques. For such equations, a graphing utility can be very helpful.

Graphical Approximations of Solutions of an Equation

1. Write the equation in *general form*, $y = 0$, with the nonzero terms on one side of the equation and zero on the other side.

2. Use a graphing utility to graph the equation. Be sure the viewing window shows all the relevant features of the graph.

3. Use the *zero* or *root* feature or the *zoom* and *trace* features of the graphing utility to approximate each of the x-intercepts of the graph. Remember that a graph can have more than one x-intercept, so you may need to change the viewing window a few times.

A polynomial equation of degree n cannot have more than n different solutions.

EXAMPLE 4   Finding Solutions of an Equation Graphically

Use a graphing utility to approximate the solutions of $2x^3 - 3x + 2 = 0$.

Solution
Begin by graphing the equation $y = 2x^3 - 3x + 2$, as shown in Figure P.31. You can see from the graph that there is only one x-intercept. It lies between -1 and -2 and is approximately -1.5. By using the *zero* or *root* feature of a graphing utility you can improve the approximation. To three-decimal-place accuracy, the solution is $x \approx -1.476$. Check this approximation on your calculator. You will find that the value of y is $y = 2(-1.476)^3 - 3(-1.476) + 2 \approx -0.003$.

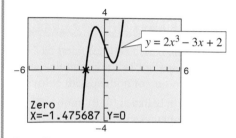

Figure P.31

Exploration

In Chapter 4 you will learn that a cubic equation such as

$$24x^3 - 36x + 17 = 0$$

can have up to 3 real solutions. Use a graphing utility to graph

$$y = 24x^3 - 36x + 17.$$

Describe a viewing window that allows you to determine the number of real solutions of the equation

$$24x^3 - 36x + 17 = 0.$$

Use the same technique to determine the number of real solutions of

$$97x^3 - 102x^2 - 200x - 63 = 0.$$

You can also use a graphing calculator's *zoom* and *trace* features to approximate the solution of an equation. Here are some suggestions for using the *zoom-in* feature of a graphing utility.

1. With each successive zoom-in, adjust the *x*-scale (if necessary) so that the resulting viewing window shows at least the two scale marks between which the solution lies.

2. The accuracy of the approximation will always be such that the error is less than the distance between two scale marks.

3. If you have a *trace* feature on your graphing utility, you can generally add one more decimal place of accuracy without changing the viewing window.

Unless stated otherwise, this book will approximate all real solutions with an error of *at most* 0.01.

EXAMPLE 5   Approximating Solutions of an Equation Graphically

Use a graphing utility to approximate the solutions of $x^2 + 3 = 5x$.

Solution

In general form, this equation is

$$x^2 - 5x + 3 = 0. \qquad \text{Equation in general form}$$

So, you can begin by graphing

$$y = x^2 - 5x + 3 \qquad \text{Equation to be graphed}$$

as shown in Figure P.32(a). This graph has two *x*-intercepts, and by using the *zoom* and *trace* features you can approximate the corresponding solutions to be $x \approx 0.70$ and $x \approx 4.30$, as shown in Figures P.32(b) and P.32(c).

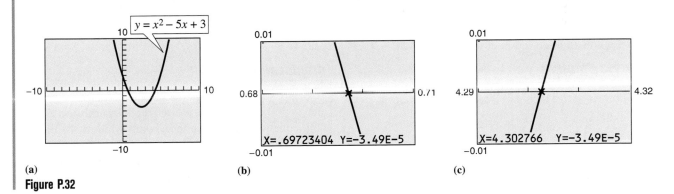

(a)

(b)

(c)

Figure P.32

The built-in *zero* or *root* programs of a graphing utility will approximate solutions of equations or approximate *x*-intercepts of graphs. If your graphing utility has such features, try using them to approximate the solutions in Example 5.

Points of Intersection of Two Graphs

An ordered pair that is a solution of two different equations is called a **point of intersection** of the graphs of the two equations. For instance, in Figure P.33 you can see that the graphs of the following equations have two points of intersection.

$$y = x + 2 \qquad \text{Equation 1}$$

$$y = x^2 - 2x - 2 \qquad \text{Equation 2}$$

The point $(-1, 1)$ is a solution of both equations, and the point $(4, 6)$ is a solution of both equations. To check this algebraically, substitute -1 and 4 into each equation.

Check that $(-1, 1)$ is a solution.

Equation 1: $y = -1 + 2 = 1$ \qquad Solution checks. ✓

Equation 2: $y = (-1)^2 - 2(-1) - 2$

$$= 1 \qquad \text{Solution checks. ✓}$$

Check that $(4, 6)$ is a solution.

Equation 1: $y = 4 + 2$

$$= 6 \qquad \text{Solution checks. ✓}$$

Equation 2: $y = (4)^2 - 2(4) - 2$

$$= 6 \qquad \text{Solution checks. ✓}$$

To find the points of intersection of the graphs of two equations, solve each equation for y (or x) and set the two results equal to each other. The resulting equation will be an equation in one variable, which can be solved using standard procedures, as shown in Example 6.

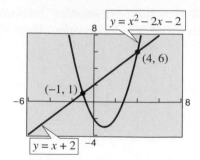

Figure P.33

STUDY T!P

The table shows some points of the graphs of the equations at the left. Find the points of intersection of the graphs by finding the values of x for which y_1 and y_2 are equal.

X	Y1	Y2
-2	0	6
-1	1	1
0	2	-2
1	3	-3
2	4	-2
3	5	1
4	6	6

X=-2

EXAMPLE 6   Finding Points of Intersection

Find the points of intersection of the graphs of $2x - 3y = -2$ and $4x - y = 6$.

Algebraic Solution

To begin, solve each equation for y to obtain

$$y = \frac{2}{3}x + \frac{2}{3} \quad \text{and} \quad y = 4x - 6.$$

Next, set the two expressions for y equal to each other and solve the resulting equation for x, as follows.

$$\frac{2}{3}x + \frac{2}{3} = 4x - 6 \qquad \text{Equate expressions for } y.$$

$$2x + 2 = 12x - 18 \qquad \text{Multiply each side by 3.}$$

$$-10x = -20 \qquad \text{Subtract } 12x \text{ and 2 from each side.}$$

$$x = 2 \qquad \text{Divide each side by } -10.$$

When $x = 2$, the y-value of each of the given equations is 2. So, the point of intersection is $(2, 2)$.

Graphical Solution

To begin, solve each equation for y to obtain $y_1 = \frac{2}{3}x + \frac{2}{3}$ and $y_2 = 4x - 6$. Then use a graphing utility to graph both equations in the same viewing window. In Figure P.34, the graphs appear to have one point of intersection. Use the *intersect* feature of the graphing utility to approximate the point of intersection to be $(2, 2)$.

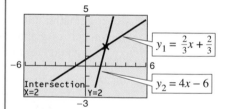

Figure P.34

Another way to approximate points of intersection of two graphs is to graph both equations and use the *zoom* and *trace* features to find the point or points at which the two graphs intersect.

EXAMPLE 7 Approximating Points of Intersection Graphically

Approximate the point(s) of intersection of the graphs of the following equations.

$$y = x^2 - 3x - 4 \qquad\qquad \text{Equation 1 (quadratic equation)}$$

$$y = x^3 + 3x^2 - 2x - 1 \qquad\qquad \text{Equation 2 (cubic equation)}$$

Solution

Begin by using a graphing utility to graph both equations, as shown in Figure P.35. From this display, you can see that the two graphs have only one point of intersection. Then, using the *zoom* and *trace* features, approximate the point of intersection to be $(-2.17, 7.25)$. To test the reasonableness of this approximation, you can evaluate both equations when $x = -2.17$.

Quadratic Equation:

$$y = (-2.17)^2 - 3(-2.17) - 4$$

$$\approx 7.22$$

Cubic Equation:

$$y = (-2.17)^3 + 3(-2.17)^2 - 2(-2.17) - 1$$

$$\approx 7.25$$

Because both equations yield approximately the same y-value, you can conclude that the approximate coordinates of the point of intersection are $x \approx -2.17$ and $y \approx 7.25$.

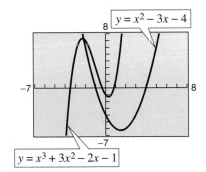

Figure P.35

The method shown in Example 7 gives a nice graphical picture of the points of intersection of two graphs. However, for actual approximation purposes, it is better to use the algebraic procedure described in Example 6. That is, the point of intersection of $y = x^2 - 3x - 4$ and $y = x^3 + 3x^2 - 2x - 1$ coincides with the solution of the equation

$$x^3 + 3x^2 - 2x - 1 = x^2 - 3x - 4 \qquad \text{Equate } y\text{-values.}$$

$$x^3 + 2x^2 + x + 3 = 0. \qquad \text{Write in general form.}$$

By graphing $y = x^3 + 2x^2 + x + 3$ on a graphing utility and using the *zoom* and *trace* features (or the *zero* or *root* feature), you can approximate the solution of this equation to be $x \approx -2.17$. The corresponding y-value for *both* of the equations given in Example 7 is $y \approx 7.25$.

Solving Polynomial Equations Algebraically

Polynomial equations can be classified by their degree.

Degree	Name	Example
First degree	Linear equation	$6x + 2 = 4$
Second degree	Quadratic equation	$2x^2 - 5x + 3 = 0$
Third degree	Cubic equation	$x^3 - x = 0$
Fourth degree	Quartic equation	$x^4 - 3x^2 + 2 = 0$
Fifth degree	Quintic equation	$x^5 - 12x^2 + 7x + 4 = 0$

In general, the higher the degree, the more difficult it is to solve the equation—either algebraically or graphically.

You should be familiar with the following four methods for solving quadratic equations *algebraically*.

Solving a Quadratic Equation

Method

Example

Factoring: If $ab = 0$, then $a = 0$ or $b = 0$.

$$x^2 - x - 6 = 0$$
$$(x - 3)(x + 2) = 0$$
$$x - 3 = 0 \implies x = 3$$
$$x + 2 = 0 \implies x = -2$$

Square Root Principle: If $u^2 = c$, where $c > 0$, then $u = \pm\sqrt{c}$.

$$(x + 3)^2 = 16$$
$$x + 3 = \pm 4$$
$$x = -3 \pm 4$$
$$x = 1 \quad \text{or} \quad x = -7$$

Completing the Square: If $x^2 + bx = c$, then

$$x^2 + bx + \left(\frac{b}{2}\right)^2 = c + \left(\frac{b}{2}\right)^2$$

$$\left(x + \frac{b}{2}\right)^2 = c + \frac{b^2}{4}.$$

$$x^2 + 6x = 5$$
$$x^2 + 6x + 3^2 = 5 + 3^2$$
$$(x + 3)^2 = 14$$
$$x + 3 = \pm\sqrt{14}$$
$$x = -3 \pm \sqrt{14}$$

Quadratic Formula: If $ax^2 + bx + c = 0$, then

$$x = \frac{-b \pm \sqrt{b^2 - 4ac}}{2a}.$$

$$2x^2 + 3x - 1 = 0$$
$$x = \frac{-3 \pm \sqrt{3^2 - 4(2)(-1)}}{2(2)}$$
$$= \frac{-3 \pm \sqrt{17}}{4}$$

The methods used to solve quadratic equations can sometimes be extended to polynomial equations of higher degree, as shown in the next two examples.

EXAMPLE 8   Solving an Equation of Quadratic Type

Solve the equation $x^4 - 3x^2 + 2 = 0$.

Solution

The expression $x^4 - 3x^2 + 2$ is said to be in *quadratic form* because it is written in the form $au^2 + bu + c$, where u is any expression in x, namely x^2. You can use factoring to solve the equation as follows.

$$x^4 - 3x^2 + 2 = 0 \qquad \text{Write original equation.}$$
$$(x^2)^2 - 3(x^2) + 2 = 0 \qquad \text{Write in quadratic form in } x^2.$$
$$(x^2 - 1)(x^2 - 2) = 0 \qquad \text{Partially factor.}$$
$$(x + 1)(x - 1)(x^2 - 2) = 0 \qquad \text{Factor.}$$
$$x + 1 = 0 \implies x = -1$$
$$x - 1 = 0 \implies x = 1$$
$$x^2 - 2 = 0 \implies x = \pm\sqrt{2}$$

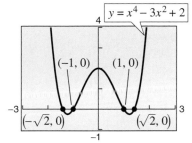

Figure P.36

The equation has four solutions: -1, 1, $\sqrt{2}$, and $-\sqrt{2}$. Check these solutions in the original equation. The graph of $y = x^4 - 3x^2 + 2$, as shown in Figure P.36, verifies the solutions graphically.

EXAMPLE 9   Solving a Polynomial Equation by Factoring

Solve the equation $2x^3 - 6x^2 - 6x + 18 = 0$.

Solution

This equation has a common factor of 2. To simplify things, first divide each side of the equation by 2.

$$2x^3 - 6x^2 - 6x + 18 = 0 \qquad \text{Write original equation.}$$
$$x^3 - 3x^2 - 3x + 9 = 0 \qquad \text{Divide each side by 2.}$$
$$x^2(x - 3) - 3(x - 3) = 0 \qquad \text{Factor out common monomial factors.}$$
$$(x - 3)(x^2 - 3) = 0 \qquad \text{Factor by grouping.}$$
$$x - 3 = 0 \implies x = 3$$
$$x^2 - 3 = 0 \implies x = \pm\sqrt{3}$$

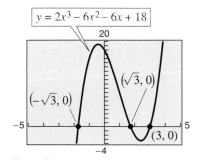

Figure P.37

The equation has three solutions: 3, $\sqrt{3}$, and $-\sqrt{3}$. Check these solutions in the original equation. The graph of $y = 2x^3 - 6x^2 - 6x + 18$, as shown in Figure P.37, verifies the solutions graphically.

Other Types of Equations

An equation involving a radical expression can often be cleared of radicals by raising both sides of the equation to an appropriate power. When using this procedure, remember that it can introduce extraneous solutions—so be sure to check each solution in the original equation.

EXAMPLE 10   Solving an Equation Involving a Radical

Solve $\sqrt{2x + 7} - x = 2$.

Algebraic Solution

$\sqrt{2x + 7} - x = 2$	Write original equation.
$\sqrt{2x + 7} = x + 2$	Isolate radical.
$2x + 7 = x^2 + 4x + 4$	Square each side.
$x^2 + 2x - 3 = 0$	Write in general form.
$(x + 3)(x - 1) = 0$	Factor.
$x + 3 = 0 \implies x = -3$	Set 1st factor equal to 0.
$x - 1 = 0 \implies x = 1$	Set 2nd factor equal to 0.

By substituting into the original equation, you can determine that -3 is extraneous, whereas 1 is valid. So, the equation has only one real solution: $x = 1$.

Graphical Solution

First rewrite the equation as $\sqrt{2x + 7} - x - 2 = 0$. Then use a graphing utility to graph $y = \sqrt{2x + 7} - x - 2$, as shown in Figure P.38(a). Notice that the domain is $x \geq -\frac{7}{2}$ because the expression under the radical cannot be negative. There appears to be one solution near $x = 1$. Use the *zoom* and *trace* features, as shown in Figure P.38(b), to approximate the only solution to be $x = 1$.

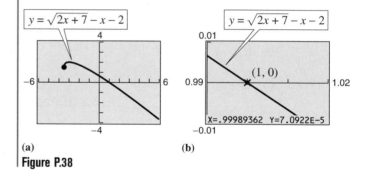

(a)   (b)

Figure P.38

EXAMPLE 11   Solving an Equation Involving Two Radicals

Solve the equation $\sqrt{2x + 6} - \sqrt{x + 4} = 1$.

Solution

$\sqrt{2x + 6} - \sqrt{x + 4} = 1$	Write original equation.
$2x + 6 = 1 + 2\sqrt{x + 4} + (x + 4)$	Isolate radical and square each side.
$x + 1 = 2\sqrt{x + 4}$	Isolate radical.
$x^2 + 2x + 1 = 4(x + 4)$	Square each side.
$x^2 - 2x - 15 = 0$	Write in general form.
$(x - 5)(x + 3) = 0$	Factor.
$x - 5 = 0 \implies x = 5$	Set 1st factor equal to 0.
$x + 3 = 0 \implies x = -3$	Set 2nd factor equal to 0.

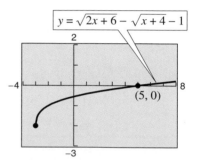

By substituting into the original equation, you can determine that -3 is extraneous, whereas 5 is valid. Figure P.39 verifies that $x = 5$ is the only solution.

Figure P.39

Sometimes radicals in equations are represented by fractional exponents.

EXAMPLE 12   Solving an Equation with Rational Exponents

Solve $(x + 1)^{2/3} = 4$.

Algebraic Solution

$$(x + 1)^{2/3} = 4 \qquad \text{Write original equation.}$$

$$\sqrt[3]{(x + 1)^2} = 4 \qquad \text{Rewrite with radical sign.}$$

$$(x + 1)^2 = 64 \qquad \text{Cube each side.}$$

$$x + 1 = \pm 8 \qquad \text{Take square root of each side.}$$

$$x = -9, x = 7 \qquad \text{Subtract 1 from each side.}$$

Substitute $x = -9$ and $x = 7$ into the original equation to determine that both are valid solutions.

Graphical Solution

Use a graphing utility to graph $y_1 = \sqrt[3]{(x + 1)^2}$ and $y_2 = 4$ in the same viewing window, as shown in Figure P.40. Use the *intersect* feature of the graphing utility to approximate the solutions to be $x = -9$ and $x = 7$.

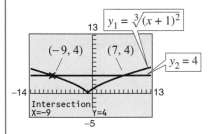

Figure P.40

As demonstrated in Example 1, you can algebraically solve an equation involving fractions by multiplying both sides of the equation by the least common denominator of each term in the equation to "clear an equation of fractions."

EXAMPLE 13   Solving an Equation Involving Fractions

Solve $\dfrac{2}{x} = \dfrac{3}{x - 2} - 1$.

Solution

For this equation, the least common denominator of the three terms is $x(x - 2)$, so you can begin by multiplying each term in the equation by this expression.

$$\frac{2}{x} = \frac{3}{x - 2} - 1$$

$$x(x - 2)\frac{2}{x} = x(x - 2)\frac{3}{x - 2} - x(x - 2)(1)$$

$$2(x - 2) = 3x - x(x - 2), \qquad x \neq 0, 2$$

$$2x - 4 = -x^2 + 5x$$

$$x^2 - 3x - 4 = 0$$

$$(x - 4)(x + 1) = 0$$

$$x - 4 = 0 \implies x = 4$$

$$x + 1 = 0 \implies x = -1$$

The equation has two solutions: 4 and -1. Check these solutions in the original equation. Use a graphing utility to verify these solutions graphically.

Exploration

Using *dot* mode, graph the two equations

$$y_1 = \frac{2}{x}$$

$$y_2 = \frac{3}{x - 2} - 1$$

in the same viewing window. How many times do the graphs of the equations intersect? What does this tell you about the solution to Example 13?

STUDY TIP

Graphs of equations involving variable denominators can be tricky because of the way graphing utilities skip over points where the denominator is zero.

EXAMPLE 14   Solving an Equation Involving Absolute Value

Solve $|x^2 - 3x| = -4x + 6$.

Solution

Begin by writing the equation as $|x^2 - 3x| + 4x - 6 = 0$. From the graph of $y = |x^2 - 3x| + 4x - 6$ in Figure P.41, you can estimate the solutions to be -3 and 1. These can be verified by substitution into the equation. To solve an equation involving an absolute value *algebraically*, you must consider the fact that the expression inside the absolute value symbols can be positive or negative. This consideration results in *two* separate equations, each of which must be solved.

First Equation:

$$x^2 - 3x = -4x + 6 \qquad \text{Use positive expression.}$$

$$x^2 + x - 6 = 0 \qquad \text{Write in general form.}$$

$$(x + 3)(x - 2) = 0 \qquad \text{Factor.}$$

$$x + 3 = 0 \quad \Longrightarrow \quad x = -3 \qquad \text{Set 1st factor equal to 0.}$$

$$x - 2 = 0 \quad \Longrightarrow \quad x = 2 \qquad \text{Set 2nd factor equal to 0.}$$

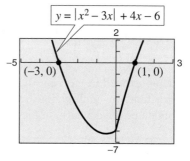

Figure P.41

Second Equation:

$$-(x^2 - 3x) = -4x + 6 \qquad \text{Use negative expression.}$$

$$x^2 - 7x + 6 = 0 \qquad \text{Write in general form.}$$

$$(x - 1)(x - 6) = 0 \qquad \text{Factor.}$$

$$x - 1 = 0 \quad \Longrightarrow \quad x = 1 \qquad \text{Set 1st factor equal to 0.}$$

$$x - 6 = 0 \quad \Longrightarrow \quad x = 6 \qquad \text{Set 2nd factor equal to 0.}$$

Check

$$|(-3)^2 - 3(-3)| \stackrel{?}{=} -4(-3) + 6 \qquad \text{Substitute } -3 \text{ for } x.$$

$$18 = 18 \qquad -3 \text{ checks. } \checkmark$$

$$|2^2 - 3(2)| \stackrel{?}{=} -4(2) + 6 \qquad \text{Substitute 2 for } x.$$

$$2 \neq -2 \qquad 2 \text{ does not check.}$$

$$|1^2 - 3(1)| \stackrel{?}{=} -4(1) + 6 \qquad \text{Substitute 1 for } x.$$

$$2 = 2 \qquad 1 \text{ checks. } \checkmark$$

$$|6^2 - 3(6)| \stackrel{?}{=} -4(6) + 6 \qquad \text{Substitute 6 for } x.$$

$$18 \neq -18 \qquad 6 \text{ does not check.}$$

The equation has only two solutions: -3 and 1, just as you obtained by graphing.

In Figure P.41, the graph of $y = |x^2 - 3x| + 4x - 6$ appears to be a straight line to the right of the *y*-axis. Is it? Explain how you decided.

P.3 E x e r c i s e s

In Exercises 1–6, determine whether the given values of x are solutions of the equation.

Equation	*Values*

1. $\dfrac{5}{2x} - \dfrac{4}{x} = 3$
  (a) $x = -\dfrac{1}{2}$ (b) $x = 4$
  (c) $x = 0$ (d) $x = \dfrac{1}{4}$

2. $\dfrac{x}{2} + \dfrac{6x}{7} = \dfrac{19}{14}$
  (a) $x = -2$ (b) $x = 1$
  (c) $x = \dfrac{1}{2}$ (d) $x = 7$

3. $3 + \dfrac{1}{x + 2} = 4$
  (a) $x = -1$ (b) $x = -2$
  (c) $x = 0$ (d) $x = 5$

4. $\dfrac{(x + 5)(x - 3)}{2} = 24$
  (a) $x = 3$ (b) $x = -2$
  (c) $x = 7$ (d) $x = 9$

5. $\dfrac{\sqrt{x + 4}}{6} + 3 = 4$
  (a) $x = -3$ (b) $x = 0$
  (c) $x = 21$ (d) $x = 32$

6. $\dfrac{\sqrt[3]{x - 8}}{3} = -\dfrac{2}{3}$
  (a) $x = -16$ (b) $x = 0$
  (c) $x = 9$ (d) $x = 16$

In Exercises 7–12, determine whether the equation is an identity or a conditional equation.

7. $2(x - 1) = 2x - 2$

8. $-7(x - 3) + 4x = 3(7 - x)$

9. $x^2 - 8x + 5 = (x - 4)^2 - 11$

10. $x^2 + 2(3x - 2) = x^2 + 6x - 4$

11. $3 + \dfrac{1}{x + 1} = \dfrac{4x}{x + 1}$ **12.** $\dfrac{5}{x} + \dfrac{3}{x} = 24$

In Exercises 13 and 14, solve the equation in two ways. Then explain which way was easier for you.

13. $\dfrac{3x}{8} - \dfrac{4x}{3} = 4$ **14.** $\dfrac{3z}{8} - \dfrac{z}{10} = 6$

In Exercises 15–30, solve the equation (if possible). Then use a graphing utility to verify your solution.

15. $\dfrac{5x}{4} + \dfrac{1}{2} = x - \dfrac{1}{2}$ **16.** $\dfrac{x}{5} - \dfrac{x}{2} = 3$

17. $\dfrac{3}{2}(z + 5) - \dfrac{1}{4}(z + 24) = 0$

18. $\dfrac{3x}{2} + \dfrac{1}{4}(x - 2) = 10$

19. $\dfrac{100 - 4u}{3} = \dfrac{5u + 6}{4} + 6$

20. $\dfrac{17 + y}{y} + \dfrac{32 + y}{y} = 100$

21. $\dfrac{5x - 4}{5x + 4} = \dfrac{2}{3}$ **22.** $\dfrac{15}{x} - 4 = \dfrac{6}{x} + 3$

23. $\dfrac{1}{x - 3} + \dfrac{1}{x + 3} = \dfrac{10}{x^2 - 9}$

24. $\dfrac{1}{x - 2} + \dfrac{3}{x + 3} = \dfrac{4}{x^2 + x - 6}$

25. $\dfrac{7}{2x + 1} - \dfrac{8x}{2x - 1} = -4$

26. $\dfrac{4}{u - 1} + \dfrac{6}{3u + 1} = \dfrac{15}{3u + 1}$

27. $\dfrac{1}{x} + \dfrac{2}{x - 5} = 0$ **28.** $\dfrac{6}{x} - \dfrac{2}{x + 3} = \dfrac{3(x + 5)}{x(x + 3)}$

29. $\dfrac{3}{x(x - 3)} + \dfrac{4}{x} = \dfrac{1}{x - 3}$

30. $3 = 2 + \dfrac{2}{z + 2}$

In Exercises 31–40, find the x- and y-intercepts of the graph of the equation.

31. $y = x - 5$ **32.** $y = -\dfrac{3}{4}x - 3$

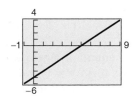   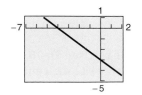

33. $y = x^2 + x - 2$ **34.** $y = 4 - x^2$

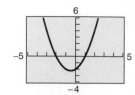   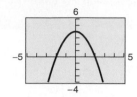

35. $y = x\sqrt{x + 2}$ **36.** $y = -\frac{1}{2}x\sqrt{x + 3} + 1$

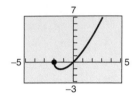   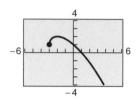

37. $y = |x - 2| - 4$ **38.** $y = 3 - \frac{1}{2}|x + 1|$

39. $xy - 2y - x + 1 = 0$

40. $x^2y - x^2 + 4y = 0$

In Exercises 41–46, the solution(s) of the equation are given. Verify the solution(s) both algebraically and graphically.

Equation	Solution(s)
41. $y = 12 - 4x = 0$	$x = 3$
42. $y = 3(x - 5) + 9 = 0$	$x = 2$
43. $y = x^2 - 2.5x - 6 = 0$	$x = -1.5, 4$
44. $y = x^3 - 9x^2 + 18x = 0$	$x = 0, 3, 6$
45. $y = \dfrac{x + 2}{3} - \dfrac{x - 1}{5} - 1 = 0$	$x = 1$
46. $y = x - 3 - \dfrac{10}{x} = 0$	$x = -2, 5$

Graphical Analysis   In Exercises 47–50, use a graphing utility to graph the equation and approximate any x-intercepts. Set $y = 0$ and solve the resulting equation. Compare the results with the x-intercepts of the graph.

47. $y = 2(x - 1) - 4$ **48.** $y = \frac{4}{3}x + 2$

49. $y = 20 - (3x - 10)$ **50.** $y = 10 + 2(x - 2)$

In Exercises 51–70, solve the equation algebraically. Then write the equation in the form $y = 0$ and use a graphing utility to verify the algebraic solution.

51. $27 - 4x = 12$ **52.** $3.5x - 8 = 0.5x$

53. $25(x - 3) = 12(x + 2) - 10$

54. $1200 = 300 + 2(x - 500)$

55. $\dfrac{3x}{2} + \dfrac{1}{4}(x - 2) = 10$

56. $0.60x + 0.40(100 - x) = 50$

57. $\dfrac{2x}{3} = 10 - \dfrac{24}{x}$ **58.** $\dfrac{x - 3}{25} = \dfrac{x - 5}{12}$

59. $\dfrac{3}{x + 2} - \dfrac{4}{x - 2} = 5$ **60.** $\dfrac{6}{x} + \dfrac{8}{x + 5} = 3$

61. $3(x + 3) = 5(1 - x) - 1$

62. $(x + 1)^2 + 2(x - 2) = (x + 1)(x - 2)$

63. $2x^3 - x^2 - 18x + 9 = 0$

64. $4x^3 + 12x^2 - 26x - 24 = 0$

65. $x^4 = 2x^3 + 1$ **66.** $x^5 = 3 + 2x^3$

67. $\dfrac{2}{x + 2} = 3$ **68.** $\dfrac{5}{x} = 1 + \dfrac{3}{x + 2}$

69. $|x - 3| = 4$ **70.** $\sqrt{x - 2} = 3$

In Exercises 71–74, determine any point(s) of intersection algebraically. Then verify your result numerically by creating a table of values for each equation.

71. $y = 2 - x$ **72.** $2x + y = 6$
$\quad\ y = 2x - 1$ $\qquad\ -x + y = 0$

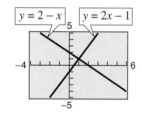   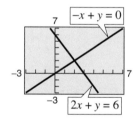

73. $x - y = -4$ **74.** $3x + y = 2$
$\quad\ x^2 - y = -2$ $\qquad\ x^3 + y = 0$

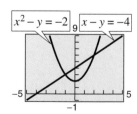   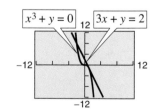

In Exercises 75–80, use a graphing utility to approximate any points of intersection (accurate to three decimal places) of the graphs of equations. Verify your results algebraically.

75. $y = 9 - 2x$
$y = x - 3$

76. $y = \frac{1}{3}x + 2$
$y = \frac{5}{2}x - 11$

77. $y = 4 - x^2$
$y = 2x - 1$

78. $y = -x$
$y = 2x - x^2$

79. $y = 2x^2$
$y = x^4 - 2x^2$

80. $y = x^3 - 3$
$y = 5 - 2x$

In Exercises 81–88, solve the quadratic equation by factoring. Check your solutions in the original equation.

81. $6x^2 + 3x = 0$

82. $9x^2 - 1 = 0$

83. $x^2 - 2x - 8 = 0$

84. $x^2 - 10x + 9 = 0$

85. $3 + 5x - 2x^2 = 0$

86. $2x^2 = 19x + 33$

87. $x^2 + 4x = 12$

88. $-x^2 + 8x = 12$

In Exercises 89–96, solve the equation by extracting square roots. List both the exact solution and the decimal solution rounded to two decimal places.

89. $x^2 = 7$

90. $9x^2 = 25$

91. $(x - 12)^2 = 18$

92. $(x - 5)^2 = 20$

93. $(2x - 1)^2 = 18$

94. $(4x + 7)^2 = 44$

95. $(x - 7)^2 = (x + 3)^2$

96. $(x + 5)^2 = (x + 4)^2$

In Exercises 97–102, solve the quadratic equation by completing the square. Verify your answer graphically.

97. $x^2 + 4x - 32 = 0$

98. $x^2 - 2x - 3 = 0$

99. $x^2 + 6x + 2 = 0$

100. $x^2 + 8x + 14 = 0$

101. $9x^2 - 18x + 3 = 0$

102. $9x^2 - 12x - 14 = 0$

In Exercises 103–108, use the Quadratic Formula to solve the equation. Use a graphing utility to verify your solutions graphically.

103. $2 + 2x - x^2 = 0$

104. $x^2 - 10x + 22 = 0$

105. $x^2 + 8x - 4 = 0$

106. $4x^2 - 4x - 4 = 0$

107. $4x^2 + 16x + 15 = 0$

108. $9x^2 - 6x - 35 = 0$

In Exercises 109–114, solve the equation by any convenient method.

109. $x^2 - 2x - 1 = 0$

110. $11x^2 + 33x = 0$

111. $(x + 3)^2 = 81$

112. $x^2 - 14x + 49 = 0$

113. $x^2 - x - \frac{11}{4} = 0$

114. $x^2 + 3x - \frac{3}{4} = 0$

In Exercises 115–130, find all solutions of the equation. Use a graphing utility to verify the solutions graphically.

115. $4x^4 - 18x^2 = 0$

116. $20x^3 - 125x = 0$

117. $x^4 - 81 = 0$

118. $x^6 - 64 = 0$

119. $5x^3 + 30x^2 + 45x = 0$

120. $9x^4 - 24x^3 + 16x^2 = 0$

121. $x^3 - 3x^2 - x + 3 = 0$

122. $x^4 + 2x^3 - 8x - 16 = 0$

123. $x^4 - 4x^2 + 3 = 0$

124. $x^4 + 5x^2 - 36 = 0$

125. $4x^4 - 65x^2 + 16 = 0$

126. $36t^4 + 29t^2 - 7 = 0$

127. $\frac{1}{t^2} + \frac{8}{t} + 15 = 0$

128. $6\left(\frac{s}{s + 1}\right)^2 + 5\left(\frac{s}{s + 1}\right) - 6 = 0$

129. $2x + 9\sqrt{x} - 5 = 0$

130. $3x^{1/3} + 2x^{2/3} = 5$

In Exercises 131–150, find all solutions of the equation algebraically. Check your solutions both algebraically and graphically.

131. $\sqrt{x - 10} - 4 = 0$

132. $\sqrt[3]{2x + 5} + 3 = 0$

133. $\sqrt{x + 1} - 3x = 1$

134. $\sqrt{x + 5} = \sqrt{x - 5}$

135. $\sqrt{x} - \sqrt{x - 5} = 1$

136. $\sqrt{x} + \sqrt{x - 20} = 10$

137. $(x - 5)^{2/3} = 16$

138. $(x^2 - x - 22)^{4/3} = 16$

139. $3x(x - 1)^{1/2} + 2(x - 1)^{3/2} = 0$

140. $4x^2(x - 1)^{1/3} + 6x(x - 1)^{4/3} = 0$

141. $\frac{20 - x}{x} = x$

142. $\frac{4}{x} - \frac{5}{3} = \frac{x}{6}$

143. $\frac{1}{x} - \frac{1}{x + 1} = 3$

144. $\frac{x}{x^2 - 4} + \frac{1}{x + 2} = 3$

145. $x = \frac{3}{x} + \frac{1}{2}$

146. $4x + 1 = \frac{3}{x}$

147. $|2x - 1| = 5$

148. $|3x + 2| = 7$

149. $|x| = x^2 + x - 3$

150. $|x - 10| = x^2 - 10x$

Graphical Analysis In Exercises 151–158, (a) use a graphing utility to graph the equation, (b) use the graph to approximate any *x*-intercepts of the graph, (c) set $y = 0$ and solve the resulting equation, and (d) compare the result with the *x*-intercepts of the graph.

151. $y = x^3 - 2x^2 - 3x$ **152.** $y = x^4 - 10x^2 + 9$

153. $y = \sqrt{11x - 30} - x$ **154.** $y = 2x - \sqrt{15 - 4x}$

155. $y = \dfrac{1}{x} - \dfrac{4}{x - 1} - 1$ **156.** $y = x + \dfrac{9}{x + 1} - 5$

157. $y = |x + 1| - 2$ **158.** $y = |x - 2| - 3$

159. *Per Capita Utilization* The per capita utilization (in pounds) of nectarines and peaches *N* and cucumbers *C* from 1991 through 1996 can be modeled by

$$N = -0.37t + 6.88$$

$$C = 0.27t + 4.42$$

where $t = 1$ represents 1991. (Source: U.S. Department of Agriculture)

(a) What does the intersection of the graphs of these equations represent?

(b) Find the point of intersection of the graphs algebraically.

(c) Verify your answer to part (b) using the *zoom* and *trace* features of a graphing utility.

160. *Saturated Steam* The temperature *T* (in degrees Fahrenheit) of saturated steam increases as pressure increases. This relationship is approximated by

$$T = 75.82 - 2.11x + 43.51\sqrt{x}, \quad 5 \le x \le 40$$

where *x* is the absolute pressure in pounds per square inch.

(a) Use a graphing utility to graph the temperature equation over the specified domain.

(b) The temperature of steam at sea level $(x = 14.696)$ is 212°F. Evaluate the model at this pressure and verify the result graphically.

(c) Use the model to approximate the pressure for a steam temperature of 240°F.

161. *Demand Equation* The marketing department at a publisher is asked to determine the price of a book. The department determines that the demand for the book depends on the price of the book according to the formula

$$p = 40 - \sqrt{0.0001x + 1}, \quad x \ge 0$$

where *p* is the price per book in dollars and *x* is the number of books sold at the given price. For instance, if the price were \$39, then (according to the model) no one would be willing to buy the book. On the other hand, if the price were \$17.60, 5 million copies could be sold.

(a) Use a graphing utility to graph the demand equation over the specified domain.

(b) If the publisher set the price at \$12.95, how many copies would be sold?

Synthesis

True or False? In Exercises 162 and 163, determine whether the statement is true or false. Justify your answer.

162. An equation can never have more than one extraneous solution.

163. Two linear equations can have either one point of intersection or no points of intersection.

164. *Exploration* Given that *a* and *b* are nonzero real numbers, determine the solutions of the equations.

(a) $ax^2 + bx = 0$ (b) $ax^2 - ax = 0$

<table>
<tr><td>

P.4   Lines in the Plane

The Slope of a Line

In this section, you will study lines and their equations. The **slope** of a nonvertical line represents the number of units a line rises or falls vertically for each unit of horizontal change from left to right. For instance, consider the two points (x_1, y_1) and (x_2, y_2) on the line shown in Figure P.42. As you move from left to right along this line, a change of $(y_2 - y_1)$ units in the vertical direction corresponds to a change of $(x_2 - x_1)$ units in the horizontal direction. That is,

$$y_2 - y_1 = \text{the change in } y$$

and

$$x_2 - x_1 = \text{the change in } x.$$

The slope of the line is given by the ratio of these two changes.

</td><td>

What You Should Learn:

- How to find the slopes of lines
- How to write linear equations given points on lines and their slopes
- How to use slope-intercept forms of linear equations to sketch graphs of lines
- How to use slope to identify parallel and perpendicular lines

Why You Should Learn It:

Linear equations can be used to model and solve real-life problems. For instance, Exercise 95 on page 55 shows how to use a linear equation to model the average annual salaries of Major League Baseball players from 1988 to 1998.

</td></tr>
</table>

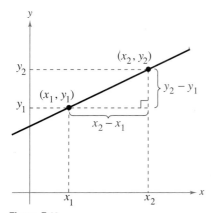

Figure P.42

Allsport

Definition of the Slope of a Line

The **slope** m of the nonvertical line through (x_1, y_1) and (x_2, y_2) is

$$m = \frac{y_2 - y_1}{x_2 - x_1} = \frac{\text{change in } y}{\text{change in } x}$$

where $x_1 \neq x_2$.

When this formula is used, the *order of subtraction* is important. Given two points on a line, you are free to label either one of them as (x_1, y_1) and the other as (x_2, y_2). However, once this has been done, you must form the numerator and denominator using the same order of subtraction.

$$m = \underbrace{\frac{y_2 - y_1}{x_2 - x_1}}_{\text{Correct}} \qquad m = \underbrace{\frac{y_1 - y_2}{x_1 - x_2}}_{\text{Correct}} \qquad m = \underbrace{\frac{y_2 - y_1}{x_1 - x_2}}_{\text{Incorrect}}$$

Throughout this text, the term *line* always means a *straight* line.

EXAMPLE 1   Finding the Slope of a Line

Find the slope of the line passing through each pair of points.

a. $(-2, 0)$ and $(3, 1)$ **b.** $(-1, 2)$ and $(2, 2)$ **c.** $(0, 4)$ and $(1, -1)$

Solution

Difference in *y*-values

a. $m = \dfrac{y_2 - y_1}{x_2 - x_1} = \dfrac{1 - 0}{3 - (-2)} = \dfrac{1}{3 + 2} = \dfrac{1}{5}$

Difference in *x*-values

b. $m = \dfrac{2 - 2}{2 - (-1)} = \dfrac{0}{3} = 0$

c. $m = \dfrac{-1 - 4}{1 - 0} = \dfrac{-5}{1} = -5$

The graphs of the three lines are shown in Figure P.43. Note that the square setting gives the correct "steepness" of the lines.

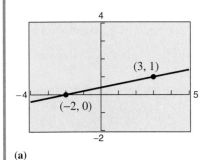
(a)

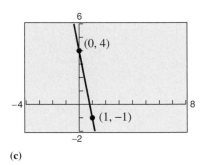
(b)

(c)

Figure P.43

The definition of slope does not apply to vertical lines. For instance, consider the points $(3, 4)$ and $(3, 1)$ on the vertical line shown in Figure P.44. Applying the formula for slope, you obtain

$$m = \frac{4 - 1}{3 - 3} = \frac{3}{0}. \qquad \text{Undefined}$$

Because division by zero is undefined, the slope of a vertical line is undefined.

From the slopes of the lines shown in Figures P.43 and P.44, you can make the following generalizations about the slope of a line.

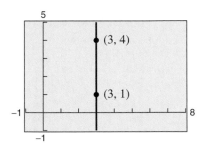
Figure P.44

The Slope of a Line

1. A line with positive slope $(m > 0)$ *rises* from left to right.
2. A line with negative slope $(m < 0)$ *falls* from left to right.
3. A line with zero slope $(m = 0)$ is *horizontal*.
4. A line with undefined slope is *vertical*.

> **Exploration**
>
> Use a graphing utility to compare the slopes of the lines $y = 0.5x$, $y = x$, $y = 2x$, and $y = 4x$. What do you observe about these lines? Compare the slopes of the lines $y = -0.5x$, $y = -x$, $y = -2x$, and $y = -4x$. What do you observe about these lines? (*Hint:* Use a square setting to guarantee a true geometric perspective.)

The Point-Slope Form of the Equation of a Line

If you know the slope of a line *and* you also know the coordinates of one point on the line, you can find an equation for the line. For instance, in Figure P.45, let (x_1, y_1) be a given point on the line whose slope is m. If (x, y) is any *other* point on the line, it follows that

$$\frac{y - y_1}{x - x_1} = m.$$

This equation in the variables x and y can be rewritten in the **point-slope form** of the equation of a line.

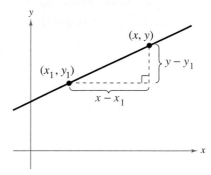

Figure P.45

Point-Slope Form of the Equation of a Line

The **point-slope form** of the equation of the line that passes through the point (x_1, y_1) and has a slope of m is

$$y - y_1 = m(x - x_1).$$

EXAMPLE 2   The Point-Slope Form of the Equation of a Line

Find an equation of the line that passes through the point $(1, -2)$ and has a slope of 3.

Solution

$$
\begin{aligned}
y - y_1 &= m(x - x_1) && \text{Point-slope form} \\
y - (-2) &= 3(x - 1) && \text{Substitute for } y_1, m, \text{ and } x_1. \\
y + 2 &= 3x - 3 \\
y &= 3x - 5
\end{aligned}
$$

This line is shown in Figure P.46.

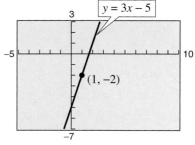

Figure P.46

The point-slope form can be used to find an equation of a nonvertical line passing through two points (x_1, y_1) and (x_2, y_2). First, use the formula for the slope of the line passing through two points.

$$m = \frac{y_2 - y_1}{x_2 - x_1}$$

Then, once you know the slope, use the point-slope form to obtain the equation

$$
\begin{aligned}
y - y_1 &= m(x - x_1) \\
&= \frac{y_2 - y_1}{x_2 - x_1}(x - x_1).
\end{aligned}
$$

This is sometimes called the **two-point form** of the equation of a line.

EXAMPLE 3   A Linear Model for Sales Prediction

During 1997, Barnes & Noble's net sales were $2.8 billion, and in 1998 net sales were $3.0 billion.   (Source: Barnes & Noble, Inc.)

a. Write a linear equation giving the net sales y in terms of the year x.

b. Use the equation to estimate the net sales during 2000.

Solution

a. Let $x = 7$ represent 1997. In Figure P.47, let $(7, 2.8)$ and $(8, 3.0)$ be two points on the line representing the net sales. The slope of the line passing through these two points is

$$m = \frac{3.0 - 2.8}{8 - 7} = 0.2. \qquad m = \frac{y_2 - y_1}{x_2 - x_1}$$

By the point-slope form, the equation of the line is as follows.

$$y - y_1 = m(x - x_1) \qquad \text{Point-slope form}$$

$$y - 2.8 = 0.2(x - 7) \qquad \text{Substitute for } y_1, m, \text{ and } x_1.$$

$$y = 0.2x - 1.4 + 2.8$$

$$y = 0.2x + 1.4 \qquad \text{Simplify.}$$

b. Using the equation from part (a), estimate the 2000 net sales $(x = 10)$ to be

$$y = 0.2(10) + 1.4$$

$$= 2 + 1.4$$

$$= \$3.4 \text{ billion.}$$

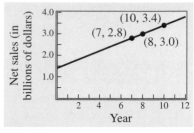

Figure P.47

The approximation method illustrated in Example 3 is **linear extrapolation.** Note in Figure P.48 that for linear extrapolation, the estimated point lies outside of the given points. When the estimated point lies *between* two given points, the procedure is called **linear interpolation.**

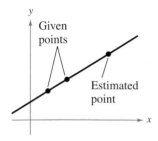

Linear Extrapolation
Figure P.48

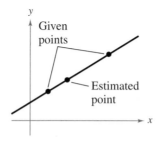

Linear Interpolation

Library of Functions

In the next section, you will be introduced to the precise meaning of the term *function*. The simplest type of function is a linear function and has the form

$$y = mx + b.$$

As its name implies, the graph of a linear function is a line that has a slope of m and a y-intercept at $(0, b)$.

Consult the Library of Functions Summary inside the front cover for a description of the linear function.

Sketching Graphs of Lines

Many problems in coordinate geometry can be classified as follows.

1. Given a graph (or parts of it), find its equation.

2. Given an equation, sketch its graph.

For lines, the first problem is solved easily by using the point-slope form. This formula, however, is not particularly useful for solving the second type of problem. The form that is better suited to graphing linear equations is the **slope-intercept form,** $y = mx + b$, of the equation of a line.

Slope-Intercept Form of the Equation of a Line

The graph of the equation

$$y = mx + b$$

is a line whose slope is m and whose y-intercept is $(0, b)$.

A computer simulation of this concept appears in the *Interactive* CD-ROM and *Internet* versions of this text.

EXAMPLE 1   Using the Slope-Intercept Form

Determine the slope and y-intercept of each linear equation. Then describe its graph.

a. $x + y = 2$ **b.** $y = 2$

Algebraic Solution

a. Begin by writing the equation in slope-intercept form.

$$x + y = 2 \qquad \text{Write original equation.}$$

$$y = 2 - x \qquad \text{Subtract } x \text{ from each side.}$$

$$y = -x + 2 \qquad \text{Slope-intercept form}$$

From the slope-intercept form of the equation, the slope is -1 and the y-intercept is $(0, 2)$. Because the slope is negative, you know that the graph of the equation is a line that falls one unit for every unit it moves to the right.

b. By writing the equation $y = 2$ in slope-intercept form

$$y = (0)x + 2$$

you can see that the slope is 0 and the y-intercept is $(0, 2)$. A zero slope implies that the line is horizontal.

Graphical Solution

a. Solve the equation for y to obtain $y = 2 - x$. Enter this equation in your graphing utility. Use a decimal viewing window to graph the equation as shown in Figure P.49(a).

To find the y-intercept, use the *value* or *trace* feature. When $x = 0$, $y = 2$. So, the y-intercept is $(0, 2)$. To find the slope, continue to use the *trace* feature. Move the cursor along the line until $x = 1$. At this point, $y = 1$. So the graph falls 1 unit for every unit it moves to the right, and the slope is -1.

b. Enter the equation $y = 2$ in your graphing utility and graph the equation as shown in Figure P.49(b). Use the *trace* feature to verify the y-intercept $(0, 2)$ and to see that the value of y is the same for all values of x. So, the slope of the horizontal line is 0.

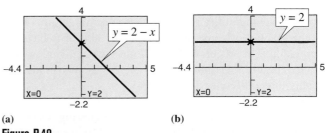

(a)   (b)

Figure P.49

From the slope-intercept form of the equation of a line, you can see that a horizontal line ($m = 0$) has an equation of the form $y = b$. This is consistent with the fact that each point on a horizontal line through $(0, b)$ has a y-coordinate of b.

Similarly, each point on a vertical line through $(a, 0)$ has an x-coordinate of a. So, a vertical line has an equation of the form $x = a$. This equation cannot be written in the slope-intercept form, because the slope of a vertical line is undefined. However, *every* line has an equation that can be written in the **general form**

$$Ax + By + C = 0 \qquad \text{General form of the equation of a line}$$

where A and B are not *both* zero.

Summary of Equations of Lines

1. General form: $Ax + By + C = 0$
2. Vertical line: $x = a$
3. Horizontal line: $y = b$
4. Slope-intercept form: $y = mx + b$
5. Point-slope form: $y - y_1 = m(x - x_1)$

Exploration

Graph the lines $y = 2x + 1$, $y = \frac{1}{2}x + 1$, and $y = -2x + 1$ in the same viewing window. What do you observe?

Graph the lines $y = 2x + 1$, $y = 2x$, and $y = 2x - 1$ in the same viewing window. What do

EXAMPLE 5   Different Viewing Windows

The graphs of the two lines

$$y = -x - 1 \qquad \text{and} \qquad y = -10x - 1$$

are shown in Figure P.50. Even though the slopes of these lines are quite different (-1 and -10, respectively), the graphs seem misleadingly similar because the viewing windows are different.

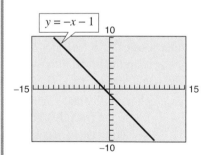

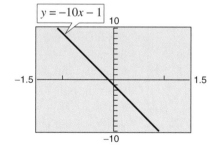

Figure P.50

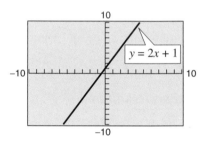

(a)

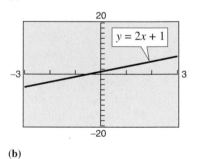

(b)

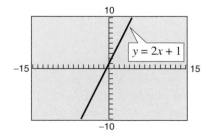

(c)
Figure P.51

When a graphing utility is used to sketch a straight line, it is important to realize that the graph of the line may not visually appear to have the slope indicated by its equation. This occurs because of the viewing window used for the graph. For instance, Figure P.51 shows graphs of $y = 2x + 1$ produced on a graphing utility using three different viewing windows.

Notice that the slopes in Figure P.51(a) and (b) do not visually appear to be equal to 2. However, if you use the *square* viewing window, as in Figure P.51(c), the slope visually appears to be 2. In general, two graphs of the same equation can appear to be quite different depending on the viewing window selected.

Parallel and Perpendicular Lines

The slope of a line is a convenient tool for determining whether two lines are parallel or perpendicular.

Parallel Lines

Two distinct nonvertical lines are **parallel** if and only if their slopes are equal.

EXAMPLE 6   Equations of Parallel Lines

Find the slope-intercept form of the equation of the line that passes through the point $(2, -1)$ and is parallel to the line $2x - 3y = 5$.

Solution

Begin by finding the slope of the given line.

$$2x - 3y = 5 \qquad \text{Write original equation.}$$

$$-2x + 3y = -5 \qquad \text{Multiply by } -1.$$

$$3y = 2x - 5 \qquad \text{Add } 2x \text{ to each side.}$$

$$y = \frac{2}{3}x - \frac{5}{3} \qquad \text{Write in slope-intercept form.}$$

Therefore, the given line has a slope of $m = \frac{2}{3}$. Because any line parallel to the given line must also have a slope of $\frac{2}{3}$, the required line through $(2, -1)$ has the following equation.

$$y - (-1) = \frac{2}{3}(x - 2) \qquad \text{Substitute for } m, x_1, \text{ and } y_1 \text{ in point-slope form.}$$

$$y + 1 = \frac{2}{3}x - \frac{4}{3} \qquad \text{Simplify.}$$

$$y = \frac{2}{3}x - \frac{4}{3} - 1 \qquad \text{Subtract 1 from each side.}$$

$$y = \frac{2}{3}x - \frac{7}{3} \qquad \text{Write in slope-intercept form.}$$

Notice the similarity between the slope-intercept form of the original equation and the slope-intercept form of the parallel equation. The graphs of both equations are shown in Figure P.52.

STUDY T!P

Be careful when you graph equations such as $y = \frac{2}{3}x - \frac{7}{3}$ on your graphing utility. A common mistake is to type it in as

$$Y_1 = 2/3X - 7/3,$$

which may not be interpreted as the original equation by your graphing utility. You should use one of the following formulas.

$$Y_1 = 2X/3 - 7/3$$

$$Y_1 = (2/3)X - 7/3$$

Do you see why?

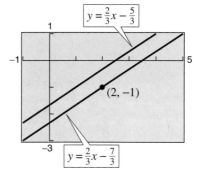

Figure P.52

Perpendicular Lines

Two nonvertical lines are **perpendicular** if and only if their slopes are negative reciprocals of each other. That is,

$$m_1 = -\frac{1}{m_2}.$$

EXAMPLE 7   Equations of Perpendicular Lines

Find an equation of the line that passes through the point $(2, -1)$ and is perpendicular to the line $2x - 3y = 5$.

Solution

By writing the given line in the form $y = \frac{2}{3}x - \frac{5}{3}$, you can see that the line has a slope of $\frac{2}{3}$. So, any line that is perpendicular to this line must have a slope of $-\frac{3}{2}$ (because $-\frac{3}{2}$ is the negative reciprocal of $\frac{2}{3}$). Therefore, the required line through the point $(2, -1)$ has the following equation.

$$y - (-1) = -\frac{3}{2}(x - 2) \qquad \text{Substitute for } m, x_1, \text{ and } y_1 \text{ in point-slope form.}$$

$$y + 1 = -\frac{3}{2}x + 3 \qquad \text{Simplify.}$$

$$y = -\frac{3}{2}x + 3 - 1 \qquad \text{Subtract 1 from each side.}$$

$$y = -\frac{3}{2}x + 2 \qquad \text{Slope-intercept form}$$

The graphs of both equations are shown in Figure P.53.

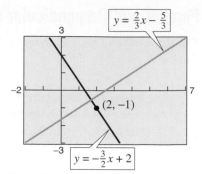

Figure P.53

EXAMPLE 8   Graphs of Perpendicular Lines

Use a graphing utility to graph the lines

$$y = x + 1$$

and

$$y = -x + 3.$$

Display *both* graphs in the same viewing window. The lines are supposed to be perpendicular (they have slopes of $m_1 = 1$ and $m_2 = -1$). Do they appear to be perpendicular on the display?

Solution

If the viewing window is nonsquare, as in Figure P.54(a), the two lines will not appear perpendicular. If, however, the viewing window is square, as in Figure P.54(b), the lines will appear perpendicular.

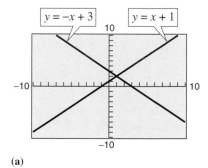

(a)

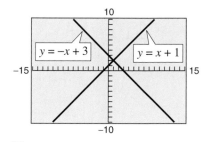

(b)

Figure P.54

Writing About Math   *An Application of Slope*

In 1990, a college had an enrollment of 5500 students. By 2000, the enrollment had increased to 7000 students.

a. What was the average annual change in enrollment from 1990 to 2000?

b. Use the average annual change in enrollment to estimate the enrollments in 1993, 1997, and 1999.

c. Write the equation of the line that represents the data in part (b). What is its slope? Interpret the slope in the context of the problem.

d. Write a short paragraph discussing the concepts of *slope* and *average rate of change*.

P.4 Exercises

In Exercises 1 and 2, identify the line that has the specified slope.

1. (a) $m = \frac{2}{3}$ (b) m is undefined.   (c) $m = -2$

2. (a) $m = 0$ (b) $m = -\frac{3}{4}$ (c) $m = 1$

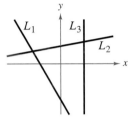

FIGURE FOR 1

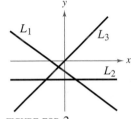

FIGURE FOR 2

In Exercises 3 and 4, sketch the line through the point with each indicated slope on the same set of coordinate axes.

Point	Slopes
3. $(2, 3)$	(a) 0 (b) 1 (c) 2 (d) -3
4. $(-4, 1)$	(a) 3 (b) -3 (c) $\frac{1}{2}$ (d) Undefined

In Exercises 5–10, estimate the slope of the line.

5.
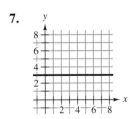

6.
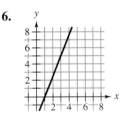

7.
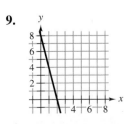

8.
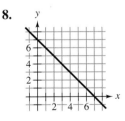

9.
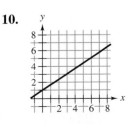

10.

In Exercises 11–14, plot the points and find the slope of the line passing through the points. Verify the slope using the *draw* feature on your graphing utility to graph the line segment connecting the two points. (Use a square setting.)

11. $(0, -10), (-4, 0)$ **12.** $(2, 4), (4, -4)$

13. $(-6, -1), (-6, 4)$ **14.** $(-3, -2), (1, 6)$

In Exercises 15–20, you are given the slope of the line and a point on the line. Find three additional points through which the line passes. (There are many correct answers.)

	Point	Slope
15.	$(2, 1)$	$m = 0$
16.	$(-4, 1)$	m is undefined.
17.	$(-5, 4)$	$m = 2$
18.	$(0, -9)$	$m = -2$
19.	$(7, -2)$	$m = \frac{1}{2}$
20.	$(-1, -6)$	$m = -\frac{1}{2}$

In Exercises 21–24, determine whether the lines L_1 and L_2 passing through the pairs of points are parallel, perpendicular, or neither. Use a graphing utility to graph the line segments connecting the pairs of points on the respective lines. (Use a square setting.)

21. L_1: $(0, -1), (5, 9)$ **22.** L_1: $(-2, -1), (1, 5)$
  L_2: $(0, 3), (4, 1)$ L_2: $(1, 3), (5, -5)$

23. L_1: $(3, 6), (-6, 0)$ **24.** L_1: $(4, 8), (-4, 2)$
  L_2: $(0, -1), \left(5, \frac{7}{3}\right)$ L_2: $(3, -5), \left(-1, \frac{1}{3}\right)$

In Exercises 25–32, (a) find the slope and y-intercept (if possible) of the equation of the line algebraically, (b) sketch the line by hand, and (c) use a graphing utility to verify your answers to parts (a) and (b).

25. $5x - y + 3 = 0$ **26.** $2x + 3y - 9 = 0$

27. $5x - 2 = 0$ **28.** $3x + 7 = 0$

29. $3y + 5 = 0$ **30.** $-11 - 8y = 0$

31. $7x + 6y - 30 = 0$ **32.** $x - y - 10 = 0$

In Exercises 33–42, find the general form of the equation of the line that passes through the given point and has the indicated slope. Sketch the line by hand. Use a graphing utility to verify your sketch.

	Point	*Slope*
33.	$(0, -2)$	$m = 3$
34.	$(0, 10)$	$m = -1$
35.	$(-3, 6)$	$m = -2$
36.	$(0, 0)$	$m = 4$
37.	$(4, 0)$	$m = -\frac{1}{3}$
38.	$(-2, -5)$	$m = \frac{3}{4}$
39.	$(6, -1)$	m is undefined.
40.	$(-10, 4)$	$m = 0$
41.	$\left(-\frac{1}{2}, \frac{3}{2}\right)$	$m = -3$
42.	$(2.3, -8.5)$	$m = -\frac{5}{2}$

In Exercises 43–52, find the general form of the equation of the line that passes through the points. Use a graphing utility to sketch the line.

43. $(5, -1), (-5, 5)$ **44.** $(4, 3), (-4, -4)$

45. $(-8, 1), (-8, 7)$ **46.** $(-1, 4), (6, 4)$

47. $\left(2, \frac{1}{2}\right), \left(\frac{1}{2}, \frac{5}{4}\right)$ **48.** $\left(1, 1\right), \left(6, -\frac{2}{3}\right)$

49. $\left(-\frac{1}{10}, -\frac{3}{5}\right), \left(\frac{9}{10}, -\frac{9}{5}\right)$ **50.** $\left(\frac{3}{4}, \frac{3}{2}\right), \left(-\frac{4}{3}, \frac{7}{4}\right)$

51. $(1, 0.6), (-2, -0.6)$ **52.** $(-8, 0.6), (2, -2.4)$

Exploration   **In Exercises 53 and 54, use the values of** *a* **and** *b* **and a graphing utility to graph the equation of the line**

$$\frac{x}{a} + \frac{y}{b} = 1, \qquad a \neq 0, b \neq 0.$$

Use the graphs to make a conjecture about what *a* **and** *b* **represent. Verify your conjecture.**

53. $a = 5, \quad b = -3$ **54.** $a = -6, \quad b = 2$

In Exercises 55–58, use the results of Exercises 53 and 54 to write an equation of the line that passes through the points.

55. x-intercept: $(2, 0)$ **56.** x-intercept: $(-5, 0)$
   y-intercept: $(0, 3)$ y-intercept: $(0, -4)$

57. x-intercept: $\left(-\frac{1}{6}, 0\right)$ **58.** x-intercept: $\left(\frac{3}{4}, 0\right)$
   y-intercept: $\left(0, -\frac{2}{3}\right)$ y-intercept: $\left(0, \frac{4}{5}\right)$

In Exercises 59 and 60, use a graphing utility to graph the equation using each of the suggested viewing windows. Describe the difference between the two graphs.

59. $y = 0.5x - 3$

Xmin = -5		Xmin = -2
Xmax = 10		Xmax = 10
Xscl = 1		Xscl = 1
Ymin = -1		Ymin = -4
Ymax = 10		Ymax = 1
Yscl = 1		Yscl = 1

60. $y = -8x + 5$

Xmin = -5		Xmin = -5
Xmax = 5		Xmax = 10
Xscl = 1		Xscl = 1
Ymin = -10		Ymin = -80
Ymax = 10		Ymax = 80
Yscl = 1		Yscl = 20

Graphical Analysis   **In Exercises 61–64, use a graphing utility to graph the three equations in the same viewing window. Adjust the viewing window so that the slope appears visually correct. Identify any relationships that exist among the lines. Use the slope of the lines to verify your results.**

61. (a) $y = 2x$ (b) $y = -2x$ (c) $y = \frac{1}{2}x$

62. (a) $y = \frac{2}{3}x$ (b) $y = -\frac{3}{2}x$ (c) $y = \frac{2}{3}x + 2$

63. (a) $y = -\frac{1}{2}x$ (b) $y = -\frac{1}{2}x + 3$ (c) $y = 2x - 4$

64. (a) $y = x - 8$ (b) $y = x + 1$ (c) $y = -x + 3$

In Exercises 65–70, write equations of the lines through the given point (a) parallel to the given line and (b) perpendicular to the given line.

	Point	*Line*
65.	$(2, 1)$	$4x - 2y = 3$
66.	$(-3, 2)$	$x + y = 7$
67.	$\left(-\frac{2}{3}, \frac{7}{8}\right)$	$3x + 4y = 7$
68.	$\left(\frac{7}{8}, \frac{3}{4}\right)$	$5x + 3y = 0$
69.	$(2.5, 6.8)$	$x - y = 4$
70.	$(-3.9, -1.4)$	$6x + 2y = 9$

In Exercises 71 and 72, find a relationship between x **and** y **such that** (x, y) **is equidistant from the two points.**

71. $(4, -1), (-2, 3)$ **72.** $(3, -2), (-7, 1)$

73. *Business* The slopes are the slopes of lines representing annual sales y in terms of time x in years. Use each slope to interpret any change in annual sales for a 1-year increase in time.

(a) The line has a slope of $m = 135$.

(b) The line has a slope of $m = 0$.

(c) The line has a slope of $m = -40$.

74. *Business* The slopes are the slopes of lines representing daily revenues y in terms of time x in days. Use each slope to interpret any change in daily revenues for a 1-day increase in time.

(a) The line has a slope of $m = 400$.

(b) The line has a slope of $m = 100$.

(c) The line has a slope of $m = 0$.

75. *Business* The graph shows the earnings per share of stock for the Kellogg Company for the years 1988 through 1998. (Source: Kellogg Company)

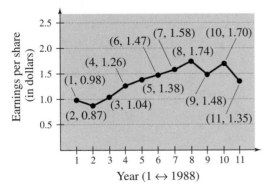

(a) Use the slopes to determine the year(s) when the earnings per share showed the greatest increase and decrease.

(b) Find the equation of the line between the years 1988 and 1998.

(c) Interpret the meaning of the slope in the equation from part (b) in the context of the problem.

(d) Use the equation from part (b) to estimate the earnings per share of stock for the year 2001. Do you think this is an accurate estimation? Explain.

76. *Business* The graph shows the dividends declared per share of stock for the Colgate-Palmolive Company for the years 1988 through 1998. (Source: Colgate-Palmolive Company)

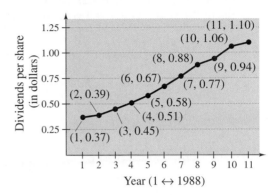

(a) Use the slopes to determine the years when the dividends declared per share showed the greatest increase and the smallest increase.

(b) Find the equation of the line between the years 1988 and 1998.

(c) Interpret the meaning of the slope in the equation from part (b) in the context of the problem.

(d) Use the equation from part (b) to estimate the dividends declared per share for the year 2001. Do you think this is an accurate estimation? Explain.

77. *Driving* When driving down a mountain road, you notice warning signs indicating that it is a "12% grade." This means that the slope of the road is $-\frac{12}{100}$. Approximate the amount of horizontal change in your position if you note from elevation markers that you have descended 2000 feet vertically.

78. *Attic Height* The "rise to run" ratio of the roof of a house determines the steepness of the roof. Suppose the rise to run ratio of a roof is 3 to 4. Determine the maximum height in the attic of the house if the house is 32 feet wide.

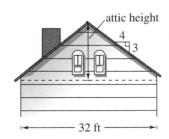

Rate of Change In Exercises 79–82, you are given the dollar value of a product in 2001 and the rate at which the value of the product is expected to change during the next 5 years. Use this information to write a linear equation that gives the dollar value V of the item in terms of the year t. (Let $t = 1$ represent 2001.)

	2001 Value	Rate
79.	$2540	$125 increase per year
80.	$156	$4.50 increase per year
81.	$20,400	$2000 decrease per year
82.	$245,000	$5600 decrease per year

Graphical Interpretation In Exercises 83–86, match the description with its graph. Also determine the slope and how it is interpreted in the situation. [The graphs are labeled (a), (b), (c), and (d).]

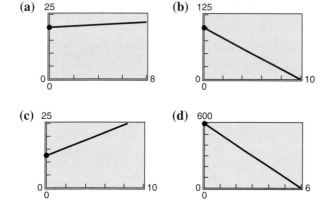

83. A person is paying $10 per week to a friend to repay a $100 loan.

84. An employee is paid $12.50 per hour plus $1.50 for each unit produced per hour.

85. A sales representative receives $20 per day for food plus $0.25 for each mile traveled.

86. A word processor that was purchased for $600 depreciates $100 per year.

87. *Temperature* Find the equation of the line that shows the relationship between the temperature in degrees Celsius C and degrees Fahrenheit F. Remember that water freezes at 0°C (32°F) and boils at 100°C (212°F).

88. *Temperature* Use the result of Exercise 87 to complete the table.

C		−10°	10°			177°
F	0°			68°	90°	

89. *Finance* Your salary was $28,500 in 1998 and $32,900 in 2000. If your salary follows a linear growth pattern, what will your salary be in 2003?

90. *College Enrollment* A small college had 2546 students in 1998 and 2702 students in 2000. If the enrollment follows a linear growth pattern, how many students will the college have in 2004?

91. *Business* A small business purchases a fax machine for $875. After 5 years, the fax machine will be outdated and have no value.

(a) Write a linear equation giving the value V of the fax machine during the 5 years it will be used.

(b) Use a graphing utility to graph the linear equation representing the depreciation of the fax machine, and use the *value* or *trace* feature to complete the table.

t	0	1	2	3	4	5
V						

(c) Verify your answers in part (b) algebraically by using the equation you found in part (a).

92. *Business* A small business purchases a computer network system for $25,000. After 10 years, the system will have to be replaced. Its value at that time is expected to be $2000.

(a) Write a linear equation giving the value V of the system during the 10 years it will be used.

(b) Use a graphing utility to graph the linear equation representing the depreciation of the system, and use the *value* or *trace* feature to complete the table.

t	0	1	2	3	4	5	6	7	8	9	10
V											

(c) Verify your answers in part (b) algebraically by using the equation you found in part (a).

93. *Business* A contractor purchases a bulldozer for $36,500. The bulldozer requires an average expenditure of $5.25 per hour for fuel and maintenance, and the operator is paid $11.50 per hour.

(a) Write a linear equation giving the total cost C of operating the bulldozer for t hours. (Include the purchase cost of the bulldozer.)

(b) Assuming that customers are charged $27 per hour of bulldozer use, write an equation for the revenue R derived from t hours of use.

(c) Use the formula for profit ($P = R - C$) to write an equation for the profit derived from t hours of use.

(d) Use the result of part (c) to find the break-even point (the number of hours the bulldozer must be used to yield a profit of 0 dollars).

94. *Real Estate Purchase*   A real estate office handles an apartment complex with 50 units. When the rent per unit is $580 per month, all 50 units are occupied. However, when the rent is $625 per month, the average number of occupied units drops to 47. Assume that the relationship between the monthly rent p and the demand x is linear.

(a) Write the equation of the line giving the demand x in terms of the rent p.

(b) Use a graphing utility to graph the demand equation and use the *trace* feature to estimate the number of units occupied if the rent is raised to $655. Verify your answer algebraically.

(c) Use the demand equation to estimate the number of units occupied if the rent is lowered to $595. Verify your answer graphically.

95. *Sports*   The average annual salaries of Major League Baseball players (in thousands of dollars) from 1988 to 1998 are shown in the scatter plot. Let y represent the average salary and let t represent the year, with $t = 0$ corresponding to 1988. (Source: Major League Baseball Player Relations Committee)

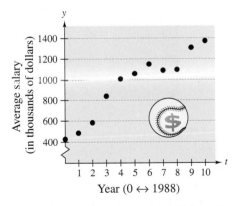

Year (0 ↔ 1988)

(a) Use the regression capabilities of a graphing utility to find the line that best fits the data.

(b) Use the regression line to estimate the average salary in the year 2000.

(c) Interpret the meaning of the slope of the regression line.

Synthesis

True or False?   **In Exercises 96 and 97, determine whether the statement is true or false. Justify your answer.**

96. The line through $(-8, 2)$ and $(-1, 4)$ and the line through $(0, -4)$ and $(-7, 7)$ are parallel.

97. If the points $(10, -3)$ and $(2, -9)$ lie on the same line, then the point $\left(-12, -\frac{37}{2}\right)$ also lies on that line.

98. *Writing*   Explain how you could show that the points $A(2, 3)$, $B(2, 9)$, and $C(7, 3)$ are the vertices of a right triangle.

99. *Think About It*   The slopes of two lines are -4 and $\frac{5}{2}$. Which is steeper?

100. *Writing*   Write a brief paragraph explaining whether or not any pair of points on a line can be used to calculate the slope of the line.

101. *Think About It*   Is it possible for two lines with positive slopes to be perpendicular? Explain.

P.5 Functions

Introduction to Functions

Many everyday phenomena involve pairs of quantities that are related to each other by some rule of correspondence. The mathematical term for such a rule of correspondence is a **relation.** Here are two examples.

1. The simple interest I earned on an investment of $1000 for 1 year is related to the annual interest rate r by the formula $I = 1000r$.

2. The area A of a circle is related to its radius r by the formula $A = \pi r^2$.

Not all relations have simple mathematical formulas. For instance, people commonly match up NFL starting quarterbacks with touchdown passes, and hours of the day with temperature. In each of these cases, however, there is some relation that matches each item from one set with exactly one item from a different set. Such a relation is called a **function.**

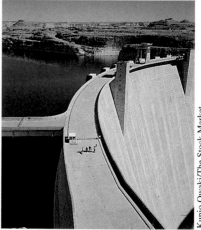

> ## Definition of a Function
>
> A **function** f from a set A to a set B is a relation that assigns to each element x in the set A exactly one element y in the set B. The set A is the **domain** (or set of inputs) of the function f, and the set B contains the **range** (or set of outputs).

To help understand this definition, look at the function in Figure P.55.

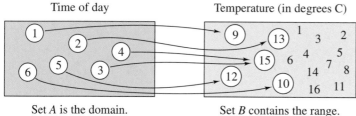

Set A is the domain.
Inputs: 1, 2, 3, 4, 5, 6

Set B contains the range.
Outputs: 9, 10, 12, 13, 15

Figure P.55

This function can be represented by the ordered pairs $\{(1, 9°), (2, 13°), (3, 15°), (4, 15°), (5, 12°), (6, 10°)\}$. In each ordered pair, the first coordinate is the **input** and the second coordinate is the **output.** In this example, note the following characteristics of a function.

1. Each element in A must be matched with an element of B.

2. Some elements in B may not be matched with any element in A.

3. Two or more elements of A may be matched with the same element of B.

The converse of the third statement is not true. That is, an element of A (the domain) cannot be matched with two different elements of B. In other words, each element of A must be matched with only one element of B.

Functions are commonly represented in four ways.

1. *Verbally* by a sentence that describes how the input variable is related to the output variable
2. *Numerically* by a table or a list of ordered pairs that matches input values with output values
3. *Graphically* by points on a graph in a coordinate plane in which the input values are represented by the horizontal axis and the output values are represented by the vertical axis
4. *Algebraically* by an equation in two variables

In the following example, you are asked to decide whether the given relation is a function. To do this, you must decide whether each input value is matched with exactly one output value. If any input value is matched with two or more output values, the relation is not a function.

Library of Functions

Many functions do not have simple mathematical formulas but are defined by real-life data. Such functions arise when you are using collections of data to model real-life applications. You will see that it is often convenient to approximate the data using a mathematical model or formula.

Consult the Library of Functions Summary inside the front cover for a description of functions used to model data.

EXAMPLE 1   Testing for Functions

Decide whether the description represents *y* as a function of *x*.

a. The input value *x* is the number of representatives from a state and the output value *y* is the number of senators.

b.

Input x	2	2	3	4	5
Output y	11	10	8	5	1

c.

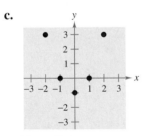

Figure P.56

STUDY T!P

Be sure you see that the *range* of a function is not the same as the use of *range* relating to the viewing window.

Solution

a. This verbal description does describe *y* as a function of *x*. Regardless of the value of *x*, the value of *y* is always 2. Such functions are called *constant functions.*

b. This table does not describe *y* as a function of *x*. The input value 2 is mapped to two different *y*-values.

c. The graph in Figure P.56 does describe *y* as a function of *x*. No input value is mapped to two output values.

Representing functions by sets of ordered pairs is common in *discrete mathematics.* For instance, a function that relates the number of units produced or sold in business to the profit could be represented by sets of ordered pairs. In algebra, however, it is more common to assume that the variables can be any real value and represent functions by equations or formulas involving two variables.

For instance, the equation

$$y = x^2 \qquad \text{\small \emph{y} is a function of \emph{x}.}$$

represents the variable y as a function of the variable x. In this equation, x is the **independent variable** and y is the **dependent variable.** The domain of the function is the set of all values taken on by the independent variable x, and the range of the function is the set of all values taken on by the dependent variable y.

EXAMPLE 2   Testing for Functions Represented Algebraically

Which of the equations represents y as a function of x?

a. $x^2 + y = 1$ **b.** $-x + y^2 = 1$

Solution
To determine whether y is a function of x, try to solve for y in terms of x.
a. Solving for y yields

$$x^2 + y = 1 \qquad \text{\small Write original equation.}$$

$$y = 1 - x^2. \qquad \text{\small Solve for \emph{y}.}$$

To each value of x there corresponds exactly one value of y. So, y *is* a function of x.

b. Solving for y yields

$$-x + y^2 = 1 \qquad \text{\small Write original equation.}$$

$$y^2 = 1 + x \qquad \text{\small Add \emph{x} to each side.}$$

$$y = \pm\sqrt{1 + x}. \qquad \text{\small Solve for \emph{y}.}$$

The $\pm$ indicates that for a given value of x there correspond two values of y. So, y *is not* a function of x.

Function Notation

When an equation is used to represent a function, it is convenient to name the function so that it can be referenced easily. For example, you know that the equation $y = 1 - x^2$ describes y as a function of x. Suppose you give this function the name "f." Then you can use the following **function notation.**

Input	Output	Equation
x	$f(x)$	$f(x) = 1 - x^2$

The symbol $f(x)$ is read as the *value of f at x* or simply *f of x.* The symbol $f(x)$ corresponds to the y-value for a given x. So, you can write $y = f(x)$. Keep in mind that f is the *name* of the function, whereas $f(x)$ is the *output value* of the function at the *input value x.* In function notation, the *input* is the independent variable and the *output* is the dependent variable. For instance, the function $f(x) = 3 - 2x$ has *function values* denoted by $f(-1), f(0), f(2)$, and so on. To find these values, substitute the specified input values into the given equation.

For $x = -1$, $f(-1) = 3 - 2(-1) = 3 + 2 = 5.$
For $x = 0$, $f(0) = 3 - 2(0) = 3 - 0 = 3.$
For $x = 2$, $f(2) = 3 - 2(2) = 3 - 4 = -1.$

Exploration

Use a graphing utility to graph $x^2 + y = 1$. Then use the graph to write a convincing argument that each x-value has at most one y-value.

Use a graphing utility to graph $-x + y^2 = 1$. (*Hint:* You will need to use two equations.) Then use the graph to find an x-value that corresponds to two y-values. Does the graph represent y as a function of x? Explain.

STUDY T!P

You can use a graphing utility to evaluate a function. On the website *college.hmco.com*, you will find the graphing utility program EVALUATE. The program will prompt you for a value of x, and then evaluate the expression in the equation editor for that value of x. Try using the program to evaluate several different functions of x.

Although f is often used as a convenient function name and x is often used as the independent variable, you can use other letters. For instance,

$$f(x) = x^2 - 4x + 7, \quad f(t) = t^2 - 4t + 7, \quad \text{and} \quad g(s) = s^2 - 4s + 7$$

all define the same function. In fact, the role of the independent variable is that of a "placeholder." Consequently, the function could be described by

$$f(\quad) = (\quad)^2 - 4(\quad) + 7.$$

EXAMPLE 3   Evaluating a Function

Let $g(x) = -x^2 + 4x + 1$ and find

a. $g(2)$ **b.** $g(t)$ **c.** $g(x + 2)$.

Solution
a. Replacing x with 2 in $g(x) = -x^2 + 4x + 1$ yields the following.

$$g(2) = -(2)^2 + 4(2) + 1 = -4 + 8 + 1 = 5$$

b. Replacing x with t yields the following.

$$g(t) = -(t)^2 + 4(t) + 1 = -t^2 + 4t + 1$$

c. Replacing x with $x + 2$ yields the following.

$$
\begin{aligned}
g(x + 2) &= -(x + 2)^2 + 4(x + 2) + 1 & &\text{Substitute } x + 2 \text{ for } x.\\
&= -(x^2 + 4x + 4) + 4x + 8 + 1 & &\text{Multiply.}\\
&= -x^2 - 4x - 4 + 4x + 8 + 1 & &\text{Distributive Property}\\
&= -x^2 + 5 & &\text{Simplify.}
\end{aligned}
$$

In Example 3, note that $g(x + 2)$ is not equal to $g(x) + g(2)$. In general, $g(u + v) \neq g(u) + g(v)$.

Sometimes a function is defined by more than one equation. An illustration of this is given in Example 4.

EXAMPLE 4   A Piecewise-Defined Function

Evaluate the function when $x = -1, 0,$ and 1.

$$f(x) = \begin{cases} x^2 + 1, & x < 0 \\ x - 1, & x \geq 0 \end{cases}$$

Solution
Because $x = -1$ is less than 0, use $f(x) = x^2 + 1$ to obtain

$$f(-1) = (-1)^2 + 1 = 2.$$

For $x = 0$, use $f(x) = x - 1$ to obtain

$$f(0) = (0) - 1 = -1.$$

For $x = 1$, use $f(x) = x - 1$ to obtain

$$f(1) = (1) - 1 = 0.$$

Library of Functions

The function in Example 4 is a *piecewise-defined* function. This means that the function is defined by two or more equations over a specified domain. In Example 4, you use the top equation for all x-values less than 0, and the bottom equation for all x-values greater than or equal to 0.

Consult the Library of Functions Summary inside the front cover for a description of the piecewise-defined function.

STUDY TIP

Most graphing utilities can graph functions that are defined piecewise. For instructions on how to enter a piecewise-defined function into your graphing utility, consult your user's manual. You may find it helpful to set your graphing utility to *dot mode* before graphing.

The Domain of a Function

The domain of a function is the set of all values of the independent variable for which the function is defined. If x is in the domain of f, f is said to be *defined* at x. If x is not in the domain of f, f is said to be *undefined* at x.

The domain of a function can be described explicitly or it can be *implied* by the expression used to define the function. The **implied domain** is the set of all real numbers for which the expression is defined. For instance, the function

$$f(x) = \frac{1}{x^2 - 4}$$ Domain excludes x-values that result in division by zero.

has an implied domain that consists of all real x other than $x = \pm 2$. These two values are excluded from the domain because division by zero is undefined. Another common type of implied domain is that used to avoid even roots of negative numbers. For example, the function

$$f(x) = \sqrt{x}$$ Domain excludes x-values that result in even roots of negative numbers.

is defined only for $x \geq 0$. So, its implied domain is the interval $[0, \infty)$. In general, the domain of a function *excludes* values that would cause division by zero *or* result in the even root of a negative number.

EXAMPLE 5   Finding the Domain of a Function

Find the domain of each function.

a. $f: \{(-3, 0), (-1, 4), (0, 2), (2, 2), (4, -1)\}$

b. $g(x) = -3x^2 + 4x + 5$ **c.** $h(x) = \dfrac{1}{x + 5}$

d. Volume of a sphere: $V = \frac{4}{3}\pi r^3$ **e.** $k(x) = \sqrt{4 - 3x}$

Solution

a. The domain of f consists of all first coordinates in the set of ordered pairs.

   Domain $= \{-3, -1, 0, 2, 4\}$.

b. The domain of g is the set of all real numbers.

c. Excluding x-values that yield zero in the denominator, the domain of h is the set of all real numbers $x \neq -5$.

d. Because this function represents the volume of a sphere, the values of the radius r must be positive. So, the domain is the set of all real numbers r such that $r > 0$.

e. This function is defined only for x-values for which

   $4 - 3x \geq 0$.

   The domain of k is all real numbers that are less than or equal to $\frac{4}{3}$.

In Example 5(d), note that the *domain of a function may be implied by the physical context.* For instance, from the equation $V = \frac{4}{3}\pi r^3$, you would have no reason to restrict r to positive values, but the physical context implies that a sphere cannot have a negative or zero radius.

Exploration

Use a graphing utility to graph $y = \sqrt{4 - x^2}$. What is the domain of this function? Then graph $y = \sqrt{x^2 - 4}$. What is the domain of this function? Do the domains of these two functions overlap? If so, for what values?

Library of Functions

The *square root* function $f(x) = \sqrt{x}$ is not defined for $x < 0$. This means that you must be careful when analyzing the domain of complicated functions involving the square root symbol.

Consult the Library of Functions Summary inside the front cover for a description of the square root or radical function.

STUDY T!P

In Example 5(e), $4 - 3x \geq 0$ is a *linear inequality*. For help with solving linear inequalities see Appendix C.

Applications

$$\frac{h}{r} = 4$$

|← r →|

Figure P.57

EXAMPLE 6 The Dimensions of a Container

You work in the marketing department of a soft-drink company and are experimenting with a new soft-drink can that is slightly narrower and taller than a standard can. For your experimental can, the ratio of the height to the radius is 4, as shown in Figure P.57.

a. Express the volume of the can as a function of the radius r.

b. Express the volume of the can as a function of the height h.

Solution

The volume of a right circular cylinder is $V = \pi r^2 h$.

a. To write the volume as a function of the radius, use the fact that $h = 4r$.

$$V(r) = \pi r^2 h = \pi r^2 (4r) = 4\pi r^3 \qquad \text{Write } V \text{ as a function of } r.$$

b. To write the volume as a function of the height, use the fact that $r = h/4$.

$$V(h) = \pi r^2 h = \pi \left(\frac{h}{4}\right)^2 h = \frac{\pi h^3}{16} \qquad \text{Write } V \text{ as a function of } h.$$

EXAMPLE 7 The Path of a Baseball

A baseball is hit at a point 3 feet above ground at a velocity of 100 feet per second and an angle of 45°. The path of the baseball is given by the function

$$f(x) = -0.0032x^2 + x + 3$$

where y and x are measured in feet. Will the baseball clear a 10-foot fence located 300 feet from home plate?

Algebraic Solution

The height of the baseball is a function of the horizontal distance from home plate. When $x = 300$, you can find the height of the baseball as follows.

$$f(x) = -0.0032x^2 + x + 3 \qquad \text{Write original equation.}$$

$$f(300) = -0.0032(300)^2 + 300 + 3 \qquad \text{Substitute 300 for } x.$$

$$= 15 \qquad \text{Simplify.}$$

When $x = 300$, the height of the baseball is 15 feet, so the baseball will clear a 10-foot fence.

Graphical Solution

Use a graphing utility to graph the function $y = -0.0032x^2 + x + 3$, as shown in Figure P.58.

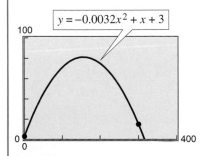

$$y = -0.0032x^2 + x + 3$$

Figure P.58

Use the *value* feature or *zoom* and *trace* features of the graphing utility to estimate that $y = 15$ when $x = 300$. So, the ball will clear a 10-foot fence.

EXAMPLE 8 Direct Mail Advertising

The money C (in billions of dollars) spent for direct mail advertising in the United States increased in a linear pattern from 1990 to 1992, as shown in Figure P.59. Then, in 1993, the money spent took a jump and, until 1996, increased in a *different* linear pattern. These two patterns can be approximated by the function

$$C(t) = \begin{cases} 23.40 + 1.01t & 0 \leq t \leq 2 \\ 19.85 + 2.50t & 3 \leq t \leq 6 \end{cases}$$

where $t = 0$ represents 1990. Use this function to approximate the total amount spent for direct mail advertising between 1990 and 1996. (Source: McCann-Erickson)

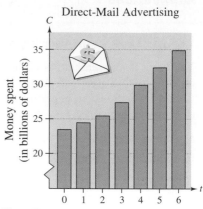

Figure P.59

Solution

From 1990 to 1992, use the formula $C(t) = 23.40 + 1.01t$.

$$\underbrace{\$23.40,}_{1990} \quad \underbrace{\$24.41,}_{1991} \quad \underbrace{\$25.42}_{1992}$$

From 1993 to 1996, use the formula $C(t) = 19.85 + 2.50t$.

$$\underbrace{\$27.35,}_{1993} \quad \underbrace{\$29.85,}_{1994} \quad \underbrace{\$32.35,}_{1995} \quad \underbrace{\$34.85}_{1996}$$

The total of these seven amounts is $197.63, which implies that the total amount spent was approximately $197,630,000,000.

One of the basic definitions in calculus employs the ratio

$$\frac{f(x + h) - f(x)}{h}, \qquad h \neq 0.$$

This ratio is called a **difference quotient,** as illustrated in Examples 9 and 10.

EXAMPLE 9 Evaluating a Difference Quotient

For $f(x) = -2x + 4$, find $\dfrac{f(x + h) - f(x)}{h}$.

Solution

$$\frac{f(x + h) - f(x)}{h} = \frac{[-2(x + h) + 4] - (-2x + 4)}{h} \qquad \text{Substitute } x + h \text{ for } x.$$

$$= \frac{-2x - 2h + 4 + 2x - 4}{h} \qquad \text{Distributive Property}$$

$$= \frac{-2h}{h} = -2, \quad h \neq 0 \qquad \text{Simplify and divide out common factor.}$$

The symbol 🚧 indicates an example or exercise that highlights algebraic techniques specifically used in calculus.

EXAMPLE 10   Evaluating a Difference Quotient

For $f(x) = x^2 - 4x + 7$, find $\dfrac{f(x + h) - f(x)}{h}$.

Solution

$$\frac{f(x + h) - f(x)}{h} = \frac{[(x + h)^2 - 4(x + h) + 7] - (x^2 - 4x + 7)}{h}$$

$$= \frac{x^2 + 2xh + h^2 - 4x - 4h + 7 - x^2 + 4x - 7}{h}$$

$$= \frac{2xh + h^2 - 4h}{h}$$

$$= \frac{h(2x + h - 4)}{h} = 2x + h - 4, \quad h \neq 0$$

Summary of Function Terminology

Function: A **function** is a relationship between two variables such that to each value of the independent variable there corresponds exactly one value of the dependent variable.

Function Notation: $y = f(x)$
    f is the *name* of the function.
    y is the **dependent variable,** or output value.
    x is the **independent variable,** or input value.
    $f(x)$ is the *value of the function at x.*

Domain: The **domain** of a function is the set of all values (inputs) of the independent variable for which the function is defined.

Range: The **range** of a function is the set of all values (outputs) assumed by the dependent variable (that is, the set of all function values).

Implied Domain: If f is defined by an algebraic expression and the domain is not specified, the **implied domain** consists of all real numbers for which the expression is defined.

Leonhard Euler (1707–1783), a Swiss mathematician, is considered to have been the most prolific and productive mathematician in history. One of his greatest influences on mathematics was his use of symbols, or notation. The function notation $y = f(x)$ was introduced by Euler.

The Granger Collection

Writing About Math   *Modeling with Piecewise-Defined Functions*

x	y		x	y
1	5.2		7	12.8
2	5.6		8	10.1
3	6.6		9	8.6
4	8.3		10	6.9
5	11.5		11	4.5
6	15.8		12	2.7

The table at the left shows the monthly revenue y (in thousands of dollars) for one year of a landscaping business, with $x = 1$ representing January.

A mathematical model that represents this data is

$$f(x) = \begin{cases} -1.97x + 26.33 \\ 0.51x^2 - 1.47x + 6.31. \end{cases}$$

What is the domain of each part of the piecewise-defined function? How can you tell? Explain your reasoning.

Find $f(5)$ and $f(11)$, and interpret your results in the context of the problem. How do these model values compare with the actual data values?

P.5 Exercises

In Exercises 1–4, is the relationship a function?

1. *Domain*   *Range*   **2.** *Domain*   *Range*

3. *Domain*   *Range*   **4.** *Domain*   *Range*

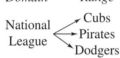
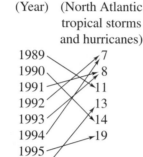

In Exercises 5–8, does the table describe a function? Explain your reasoning.

5.

Input Value	-2	-1	0	1	2
Output Value	-8	-1	0	1	8

6.

Input Value	0	1	2	1	0
Output Value	-4	-2	0	2	4

7.

Input Value	10	7	4	7	10
Output Value	3	6	9	12	15

8.

Input Value	0	3	9	12	15
Output Value	3	3	3	3	3

In Exercises 9 and 10, which sets of ordered pairs represent function(s) from *A* to *B*? Explain.

9. $A = \{0, 1, 2, 3\}$ and $B = \{-2, -1, 0, 1, 2\}$

(a) $\{(0, 1), (1, -2), (2, 0), (3, 2)\}$

(b) $\{(0, -1), (2, 2), (1, -2), (3, 0), (1, 1)\}$

(c) $\{(0, 0), (1, 0), (2, 0), (3, 0)\}$

(d) $\{(0, 2), (3, 0) (1, 1)\}$

10. $A = \{a, b, c\}$ and $B = \{0, 1, 2, 3\}$

(a) $\{(a, 1), (c, 2), (c, 3), (b, 3)\}$

(b) $\{(a, 1), (b, 2), (c, 3)\}$

(c) $\{(1, a), (0, a), (2, c), (3, b)\}$

(d) $\{(c, 0), (b, 0), (a, 3)\}$

Circulation of Newspapers **In Exercises 11 and 12, use the graph, which shows the circulation (in millions) of daily newspapers in the United States.**
(Source: Editor & Publisher Company)

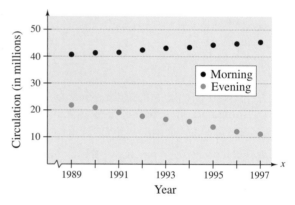

11. Is the circulation of morning newspapers a function of the year? Is the circulation of evening newspapers a function of the year? Explain.

12. Let $f(x)$ represent the circulation of evening newspapers in year x. Find $f(1994)$.

In Exercises 13–24, determine if the equation represents *y* as a function of *x*.

13. $x^2 + y^2 = 4$ **14.** $x = y^2$

15. $x^2 + y = 4$ **16.** $x + y^2 = 4$

17. $2x + 3y = 4$ **18.** $x = -y + 5$

19. $y^2 = x^2 - 1$ **20.** $y = \sqrt{x + 5}$

21. $y = |4 - x|$ **22.** $|y| = 4 - x$

23. $x = 4$ **24.** $y = -2$

In Exercises 25 and 26, fill in the blanks using the specified function and the given values of the independent variable. Simplify the result.

25. $f(x) = \dfrac{1}{x + 1}$

(a) $f(4) = \dfrac{1}{(\quad) + 1}$

(b) $f(0) = \dfrac{1}{(\quad) + 1}$

(c) $f(4t) = \dfrac{1}{(\quad) + 1}$

(d) $f(x + c) = \dfrac{1}{(\quad) + 1}$

26. $g(x) = x^2 - 2x$

(a) $g(2) = (\quad)^2 - 2(\quad)$

(b) $g(-3) = (\quad)^2 - 2(\quad)$

(c) $g(t + 1) = (\quad)^2 - 2(\quad)$

(d) $g(x + c) = (\quad)^2 - 2(\quad)$

In Exercises 27–38, evaluate the function at each specified value of the independent variable and simplify.

27. $f(x) = 2x - 3$

(a) $f(1)$ (b) $f(-3)$ (c) $f(x - 1)$

28. $g(y) = 7 - 3y$

(a) $g(0)$ (b) $g\left(\frac{7}{3}\right)$ (c) $g(s + 2)$

29. $h(t) = t^2 - 2t$

(a) $h(2)$ (b) $h(1.5)$ (c) $h(x + 2)$

30. $V(r) = \frac{4}{3}\pi r^3$

(a) $V(3)$ (b) $V\left(\frac{3}{2}\right)$ (c) $V(2r)$

31. $f(y) = 3 - \sqrt{y}$

(a) $f(4)$ (b) $f(0.25)$ (c) $f(4x^2)$

32. $f(x) = \sqrt{x + 8} + 2$

(a) $f(-8)$ (b) $f(1)$ (c) $f(x - 8)$

33. $q(x) = \dfrac{1}{x^2 - 9}$

(a) $q(0)$ (b) $q(3)$ (c) $q(y + 3)$

34. $q(t) = \dfrac{2t^2 + 3}{t^2}$

(a) $q(2)$ (b) $q(0)$ (c) $q(-x)$

35. $f(x) = \dfrac{|x|}{x}$

(a) $f(2)$ (b) $f(-2)$ (c) $f(x^2)$

36. $f(x) = |x| + 4$

(a) $f(2)$ (b) $f(-2)$ (c) $f(x^2)$

37. $f(x) = \begin{cases} 2x + 1, & x < 0 \\ 2x + 2, & x \geq 0 \end{cases}$

(a) $f(-1)$ (b) $f(0)$ (c) $f(2)$

38. $f(x) = \begin{cases} x^2 + 2, & x \leq 1 \\ 2x^2 + 2, & x > 1 \end{cases}$

(a) $f(-2)$ (b) $f(1)$ (c) $f(2)$

In Exercises 39–44, complete the table.

39. $f(x) = x^2 - 3$

x	-2	-1	0	1	2
$f(x)$					

40. $g(x) = \sqrt{x - 3}$

x	3	4	5	6	7
$g(x)$					

41. $h(t) = \frac{1}{2}|t + 3|$

t	-5	-4	-3	-2	-1
$h(t)$					

42. $f(s) = \dfrac{|s - 2|}{s - 2}$

s	0	1	$\frac{3}{2}$	$\frac{5}{2}$	4
$f(s)$					

43. $f(x) = \begin{cases} -\frac{1}{2}x + 4, & x \leq 0 \\ (x - 2)^2, & x > 0 \end{cases}$

x	-2	-1	0	1	2
$f(x)$					

44. $h(x) = \begin{cases} 9 - x^2, & x < 3 \\ x - 3, & x \geq 3 \end{cases}$

x	1	2	3	4	5
$h(x)$					

In Exercises 45–52, find all real values of x such that $f(x) = 0$.

45. $f(x) = 15 - 3x$

46. $f(x) = 5x + 1$

47. $f(x) = \dfrac{3x - 4}{5}$

48. $f(x) = \dfrac{12 - x^2}{5}$

49. $f(x) = x^2 - 9$

50. $f(x) = x^3 - x$

51. $f(x) = \sqrt{x^2 - 16}$

52. $f(x) = \sqrt{4x^2 - x}$

In Exercises 53–56, find the value(s) of x for which $f(x) = g(x)$.

53. $f(x) = x^2$, $\quad g(x) = x + 2$

54. $f(x) = x^2 + 2x + 1$, $\quad g(x) = 3x + 3$

55. $f(x) = \sqrt{3x} + 1$, $\quad g(x) = x + 1$

56. $f(x) = x^4 - 2x^2$, $\quad g(x) = 2x^2$

In Exercises 57–70, determine the domain of the function.

57. $f(x) = 5x^2 + 2x - 1$

58. $g(x) = 1 - 2x^2$

59. $h(t) = \dfrac{4}{t}$

60. $s(y) = \dfrac{3y}{y + 5}$

61. $g(y) = \sqrt{y - 10}$

62. $f(t) = \sqrt[3]{t + 4}$

63. $f(x) = \sqrt[4]{1 - x^2}$

64. $f(x) = \sqrt[4]{x^2 + 3x}$

65. $g(x) = \dfrac{1}{x} - \dfrac{3}{x + 2}$

66. $h(x) = \dfrac{10}{x^2 - 2x}$

67. $f(s) = \dfrac{\sqrt{s - 1}}{s - 4}$

68. $f(x) = \dfrac{\sqrt{x + 6}}{6 + x}$

69. $f(x) = \dfrac{\sqrt[3]{x - 4}}{x}$

70. $f(x) = \dfrac{x - 5}{\sqrt[4]{x^2 - 9}}$

In Exercises 71–74, assume that the domain of f is the set $A = \{-2, -1, 0, 1, 2\}$. Determine the set of ordered pairs representing the function f.

71. $f(x) = x^2$

72. $f(x) = \dfrac{2x}{x^2 + 1}$

73. $f(x) = \sqrt{x + 2}$

74. $f(x) = |x + 1|$

Exploration **In Exercises 75–78, select a function from $f(x) = cx$, $g(x) = cx^2$, $h(x) = c\sqrt{|x|}$, or $r(x) = c/x$ and determine the value of the constant c such that the function fits the data given in the table.**

75.

x	-4	-1	0	1	4
y	-32	-2	0	-2	-32

76.

x	-4	-1	0	1	4
y	-1	$-\frac{1}{4}$	0	$\frac{1}{4}$	1

77.

x	-4	-1	0	1	4
y	-8	-32	Undef.	32	8

78.

x	-4	-1	0	1	4
y	6	3	0	3	6

In Exercises 79–86, find the difference quotient and simplify your answer.

79. $f(x) = 2x$, $\quad \dfrac{f(x + c) - f(x)}{c}, \, c \neq 0$

80. $g(x) = 3x - 1$, $\quad \dfrac{g(x + h) - g(x)}{h}, \, h \neq 0$

81. $f(x) = x^2 - x + 1$, $\quad \dfrac{f(2 + h) - f(2)}{h}, \, h \neq 0$

82. $f(x) = 5x - x^2$, $\quad \dfrac{f(5 + h) - f(5)}{h}, \, h \neq 0$

83. $f(x) = x^3$, $\quad \dfrac{f(x + c) - f(x)}{c}, \, c \neq 0$

84. $f(x) = x^3 + x$, $\quad \dfrac{f(x + h) - f(x)}{h}, \, h \neq 0$

85. $f(t) = \dfrac{1}{t}$, $\quad \dfrac{f(t) - f(1)}{t - 1}, \, t \neq 1$

86. $f(x) = \dfrac{4}{x + 1}$, $\quad \dfrac{f(x) - f(7)}{x - 7}, \, x \neq 7$

87. *Geometry* Express the area A of a circle as a function of its circumference C.

88. *Geometry* Express the area A of an equilateral triangle as a function of the length s of its sides.

89. *Geometry* Express the area A of an isosceles right triangle as a function of the length s of one of its two equal sides.

90. *Geometry* Express the area A of an equilateral triangle as a function of the height of the triangle.

The symbol 🔩 indicates an example or exercise that highlights algebraic techniques specifically used in calculus.

91. *Exploration* An open box of maximum volume is to be made from a square piece of material, 24 centimeters on a side, by cutting equal squares from the corners and turning up the sides.

(a) Use the *table* feature of a graphing utility to complete six rows of the table. Use the result to guess the maximum volume.

Height, x	Width	Volume, V
1	$24 - 2(1)$	$1[24 - 2(1)]^2 = 484$
2	$24 - 2(2)$	$2[24 - 2(2)]^2 = 800$

(b) Write the volume V as a function of x, and determine its domain. Use a graphing utility to graph the function.

(c) Use the *value* feature or *zoom* and *trace* features of the graphing utility to approximate V when $x = 9$ and when $x = 10$.

(d) Verify your answers to part (c) algebraically.

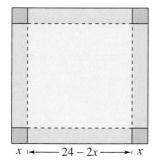

92. *Exploration* The cost per unit of a certain radio model is $60. The manufacturer charges $90 per unit for orders of 100 or less. To encourage large orders, the manufacturer reduces the charge by $0.15 per radio for each unit ordered in excess of 100 (i.e., there would be a charge of $87 per radio for an order size of 120).

(a) Use the *table* feature of a graphing utility to complete six rows of the table. Use the result to estimate the maximum profit.

Units, x	Price, p	Profit, P
110	$90 - 10(0.15)$	$xp - 110(60)$
120	$90 - 20(0.15)$	$xp - 120(60)$

(b) Write the profit P as a function of x, and determine its domain. Use a graphing utility to graph the function.

(c) Use the *value* feature or *zoom* and *trace* features of the graphing utility to approximate P when $x = 120$, 130, and 140.

(d) Verify your answers to part (c) algebraically.

93. *Geometry* A right triangle is formed in the first quadrant by the x- and y-axes and a line through the point (2, 1). Write the area of the triangle as a function of x, and determine the domain of the function.

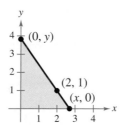

94. *Geometry* A rectangle is bounded by the x-axis and the semicircle $y = \sqrt{36 - x^2}$. Write the area of the rectangle as a function of x, and determine the domain of the function.

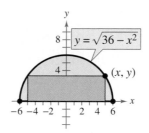

95. *Geometry* A rectangular package to be sent by the U.S. Postal Service can have a maximum combined length and girth (perimeter of a cross section) of 108 inches.

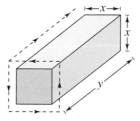

(a) Write the volume of the package as a function of x.

(b) What is the domain of the function?

(c) Use a graphing utility to graph the function. Be sure to use the appropriate viewing window.

(d) What dimensions will maximize the volume of the package? Explain your answer.

96. Mobile Home Prices   The average price p (in thousands of dollars) of a new mobile home in the United States from 1974 to 1997 can be approximated by the model

$$p(t) = \begin{cases} 17.27 + 1.036t, & -6 \le t \le 11 \\ -4.807 + 2.882t + 0.011t^2, & 12 \le t \le 17 \end{cases}$$

where $t = 0$ represents 1980. Use a graphing utility to graph this model. Then use the *value* feature or *zoom* and *trace* features to find the average price of a mobile home in 1978, 1988, 1993, and 1997. (Source: U.S. Bureau of the Census, *Construction Reports*)

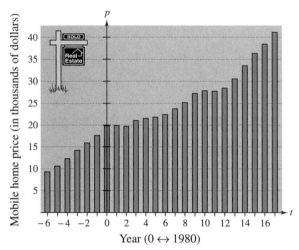

Year $(0 \leftrightarrow 1980)$

97. Business   A company produces a toy for which the variable cost is $12.30 per unit and the fixed costs are $98,000. The toy sells for $17.98. Let x be the number of units produced and sold.

(a) Write the total cost C as a function of the number of units produced.

(b) Write the revenue R as a function of the number of units sold.

(c) Write the profit P as a function of the number of units sold. (*Note:* $P = R - C$.)

98. Business   The inventor of a new game believes that the variable cost for producing the game is $0.95 per unit and the fixed costs are $6000. The inventor sells each game for $1.69. Let x be the number of games sold.

(a) Write the total cost C as a function of the number of games sold.

(b) Write the average cost per unit $\overline{C} = C/x$ as a function of x.

99. Charter Bus Fares   For groups of 80 or more people, a charter bus company determines the rate per person according to the formula

$$\text{Rate} = 8 - 0.05(n - 80), \quad n \ge 80$$

where the rate is given in dollars and n is the number of people.

(a) Express the revenue R for the bus company as a function of n.

(b) Use the function from part (a) to complete the table. What can you conclude?

n	90	100	110	120	130	140	150
$R(n)$							

(c) Use a graphing utility to graph R and determine the number of people that will produce a maximum revenue. Compare the result with your conclusion from part (b).

100. Physics   The force F (in tons) of water against the face of a dam is a function $F(y) = 149.76\sqrt{10}y^{5/2}$, where y is the depth of the water in feet. Complete the table.

y	5	10	20	30	40
$F(y)$					

(a) What can you conclude from the table?

(b) Use a graphing utility to graph the function. Describe your viewing window.

(c) Use the table to approximate the depth at which the force against the dam is 1,000,000 tons. How could you find a better estimate?

(d) Verify your answer in part (c) graphically.

101. Height of a Balloon   A balloon carrying a transmitter ascends vertically from a point 3000 feet from the receiving station.

(a) Draw a diagram to represent the problem. Let h represent the height of the balloon and let d represent the distance between the balloon and the receiving station.

(b) Express the height of the balloon as a function of d. What is the domain of the function?

(c) Use a graphing utility to graph the function in part (b). Describe your viewing window.

(d) Graphically find the height of the balloon when $d = 10,000$ feet. Verify your answer algebraically.

102. *Biology* The graph below shows the lynx population from 1988 through 1995 in a 350-square-kilometer region of the Yukon territory in Canada. Let $f(t)$ represent the number of lynx in year t. (Source: Kluane Boreal Forest Ecosystem Project)

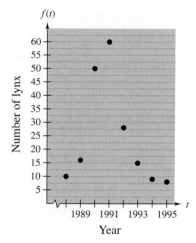

(a) Find $f(1992)$.

(b) Find $\dfrac{f(1994) - f(1991)}{1994 - 1991}$

and interpret the result in the context of the problem.

(c) An approximate model for the function is

$$N(t) = \frac{434t + 4387}{45t^2 - 55t + 100}$$

where N is the number of lynx and t is the time in years, with $t = 0$ corresponding to 1990. Complete the table and compare the result with the data. Use a graphing utility to graph the model and data in the same viewing window. Comment on the validity of the model.

t	1988	1989	1990	1991
N				

t	1992	1993	1994	1995
N				

Synthesis

True or False? **In Exercises 103 and 104, determine whether the statement is true or false. Justify your answer.**

103. The domain of the function $f(x) = x^4 - 1$ is $(-\infty, \infty)$, and the range of $f(x)$ is $(0, \infty)$.

104. The set of ordered pairs $\{(-8, -2), (-6, 0), (-4, 0), (-2, 2), (0, 4), (2, -2)\}$ represents a function.

105. *Think About It* Does the relationship shown in the figure represent a function from set A to set B? Explain.

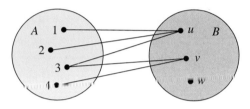

106. *Writing* In your own words, explain the meaning of domain and range.

107. *Think About It* Describe an advantage of function notation.

P.6 Graphs of Functions

The Graph of a Function

In Section P.5 functions were represented graphically by points on a graph in a coordinate plane in which the input values are represented by the horizontal axis and the output values are represented by the vertical axis. The **graph of a function** f is the collection of ordered pairs $(x, f(x))$ such that x is in the domain of f. As you study this section, remember the following geometrical interpretation of x and $f(x)$.

x = the directed distance from the y-axis

$f(x)$ = the directed distance from the x-axis

Example 1 shows how to use the graph of a function to find the domain and range of the function.

EXAMPLE 1 Finding the Domain and Range of a Function

Use the graph of the function f shown in Figure P.60 to find (a) the domain of f, (b) the function values $f(-1)$ and $f(2)$, and (c) the range of f.

Solution

a. The closed dot (on the left) indicates that $x = -1$ is in the domain of f, whereas the open dot (on the right) indicates that $x = 4$ is not in the domain. So, the domain of f is all x in the interval $[-1, 4)$.

b. Because $(-1, -5)$ is a point on the graph of f, it follows that

$$f(-1) = -5.$$

Similarly, because $(2, 4)$ is a point on the graph of f, it follows that

$$f(2) = 4.$$

c. Because the graph does not extend below $f(-1) = -5$ or above $f(2) = 4$, the range of f is the interval $[-5, 4]$.

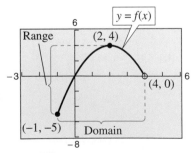

Figure P.60

What You Should Learn:

- How to find the domains and ranges of functions
- How to use the Vertical Line Test for functions
- How to determine intervals on which functions are increasing or decreasing
- How to determine relative maximum and relative minimum values of functions
- How to identify and graph step functions and other piecewise-defined functions
- How to identify even and odd functions

Why You Should Learn It:

Graphs of functions provide a visual relationship between two variables. Exercise 87 on page 80 shows how the graph of a step function can represent the cost of a telephone call.

The use of dots (open or closed) at the extreme left and right points of a graph indicates that the graph does not extend beyond these points. If no such dots are shown, assume that the graph extends beyond these points.

EXAMPLE 2   Finding the Domain and Range of a Function

Find the domain and range of $f(x) = \sqrt{x - 4}$.

Algebraic Solution

Because the expression under a radical cannot be negative, the domain of $f(x) = \sqrt{x - 4}$ is the set of all real numbers such that $x - 4 \geq 0$. Solve this linear inequality for x as follows. (For help with solving linear inequalities, see Appendix C.)

$x - 4 \geq 0$ Write original inequality.

$x \geq 4$ Add 4 to each side.

So, the domain is the set of all real numbers greater than or equal to 4. Because the value of a radical expression is never negative, the range of $f(x) = \sqrt{x - 4}$ is the set of all nonnegative real numbers.

Graphical Solution

Use a graphing utility to graph the equation $y = \sqrt{x - 4}$, as shown in Figure P.61. Use the *trace* feature to determine that the x-coordinates of points on the graph extend from 4 to the right. When x is greater than or equal to 4, the expression under the radical is nonnegative. So, you can conclude that the domain is the set of all real numbers greater than or equal to 4. From the graph, you can see that the y-coordinates of points on the graph extend from 0 upwards. So you can estimate the range to be the set of all nonnegative real numbers.

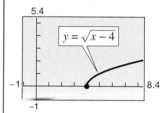

Figure P.61

By the definition of a function, at most one y-value corresponds to a given x-value. It follows, then, that a vertical line can intersect the graph of a function at most once. This leads to the **Vertical Line Test** for functions.

Vertical Line Test for Functions

A set of points in a coordinate plane is the graph of y as a function of x if and only if no vertical line intersects the graph at more than one point.

EXAMPLE 3   Vertical Line Test for Functions

Which of the graphs in Figure P.62 represent y as a function of x?

Solution

a. This *is not* a graph of y as a function of x because you can find a vertical line that intersects the graph twice.

b. This *is* a graph of y as a function of x because every vertical line intersects the graph at most once.

c. This *is* a graph of y as a function of x. (Note that if a vertical line does not intersect the graph, it simply means that the function is undefined for that particular value of x.)

Most graphing utilities are designed to graph functions of x more easily than other types of equations. For instance, the graph shown in Figure P.62(a) represents the equation $x - (y - 1)^2 = 0$. To use a graphing utility to duplicate this graph you must first solve the equation for y to obtain $y = 1 \pm \sqrt{x}$, and then graph the two equations $y_1 = 1 + \sqrt{x}$ and $y_2 = 1 - \sqrt{x}$ in the same viewing window.

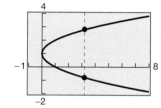

(a)

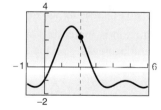

(b)

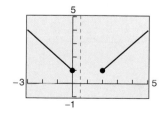

(c)

Figure P.62

Increasing and Decreasing Functions

The more you know about the graph of a function, the more you know about the function itself. Consider the graph shown in Figure P.63. Moving from *left to right,* this graph falls from $x = -2$ to $x = 0$, is constant from $x = 0$ to $x = 2$, and rises from $x = 2$ to $x = 4$.

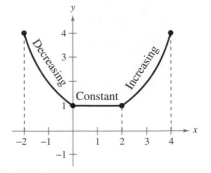

Figure P.63

Increasing, Decreasing, and Constant Functions

A function f is **increasing** on an interval if, for any x_1 and x_2 in the interval,

$x_1 < x_2$ implies $f(x_1) < f(x_2)$.

A function f is **decreasing** on an interval if, for any x_1 and x_2 in the interval,

$x_1 < x_2$ implies $f(x_1) > f(x_2)$.

A function f is **constant** on an interval if, for any x_1 and x_2 in the interval,

$f(x_1) = f(x_2)$.

EXAMPLE 4   Increasing and Decreasing Functions

In Figure P.64, determine the open intervals on which each function is increasing, decreasing, or constant.

Solution

a. Although it might appear that there is an interval in which this function is constant, you can see that if $x_1 < x_2$, then $(x_1)^3 < (x_2)^3$, which implies that $f(x_1) < f(x_2)$. So, the function is increasing over the entire real line.

b. This function is increasing on the interval $(-\infty, -1)$, decreasing on the interval $(-1, 1)$, and increasing on the interval $(1, \infty)$.

c. This function is increasing on the interval $(-\infty, 0)$, constant on the interval $(0, 2)$, and decreasing on the interval $(2, \infty)$.

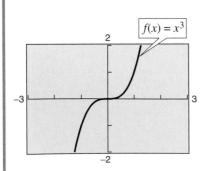

(a)

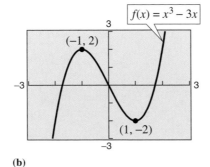

(b)

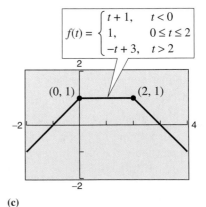

(c)

Figure P.64

Relative Minimum and Maximum Values

The points at which a function changes its increasing, decreasing, or constant behavior are helpful in determining the relative maximum or relative minimum values of the function.

> ## Definition of Relative Minimum and Relative Maximum
>
> A function value $f(a)$ is called a **relative minimum** of f if there exists an interval (x_1, x_2) that contains a such that
>
> $$x_1 < x < x_2 \quad \text{implies} \quad f(a) \le f(x).$$
>
> A function value $f(a)$ is called a **relative maximum** of f if there exists an interval (x_1, x_2) that contains a such that
>
> $$x_1 < x < x_2 \quad \text{implies} \quad f(a) \ge f(x).$$

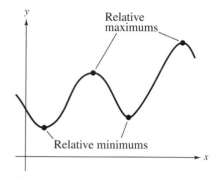

Figure P.65

Figure P.65 shows several different examples of relative minimums and relative maximums. Although there is an algebraic technique for finding the *exact points* at which a second-degree polynomial function has a relative minimum or relative maximum, you can use a graphing utility to find reasonable approximations of these points.

EXAMPLE 5   Approximating a Relative Minimum

Use a graphing utility to approximate the relative minimum of the function $f(x) = 3x^2 - 4x - 2$.

Solution

The graph of f is shown in Figure P.66. By using the *zoom* and *trace* features of a graphing utility, you can estimate that the function has a relative minimum at the point

$$(0.67, -3.33). \qquad \text{Approximate relative minimum}$$

The exact point at which the relative minimum occurs is $\left(\frac{2}{3}, -\frac{10}{3}\right)$, which can be determined algebraically.

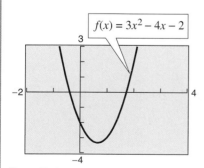

Figure P.66

> ## STUDY T!P
>
> When you use a graphing utility to estimate x- and y-values of a relative minimum or relative maximum, the automatic *zoom* feature will often produce graphs that are nearly flat. To overcome this problem, you can manually change the vertical setting of the viewing window. The graph will vertically stretch if the values of Ymin and Ymax are closer together.

EXAMPLE 6 Approximating Relative Minimums and Maximums

Use a graphing utility to approximate the relative minimum and relative maximum of the function $f(x) = -x^3 + x$.

Solution
A sketch of the graph of f is shown in Figure P.67. By using the *zoom* and *trace* features of the graphing utility, you can estimate that the function has a relative minimum at the point

$(-0.58, -0.38)$ Approximate relative minimum

and a relative maximum at the point

$(0.58, 0.38)$.   Approximate relative maximum

If you go on to take a course in calculus, you will learn a technique for finding the exact points at which this function has a relative minimum and a relative maximum.

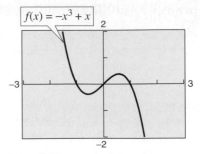

Figure P.67

EXAMPLE 7 Bowling Equipment Sales

During the 1990s, the sales of bowling equipment in the United States increased and then decreased according to the model

$C = 0.165t^3 - 7.16t^2 + 100.6t - 303.1, \qquad 8 \le t \le 16$

where C is the sales of bowling equipment (in millions of dollars) and t represents the year, with $t = 8$ corresponding to 1988. According to this model, during which years were bowling equipment sales increasing? During which years were bowling equipment sales decreasing? Approximate the maximum amount of bowling equipment sales between 1988 and 1996.   (Source: National Sporting Goods Association)

Solution
To solve this problem graph the function, as shown in Figure P.68. From the graph, you can see that the bowling equipment sales increased from 1988 until 1992. Then, from 1992 to 1996, the sales decreased. The maximum amount of bowling equipment sales during the 8-year period was approximately $158 million.

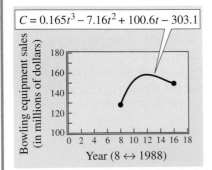

Figure P.68

Graphing Step Functions and Piecewise-Defined Functions

The **greatest integer function** is denoted by $[\![x]\!]$ and is defined by

$$f(x) = [\![x]\!] = \text{the greatest integer less than or equal to } x.$$

The graph of this function is shown in Figure P.69.

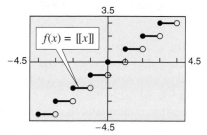

Figure P.69

The graph of the greatest integer function jumps vertically one unit at each integer and is constant (a horizontal line segment) between each pair of consecutive integers. Because of these jumps, the greatest integer function is an example of a **step function** whose graph resembles a set of stair steps. Some values of the greatest integer function are as follows.

$$[\![-2]\!] = -2 \qquad [\![-1.5]\!] = -2 \qquad [\![-1]\!] = -1$$
$$[\![-0.5]\!] = -1 \qquad [\![0]\!] = 0 \qquad [\![0.5]\!] = 0$$
$$[\![1]\!] = 1 \qquad [\![1.5]\!] = 1 \qquad [\![2]\!] = 2$$

The range of the greatest integer function is the set of all integers.

In Section P.5, you learned that a piecewise-defined function is a function that is defined by two or more equations over a specified domain. To sketch the graph of a piecewise-defined function, you need to sketch the graph of each equation on the appropriate portion of the domain.

EXAMPLE 8   Graphing a Piecewise-Defined Function

Sketch the graph of $f(x) = \begin{cases} 2x + 3, & x \le 1 \\ -x + 4, & x > 1 \end{cases}$.

Solution

This piecewise-defined function is composed of two linear functions. To the left of $x = 1$, the graph is the line given by $y = 2x + 3$. To the right of $x = 1$, the graph is the line given by $y = -x + 4$ (see Figure P.70). Notice that the point $(1, 5)$ is a solid dot and the point $(1, 3)$ is an open dot. This is because $f(1) = 5$.

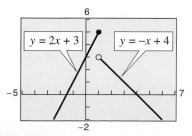

Figure P.70

STUDY T!P

Most graphing utilities display graphs in *connected mode*, which means that the graph has no breaks. When you are sketching graphs that do have breaks, it is better to use *dot mode*. Graph the greatest integer function [often called Int (x)] in connected and dot modes, and compare the two results.

Even and Odd Functions

A graph has *symmetry with respect to the y-axis* if whenever (x, y) is on the graph, so is the point $(-x, y)$. A graph has *symmetry with respect to the origin* if whenever (x, y) is on the graph, so is the point $(-x, -y)$. A graph has *symmetry with respect to the x-axis* if whenever (x, y) is on the graph, so is the point $(x, -y)$. A function whose graph is symmetric with respect to the y-axis is an **even** function. A function whose graph is symmetric with respect to the origin is an **odd** function. The graph of a (nonzero) function cannot be symmetric with respect to the x-axis. These three types of symmetry are illustrated in Figure P.71.

A computer animation of this concept appears in the *Interactive* CD-ROM and *Internet* versions of this text.

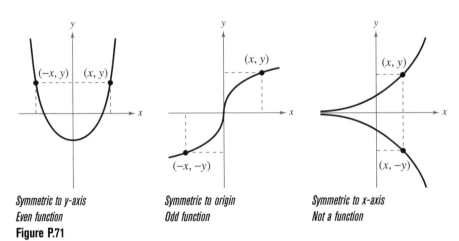

Symmetric to y-axis
Even function
Figure P.71

Symmetric to origin
Odd function

Symmetric to x-axis
Not a function

Test for Even and Odd Functions

A function f is **even** if, for each x in the domain of f, $f(-x) = f(x)$.

A function f is **odd** if, for each x in the domain of f, $f(-x) = -f(x)$.

Library of Functions

The *absolute value* function can be expressed as a piecewise-defined function.

$$|x| = \begin{cases} -x, & x < 0 \\ x, & x \geq 0 \end{cases}$$

Use this definition to show that $|-5| = 5$ and $|0| = 0$.

Consult the Library of Functions Summary inside the front cover for a description of the absolute value function.

EXAMPLE 9 Testing for Evenness and Oddness

Is the function $f(x) = |x|$ even, odd, or neither?

Algebraic Solution

This function is even because

$$f(-x) = |-x|$$
$$= |x|$$
$$= f(x).$$

Graphical Solution

Use a graphing utility to graph $y = |x|$, as shown in Figure P.72. You can see that the graph appears to be symmetric about the y-axis. So, the function is even.

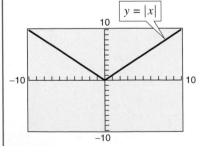

Figure P.72

EXAMPLE 10   Even and Odd Functions

Determine whether each function is even, odd, or neither.

a. $g(x) = x^3 - x$ **b.** $h(x) = x^2 + 1$ **c.** $f(x) = x^3 - 1$

Algebraic Solution

a. This function is odd because

$$g(-x) = (-x)^3 - (-x)$$
$$= -x^3 + x$$
$$= -(x^3 - x)$$
$$= -g(x).$$

b. This function is even because

$$h(-x) = (-x)^2 + 1$$
$$= x^2 + 1$$
$$= h(x).$$

c. Substituting $-x$ for x produces

$$f(-x) = (-x)^3 - 1$$
$$= -x^3 - 1.$$

Because $f(x) = x^3 - 1$ and $-f(x) = -x^3 + 1$, you can conclude that $f(-x) \neq f(x)$ and $f(-x) \neq -f(x)$. So, the function is neither even nor odd.

Graphical Solution

a. In Figure P.73(a), the graph is symmetric with respect to the origin. So, this function is odd.

b. In Figure P.73(b), the graph is symmetric with respect to the *y*-axis. So, this function is even.

c. In Figure P.73(c), the graph is neither symmetric to the origin nor to the *y*-axis. So, this function is neither even nor odd.

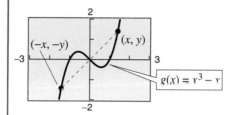

(a)

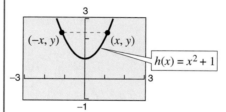

(b)

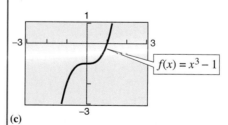

(c)
Figure P.73

W r i t i n g   A b o u t   M a t h   *Increasing and Decreasing Functions*

Write a short paragraph describing three different functions that represent quantities between 1985 and 2000. Describe one that decreased during this time, one that increased, and one that was constant. For instance, the value of the dollar decreased, the cost of first-class postage increased, and the land size of the United States remained constant. Present your results graphically.

P.6 Exercises

In Exercises 1–6, find the domain and range of the function.

1. $f(x) = 1 - x^2$

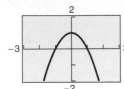

2. $f(x) = x^3 - 3x + 2$

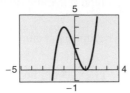

3. $f(x) = \sqrt{x^2 - 1}$

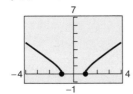

4. $h(x) = \sqrt{16 - x^2}$

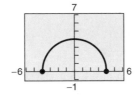

5. $f(x) = \frac{1}{2}|x - 2|$

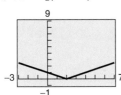

6. $g(x) = -|x - 1|$

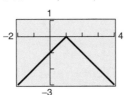

In Exercises 7–12, use a graphing utility to graph the function and estimate its domain and range. Then find the domain and range algebraically.

7. $f(x) = 2x^2 + 3$

8. $f(x) = -x^2 - 1$

9. $f(x) = \sqrt{x - 1}$

10. $h(t) = \sqrt{4 - t^2}$

11. $f(x) = |x + 3|$

12. $f(x) = -\frac{1}{4}|x - 5|$

In Exercises 13–18, use the Vertical Line Test to determine whether y is a function of x. Describe how you can use a graphing utility to produce the given graph.

13. $y = \frac{1}{2}x^2$

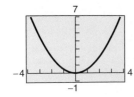

14. $y = \frac{1}{4}x^3$

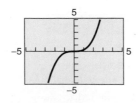

15. $x - y^2 = 1$

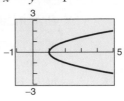

16. $x^2 + y^2 = 25$

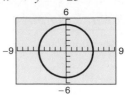

17. $x^2 = 2xy - 1$

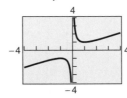

18. $x = |y + 2|$

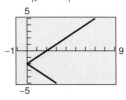

In Exercises 19–22, (a) determine the intervals over which the function is increasing, decreasing, or constant and (b) determine whether the function is even, odd, or neither.

19. $f(x) = \frac{3}{2}x$

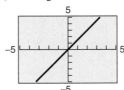

20. $f(x) = x^2 - 4x$

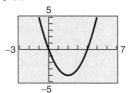

21. $f(x) = x^3 - 3x^2 + 2$

22. $f(x) = \sqrt{x^2 - 1}$

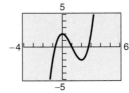 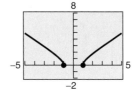

In Exercises 23–30, (a) use a graphing utility to graph the function; (b) determine the intervals over which the function is increasing, decreasing, or constant; and (c) determine whether the function is even, odd, or neither.

23. $f(x) = 3x^4 - 6x^2$

24. $f(x) = -x^6 - 2x^4$

25. $f(x) = x^{2/3}$

26. $f(x) = -x^{3/4}$

27. $f(x) = x\sqrt{x + 3}$ **28.** $f(x) = x(x^2 + 1)^{1/2}$

29. $f(x) = |x + 1| + |x - 1|$

30. $f(x) = -|x + 4| - |x + 1|$

In Exercises 31–36, use a graphing utility to approximate (to two decimal places) any relative minimum or maximum values of the function.

31. $f(x) = x^2 - 6x$ **32.** $f(x) = 3x^2 - 2x - 5$

33. $y = 2x^3 + 3x^2 - 12x$ **34.** $y = x^3 - 6x^2 + 15$

35. $h(x) = (x - 1)\sqrt{x}$ **36.** $g(x) = x\sqrt{4 - x}$

In Exercises 37–42, (a) approximate the relative minimum or maximum values of the function by sketching its graph using the point-plotting method, (b) use a graphing utility to approximate (to two decimal places) any relative minimum or maximum values, and (c) compare your answers from parts (a) and (b).

37. $f(x) = x^2 - 4x - 5$ **38.** $f(x) = 3x^2 - 12$

39. $f(x) = x^3 - 8x$ **40.** $f(x) = -x^3 + 1x$

41. $f(x) = (x - 4)^{2/3}$ **42.** $f(x) = \sqrt{4x^2 + 1}$

In Exercises 43–46, use a graphing utility to graph the piecewise-defined function.

43. $f(x) = \begin{cases} 2x + 3, & x < 0 \\ 3 - x, & x \geq 0 \end{cases}$

44. $f(x) = \begin{cases} x^2 + 5, & x \leq 1 \\ -x^2 + 4x + 3, & x > 1 \end{cases}$

45. $f(x) = \begin{cases} \sqrt{4 + x}, & x < 0 \\ \sqrt{4 - x}, & x \geq 0 \end{cases}$

46. $f(x) = \begin{cases} 1 - (x - 1)^2, & x \leq 2 \\ \sqrt{x - 2}, & x > 2 \end{cases}$

In Exercises 47–54, algebraically determine whether the function is even, odd, or neither. Verify your answer using a graphing utility.

47. $f(t) = t^2 + 2t - 3$ **48.** $f(x) = x^6 - 2x^2 + 3$

49. $g(x) = x^3 - 5x$ **50.** $h(x) = x^3 - 5$

51. $f(x) = x\sqrt{1 - x^2}$ **52.** $f(x) = x\sqrt{x + 5}$

53. $g(s) = 4s^{2/3}$ **54.** $f(s) = 4s^{3/2}$

Think About It   **In Exercises 55–60, find the coordinates of a second point on the graph of a function f if the given point is on the graph and the function is (a) even and (b) odd.**

55. $\left(-\frac{3}{2}, 4\right)$ **56.** $\left(-\frac{5}{3}, -7\right)$

57. $(4, 9)$ **58.** $(5, -1)$

59. $(x, -y)$ **60.** $(2a, 2c)$

In Exercises 61–72, use a graphing utility to graph the function and determine whether it is even, odd, or neither. Verify your answer algebraically.

61. $f(x) = 5$ **62.** $f(x) = -9$

63. $f(x) = 3x - 2$ **64.** $f(x) = 5 - 3x$

65. $h(x) = x^2 - 4$ **66.** $f(x) = -x^2 - 8$

67. $f(x) = \sqrt{1 - x}$ **68.** $g(t) = \sqrt[3]{t - 1}$

69. $f(x) = |x + 2|$ **70.** $f(x) = -|x - 5|$

71. $f(x) = \begin{cases} x + 3, & x \leq 0 \\ 3, & 0 < x \leq 2 \\ 2x - 1, & x > 2 \end{cases}$

72. $f(x) = \begin{cases} 2x + 1, & x \leq -1 \\ x^2 - 2, & x > -1 \end{cases}$

In Exercises 73–82, graph the function and determine the interval(s) (if any) on the real axis for which $f(x) \geq 0$. Use a graphing utility to verify your results.

73. $f(x) = 4 - x$ **74.** $f(x) = 4x + 2$

75. $f(x) = x^2 - 9$ **76.** $f(x) = x^2 - 4x$

77. $f(x) = 1 - x^4$ **78.** $f(x) = x^2 + 1$

79. $f(x) = \sqrt{x + 2}$ **80.** $f(x) = -2\sqrt{x - 3}$

81. $f(x) = -\left(1 + |x|\right)$ **82.** $f(x) = \frac{1}{2}(2 + |x|)$

In Exercises 83 and 84, use a graphing utility to graph the function. State the domain and range of the function. Describe the pattern of the graph.

83. $s(x) = 2\left(\frac{1}{4}x - \left[\!\left[\frac{1}{4}x\right]\!\right]\right)$ **84.** $g(x) = 2\left(\frac{1}{4}x - \left[\!\left[\frac{1}{4}x\right]\!\right]\right)^2$

85. *Geometry*   The perimeter of a rectangle is 100 meters.

(a) Show that the area of the rectangle is $A = x(50 - x)$, where x is its length.

(b) Use a graphing utility to graph the area function.

(c) Use a graphing utility to approximate the maximum area of the rectangle and the dimensions that yield the maximum area.

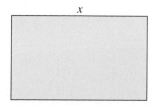

86. *Business*   The marketing department of a company estimates that the demand for a product is

$$p = 100 - 0.0001x$$

where p is the price per unit and x is the number of units. The cost of producing x units is

$$C = 350,000 + 30x$$

and the profit for producing and selling x units is

$$P = R - C = xp - C.$$

Use a graphing utility to graph the profit function and estimate the number of units that would produce a maximum profit.

87. *Communications*   The cost of using a telephone calling card is \$1.05 for the first minute and \$0.38 for each additional minute.

(a) Which of the following is the appropriate model for the cost C of a telephone call lasting t minutes? Explain.

$$C_1(t) = 1.05 + 0.38[\![t - 1]\!]$$

$$C_2(t) = 1.05 - 0.38[\![-(t - 1)]\!]$$

(b) Use a graphing utility to graph the appropriate model. Use the *value* feature or *zoom* and *trace* features to estimate the cost of a call lasting 18 minutes and 45 seconds.

88. *Delivery Service*   Suppose that the cost of sending an overnight package from New York to Atlanta is \$9.80 for under one pound and \$2.50 for each additional pound. Use the greatest integer function to create a model for the cost C of overnight delivery of a package weighing x pounds where $x > 0$. Sketch the graph of the function.

In Exercises 89–92, write the height h of the rectangle as a function of x.

89.

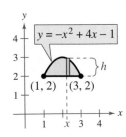

90.

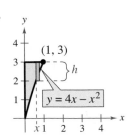

91.

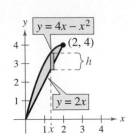

92.

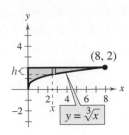

In Exercises 93 and 94, write the length L of the rectangle as a function of y.

93.

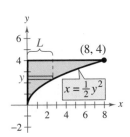

94.

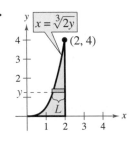

95. *Data Analysis*   The table shows the amount y (in billions of dollars) of the merchandise trade balance of the United States for the years 1990 through 1997. (Source: U.S. International Trade Administration)

Year	1990	1991	1992	1993
y	-101.7	-66.7	-84.5	-115.6

Year	1994	1995	1996	1997
y	-150.6	-158.7	-170.2	-181.5

(a) Use the regression capabilities of a graphing utility to find a cubic model for the data. Let x be the time (in years), with $x = 0$ corresponding to 1990.

(b) What is the domain of the model?

(c) Use a graphing utility to graph the data and the model on the same viewing window.

(d) For which year does the model most accurately estimate the actual data? During which year is it least accurate?

(e) If this model remained valid in the future, would the economy show improvement? Explain your answer.

96. *Fluid Flow*   The intake pipe of a 100-gallon tank has a flow rate of 10 gallons per minute, and two drain pipes have a flow rate of 5 gallons per minute each. The graph shows the volume V of fluid in the tank as a function of time t. Determine the pipes in which the fluid is flowing in specific subintervals of the 1 hour of time shown on the graph. (There are many correct answers.)

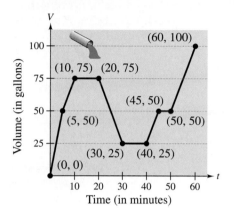

Synthesis

True or False?   **In Exercises 97 and 98, determine whether the statement is true or false. Justify your answer.**

97. A function with a square root cannot have a domain that is the set of all real numbers.

98. It is possible for an odd function to have the interval $[0, \infty)$ as its domain.

99. Prove that a function of the following form is odd.

$$y = a_{2n+1}x^{2n+1} + a_{2n-1}x^{2n-1} + \cdots + a_3x^3 + a_1x$$

100. Prove that a function of the following form is even.

$$y = a_{2n}x^{2n} + a_{2n-2}x^{2n-2} + \cdots + a_2x^2 + a_0$$

101. If f is an even function, determine if g is even, odd, or neither. Explain.

(a) $g(x) = -f(x)$ (b) $g(x) = f(-x)$

(c) $g(x) = f(x) - 2$ (d) $g(x) = -f(x - 2)$

102. *Think About It*   Does the graph of $x - y^2 = 1$ represent x as a function of y? Explain.

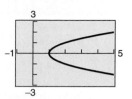

103. *Think About It*   Does the graph of $x^2 + y^2 = 25$ represent x as a function of y? Explain.

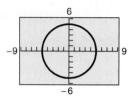

P.7 Shifting, Reflecting, and Stretching Graphs

Summary of Graphs of Common Functions

One of the goals of this text is to enable you to build your intuition for the basic shapes of the graphs of different types of functions. For instance, from your study of lines in Section P.4, you can determine the basic shape of the graph of the linear function $f(x) = mx + b$. Specifically, you know that the graph of this function is a line whose slope is m and whose y-intercept is b.

The six graphs shown in Figure P.74 represent the most commonly used functions in algebra. Familiarity with the basic characteristics of these simple graphs will help you analyze the shapes of more complicated graphs.

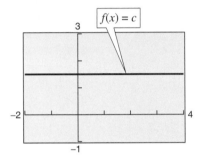

$f(x) = c$

(a) Constant Function

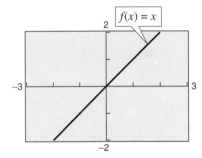

$f(x) = x$

(b) Identity Function

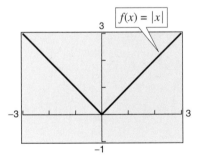

$f(x) = |x|$

(c) Absolute Value Function

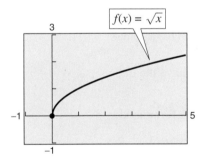

$f(x) = \sqrt{x}$

(d) Square Root Function

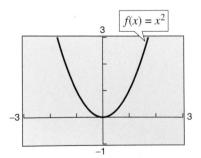

$f(x) = x^2$

(e) Quadratic Function

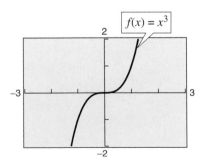

$f(x) = x^3$

(f) Cubic Function

Figure P.74

What **You Should Learn:**

- How to recognize graphs of common functions
- How to use vertical and horizontal shifts and reflections to sketch graphs of functions
- How to use nonrigid transformations to sketch graphs of functions

Why **You Should Learn It:**

Knowing the graphs of common functions and knowing how to shift, reflect, and stretch graphs of functions can help you sketch a wide variety of simple functions by hand. This skill is useful in sketching graphs of functions that model real-life data, such as in Exercise 77 on page 90, where you are asked to sketch a function that models the amount of fuel used by trucks from 1980 through 1996.

Vertical and Horizontal Shifts

Many functions have graphs that are simple transformations of the common graphs summarized in Figure P.74. For example, you can obtain the graph of

$$h(x) = x^2 + 2$$

by shifting the graph of $f(x) = x^2$ *up* two units, as shown in Figure P.75. In function notation, h and f are related as follows.

$$h(x) = x^2 + 2$$
$$\quad = f(x) + 2 \qquad \text{Upward shift of 2}$$

Similarly, you can obtain the graph of

$$g(x) = (x - 2)^2$$

by shifting the graph of $f(x) = x^2$ to the *right* two units, as shown in Figure P.76. In this case, the functions g and f have the following relationship.

$$g(x) = (x - 2)^2$$
$$\quad = f(x - 2) \qquad \text{Right shift of 2}$$

Exploration

Use a graphing utility to display the graphs of $y = x^2 + c$, where $c = -4, -2, 0, 2$, and 4. Use the result to describe the effect that c has on the graph. Use a graphing utility to display the graphs of $y = (x + c)^2$, where $c = -2, 0, 2$, and 4. Use the result to describe the effect that c has on the graph.

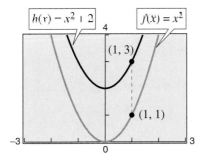

Figure P.75 *Vertical shift upward: two units*

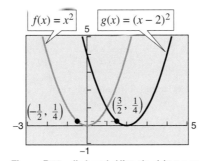

Figure P.76 *Horizontal shift to the right: two units*

The following list summarizes this discussion about horizontal and vertical shifts.

Vertical and Horizontal Shifts

Let c be a positive real number. **Vertical and horizontal shifts** in the graph of $y = f(x)$ are represented as follows.

1. Vertical shift c units *upward:* $h(x) = f(x) + c$
2. Vertical shift c units *downward:* $h(x) = f(x) - c$
3. Horizontal shift c units to the *right:* $h(x) = f(x - c)$
4. Horizontal shift c units to the *left:* $h(x) = f(x + c)$

In items 3 and 4, be sure you see that $h(x) = f(x - c)$ corresponds to a *right* shift and $h(x) = f(x + c)$ corresponds to a *left* shift for $c > 0$.

EXAMPLE 1   Shifts in the Graph of a Function

Compare the graph of each function with the graph of $f(x) = x^3$.

a. $g(x) = x^3 - 1$ **b.** $h(x) = (x - 1)^3$ **c.** $k(x) = (x + 2)^3 + 1$

Solution

a. Graph $f(x) = x^3$ and $g(x) = x^3 - 1$ [see Figure P.77(a)]. You can see that you can obtain the graph of g by shifting the graph of f one unit down.

b. Graph $f(x) = x^3$ and $h(x) = (x - 1)^3$ [see Figure P.77(b)]. You can obtain the graph of h by shifting the graph of f one unit to the right.

c. Graph $f(x) = x^3$ and $k(x) = (x + 2)^3 + 1$ [see Figure P.77(c)]. You can see that you can obtain the graph of k by shifting the graph of f two units to the left and then one unit up.

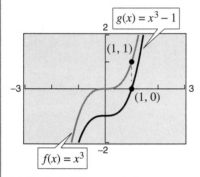
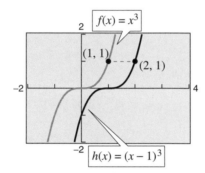
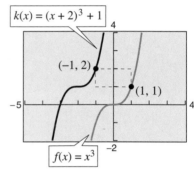

(a) **Vertical shift: one unit down**   (b) **Horizontal shift: one unit right**   (c) **Two units left and one unit up**

Figure P.77

EXAMPLE 2   Finding Equations from Graphs

The graphs shown in Figures P.78 (b) and (c) are shifts of the graph of $f(x) = x^2$ shown in Figure P.78 (a). Find equations for g and h.

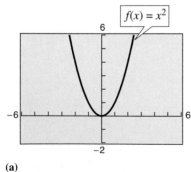
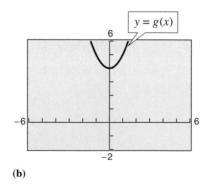
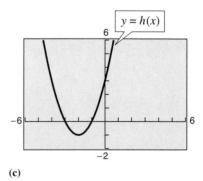

(a)   (b)   (c)

Figure P.78

Solution

a. The graph of g is a vertical shift of four units upward of the graph of $f(x) = x^2$. So, the equation for g is $g(x) = x^2 + 4$.

b. The graph of h is a horizontal shift of two units to the left of the graph of $f(x) = x^2$ and one unit downward. So, the equation for h is $h(x) = (x + 2)^2 - 1$.

Reflecting Graphs

The second common type of transformation is called a **reflection.** For instance, if you consider the x-axis to be a mirror, the graph of $h(x) = -x^2$ is the mirror image (or reflection) of the graph of $f(x) = x^2$ (see Figure P.79).

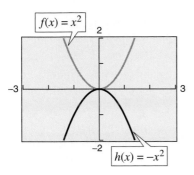

Figure P.79

Exploration

Compare the graph of each function with the graph of $f(x) = x^2$ by using a graphing utility to graph the function and f in the same viewing window. Describe the transformation.

a. $g(x) = -x^2$

b. $h(x) = (-x)^2$

A computer animation of this concept appears in the *Interactive* CD-ROM and *Internet* versions of this text.

Reflections in the Coordinate Axes

Reflections in the coordinate axes of the graph of $y = f(x)$ are represented as follows.

1. Reflection in the x-axis: $\quad h(x) = -f(x)$

2. Reflection in the y-axis: $\quad h(x) = f(-x)$

EXAMPLE 3   Finding Equations from Graphs

Each of the graphs shown in Figure P.81 is a transformation of the graph of $f(x) = x^4$ (see Figure P.80). Find an equation of each function.

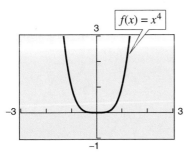

Figure P.80

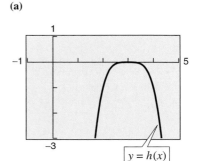

(a)

(b)

Figure P.81

Solution

a. The graph of g is a reflection in the x-axis followed by an upward shift of two units of the graph of $f(x) = x^4$. So, the equation for g is $g(x) = -x^4 + 2$.

b. The graph of h is a horizontal shift of three units to the right followed by a reflection in the x-axis of the graph of $f(x) = x^4$. So, the equation for h is $h(x) = -(x - 3)^4$.

EXAMPLE 4   Reflections and Shifts

Compare the graph of each function with the graph of $f(x) = \sqrt{x}$.

a. $g(x) = -\sqrt{x}$ **b.** $h(x) = \sqrt{-x}$ **c.** $k(x) = -\sqrt{x+2}$

Algebraic Solution

a. Relative to the graph of $f(x) = \sqrt{x}$, the graph of g is a reflection in the x-axis because

$$g(x) = -\sqrt{x}$$

$$= -f(x).$$

b. The graph of h is a reflection of the graph of $f(x) = \sqrt{x}$ in the y-axis because

$$h(x) = \sqrt{-x}$$

$$= f(-x).$$

c. From the equation

$$k(x) = -\sqrt{x+2}$$

$$= -f(x+2)$$

you can conclude that the graph of k is a left shift of two units, followed by a reflection in the x-axis.

Graphical Solution

a. Use a graphing utility to graph f and g in the same viewing window. From the graph in Figure P.82(a), you can see that the graph of g is a reflection of the graph of f in the x-axis.

b. Use a graphing utility to graph f and h in the same viewing window. From the graph in Figure P.82(b), you can see that the graph of h is a reflection of the graph of f in the y-axis.

c. Use a graphing utility to graph f and k in the same viewing window. From the graph in Figure P.82(c), you can see that the graph of k is a left shift of the graph of f of two units, followed by a reflection in the x-axis.

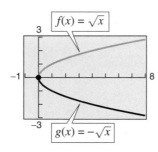

(a)

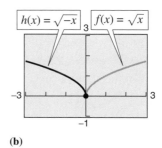

(b)

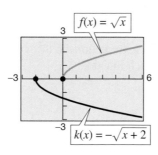

(c)

Figure P.82

When sketching the graphs of functions involving square roots, remember that the domain must be restricted to exclude negative numbers inside the radical. For instance, here are the domains of the functions in Example 4.

Domain of $g(x) = -\sqrt{x}$: $x \geq 0$

Domain of $h(x) = \sqrt{-x}$: $x \leq 0$

Domain of $k(x) = -\sqrt{x+2}$: $x \geq -2$

Nonrigid Transformations

Horizontal shifts, vertical shifts, and reflections are called **rigid transformations** because the basic shape of the graph is unchanged. These transformations change only the *position* of the graph in the *xy*-plane. **Nonrigid transformations** are those that cause a *distortion*—a change in the shape of the original graph. For instance, a nonrigid transformation of the graph of $y = f(x)$ is represented by $y = cf(x)$, where the transformation is a **vertical stretch** if $c > 1$ and a **vertical shrink** if $0 < c < 1$.

A computer animation of this concept appears in the *Interactive* CD-ROM and *Internet* versions of this text.

EXAMPLE 5   Nonrigid Transformations

Compare the graph of each function with the graph of $f(x) = |x|$.

a. $h(x) = 3|x|$ **b.** $g(x) = \dfrac{1}{3}|x|$

Solution

a. Relative to the graph of $f(x) = |x|$, the graph of

$$h(x) = 3|x|$$
$$= 3f(x)$$

is a vertical stretch (multiply each *y*-value by 3) of the graph of *f*.

b. Similarly, the equation

$$g(x) = \frac{1}{3}|x|$$
$$= \frac{1}{3}f(x)$$

indicates that the graph of *g* is a vertical shrink $\left(\text{multiply each } y\text{-value by } \frac{1}{3}\right)$ of the graph of *f*.

The graphs of all three functions are shown in Figure P.83.

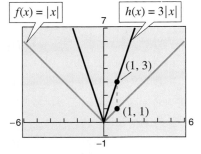

(a)

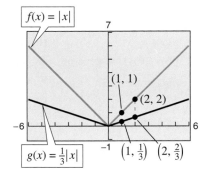

(b)

Figure P.83

EXAMPLE 6   Sequences of Nonrigid Transformations

Use a graphing utility to graph the two functions

$$g(x) = 5(x^2 - 2) \qquad \text{and} \qquad h(x) = 5x^2 - 2$$

in the same viewing window. Describe how each function was obtained from $f(x) = x^2$ as a sequence of shifts and stretches.

Solution

Notice that $g(x)$ and $h(x)$ in Figure P.84 are different. The graph of *g* is a downward shift of *f* of two units followed by a vertical stretch, whereas *h* is a vertical stretch of *f* followed by a downward shift of two units. So, the order of applying the transformations that include nonrigid transformations is important.

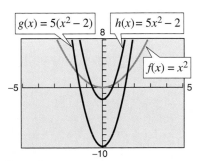

Figure P.84

P.7 Exercises

In Exercises 1–12, sketch the graphs of the three functions by hand on the same rectangular coordinate system. Verify your result with a graphing utility.

1. $f(x) = x$
   $g(x) = x - 4$
   $h(x) = 3x$

2. $f(x) = \frac{1}{2}x$
   $g(x) = \frac{1}{2}x + 2$
   $h(x) = \frac{1}{2}(x - 2)$

3. $f(x) = x^2$
   $g(x) = x^2 + 2$
   $h(x) = (x - 2)^2$

4. $f(x) = x^2$
   $g(x) = x^2 - 4$
   $h(x) = (x + 2)^2 + 1$

5. $f(x) = -x^2$
   $g(x) = -x^2 + 1$
   $h(x) = -(x - 2)^2$

6. $f(x) = (x - 2)^2$
   $g(x) = (x - 2)^2 + 2$
   $h(x) = -(x - 2)^2 + 4$

7. $f(x) = x^2$
   $g(x) = \frac{1}{2}x^2$
   $h(x) = 2x^2$

8. $f(x) = x^2$
   $g(x) = \frac{1}{4}x^2 + 2$
   $h(x) = -\frac{1}{4}x^2$

9. $f(x) = |x|$
   $g(x) = |x| - 1$
   $h(x) = |x - 3|$

10. $f(x) = |x|$
    $g(x) = 2|x|$
    $h(x) = -2|x + 2| - 1$

11. $f(x) = \sqrt{x}$
    $g(x) = \sqrt{x + 1}$
    $h(x) = \sqrt{x - 2} + 1$

12. $f(x) = \sqrt{x}$
    $g(x) = \frac{1}{2}\sqrt{x}$
    $h(x) = -\frac{1}{2}\sqrt{x + 4}$

13. Use the graph of f to sketch each graph.
    (a) $y = f(x) + 2$
    (b) $y = -f(x)$
    (c) $y = f(x - 2)$
    (d) $y = f(x + 3)$
    (e) $y = 2f(x)$
    (f) $y = f(-x)$

14. Use the graph of f to sketch each graph.
    (a) $y = f(x) - 1$
    (b) $y = f(x + 1)$
    (c) $y = f(x - 1)$
    (d) $y = -f(x - 2)$
    (e) $y = f(-x)$
    (f) $y = \frac{1}{2}f(x)$

In Exercises 15–26, identify the common function and the transformation shown in the graph. Write the formula for the graphed function.

15.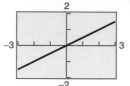

16.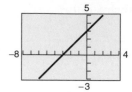

17.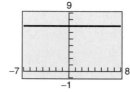

18.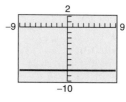

19.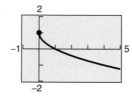

20.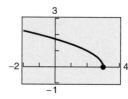

21.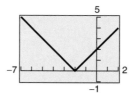

22.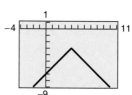

23.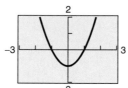

24.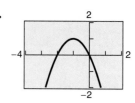

25.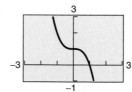

26.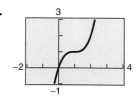

In Exercises 27–32, compare the graph of the function with the graph of $f(x) = \sqrt{x}$.

27. $y = \sqrt{x} + 2$

28. $y = -\sqrt{x} - 1$

29. $y = \sqrt{x - 2}$

30. $y = \sqrt{x + 3}$

31. $y = 2\sqrt{x}$

32. $y = \sqrt{-x + 3}$

In Exercises 33–38, compare the graph of the function with the graph of $f(x) = |x|$.

33. $y = |x + 2|$

34. $y = |x| - 3$

35. $y = -|x|$

36. $y = |-x|$

37. $y = \frac{1}{3}|x|$

38. $y = \frac{1}{2}|x|$

In Exercises 39–44, specify the sequence of transformations that will yield the graph of the given function from the graph of the function $f(x) = x^3$.

39. $g(x) = 4 - x^3$

40. $g(x) = -(x - 4)^3$

41. $h(x) = \frac{1}{4}(x + 2)^3$

42. $h(x) = -2(x - 1)^3 + 3$

43. $p(x) = \frac{1}{3}x^3 + 2$

44. $p(x) = [3(x - 2)]^3$

In Exercises 45–48, use a graphing utility to graph the three functions in the same viewing window. Describe the graphs of g and h relative to the graph of f.

45. $f(x) = x^3 - 3x^2$

$g(x) = f(x + 2)$

$h(x) = \frac{1}{2}f(x)$

46. $f(x) = x^3 - 3x^2 + 2$

$g(x) = f(x - 1)$

$h(x) = 2f(x)$

47. $f(x) = x^3 - 3x^2$

$g(x) = -\frac{1}{3}f(x)$

$h(x) = f(-x)$

48. $f(x) = x^3 - 3x^2 + 2$

$g(x) = -f(x)$

$h(x) = f(-x)$

In Exercises 49 and 50, use the graph of $f(x) = x^3 - 3x^2$ (see Exercise 45) to write a formula for the function g shown in the graph.

49.

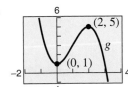

50.

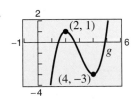

In Exercises 51–74, function g is related to one of the six common functions on page 82. (a) Identify the common function f that is related to g. (b) Describe the sequence of transformations from f to g. (c) Sketch the graph of g. (d) Use function notation to write g in terms of the common function f.

51. $g(x) = 12 - x^2$

52. $g(x) = (x - 8)^2$

53. $g(x) = 2 - (x + 5)^2$

54. $g(x) = -(x + 10)^2 + 5$

55. $g(x) = 3 + 2(x - 4)^2$

56. $g(x) = -\frac{1}{4}(x + 2)^2 - 2$

57. $g(x) = x^3 + 7$

58. $g(x) = -x^3 - 1$

59. $g(x) = (x - 1)^3 + 2$

60. $g(x) = -(x + 3)^3 - 10$

61. $g(x) = 3(x - 2)^3$

62. $g(x) = -\frac{1}{2}(x + 1)^3 - 5$

63. $g(x) = -|x| - 2$

64. $g(x) = 6 - |x + 5|$

65. $g(x) = -|x + 4| + 8$

66. $g(x) = |-x + 3| + 9$

67. $g(x) = -2|x - 1|$

68. $g(x) = \frac{1}{2}|x - 2| - 3$

69. $g(x) = \sqrt{x - 9}$

70. $g(x) = \sqrt{x + 4} + 8$

71. $g(x) = \sqrt{7 - x} - 2$

72. $g(x) = -\sqrt{x + 1} - 6$

73. $g(x) = 4\sqrt{x - 1}$

74. $g(x) = -\frac{1}{2}\sqrt{x + 3} - 1$

75. *Profit* The profit P per week on a certain product is given by the model

$$P(x) = 80 + 20x - 0.5x^2, \qquad 0 \le x \le 20$$

where x is the amount spent on advertising. In this model, x and P are both measured in hundreds of dollars.

(a) Use a graphing utility to graph the profit function.

(b) The business estimates that taxes and operating costs will increase by an average of $2500 per week during the next year. Rewrite the profit equation to reflect this expected decrease in profits. Identify the type of transformation applied to the graph of the equation.

(c) Rewrite the profit equation so that x measures advertising expenditures in dollars. [Find $P\left(\frac{x}{100}\right)$.] Identify the type of transformation applied to the graph of the profit function.

76. *Automobile Aerodynamics* The number of horsepower H required to overcome wind drag on a certain automobile is approximated by

$$H(x) = 0.002x^2 + 0.005x - 0.029, \quad 10 \le x \le 100$$

where x is the speed of the car in miles per hour.

(a) Use a graphing utility to graph the power function.

(b) Rewrite the power function so that x represents the speed in kilometers per hour. [Find $H(x/1.6)$.] Identify the type of transformation applied to the graph of the power function.

77. Energy The amount of fuel F (in billions of gallons) used by trucks from 1980 through 1996 can be approximated by the function

$$F(t) = 20.46 + 0.04t^2$$

where $t = 0$ represents 1980. (Source: U.S. Federal Highway Administration)

(a) Describe how F can be obtained from the common function $f(x) = x^2$. Then sketch the graph over the interval $0 \le t \le 16$.

(b) Rewrite the function so that $t = 0$ represents 1990. Explain how you got your answer.

78. Finance The amount of U.S. mortgage debt outstanding M (in billions of dollars) in the from 1985 through 1997 can be approximated by the function

$$M(t) = 1.5\sqrt{t} - 1.25$$

where $t = 5$ represents 1985. (Source: Board of Governors of the Federal Reserve System)

(a) Describe how M can be obtained from the common function $f(x) = \sqrt{x}$. Then sketch the graph over the interval $5 \le t \le 17$.

(b) Rewrite the function so that $t = 5$ represents 1995. Explain how you got your answer.

79. Graphical Reasoning An electronically controlled thermostat in a home is programmed to automatically lower the temperature at night (see figure). The temperature in the house T, in degrees Fahrenheit, is given in terms of t, the time (in hours) on a 24-hour clock.

(a) Explain why T is a function of t.

(b) Approximate $T(4)$ and $T(15)$.

(c) Suppose the thermostat were reprogrammed to produce a temperature H where $H(t) = T(t - 1)$. How would this change the temperature in the house? Explain.

(d) Suppose the thermostat were reprogrammed to produce a temperature H where $H(t) = T(t) - 1$. How would this change the temperature in the house? Explain.

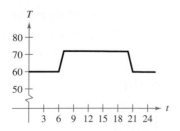

Synthesis

True or False? **In Exercises 80 and 81, determine whether the statement is true or false. Justify your answer.**

80. The graphs of $f(x) = |x| - 5$ and $g(x) = |-x| - 5$ are identical.

81. If the graph of the common function $f(x) = x^2$ is moved 6 units to the right, 3 units up, and reflected in the x-axis, then the point $(-1, 28)$ will lie on the graph of the transformation.

82. Exploration Use a graphing utility to graph each function. Describe any similarities and differences you observe among the graphs.

(a) $y = x$ (b) $y = x^2$ (c) $y = x^3$

(d) $y = x^4$ (e) $y = x^5$ (f) $y = x^6$

83. Conjecture Use the results of Exercise 82 to make a conjecture about the shapes of the graphs of $y = x^7$ and $y = x^8$. Use a graphing utility to verify your conjecture.

84. Use the results of Exercise 82 to sketch the graph of $y = (x - 3)^3$ by hand. Use a graphing utility to verify your graph.

85. Use the results of Exercise 82 to sketch the graph of $y = (x + 1)^2$ by hand. Use a graphing utility to verify your graph.

Conjecture **In Exercises 86–89, use the results of Exercise 82 to make a conjecture about the shape of the graph of the function. Use a graphing utility to verify your conjecture.**

86. $f(x) = x^2(x - 6)^2$ **87.** $f(x) = x^3(x - 6)^2$

88. $f(x) = x^2(x - 6)^3$ **89.** $f(x) = x^3(x - 6)^3$

P.8 Combinations of Functions

Arithmetic Combinations of Functions

Just as two real numbers can be combined by the operations of addition, subtraction, multiplication, and division to form other real numbers, two *functions* can be combined to create new functions. If $f(x) = 2x - 3$ and $g(x) = x^2 - 1$ you can form the sum, difference, product, and quotient of f and g as follows.

$$f(x) + g(x) = (2x - 3) + (x^2 - 1)$$
$$= x^2 + 2x - 4 \qquad \text{Sum}$$

$$f(x) - g(x) = (2x - 3) - (x^2 - 1)$$
$$= -x^2 + 2x - 2 \qquad \text{Difference}$$

$$f(x) \cdot g(x) = (2x - 3)(x^2 - 1)$$
$$= 2x^3 - 3x^2 - 2x + 3 \qquad \text{Product}$$

$$\frac{f(x)}{g(x)} = \frac{2x - 3}{x^2 - 1}, \qquad x \neq +1 \qquad \text{Quotient}$$

The domain of an **arithmetic combination** of functions f and g consists of all real numbers that are common to the domains of f and g. In the case of the quotient $f(x)/g(x)$, there is the further restriction that $g(x) \neq 0$.

Sum, Difference, Product, and Quotient of Functions

Let f and g be two functions with overlapping domains. Then, for all x common to both domains, the sum, difference, product, and quotient of f and g are defined as follows.

1. **Sum:** $(f + g)(x) = f(x) + g(x)$
2. **Difference:** $(f - g)(x) = f(x) - g(x)$
3. **Product:** $(fg)(x) = f(x) \cdot g(x)$
4. **Quotient:** $\left(\dfrac{f}{g}\right)(x) = \dfrac{f(x)}{g(x)}, \quad g(x) \neq 0$

EXAMPLE 1 Finding the Sum of Two Functions

Find $(f + g)(x)$ for the functions $f(x) = 2x + 1$ and $g(x) = x^2 + 2x - 1$. Then evaluate the sum when $x = 2$.

Solution

$$(f + g)(x) = f(x) + g(x)$$
$$= (2x + 1) + (x^2 + 2x - 1)$$
$$= x^2 + 4x$$

When $x = 2$, the value of this sum is $(f + g)(2) = 2^2 + 4(2) = 12$.

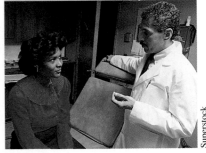

A computer animation of this concept appears in the *Interactive* CD-ROM and *Internet* versions of this text.

EXAMPLE 2   Finding the Difference of Two Functions

Evaluate $(f - g)(x)$ for the functions

$$f(x) = 2x + 1 \quad \text{and} \quad g(x) = x^2 + 2x - 1$$

when $x = 2$.

Algebraic Solution

The difference of the functions f and g is

$$(f - g)(x) = f(x) - g(x)$$

$$= (2x + 1) - (x^2 + 2x - 1)$$

$$= -x^2 + 2.$$

When $x = 2$, the value of this difference is

$$(f - g)(2) = -(2)^2 + 2$$

$$= -2.$$

Note that $(f - g)(2)$ can also be evaluated as follows.

$$(f - g)(2) = f(2) - g(2)$$

$$= [2(2) + 1] - [2^2 + 2(2) - 1]$$

$$= 5 - 7$$

$$= -2$$

Graphical Solution

You can use a graphing utility to graph the difference of two functions. Enter the functions as follows.

$$y_1 = 2x + 1$$

$$y_2 = x^2 + 2x - 1$$

$$y_3 = y_1 - y_2$$

Graph y_3 as shown in Figure P.85. Then use the *value* feature or *zoom* and *trace* features to estimate that the value of the difference when $x = 2$ is -2.

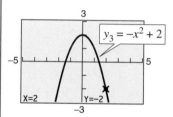

Figure P.85

In Examples 1 and 2, both f and g have domains that consist of all real numbers. So, the domain of both $(f + g)$ and $(f - g)$ is also the set of all real numbers. Remember that any restrictions on the domains of f or g must be taken into account when forming the sum, difference, product, or quotient of f and g. For instance, the domain of $f(x) = 1/x$ is all $x \neq 0$, and the domain of $g(x) = \sqrt{x}$ is $[0, \infty)$. This implies that the domain of $f + g$ is $(0, \infty)$.

EXAMPLE 3   Finding the Product of Two Functions

Given the functions $f(x) = x^2$ and $g(x) = x - 3$, find the product of f and g. Then evaluate the product when $x = 4$.

Solution

$$(fg)(x) = f(x)g(x)$$

$$= (x^2)(x - 3)$$

$$= x^3 - 3x^2$$

When $x = 4$, the value of this product is

$$(fg)(4) = 4^3 - 3(4)^2$$

$$= 16.$$

EXAMPLE 4   Finding the Quotient of Two Functions

Find $\left(\dfrac{f}{g}\right)(x)$ and $\left(\dfrac{g}{f}\right)(x)$ for the functions $f(x) = \sqrt{x}$ and $g(x) = \sqrt{4 - x^2}$. Then find the domains of f/g and g/f.

Solution

The quotient of f and g is

$$\left(\frac{f}{g}\right)(x) = \frac{f(x)}{g(x)} = \frac{\sqrt{x}}{\sqrt{4 - x^2}},$$

and the quotient of g and f is

$$\left(\frac{g}{f}\right)(x) = \frac{g(x)}{f(x)} = \frac{\sqrt{4 - x^2}}{\sqrt{x}}.$$

The domain of f is $[0, \infty)$ and the domain of g is $[-2, 2]$. The intersection of these domains is $[0, 2]$. So, the domains for f/g and g/f are as follows.

$$\text{Domain of } \frac{f}{g}: [0, 2) \qquad \text{Domain of } \frac{g}{f}: (0, 2]$$

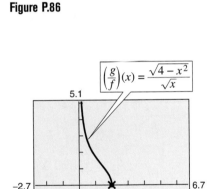

Figure P.86

Can you see why these two domains differ slightly? You can confirm the domain of f/g in Example 4 with your graphing utility by entering the three functions

$$y_1 = \sqrt{x} \qquad y_2 = \sqrt{(4 - x^2)} \qquad y_3 = \frac{y_1}{y_2}$$

and graphing y_3 as shown in Figure P.86.

Use the *trace* feature to determine that the x-coordinates of points on the graph extend from 0 to 2 but do not include 2. So, you can estimate the domain of f/g to be $[0, 2)$. You can confirm the domain of g/f in Example 4 by entering $y_4 = y_2/y_1$ and graphing y_4 as shown in Figure P.87. Use the *trace* feature to determine that the x-coordinates of points on the graph extend from 0 to 2 but do not include 0. So, you can estimate the domain of g/f to be $(0, 2]$.

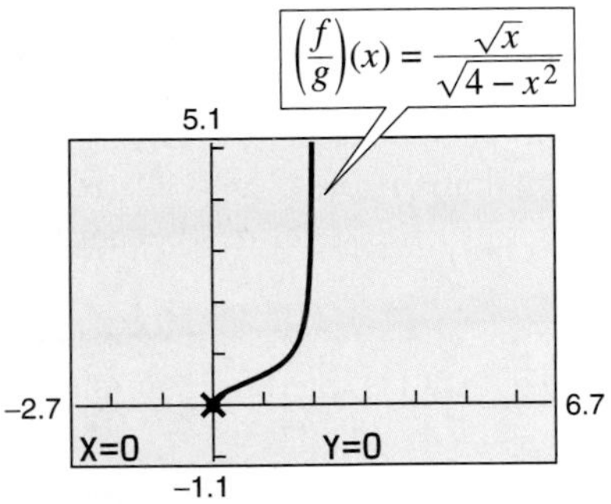

Figure P.87

Compositions of Functions

Another way of combining two functions is to form the **composition** of one with the other. For instance, if $f(x) = x^2$ and $g(x) = x + 1$, the composition of f with g is

$$f(g(x)) = f(x + 1) = (x + 1)^2.$$

This composition is denoted as $f \circ g$.

A computer animation of this concept appears in the *Interactive* CD-ROM and *Internet* versions of this text.

Definition of Composition of Two Functions

The **composition** of the function f with g is

$$(f \circ g)(x) = f(g(x)).$$

The domain of $f \circ g$ is the set of all x in the domain of g such that $g(x)$ is in the domain of f. (See Figure P.88.)

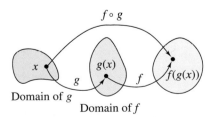

Figure P.88

EXAMPLE 5   Forming the Composition of *f* with *g*

Find $(f \circ g)(x)$ for $f(x) = \sqrt{x}, x \geq 0$, and $g(x) = x - 1, x \geq 1$. If possible, find $(f \circ g)(2)$ and $(f \circ g)(0)$.

Solution

$$(f \circ g)(x) = f(g(x)) \qquad \text{Definition of } f \circ g$$

$$= f(x - 1) \qquad \text{Definition of } g(x)$$

$$= \sqrt{x - 1}, \quad x \geq 1 \qquad \text{Definition of } f(x)$$

The domain of $f \circ g$ is $[1, \infty)$. So, $(f \circ g)(2) = \sqrt{2 - 1} = 1$ is defined, but $(f \circ g)(0)$ is not defined because 0 is not in the domain of $f \circ g$.

The composition of *f* with *g* is generally not the same as the composition of *g* with *f*. This is illustrated in Example 6.

Exploration

Let $f(x) = x + 2$ and $g(x) = 4 - x^2$. Are the compositions $f \circ g$ and $g \circ f$ equal? You can use your graphing utility to answer this question by entering and graphing the following functions.

$$Y_1 = (4 - x^2) + 2$$

$$Y_2 = 4 - (x + 2)^2$$

What do you observe? Which function represents $f \circ g$ and which represents $g \circ f$?

EXAMPLE 6   Compositions of Functions

Given $f(x) = x + 2$ and $g(x) = 4 - x^2$, evaluate the following when $x = 0, 1, 2,$ and 3.

a. $(f \circ g)(x)$ **b.** $(g \circ f)(x)$

Algebraic Solution

a. $(f \circ g)(x) = f(g(x))$ Definition of $f \circ g$

$$= f(4 - x^2) \qquad \text{Definition of } g(x)$$

$$= (4 - x^2) + 2 \qquad \text{Definition of } f(x)$$

$$= -x^2 + 6$$

$$(f \circ g)(0) = -0^2 + 6 = 6$$

$$(f \circ g)(1) = -1^2 + 6 = 5$$

$$(f \circ g)(2) = -2^2 + 6 = 2$$

$$(f \circ g)(3) = -3^2 + 6 = -3$$

b. $(g \circ f)(x) = g(f(x))$ Definition of $g \circ f$

$$= g(x + 2) \qquad \text{Definition of } f(x)$$

$$= 4 - (x + 2)^2 \qquad \text{Definition of } g(x)$$

$$= 4 - (x^2 + 4x + 4)$$

$$= -x^2 - 4x$$

$$(g \circ f)(0) = -0^2 - 4(0) = 0$$

$$(g \circ f)(1) = -1^2 - 4(1) = -5$$

$$(g \circ f)(2) = -2^2 - 4(2) = -12$$

$$(g \circ f)(3) = -3^2 - 4(3) = -21$$

Note that $(f \circ g)(x) \neq (g \circ f)(x)$.

Numerical Solution

a. You can use a table to evaluate $f \circ g$ when $x = 0, 1, 2,$ and 3. First evaluate $g(x)$ for the values of x in the table. Then evaluate $f(g(x))$ for the values of $g(x)$ in the table.

x	0	1	2	3
$g(x)$	4	3	0	-5
$f(g(x))$	6	5	2	-3

b. To evaluate $g \circ f$ when $x = 0, 1, 2,$ and 3, first evaluate $f(x)$ for the values of x in the table. Then evaluate $g(f(x))$ for the values of $f(x)$ in the table.

x	0	1	2	3
$f(x)$	2	3	4	5
$g(f(x))$	0	-5	-12	-21

From the tables you can see that

$$(f \circ g)(x) \neq (g \circ f)(x).$$

To determine the domain of a composite function $f \circ g$, you need to restrict the outputs of g so that they are in the domain of f. For instance, to find the domain of $f \circ g$ given that $f(x) = 1/x$, and $g(x) = x + 1$, consider the outputs of g. These can be any real number. However, the domain of f is restricted to all real numbers except 0. So, the outputs of g must be restricted to all real numbers except 0. This means that $g(x) = x + 1 \neq 0$, or $x \neq -1$. So, the domain of $f \circ g$ is all real numbers except $x = -1$.

EXAMPLE 7   Finding the Domain of a Composite Function

Find the domain of the composition $(f \circ g)(x)$ for the functions

$$f(x) = x^2 - 9 \quad \text{and} \quad g(x) = \sqrt{9 - x^2}.$$

Algebraic Solution

The composition of the functions is as follows.

$$(f \circ g)(x) = f(g(x))$$

$$= f\left(\sqrt{9 - x^2}\right)$$

$$= \left(\sqrt{9 - x^2}\right)^2 - 9$$

$$= 9 - x^2 - 9$$

$$= -x^2$$

From this, it might appear that the domain of the composition is the set of all real numbers. This, however, is not true because the domain of g is $-3 \leq x \leq 3$. So, the domain of $f \circ g$ is $-3 \leq x \leq 3$.

Graphical Solution

You can use a graphing utility to graph the composition of the functions $(f \circ g)(x)$ as $y = \left(\sqrt{9 - x^2}\right)^2 - 9$. Enter the functions as follows.

$$y_1 = \sqrt{(9 - x^2)}$$

$$y_2 = y_1{}^2 - 9$$

Graph y_2 as shown in Figure P.89. Use the *trace* feature to determine that the x-coordinates of points on the graph extend from -3 to 3. So, you can graphically estimate the domain of $(f \circ g)(x)$ to be $-3 \leq x \leq 3$.

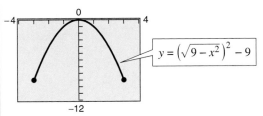

$$y = \left(\sqrt{9 - x^2}\right)^2 - 9$$

Figure P.89

EXAMPLE 8   A Case in Which $f \circ g = g \circ f$

Given $f(x) = 2x + 3$ and $g(x) = \frac{1}{2}(x - 3)$, find the following.

a. $(f \circ g)(x)$ **b.** $(g \circ f)(x)$

Solution

a. $(f \circ g)(x) = f(g(x))$

$$= f\left(\frac{1}{2}(x - 3)\right)$$

$$= 2\left(\frac{1}{2}(x - 3)\right) + 3$$

$$= x - 3 + 3$$

$$= x$$

b. $(g \circ f)(x) = g(f(x))$

$$= g(2x + 3)$$

$$= \frac{1}{2}((2x + 3) - 3)$$

$$= \frac{1}{2}(2x)$$

$$= x$$

STUDY T!P

In Example 8, note that the two composite functions $f \circ g$ and $g \circ f$ are equal, and both represent the identity function. That is,

$$(f \circ g)(x) = x$$

$$(g \circ f)(x) = x.$$

You will study this special case in the next section.

In Examples 5, 6, 7, and 8 you formed the composition of two given functions. In calculus, it is also important to be able to identify two functions that make up a given composite function. For instance, the function h

$$h(x) = (3x - 5)^3$$

is the composition of f with g, where $f(x) = x^3$ and $g(x) = 3x - 5$. That is,

$$h(x) = (3x - 5)^3 = [g(x)]^3 = f(g(x)).$$

Basically, to "decompose" a composite function, look for an "inner" and an "outer" function. In the function h above, $g(x) = 3x - 5$ is the inner function and $f(x) = x^3$ is the outer function.

EXAMPLE 9   Identifying a Composite Function

Express the function

$$h(x) = \frac{1}{(x - 2)^2}$$

as a composition of two functions.

Solution

One way to write h as a composition of two functions is to take the inner function to be $g(x) = x - 2$ and the outer function to be

$$f(x) = \frac{1}{x^2} = x^{-2}.$$

Then you can write

$$h(x) = \frac{1}{(x - 2)^2} = (x - 2)^{-2} = f(x - 2) = f(g(x)).$$

Exploration

The function in Example 9 can be decomposed in other ways. For which of the following pairs of functions is $h(x)$ equal to $f(g(x))$?

a. $g(x) = \dfrac{1}{x - 2}$ and $f(x) = x^2$

b. $g(x) = x^2$ and $f(x) = \dfrac{1}{x - 2}$

c. $g(x) = \dfrac{1}{x}$ and $f(x) = (x - 2)^2$

Application

EXAMPLE 10   Bacteria Count

The number of bacteria in a refrigerated food is

$$N(T) = 20T^2 - 80T + 500, \qquad 2 \le T \le 14$$

where T is the temperature of the food in degrees Celsius. When the food is removed from refrigeration, the temperature is

$$T(t) = 4t + 2, \qquad 0 \le t \le 3$$

where t is the time (in hours). Find the following.

a. The composite $N(T(t))$. What does this function represent?

b. The number of bacteria in the food when $t = 2$ hours

c. The time when the bacterial count reaches 2000

Solution

a. $N(T(t)) = 20(4t + 2)^2 - 80(4t + 2) + 500$

$$= 20(16t^2 + 16t + 4) - 320t - 160 + 500$$

$$= 320t^2 + 320t + 80 - 320t - 160 + 500$$

$$= 320t^2 + 420$$

This composite function $N(T(t))$ represents the number of bacteria as a function of the amount of time the food has been out of refrigeration.

b. When $t = 2$, the number of bacteria is

$$N = 320(2)^2 + 420 = 1280 + 420 = 1700.$$

c. The bacterial count will reach $N = 2000$ when $320t^2 + 420 = 2000$. You can solve this equation for t algebraically as follows.

$$320t^2 + 420 = 2000$$

$$320t^2 = 1580$$

$$t^2 = \frac{1580}{320} = \frac{79}{16}$$

$$t = \frac{\sqrt{79}}{4} \approx 2.2 \text{ hours}$$

So, the count will reach 2000 when $t \approx 2.2$ hours. When you solve this equation, note that the negative value is rejected because it is not in the domain of the composite function. You can use a graphing utility to approximate the solution, as shown in Figure P.90.

Exploration

Use a graphing utility to graph $y = 320t^2 + 420$ and $y = 2000$ in the same viewing window. (Use a viewing window in which $0 \le x \le 3$ and $420 \le y \le 4000$.) Explain how the graphs can be used to answer the question asked in Example 10(c). Compare your answer with that given in part (c). When will the bacteria count reach 3200?

Notice that the model for this bacteria count situation is valid only for a span of 3 hours. Now suppose the minimum number of bacteria in the food is reduced from 420 to 100. Will the number of bacteria still reach a level of 2000 within the 3-hour time span? Will the number of bacteria reach a level of 3200 within 3 hours?

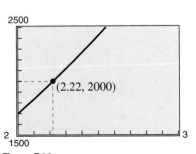

Figure P.90

P.8 E x e r c i s e s

In Exercises 1–8, find the following.

(a) $(f + g)(x)$ (b) $(f - g)(x)$

(c) $(fg)(x)$ (d) $(f/g)(x)$

(e) the domain of f/g

1. $f(x) = x + 1, \quad g(x) = x - 1$
2. $f(x) = 2x - 5, \quad g(x) = 1 - x$
3. $f(x) = x^2, \quad g(x) = 1 - x$
4. $f(x) = 2x - 5, \quad g(x) = 5$
5. $f(x) = x^2 + 5, \quad g(x) = \sqrt{1 - x}$
6. $f(x) = \sqrt{x^2 - 4}, \quad g(x) = \dfrac{x^2}{x^2 + 1}$
7. $f(x) = \dfrac{1}{x}, \quad g(x) = \dfrac{1}{x^2}$
8. $f(x) = \dfrac{x}{x + 1}, \quad g(x) = x^3$

In Exercises 9–22, evaluate the indicated function for $f(x) = x^2 + 1$ and $g(x) = x - 4$ algebraically. If possible, use a graphing utility to verify your answer.

9. $(f + g)(3)$ 10. $(f - g)(-2)$
11. $(f - g)(0)$ 12. $(f + g)(1)$
13. $(fg)(4)$ 14. $(fg)(-6)$
15. $\left(\dfrac{f}{g}\right)(5)$ 16. $\left(\dfrac{f}{g}\right)(0)$
17. $(f - g)(2t)$ 18. $(f + g)(t - 4)$
19. $(fg)(-5t)$ 20. $(fg)(3t^2)$
21. $\left(\dfrac{f}{g}\right)(-t)$ 22. $\left(\dfrac{f}{g}\right)(t + 2)$

In Exercises 23–26, use the graphs to graph $h(x) = (f + g)(x)$.

23.

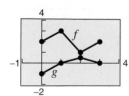

24.

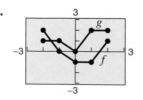

25.

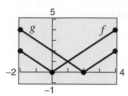

26.

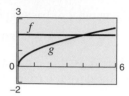

In Exercises 27–30, graph the functions f, g, and $f + g$ in the same viewing window.

27. $f(x) = \frac{1}{2}x, \quad g(x) = x - 1$
28. $f(x) = \frac{1}{3}x, \quad g(x) = -x + 4$
29. $f(x) = x^2, \quad g(x) = -2x$
30. $f(x) = 4 - x^2, \quad g(x) = x$

In Exercises 31–34, use a graphing utility to sketch the graphs of f, g, and $f + g$ in the same viewing window. Which function contributes more to the magnitude of the sum when $0 \le x \le 2$? Which function contributes more to the magnitude of the sum when $x > 6$?

31. $f(x) = 3x, \quad g(x) = -\dfrac{x^3}{10}$
32. $f(x) = \dfrac{x}{2}, \quad g(x) = \sqrt{x}$
33. $f(x) = 3x + 2, \quad g(x) = -\sqrt{x + 5}$
34. $f(x) = x^2 - \frac{1}{2}, \quad g(x) = -3x^2 - 1$

In Exercises 35–38, find (a) $f \circ g$ and (b) $g \circ f$.

35. $f(x) = x^2, \quad g(x) = x - 1$
36. $f(x) = \sqrt[3]{x - 1}, \quad g(x) = x^3 + 1$
37. $f(x) = 3x + 5, \quad g(x) = 5 - x$
38. $f(x) = x^3, \quad g(x) = \dfrac{1}{x}$

In Exercises 39–44, (a) find $f \circ g$ and $g \circ f$. (b) Use a graphing utility to graph $f \circ g$ and $g \circ f$. Determine whether $f \circ g = g \circ f$.

39. $f(x) = \sqrt{x + 4}, \quad g(x) = x^2$
40. $f(x) = \sqrt[3]{x + 1}, \quad g(x) = x^3 - 1$
41. $f(x) = \frac{1}{3}x - 3, \quad g(x) = 3x + 1$
42. $f(x) = \sqrt{x}, \quad g(x) = \sqrt{x}$
43. $f(x) = x^{2/3}, \quad g(x) = x^6$
44. $f(x) = |x|, \quad g(x) = x + 6$

In Exercises 45–50, (a) find $(f \circ g)(x)$ and $(g \circ f)(x)$, (b) determine algebraically whether $(f \circ g)(x) = (g \circ f)(x)$, and (c) verify your answer to part (b) by comparing a table of values for each composition.

45. $f(x) = 5x + 4$, $g(x) = 4 - x$

46. $f(x) = \frac{1}{4}(x - 1)$, $g(x) = 4x + 1$

47. $f(x) = \sqrt{x + 6}$, $g(x) = x^2 - 5$

48. $f(x) = x^3 - 4$, $g(x) = \sqrt[3]{x + 10}$

49. $f(x) = |x + 3|$, $g(x) = 2x - 1$

50. $f(x) = \frac{6}{3x - 5}$, $g(x) = -x$

In Exercises 51–56, use the graphs of f and g to evaluate the functions.

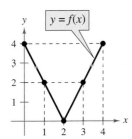
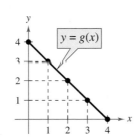

51. (a) $(f + g)(3)$ (b) $(f/g)(2)$
52. (a) $(f - g)(1)$ (b) $(fg)(4)$
53. (a) $(f \circ g)(2)$ (b) $(g \circ f)(2)$
54. (a) $(f \circ g)(1)$ (b) $(g \circ f)(3)$
55. (a) $(f \circ f)(3)$ (b) $(f \circ f)(4)$
56. (a) $(g \circ g)(1)$ (b) $(g \circ g)(0)$

In Exercises 57–64, find two functions f and g such that $(f \circ g)(x) = h(x)$. (There are many correct answers.)

57. $h(x) = (2x + 1)^2$ **58.** $h(x) = (1 - x)^3$
59. $h(x) = \sqrt[3]{x^2 - 4}$ **60.** $h(x) = \sqrt{9 - x}$
61. $h(x) = \frac{1}{x + 2}$ **62.** $h(x) = \frac{4}{(5x + 2)^2}$
63. $h(x) = (x + 4)^2 + 2(x + 4)$
64. $h(x) = (x + 3)^{3/2}$

In Exercises 65–70, determine the domains of (a) f, (b) g, and (c) $f \circ g$. Use a graphing utility to verify your answer.

65. $f(x) = \sqrt{x}$, $g(x) = x^2 + 1$

66. $f(x) = \sqrt{x + 3}$, $g(x) = \frac{x}{2}$

67. $f(x) = \frac{1}{x}$, $g(x) = x + 3$

68. $f(x) = \frac{1}{x}$, $g(x) = \frac{1}{2x}$

69. $f(x) = \frac{2}{|x|}$, $g(x) = x - 1$

70. $f(x) = \frac{3}{x^2 - 1}$, $g(x) = x + 1$

Average Rate of Change   **In Exercises 71–78, find the difference quotient**

$$\frac{f(x + h) - f(x)}{h}$$

and simplify your answer.

71. $f(x) = 3x - 4$ **72.** $f(x) = 5x + 1$
73. $f(x) = 1 - x^2$ **74.** $f(x) = x^2 + 4$
75. $f(x) = \frac{4}{x}$ **76.** $f(x) = \frac{2}{x^2}$
77. $f(x) = \sqrt{2x + 1}$ **78.** $f(x) = -\sqrt{4x}$

79. **Stopping Distance**   A car traveling x miles per hour stops quickly. The distance a car travels during the driver's reaction time is given by $R(x) = \frac{3}{4}x$. The distance traveled while braking is given by $B(x) = \frac{1}{15}x^2$.

(a) Find the stopping-distance function T.

(b) Use a graphing utility to graph the functions R, B, and T in the interval $0 \le x \le 60$.

(c) Which function contributes most to the magnitude of the sum at higher speeds? Explain.

80. **Business**   You own two restaurants. From 1995 to 2000, the sales R_1 (in thousands of dollars) for one restaurant can be modeled by

$$R_1 = 480 - 8t - 0.8t^2, \qquad t = 0, 1, 2, 3, 4, 5$$

where $t = 0$ represents 1995. During the same 6-year period, the sales R_2 (in thousands of dollars) for the other restaurant can be modeled by

$$R_2 = 254 + 0.78t, \qquad t = 0, 1, 2, 3, 4, 5.$$

(a) Write a function R_3 that represents the total sales for the two restaurants.

(b) Use a graphing utility to graph R_1, R_2, and R_3 (the total sales function) in the same viewing window.

Data Analysis   In Exercises 81 and 82, use the table, which gives the total amount spent (in billions of dollars) on health services and supplies in the United States and Puerto Rico for the years 1990 through 1996. The variables y_1, y_2, and y_3 represent out-of-pocket payments, insurance premiums, and other types of payments, respectively.   (Source: U.S. Health Care Financing Administration)

Year	1990	1991	1992	1993	1994	1995	1996
y_1	144.4	151.6	159.5	163.6	164.8	166.7	171.2
y_2	238.6	259.4	282.5	303.3	315.6	326.9	337.3
y_3	21.9	24.0	25.1	27.3	29.6	31.7	32.4

81. Use a graphing utility to find a mathematical model for each of the variables. Let $t = 0$ represent 1990. Find a quadratic model $(y = ax^2 + bx + c)$ for y_1 and linear models $(y = ax + b)$ for y_2 and y_3.

82. Use a graphing utility to graph y_1, y_2, y_3, and $y_1 + y_2 + y_3$ in the same viewing window. Use the model to estimate the total amount spent on health services and supplies in 2000.

83. *Ripples*   A pebble is dropped into a calm pond, causing ripples in the form of concentric circles. The radius (in feet) of the outer ripple is $r(t) = 0.6t$, where t is the time (in seconds) after the pebble strikes the water. The area of the circle is $A(r) = \pi r^2$. Find and interpret $(A \circ r)(t)$.

84. *Geometry*   A square concrete foundation was prepared as a base for a large cylindrical gasoline tank.

(a) Express the radius r of the tank as a function of the length x of the sides of the square.

(b) Express the area A of the circular base of the tank as a function of the radius r.

(c) Find and interpret $(A \circ r)(x)$.

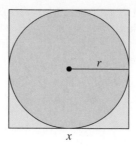

FIGURE FOR 84

85. *Business*   The weekly cost of producing x units in a manufacturing process is $C(x) = 60x + 750$. The number of units produced in t hours is $x(t) = 50t$.

(a) Find and interpret $(C \circ x)(t)$.

(b) Use a graphing utility to graph the cost as a function of time. Use the *trace* feature to estimate (to two-decimal-place accuracy) the time that must elapse until the cost increases to $15,000.

86. *Air Traffic Control*   An air traffic controller spots two planes at the same altitude flying toward each other. Their flight paths form a right angle at point P. One plane is 150 miles from point P and is moving at 450 miles per hour. The other plane is 200 miles from point P and is moving at 450 miles per hour. Write the distance s between the planes as a function of time t.

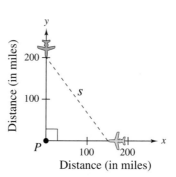

87. *Business*   You are a sales representative for an automobile manufacturer. You are paid an annual salary plus a bonus of 3% of your sales over $500,000. Consider the two functions $f(x) = x - 500,000$ and $g(x) = 0.03x$. If x is greater than $500,000, which of the following represents your bonus? Explain.

(a) $f(g(x))$ (b) $g(f(x))$

88. ***Exploration***   The suggested retail price of a new car is p dollars. The dealership advertised a factory rebate of $1200 and an 8% discount.

    (a) Write a function R in terms of p, giving the cost of the car after receiving the rebate from the factory.

    (b) Write a function S in terms of p, giving the cost of the car after receiving the dealership discount.

    (c) Form the composite functions $(R \circ S)(p)$ and $(S \circ R)(p)$, and interpret each.

    (d) Find $(R \circ S)(18,400)$ and $(S \circ R)(18,400)$. Which yields the lower cost for the car? Explain.

89. ***Data Analysis***   The data in the table shows the circulations of morning and evening newspapers in the United States for the years 1988 through 1997. The variables y_1 and y_2 represent the circulations in millions of the morning and evening papers, respectively.   (Source: Editor & Publisher Co.)

Year	1988	1989	1990	1991	1992
y_1	40.5	40.7	41.3	41.5	42.4
y_2	22.2	21.9	21.0	19.2	17.8

Year	1993	1994	1995	1996	1997
y_1	43.1	43.4	44.3	44.8	45.4
y_2	16.7	15.9	13.9	12.2	11.3

Use a graphing utility to create a scatter plot of each data set. What type of model would best fit the data? Use the regression capabilities of a graphing utility to find models for y_1 and y_2 (use $t = 8$ to represent 1988). Graph the models for $y_1, y_2,$ and $y_1 - y_2$ in the same viewing window. What does the graph of the difference of the functions indicate about newspaper circulation in general?

Synthesis

True or False?   **In Exercises 90 and 91, determine whether the statement is true or false. Justify your answer.**

90. If $f(x) = x + 1$ and $g(x) = 6x$, then

    $(f \circ g)(x) = (g \circ f)(x)$.

91. If you are given two functions $f(x)$ and $g(x)$, you can calculate $(f \circ g)(x)$ if and only if the range of g is a subset of the domain of f.

92. ***Think About It***   Prove that the product of two odd functions is an even function and the product of two even functions is an even function.

93. ***Conjecture***   Use examples to hypothesize whether the product of an odd function and an even function is even or odd. Then prove your hypothesis.

94. Given a function f, prove that $g(x)$ is even and $h(x)$ is odd where

    $g(x) = \frac{1}{2}[f(x) + f(-x)]$ and

    $h(x) = \frac{1}{2}[f(x) - f(-x)]$.

95. Use the result of Exercise 94 to prove that any function can be written as a sum of even and odd functions. (*Hint:* Add the two equations in Exercise 94.)

96. Use the result of Exercise 95 to write each function as a sum of even and odd functions.

    (a) $f(x) = x^2 - 2x + 1$ (b) $f(x) = \dfrac{1}{x + 1}$

P.9   Inverse Functions

The Inverse of a Function

Recall from Section P.5 that a function can be represented by a set of ordered pairs. For instance, the function $f(x) = x + 4$ from the set $A = \{1, 2, 3, 4\}$ to the set $B = \{5, 6, 7, 8\}$ can be written as follows.

$$f(x) = x + 4: \ \{(1, 5), (2, 6), (3, 7), (4, 8)\}$$

In this case, by interchanging the first and second coordinates of each of these ordered pairs, you can form the **inverse function** of f, which is denoted by f^{-1}. It is a function from the set B to the set A, and can be written as follows.

$$f^{-1}(x) = x - 4: \ \{(5, 1), (6, 2), (7, 3), (8, 4)\}$$

Note that the domain of f is equal to the range of f^{-1}, and vice versa, as shown in Figure P.91. Also note that the functions f and f^{-1} have the effect of "undoing" each other. In other words, when you form the composition of f with f^{-1} or the composition of f^{-1} with f, you obtain the identity function.

$$f(f^{-1}(x)) = f(x - 4) = (x - 4) + 4 = x$$

$$f^{-1}(f(x)) = f^{-1}(x + 4) = (x + 4) - 4 = x$$

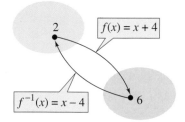

Figure P.91

EXAMPLE 1   Finding Inverse Functions Informally

Find the inverse of $f(x) = 4x$. Then verify that both $f(f^{-1}(x))$ and $f^{-1}(f(x))$ are equal to the identity function.

Solution

The given function *multiplies* each input by 4. To "undo" this function, you need to *divide* each input by 4. So, the inverse function of $f(x) = 4x$ is

$$f^{-1}(x) = \frac{x}{4}.$$

You can verify that both $f(f^{-1}(x))$ and $f^{-1}(f(x))$ are equal to the identity function as follows.

$$f(f^{-1}(x)) = f\left(\frac{x}{4}\right) = 4\left(\frac{x}{4}\right) = x$$

$$f^{-1}(f(x)) = f^{-1}(4x) = \frac{4x}{4} = x$$

What You Should Learn:

- How to find inverse functions informally and verify that two functions are inverses of each other
- How to use graphs of functions to decide whether functions have inverses
- How to find inverse functions algebraically

Why You Should Learn It:

Inverse functions can be helpful in further exploring how two variables relate to each other. Exercise 90 on page 111 investigates the relationship between the exhaust temperature and the percent load on a diesel engine.

David J. Sams/Stock Boston

EXAMPLE 2   Finding Inverse Functions Informally

Find the inverse of $f(x) = x - 6$. Then verify that both $f(f^{-1}(x))$ and $f^{-1}(f(x))$ are equal to the identity function.

Solution

The given function *subtracts* 6 from each input. To "undo" this function, you need to *add* 6 to each input. So, the inverse function of $f(x) = x - 6$ is

$$f^{-1}(x) = x + 6.$$

You can verify that both $f(f^{-1}(x))$ and $f^{-1}(f(x))$ are equal to the identity function as follows.

$$f(f^{-1}(x)) = f(x + 6) = (x + 6) - 6 = x$$
$$f^{-1}(f(x)) = f^{-1}(x - 6) = (x - 6) + 6 = x$$

A table of values can help you understand inverse functions. For instance, the following table shows several values of the function in Example 2. Interchange the rows of this table to obtain values of the inverse function.

x	-2	-1	0	1	2
$f(x)$	-8	-7	-6	-5	-4

x	-8	-7	-6	-5	-4
$f^{-1}(x)$	-2	-1	0	1	2

In the table at the left, each output is 6 less than the input, and in the table at the right, each output is 6 more than the input.

The formal definition of the inverse of a function is as follows.

Definition of the Inverse of a Function

Let f and g be two functions such that

$$f(g(x)) = x \quad \text{for every } x \text{ in the domain of } g$$

and

$$g(f(x)) = x \quad \text{for every } x \text{ in the domain of } f.$$

Under these conditions, the function g is the **inverse** of the function f. The function g is denoted by f^{-1} (read "f-inverse"). So,

$$f(f^{-1}(x)) = x \quad \text{and} \quad f^{-1}(f(x)) = x.$$

The domain of f must be equal to the range of f^{-1}, and the range of f must be equal to the domain of f^{-1}.

STUDY T!P

Don't be confused by the use of -1 to denote the inverse function f^{-1}. In this text, whenever f^{-1} is written, it *always* refers to the inverse of the function f and *not* to the reciprocal of $f(x)$, which is

$$\frac{1}{f(x)}.$$

If the function g is the inverse of the function f, it must also be true that the function f is the inverse of the function g. For this reason, you can say that the functions f and g are *inverses of each other.*

EXAMPLE 3   Verifying Inverse Functions Algebraically

Show that the functions are inverses of each other.

$$f(x) = 2x^3 - 1 \quad \text{and} \quad g(x) = \sqrt[3]{\frac{x+1}{2}}$$

Solution

$$f(g(x)) = f\left(\sqrt[3]{\frac{x+1}{2}}\right) = 2\left(\sqrt[3]{\frac{x+1}{2}}\right)^3 - 1$$

$$= 2\left(\frac{x+1}{2}\right) - 1$$

$$= x + 1 - 1$$

$$= x$$

$$g(f(x)) = g(2x^3 - 1) = \sqrt[3]{\frac{(2x^3 - 1) + 1}{2}}$$

$$= \sqrt[3]{\frac{2x^3}{2}}$$

$$= \sqrt[3]{x^3}$$

$$= x$$

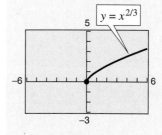

STUDY TIP

Most graphing utilities can graph $y = x^{1/3}$ in two ways:

$$y_1 = x \wedge (1/3) \quad \text{or}$$

$$y_1 = \sqrt[3]{x}.$$

However, you may not be able to obtain the complete graph of $y = x^{2/3}$ by entering $y_1 = x \wedge (2/3)$. If not, you should use

$$y_1 = (x \wedge (1/3))^2 \quad \text{or}$$

$$y_1 = \sqrt[3]{x^2}.$$

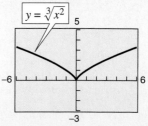

EXAMPLE 4   Verifying Inverse Functions Algebraically

Which of the functions is the inverse of $f(x) = \dfrac{5}{x-2}$?

$$g(x) = \frac{x-2}{5} \quad \text{or} \quad h(x) = \frac{5}{x} + 2$$

Solution

By forming the composition of f with g, you have

$$f(g(x)) = f\left(\frac{x-2}{5}\right) = \frac{5}{\dfrac{x-2}{5} - 2} = \frac{25}{x - 12} \neq x.$$

Because this composition is not equal to the identity function x, it follows that g *is not* the inverse of f. By forming the composition of f with h, you have

$$f(h(x)) = f\left(\frac{5}{x} + 2\right) = \frac{5}{\dfrac{5}{x} + 2 - 2} = \frac{5}{5/x} = x.$$

So, it appears that h is the inverse of f. You can confirm this by showing that the composition of h with f is also equal to the identity function.

The Graph of an Inverse Function

The graphs of f and f^{-1} are related to each other in the following way. If the point (a, b) lies on the graph of f, then the point (b, a) lies on the graph of f^{-1} and vice versa. This means that the graph of f^{-1} is a reflection of the graph of f in the line $y = x$, as shown in Figure P.92.

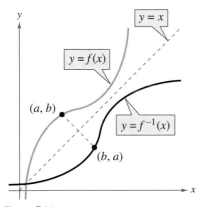

Figure P.92

EXAMPLE 5   Verifying Inverse Functions Graphically and Numerically

Example 3 shows how to verify *algebraically* that the functions

$$f(x) = 2x^3 - 1 \quad \text{and} \quad g(x) = \sqrt[3]{\frac{x + 1}{2}}$$

are inverses of each other. Verify that f and g are inverses of each other graphically and numerically.

Graphical Solution

You can *graphically* verify that f and g are inverses of each other by using a graphing utility to graph f and g in the same viewing window. (Be sure to use a square setting.) From the graph in Figure P.93, you can verify that the graph of g is the reflection of the graph of f in the line $y = x$.

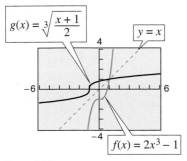

Figure P.93

Numerical Solution

You can *numerically* verify that f and g are inverses of each other by creating two tables as shown below.

x	-2	-1	0	1	2
$f(x)$	-17	-3	-1	1	15

x	-17	-3	-1	1	15
$g(x)$	-2	-1	0	1	2

Note that the entries in the tables are the same except that their rows are interchanged. From the tables, you can verify that f and g are inverses of each other.

The Existence of an Inverse Function

A function need not have an inverse function. For instance, the function $f(x) = x^2$ has no inverse [assuming a domain of $(-\infty, \infty)$]. You can observe this numerically in the table at the right.

x	-2	-1	0	1	2
$f(x)$	4	1	0	1	4

When you interchange the rows of the tables, you can find two different outputs correspond to the same input. For example, the input 4 corresponds to two different outputs, -2 and 2. So, the bottom table does not represent a function, and therefore $f(x) = x^2$ does not have an inverse.

x	4	1	0	1	4
$g(x)$	-2	-1	0	1	2

To have an inverse, a function must be **one-to-one,** which means that no two elements in the domain of f correspond to the same element in the range of f.

Definition of a One-to-One Function

A function f is **one-to-one** if, for a and b in its domain,

$$f(a) = f(b) \quad \text{implies that} \quad a = b.$$

Existence of an Inverse Function

A function f has an inverse function f^{-1} if and only if f is one-to-one.

From its graph, it is easy to tell whether a function of x is one-to-one. Simply check to see that every horizontal line intersects the graph of the function at most once. For instance, Figure P.94 shows the graph of $y = x^4$. On the graph, you can find a horizontal line that intersects the graph twice.

Two special types of functions that pass the **Horizontal Line Test** are those that are increasing or decreasing on their entire domains.

1. If f is *increasing* on its entire domain, f is one-to-one.

2. If f is *decreasing* on its entire domain, f is one-to-one.

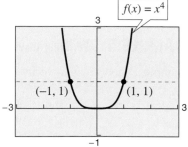

Figure P.94 $f(x) = x^4$ *is not one-to-one.*

EXAMPLE 6   Testing for One-to-One Functions

Is the function $f(x) = \sqrt{x} + 1$ one-to-one?

Algebraic Solution

Let a and b be nonnegative real numbers with $f(a) = f(b)$.

$$\sqrt{a} + 1 = \sqrt{b} + 1 \qquad \text{Set } f(a) = f(b).$$

$$\sqrt{a} = \sqrt{b}$$

$$a = b$$

Therefore,

$$f(a) = f(b)$$

implies that

$$a = b.$$

So, f is one-to-one.

Graphical Solution

Use a graphing utility to graph the function $y = \sqrt{x} + 1$. From Figure P.95, you can see that a horizontal line will intersect the graph at most once. So, f is one-to-one.

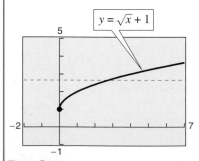

Figure P.95

Finding Inverse Functions Algebraically

For simple functions (such as the ones in Examples 1 and 2) you can find inverse functions by inspection. For instance, the inverse of $f(x) = 8x$ is $f^{-1}(x) = x/8$. For more complicated functions, however, it is best to use the following procedure for finding the inverse of a function.

Finding the Inverse of a Function

To find the inverse of f, use the following steps.

1. Use the Horizontal Line Test to decide whether f has an inverse.
2. In the equation for $f(x)$, replace $f(x)$ by y.
3. Interchange the roles of x and y, and solve for y.
4. Replace y by $f^{-1}(x)$ in the new equation.
5. Verify that f and f^{-1} are inverses of each other by showing that $f(f^{-1}(x)) = x$ and $f^{-1}(f(x)) = x$.

It is important to note that in Step 1 above, the domain of f is assumed to be the entire real line. The domain of f may be restricted so that f does have an inverse. For instance, if the domain of $f(x) = x^2$ is restricted to the nonnegative real numbers, then f does have an inverse.

EXAMPLE 7   Finding the Inverse of a Function

Find the inverse (if it exists) of $f(x) = \dfrac{5 - 3x}{2}$.

Solution
The graph of f in Figure P.96 passes the Horizontal Line Test, so you can see that f is one-to-one, and therefore has an inverse.

$f(x) = \dfrac{5 - 3x}{2}$ Write original equation.

$y = \dfrac{5 - 3x}{2}$ Replace $f(x)$ by y.

$x = \dfrac{5 - 3y}{2}$ Interchange x and y.

$2x = 5 - 3y$ Multiply each side by 2.

$3y = 5 - 2x$ Isolate the y-term.

$y = \dfrac{5 - 2x}{3}$ Solve for y.

$f^{-1}(x) = \dfrac{5 - 2x}{3}$ Replace y by $f^{-1}(x)$.

The domain and range of both f and f^{-1} consist of all real numbers. Verify that $f(f^{-1}(x)) = x$ and $f^{-1}(f(x)) = x$.

STUDY T!P

Many graphing utilities have a built-in feature to draw the inverse of a function. To see how this works, consider the function $f(x) = \sqrt{x}$. The inverse of f is $f^{-1}(x) = x^2$, $x \geq 0$. Enter the function $y_1 = \sqrt{x}$. Then graph it in the standard viewing window and use the *draw inverse* feature. You should obtain the figure below, which shows both f and its inverse f^{-1}.

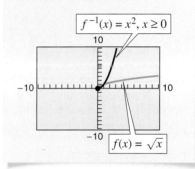

$f^{-1}(x) = x^2,\ x \geq 0$

$f(x) = \sqrt{x}$

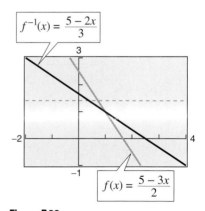

$f^{-1}(x) = \dfrac{5 - 2x}{3}$

$f(x) = \dfrac{5 - 3x}{2}$

Figure P.96

EXAMPLE 8   Finding the Inverse of a Function

Find the inverse of $f(x) = \sqrt{2x - 3}$ and sketch the graphs of f and f^{-1}.

Solution

The graph of f in Figure P.97 passes the Horizontal Line Test, so you can see that f is one-to-one and therefore has an inverse.

$f(x) = \sqrt{2x - 3}$	Write original equation.
$y = \sqrt{2x - 3}$	Replace $f(x)$ by y.
$x = \sqrt{2y - 3}$	Interchange x and y.
$x^2 = 2y - 3$	
$2y = x^2 + 3$	
$y = \dfrac{x^2 + 3}{2}$	Solve for y.
$f^{-1}(x) = \dfrac{x^2 + 3}{2}, \quad x \geq 0$	Replace y by $f^{-1}(x)$.

The graph of f^{-1} in Figure P.97 is the reflection of the graph of f in the line $y = x$. Note that the range of f is the interval $[0, \infty)$, which implies that the domain of f^{-1} is the interval $[0, \infty)$. Moreover, the domain of f is the interval $\left[\frac{3}{2}, \infty\right)$, which implies that the range of f^{-1} is the interval $\left[\frac{3}{2}, \infty\right)$.

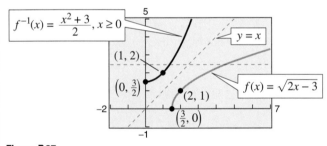

Figure P.97

Writing About Math   *The Existence of an Inverse Function*

Write a short paragraph describing why the following functions do or do not have inverse functions. Give a numerical example for each.

a. Your hourly wage is $7.50 plus $0.90 for each unit x produced per hour. Let $f(x)$ represent your weekly wage for 40 hours of work. Does this function have an inverse?

b. Let x represent the retail price of an item (in dollars), and let $f(x)$ represent the sales tax on the item. Assume that the sales tax is 7% of the retail price *and* that the sales tax is rounded to the nearest cent. Does this function have an inverse? (*Hint:* Can you undo this function? For instance, if you know that the sales tax is $0.14, can you determine *exactly* what the retail price is?)

P.9   E x e r c i s e s

In Exercises 1–4, match the graph of the function with the graph of its inverse. [The graphs of the inverse functions are labeled (a), (b), (c), and (d).]

(a)

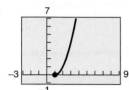

(b)

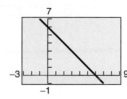

(c)

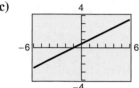

(d)

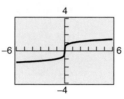

1.

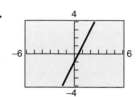

2.

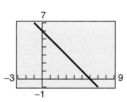

3.

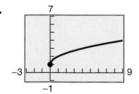

4.

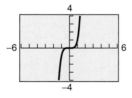

In Exercises 5–12, find the inverse of f informally. Verify that $f(f^{-1}(x)) = x$ and $f^{-1}(f(x)) = x$.

5. $f(x) = 8x$

6. $f(x) = \frac{1}{5}x$

7. $f(x) = x + 10$

8. $f(x) = x - 5$

9. $f(x) = 2x + 1$

10. $f(x) = \frac{x - 1}{4}$

11. $f(x) = \sqrt[3]{x}$

12. $f(x) = x^5$

In Exercises 13–16, show that f and g are inverse functions (a) algebraically and (b) graphically.

13. $f(x) = 2x, \quad g(x) = \frac{x}{2}$

14. $f(x) = x - 5, \quad g(x) = x + 5$

15. $f(x) = 5x + 1, \quad g(x) = \frac{x - 1}{5}$

16. $f(x) = 3 - 4x, \quad g(x) = \frac{3 - x}{4}$

In Exercises 17–22, show that f and g are inverse functions algebraically. Use a graphing utility to graph f and g in the same viewing window. Describe the relationship between the graphs.

17. $f(x) = x^3, \quad g(x) = \sqrt[3]{x}$

18. $f(x) = \frac{1}{x}, \quad g(x) = \frac{1}{x}$

19. $f(x) = \sqrt{x - 4}; \quad g(x) = x^2 + 4, \ x \geq 0$

20. $f(x) = 9 - x^2, \quad x > 0; \quad g(x) = \sqrt{9 - x}$

21. $f(x) = 1 - x^3, \quad g(x) = \sqrt[3]{1 - x}$

22. $f(x) = \frac{1}{1 + x}, \ x \geq 0; \quad g(x) = \frac{1 - x}{x}, \ 0 < x \leq 1$

In Exercises 23–28, (a) show that f and g are inverse functions algebraically and (b) verify that f and g are inverses numerically by creating a table of values for each function.

23. $f(x) = -\frac{7}{2}x - 3, \quad g(x) = -\frac{2x + 6}{7}$

24. $f(x) = \frac{x + 8}{3}, \quad g(x) = 3x - 8$

25. $f(x) = x^3 + 5, \quad g(x) = \sqrt[3]{x - 5}$

26. $f(x) = \frac{x^3}{5}, \quad g(x) = \sqrt[3]{5x}$

27. $f(x) = -\sqrt{x - 8}; \quad g(x) = 8 + x^2, \ x \leq 0$

28. $f(x) = \sqrt[3]{3x - 10}, \quad g(x) = \frac{x^3 + 10}{3}$

In Exercises 29–42, use a graphing utility to graph the function and use the Horizontal Line Test to determine whether the function is one-to-one.

29. $f(x) = 3 - \frac{1}{2}x$

30. $g(x) = \frac{4 - x}{6}$

31. $h(x) = \frac{x^2}{x^2 + 1}$

32. $f(x) = \frac{1}{8}(x + 2)^2 - 1$

33. $h(x) = \sqrt{16 - x^2}$ **34.** $f(x) = -2x\sqrt{16 - x^2}$

35. $f(x) = \sqrt{x - 2}$ **36.** $f(x) = 4 - 3x^{2/3}$

37. $f(x) = 10$ **38.** $f(x) = -0.65$

39. $g(x) = (x + 5)^3$ **40.** $f(x) = x^5 - 7$

41. $h(x) = |x + 4| - |x - 4|$

42. $f(x) = -\dfrac{|x - 6|}{|x + 6|}$

In Exercises 43–54, find the inverse of the function *f*. Use a graphing utility to graph both *f* and f^{-1} in the same viewing window. Describe the relationship between the graphs.

43. $f(x) = 2x - 3$ **44.** $f(x) = 3x$

45. $f(x) = x^5$ **46.** $f(x) = x^3 + 1$

47. $f(x) = \sqrt{x}$ **48.** $f(x) = x^2, \quad x \geq 0$

49. $f(x) = \sqrt{4 - x^2}, \quad 0 \leq x \leq 2$

50. $f(x) = \sqrt{16 - x^2}, \quad -4 \leq x \leq 0$

51. $f(x) = \sqrt[3]{x - 1}$ **52.** $f(x) = x^{3/5}$

53. $f(x) = \dfrac{4}{x}$ **54.** $f(x) = \dfrac{6}{\sqrt{x}}$

In Exercises 55–68, determine algebraically whether the function is one-to-one. If it is, find its inverse. Verify your answer graphically.

55. $f(x) = x^4$ **56.** $f(x) = \dfrac{1}{x^2}$

57. $f(x) = \dfrac{3x + 4}{5}$ **58.** $f(x) = 3x + 5$

59. $f(x) = (x + 3)^2, \quad x \geq -3$

60. $q(x) = (x - 5)^2, \quad x \leq 5$

61. $h(x) = \dfrac{4}{x^2}$

62. $f(x) = |x - 2|, \quad x \leq 2$

63. $f(x) = \sqrt{2x + 3}$ **64.** $f(x) = \sqrt{x - 2}$

65. $g(x) = x^2 - x^4$ **66.** $f(x) = \dfrac{x^2}{x^2 + 1}$

67. $f(x) = ax + b, \quad a \neq 0$

68. $f(x) = c$

Think About It   In Exercises 69–72, delete part of the graph of the function so that the part that remains is one-to-one. Find the inverse of the remaining part and give the domain of the inverse. (There are many correct answers.)

69. $f(x) = (x - 2)^2$ **70.** $f(x) = 1 - x^4$

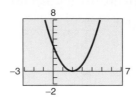   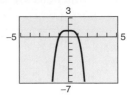

71. $f(x) = |x + 2|$ **72.** $f(x) = |x - 2|$

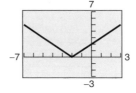   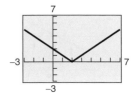

In Exercises 73 and 74, use the graph of the function *f* to complete the table and sketch the graph of f^{-1}.

73.

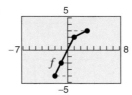

x	$f^{-1}(x)$
-4	
-2	
2	
3	

74.

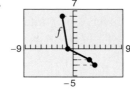

x	$f^{-1}(x)$
-3	
-2	
0	
6	

Graphical Reasoning   In Exercises 75–78, (a) use a graphing utility to graph the function, (b) use the *draw inverse* feature of the graphing utility to draw the inverse of the function, and (c) determine whether the graph of the inverse relation is an inverse function, explaining your reasoning.

75. $f(x) = x^3 + x + 1$ **76.** $h(x) = x\sqrt{4 - x^2}$

77. $g(x) = \dfrac{3x^2}{x^2 + 1}$ **78.** $f(x) = \dfrac{4x}{\sqrt{x^2 + 15}}$

In Exercises 79–84, use the functions $f(x) = \frac{1}{8}x - 3$ and $g(x) = x^3$ to find the indicated value or function.

79. $(f^{-1} \circ g^{-1})(1)$ **80.** $(g^{-1} \circ f^{-1})(-3)$

81. $(f^{-1} \circ f^{-1})(6)$ **82.** $(g^{-1} \circ g^{-1})(-4)$

83. $(f \circ g)^{-1}$ **84.** $g^{-1} \circ f^{-1}$

In Exercises 85–88, use the functions $f(x) = x + 4$ and $g(x) = 2x - 5$ to find the specified functions.

85. $g^{-1} \circ f^{-1}$ **86.** $f^{-1} \circ g^{-1}$

87. $(f \circ g)^{-1}$ **88.** $(g \circ f)^{-1}$

89. *Hourly Wage* Your wage is $8.00 per hour plus $0.75 for each unit produced per hour. So, your hourly wage y in terms of the number of units produced is $y = 8 + 0.75x$.

(a) Determine the inverse of the function. What does each variable in the inverse function represent?

(b) Use a graphing utility to graph the function and its inverse.

(c) Use the *trace* feature of your graphing utility to find the hourly wage if 10 units are produced per hour.

(d) Use the *trace* feature of your graphing utility to find the number of units produced when your hourly wage is $22.25.

90. *Diesel Engine* The function

$$y = 0.03x^2 + 254.50, \qquad 0 < x < 100$$

approximates the exhaust temperature y of a diesel engine in degrees Fahrenheit, where x is the percent load on the engine.

(a) Determine the inverse of the function. What does each variable in the inverse function represent?

(b) Use a graphing utility to graph the inverse function.

(c) Determine the percent load interval if the exhaust temperature of the engine must not exceed 500°F.

91. *Transportation* The total value of new car sales f (in billions of dollars) in the United States from 1992 through 1997 is shown in the table. The time (in years) is given by t, with $t = 2$ corresponding to 1992. (Source: National Automobile Dealers Association)

t	2	3	4	5	6	7
$f(t)$	333.8	377.3	430.6	456.2	490.0	507.5

(a) Does f^{-1} exist?

(b) If f^{-1} exists, what does it mean in the context of the problem?

(c) If f^{-1} exists, find $f^{-1}(456.2)$.

(d) If the table above was extended to 1998 and if the total value of new car sales for that year was $430.6 billion, would f^{-1} exist? Explain.

Synthesis

True or False? In Exercises 92 and 93, determine whether the statement is true or false. Justify your answer.

92. If f is an even function, f^{-1} exists.

93. If the inverse of f exists, the y-intercept of f is an x-intercept of f^{-1}.

94. Prove that if f and g are one-to-one functions, $(f \circ g)^{-1}(x) = (g^{-1} \circ f^{-1})(x)$.

95. Prove that if f is a one-to-one odd function, f^{-1} is an odd function.

96. *Think About It* The function

$$f(x) = k(2 - x - x^3)$$

is one-to-one and $f^{-1}(3) = -2$. Find k.

P Chapter Summary

What did you learn?

P   Review Exercises

P.1   In Exercises 1 and 2, determine which numbers are (a) natural numbers, (b) integers, (c) rational numbers, and (d) irrational numbers.

1. $11, -14, -\frac{8}{9}, \frac{5}{2}, \sqrt{6}, 0.4$

2. $\sqrt{15}, -22, -\frac{10}{3}, 0, 5.2, \frac{3}{7}$

In Exercises 3 and 4, use a calculator to find the decimal form of the rational number. If it is a nonterminating decimal, write the repeating pattern. Then place the correct inequality symbol (< or >) between the numbers.

3. $\frac{5}{6}, \frac{7}{8}$

4. $\frac{9}{25}, \frac{5}{7}$

In Exercises 5 and 6, give a verbal description of the real numbers that are represented by the inequality. Then sketch the inequality on the real number line.

5. $x \le 7$

6. $x > 1$

In Exercises 7 and 8, find the distance between a and b.

7. $a = -74, \quad b = 48$

8. $a = -123, \quad b = -9$

In Exercises 9–12, use absolute value notation to describe the expression.

9. The distance between x and 7 is at least 4.

10. The distance between x and 25 is no more than 10.

11. The distance between y and -30 is less than 5.

12. The distance between y and $\frac{1}{2}$ is more than 2.

In Exercises 13–16, evaluate the expression for each value of x. (If not possible, state the reason.)

	Expression		*Values*	
13.	$10x - 3$	(a) $x = -1$	(b) $x = 3$
14.	$x^2 - 11x + 24$	(a) $x = -2$	(b) $x = 2$
15.	$-2x^2 - x + 3$	(a) $x = 3$	(b) $x = -3$
16.	$\dfrac{4x}{x-1}$	(a) $x = -1$	(b) $x = 1$

In Exercises 17–20, identify the rule of algebra illustrated by the equation.

17. $2x + (3x - 10) = (2x + 3x) - 10$

18. $\dfrac{2}{y+4} \cdot \dfrac{y+4}{2} = 1, \quad y \neq -4$

19. $(t + 4)(2t) = (2t)(t + 4)$

20. $0 + (a - 5) = a - 5$

In Exercises 21–26, perform the operations. (Write fractional answers in simplest form.)

21. $\frac{2}{3} + \frac{8}{9}$

22. $\frac{3}{4} - \frac{1}{6} + \frac{1}{8}$

23. $\frac{3}{16} \div \frac{9}{2}$

24. $\frac{5}{8} \cdot \frac{2}{3}$

25. $\dfrac{x}{5} + \dfrac{7x}{12}$

26. $\dfrac{9}{x} \div \dfrac{1}{6}$

P.2   In Exercises 27–30, plot the point and determine the quadrant in which it is located.

27. $(8, -3)$

28. $(-4, -9)$

29. $\left(-\frac{5}{2}, 10\right)$

30. $(-6.5, -0.5)$

In Exercises 31 and 32, determine the quadrant(s) in which (x, y) is located so that the conditions are satisfied.

31. $x > 0$ and $y = -2$

32. $(x, y), \quad xy = 4$

33. *Patents*   The number of patents P (in thousands) issued in the United States from 1988 through 1996 is shown in the table.   (Source: U.S. Patent and Trademark Office)

Year	1988	1989	1990	1991	1992
P	84.4	102.7	99.2	106.8	107.4

Year	1993	1994	1995	1996
P	109.7	113.6	113.8	121.7

(a) Sketch a scatter plot of the data.

(b) What statement can be made about the number of patents issued in the United States?

34. *Weather* The normal daily maximum and minimum temperatures (in degrees Fahrenheit) for each month for the city of Chicago are shown in the table. Make a double line graph for the data. (Source: NOAA)

Month	Jan.	Feb.	Mar.	Apr.	May	Jun.
Max.	29.0	33.5	45.8	58.6	70.1	79.6
Min.	12.9	17.2	28.5	38.6	47.7	57.5

Month	Jul.	Aug.	Sep.	Oct.	Nov.	Dec.
Max.	83.7	81.8	74.8	63.3	48.4	34.0
Min.	62.6	61.6	53.9	42.2	31.6	19.1

35. *Business* The net profits (in millions of dollars) for the Progressive Corporation for the years 1994 through 1998 are shown in the table. Create a bar graph for the data. (Source: Progressive Corporation)

Year	1994	1995	1996	1997	1998
Profits	228.1	250.5	316.6	400.0	456.7

In Exercises 36 and 37, plot the points and find the distance between the points.

36. $(-3, 8), (1, 5)$ **37.** $(5.6, 0), (0, 8.2)$

Geometry **In Exercises 38 and 39, plot the points and verify that the points form the polygon.**

38. *Right Triangle:* $(2, 3), (13, 11), (5, 22)$

39. *Parallelogram:* $(1, 2), (8, 3), (9, 6), (2, 5)$

In Exercises 40 and 41, plot the points and find the midpoint of the line segment joining the points.

40. $(-12, 5), (4, -7)$

41. $(1.8, 7.4), (-0.6, -14.5)$

42. *Business* The Sbarro, Inc., restaurant chain had revenues of $329.5 million in 1996 and $375.2 million in 1998. (Source: Sbarro, Inc.)

(a) Without any additional information, what would you estimate the 1997 revenues to have been?

(b) The actual revenue for 1997 was $349.4 million. How accurate is your estimate?

In Exercises 43 and 44, complete the table. Use the resulting solution points to sketch the graph of the equation.

43. $y = -\frac{1}{2}x + 2$

x	-2	0	2	3	4
y					

44. $y = x^2 - 3x$

x	-1	0	1	2	3
y					

In Exercises 45–54, sketch the graph of the equation by hand. Use a graphing utility to verify the graph.

45. $y - 2x - 3 = 0$ **46.** $3x + 2y + 6 = 0$

47. $x - 5 = 0$ **48.** $y = 8 - |x|$

49. $y = \sqrt{5 - x}$ **50.** $y = \sqrt{x + 2}$

51. $y + 2x^2 = 0$ **52.** $y = x^2 - 4x$

53. $x + y^2 = 9$ **54.** $x^2 + y^2 = 10$

In Exercises 55–62, use a graphing utility to graph the equation. Approximate any intercepts.

55. $y = \frac{1}{4}(x + 1)^3$ **56.** $y = 4 - (x - 4)^2$

57. $y = \frac{1}{4}x^4 - 2x^2$ **58.** $y = \frac{1}{4}x^3 - 3x$

59. $y = x\sqrt{9 - x^2}$ **60.** $y = x\sqrt{x + 3}$

61. $y = |x - 4| - 4$ **62.** $y = |x + 2| + |3 - x|$

In Exercises 63 and 64, find the standard form of the equation of the specified circle. Then graph the equation.

63. Center: $(3, -1)$; Solution point: $(-5, 1)$

64. End points of a diameter: $(-4, 6), (10, -2)$

65. *Data Analysis* The average expenditures y for automobile insurance per insured vehicle from 1992 through 1996 can be approximated by the model

$$y = 16.7t + 584.6, \quad 2 \le t \le 6$$

where t is the time (in years), with $t = 2$ corresponding to 1992. (Source: National Association of Insurance Commissioners)

(a) Determine algebraically the value of y in 1996.

(b) Use the *zoom* and *trace* features of your graphing utility to verify your answer to part (a).

(c) Use the model to estimate the values of y for the years 2000 and 2002.

P.3 In Exercises 66–69, solve the equation (if possible) and use a graphing utility to verify your solution.

66. $14 + \dfrac{2}{x - 1} = 10$ **67.** $6 - \dfrac{11}{x} = 3 + \dfrac{7}{x}$

68. $\dfrac{9x}{3x - 1} - \dfrac{4}{3x + 1} = 3$

69. $\dfrac{5}{x - 5} + \dfrac{1}{x + 5} = \dfrac{2}{x^2 - 25}$

In Exercises 70–73, determine the x- and y-intercepts of the graph of the equation algebraically. Use a graphing utility to verify your answer.

70. $-x + y = 3$ **71.** $x - 5y = 20$

72. $y = x^2 - 9x + 8$ **73.** $y = 25 - x^2$

In Exercises 74 and 75, use a graphing utility to graph the equation and approximate the x- and y-intercepts.

74. $y = -|x + 5| - 2$ **75.** $y = 6 - 2|x - 3|$

In Exercises 76–81, use a graphing utility to approximate any solutions (accurate to three decimal places) of the equation.

76. $5(x - 2) - 1 = 0$ **77.** $12 - 5(x - 7) = 0$

78. $3x^3 - 2x + 4 = 0$ **79.** $\frac{1}{3}x^3 - x + 4 = 0$

80. $x^4 - 3x + 1 = 0$ **81.** $6 - \frac{1}{2}x^2 + \frac{5}{6}x^4 = 0$

In Exercises 82–85, determine algebraically any points of intersection of the graphs of the equations. Use a graphing utility to verify your answer(s).

82. $3x + 5y = -7$ **83.** $x - y = 3$
      $-x - 2y = 3$ $2x + y = 12$

84. $x^2 + 2y = 14$ **85.** $y = -x + 7$
      $3x + 4y = 1$ $y = 2x^3 - x + 9$

In Exercises 86–95, use any method to solve the equation. Use a graphing utility to verify your solution(s).

86. $6x = 3x^2$ **87.** $15 + x - 2x^2 = 0$

88. $(x + 4)^2 = 18$ **89.** $16x^2 = 25$

90. $x^2 - 12x + 30 = 0$ **91.** $x^2 + 6x - 3 = 0$

92. $2x^2 + 9x - 5 = 0$ **93.** $-x^2 - x + 15 = 0$

94. $x^2 - 4x - 10 = 0$ **95.** $-2x^2 - 13x = 0$

In Exercises 96–113, solve the equation (if possible) and use a graphing utility to verify your solution.

96. $3x^3 - 26x^2 + 16x = 0$ **97.** $216x^4 - x = 0$

98. $5x^4 - 12x^3 = 0$ **99.** $4x^3 - 6x^2 = 0$

100. $\sqrt{x + 4} = 3$ **101.** $\sqrt{x - 2} - 8 = 0$

102. $\sqrt{2x + 3} + \sqrt{x - 2} = 2$

103. $5\sqrt{x} - \sqrt{x - 1} = 6$

104. $(x - 1)^{2/3} - 25 = 0$ **105.** $(x + 2)^{3/4} = 27$

106. $3\left(1 - \dfrac{1}{5t}\right) = 0$ **107.** $\dfrac{1}{x - 2} = 3$

108. $\dfrac{4}{(x - 4)^2} = 1$ **109.** $\dfrac{1}{(t + 1)^2} = 1$

110. $|x - 5| = 10$ **111.** $|2x + 3| = 7$

112. $|x^2 - 3| = 2x$ **113.** $|x^2 - 6| = x$

P.4 In Exercises 114–119, plot the two points and find the slope of the line that passes through the points.

114. $(-3, 2), (8, 2)$ **115.** $(7, -1), (7, 12)$

116. $\left(\frac{3}{2}, 1\right), \left(5, \frac{5}{2}\right)$ **117.** $\left(-\frac{3}{4}, \frac{5}{6}\right), \left(\frac{1}{2}, -\frac{5}{2}\right)$

118. $(-4.5, 6), (2.1, 3)$

119. $(-2.7, -6.3), (-1, -1.2)$

In Exercises 120–123, use the concept of slope to find t such that the three points are collinear.

120. $(-2, 5), (0, t), (1, 1)$

121. $(-6, 1), (1, t), (10, 5)$

122. $(1, -4), (t, 3), (5, 10)$

123. $(-3, 3), (t, -1), (8, 6)$

In Exercises 124–135, (a) find an equation of the line that passes through the given point and has the specified slope, and (b) find three additional points through which the line passes.

	Point	*Slope*
124.	$(2, -1)$	$m = \frac{1}{4}$
125.	$(-3, 5)$	$m = -\frac{3}{2}$
126.	$(0, -5)$	$m = \frac{3}{2}$
127.	$(3, 0)$	$m = -\frac{2}{3}$

Point	Slope
128. $(-6, -5)$	$m = -2$
129. $(-7, 2)$	$m = 4$
130. $\left(\frac{1}{5}, -5\right)$	$m = -1$
131. $\left(0, \frac{7}{8}\right)$	$m = -\frac{4}{5}$
132. $(-2, 6)$	$m = 0$
133. $(-8, 8)$	$m = 0$
134. $(10, -6)$	m is undefined.
135. $(5, 4)$	m is undefined.

In Exercises 136–141, (a) find an equation of the line (in slope-intercept form) that passes through the points and (b) sketch the graph of the equation.

136. $(2, -1), (4, -1)$ **137.** $(0, 0), (0, 10)$

138. $(2, 1), (14, 6)$ **139.** $(-2, 2), (3, -10)$

140. $(-1, 0), (6, 2)$ **141.** $(1, 6), (4, 2)$

Rate of Change **In Exercises 142 and 143, you are given the dollar value of a product in 2000 *and* the rate at which the value of the item is expected to change during the 5 years following. Use this information to write a linear equation that gives the dollar value V of the product in terms of the year t. (Let $t = 0$ represent 2000.)**

2000 Value	Rate
142. $12,500	$850 increase per year
143. $72.95	$5.15 increase per year

In Exercises 144–147, write equations of the lines through the point (a) parallel to the given line and (b) perpendicular to the given line. Verify your result with a graphing utility (use a square setting).

Point	Line
144. $(3, -2)$	$5x - 4y = 8$
145. $(-8, 3)$	$2x + 3y = 5$
146. $(-6, 2)$	$x = 4$
147. $(3, -4)$	$y = 2$

P.5 **In Exercises 148 and 149, determine which of the sets of ordered pairs represents a function from A to B. Give reasons for your answers.**

148. $A = \{10, 20, 30, 40\}$ and $B = \{0, 2, 4, 6\}$

(a) $\{(20, 4), (40, 0), (20, 6), (30, 2)\}$

(b) $\{(10, 4), (20, 4), (30, 4), (40, 4)\}$

(c) $\{(40, 0), (30, 2), (20, 4), (10, 6)\}$

(d) $\{(20, 2), (10, 0), (40, 4)\}$

149. $A = \{u, v, w\}$ and $B = \{-2, -1, 0, 1, 2\}$

(a) $\{(v, -1), (u, 2), (w, 0), (u, -2)\}$

(b) $\{(u, -2), (v, 2), (w, 1)\}$

(c) $\{(u, 2), (v, 2), (w, 1), (w, 1)\}$

(d) $\{(w, -2), (v, 0), (w, 2)\}$

In Exercises 150–153, determine if the equation represents y as a function of x.

150. $16x - y^4 = 0$ **151.** $2x - y - 3 = 0$

152. $y = \sqrt{1 - x}$ **153.** $|y| = x + 2$

In Exercises 154 and 155, evaluate the function at each value of the specified variable. Simplify your answers.

154. $f(x) = x^2 + 1$

(a) $f(2)$ (b) $f(-4)$

(c) $f(t^2)$ (d) $-f(x)$

155. $g(x) = x^{4/3}$

(a) $g(8)$ (b) $g(t + 1)$

(c) $\dfrac{g(8) - g(1)}{8 - 1}$ (d) $g(-x)$

In Exercises 156–161, determine the domain of the function. Verify your result with a graphing utility.

156. $f(x) = (x - 1)(x + 2)$ **157.** $f(x) = x^2 - 4x - 32$

158. $f(x) = \sqrt{25 - x^2}$ **159.** $f(x) = \sqrt{x^2 + 8x}$

160. $g(s) = \dfrac{5}{3s - 9}$ **161.** $f(x) = \dfrac{2}{3x + 4}$

162. *Business* A company produces a product for which the variable cost is $5.35 per unit and the fixed costs are $16,000. The company sells the product for $8.20 and can sell all that it produces.

(a) Find the total cost as a function of x, the number of units produced.

(b) Find the profit as a function of x.

163. ***Boating*** The retail expenditures B (in billions of dollars) on boating in the United States from 1985 to 1996 can be represented by the piecewise-defined function

$$B(t) = \begin{cases} -0.631t^2 - 2.845t + 14.160, & -5 \le t < 2 \\ 2.088t + 5.768, & 2 \le t \le 6 \end{cases}$$

where $t = 0$ represents 1990. Use a graphing utility to graph the model and find the amount spent on boating in 1985, 1990, and 1995. (Source: National Marine Manufacturers Association)

P.6 In Exercises 164–167, find the domain and range of the function.

164. $f(x) = 3 - 2x^2$ **165.** $f(x) = \sqrt{2x^2 - 1}$

166. $h(x) = \sqrt{36 - x^2}$ **167.** $g(x) = |x + 5|$

In Exercises 168–171, (a) use a graphing utility to graph the equation and (b) use the Vertical Line Test to determine whether y is a function of x.

168. $y = \dfrac{x^2 + 3x}{6}$ **169.** $y = -\frac{2}{3}|x + 5|$

170. $3x + y^2 = 2$ **171.** $x^2 + y^2 = 49$

In Exercises 172–175, determine the open intervals over which the function is increasing, decreasing, or constant.

172. $f(x) = x^3 - 3x$ **173.** $f(x) = \sqrt{x^2 - 9}$

174. $f(x) = x\sqrt{x - 6}$ **175.** $f(x) = \dfrac{|x + 8|}{2}$

Graphical Analysis In Exercises 176–179, use a graphing utility to approximate (accurate to two decimal places) any relative maximum or minimum values of the function.

176. $f(x) = (x^2 - 4)^2$ **177.** $f(x) = x^2 - x - 1$

178. $h(x) = 4x^3 - x^4$ **179.** $f(x) = x^3 - 4x^2 - 1$

In Exercises 180 and 181, sketch the graph of the piecewise-defined function by hand. Verify using a graphing utility.

180. $f(x) = \begin{cases} 3x + 5, & x < 0 \\ x - 4, & x \ge 0 \end{cases}$

181. $f(x) = \begin{cases} x^2 + 7, & x < 1 \\ x^2 - 5x + 6, & x \ge 1 \end{cases}$

In Exercises 182 and 183, determine whether the function is even, odd, or neither.

182. $f(x) = (x^2 - 8)^2$ **183.** $f(x) = 2x^3 - x^2$

P.7 In Exercises 184–187, the graph is related to one of the common functions on page 82. Identify the common function and describe the transformation(s) shown in the graph. Write the equation for the graphed function.

184.

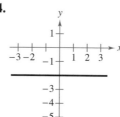

185.

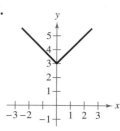

186.

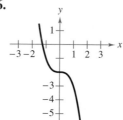

187.

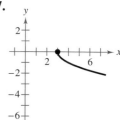

In Exercises 188–211, sketch the graph of the function by hand.

188. $f(x) = x^2 - 6$ **189.** $f(x) = (x - 3)^2 - 2$

190. $f(x) = (x - 1)^3 + 7$ **191.** $f(x) = (x + 2)^3 + 5$

192. $f(x) = \sqrt{x} - 5$ **193.** $f(x) = \sqrt{x + 5} + 4$

194. $f(x) = 7 + |x|$ **195.** $f(x) = |x + 8| - 1$

196. $f(x) = -x^2 - 3$

197. $f(x) = -(x - 2)^2 - 8$

198. $f(x) = -(x + 5)^3 - 6$

199. $f(x) = 3 - x^3$ **200.** $f(x) = -\sqrt{x + 4} - 3$

201. $f(x) = \sqrt{-x - 1} + 7$

202. $f(x) = -|x + 2| - 1$

203. $f(x) = 1 - |-x|$

204. $f(x) = -2x^2 + 3$ **205.** $f(x) = \frac{1}{2}(x - 3)^2 + 6$

206. $f(x) = -\frac{1}{4}(1 - x)^3 - 2$

207. $f(x) = 5(x - 1)^3 - 5$

208. $f(x) = -\frac{1}{2}|x| + 9$ **209.** $f(x) = -2|-x - 4|$

210. $f(x) = -3\sqrt{6 - x} + 4$

211. $f(x) = 4\sqrt{x + 1} - 8$

P.8 In Exercises 212–221, let $f(x) = 3 - 2x$, $g(x) = \sqrt{x}$, and $h(x) = 3x^2 + 2$, and find the indicated values.

212. $(f - g)(4)$

213. $(f + h)(5)$

214. $(f + g)(25)$

215. $(g - h)(1)$

216. $(fh)(1)$

217. $\left(\dfrac{g}{h}\right)(1)$

218. $(h \circ g)(7)$

219. $(g \circ f)(-2)$

220. $(f \circ h)(-4)$

221. $(g \circ h)(6)$

Data Analysis In Exercises 222 and 223, use the table, which shows the total value (in billions of dollars) of U.S. imports from China and Taiwan for the years 1992 through 1997. The variables y_1 and y_2 represent the total value of imports from China and Taiwan, respectively. (Source: U.S. Bureau of the Census)

Year	1992	1993	1994	1995	1996	1997
y_1	25.7	31.5	38.8	45.5	51.5	62.6
y_2	24.6	25.1	26.7	29.0	29.9	32.6

222. Use the regression capabilities of a graphing utility to find quadratic models for each of the variables. Let $t = 2$ represent 1992.

223. Use a graphing utility to graph y_1, y_2, and $y_1 + y_2$ in the same viewing window. Use the model to estimate the total value of U.S. imports from China and Taiwan in 2002.

P.9 In Exercises 224–227, find the inverse of f informally. Verify that $f(f^{-1}(x)) = f^{-1}(f(x)) = x$.

224. $f(x) = 6x$

225. $f(x) = \frac{1}{12}x$

226. $f(x) = x - 7$

227. $f(x) = x + 5$

In Exercises 228–235, use a graphing utility to graph f and determine whether f has an inverse. Find f^{-1}, if it exists.

228. $f(x) = \frac{1}{2}x - 3$

229. $f(x) = 5x - 7$

230. $f(x) = 2(x - 4)^2$

231. $f(x) = -x^2 + 4$

232. $f(x) = |x - 2|$

233. $f(x) = \sqrt{x + 1}$

234. $f(x) = x^3 + 2$

235. $f(x) = \frac{1}{2}|x + 4|$

In Exercises 236–241, find the inverse of f algebraically.

236. $f(x) = \dfrac{x}{12}$

237. $f(x) = \dfrac{7x + 3}{8}$

238. $f(x) = 4x^3 - 3$

239. $f(x) = x^3 - 2$

240. $f(x) = \sqrt{x + 10}$

241. $f(x) = 4\sqrt{6 - x}$

Synthesis

True or False? In Exercises 242–246, determine whether the statement is true or false. Justify your answer.

242. If $ab = 0$, then the point (a, b) lies on the x-axis or on the y-axis.

243. The graph of a function may have two distinct y-intercepts.

244. Relative to the graph of $f(x) = \sqrt{x}$, the function $h(x) = -\sqrt{x + 9} - 13$ is shifted 9 units to the left and 13 units down, then reflected in the x-axis.

245. If $f(x) = x^n$ where n is odd, f^{-1} exists.

246. There exists no function f such that $f = f^{-1}$.

In Exercises 247–250, describe the error and then make the necessary correction.

247. $\dfrac{x - 1}{1 - x} = 1$ ✗

248. $-x^2(-x^2 + 3) = x^4 + 3x^2$ ✗

249. $(2x)^4 = 2x^4$ ✗

250. $(-x)^6 = -x^6$ ✗

251. In your own words, explain the difference between an identity and a conditional equation.

252. In your own words, explain what is meant by equivalent equations. Describe the steps used to transform an equation into an equivalent equation.

253. Explain why not all equations of lines are functions.

Chapter Project *Modeling the Area of a Plot*

Many real-life problems can be analyzed from a *graphical,* a *numerical,* and an *algebraic* perspective. In this project, you will use all three strategies to determine the maximum size of a rectangular plot that can be enclosed by a fixed amount of fencing.

You have 100 meters of fencing material to enclose a rectangular plot. Your goal is to determine the dimensions of the plot such that you enclose the maximum area possible.

a.   Express the area $A(x)$ of the rectangular plot as a function of the length x of one side, as shown in the figure below.

b.   Analyze the problem *numerically* by completing the table.

x	0	5	10	15	20	25	30	35	40	45	50
$A(x)$											

According to this table, what do you think the dimensions of the plot should be to enclose the maximum area? Explain.

c.   Use a graphing utility to graph the area function. What is the domain of the function? Solve the problem *graphically* by using the *trace* feature of your graphing utility to find the value of x that yields the maximum area.

d.   Solve the problem *algebraically* by showing that the area function can be written as $A(x) = 625 - (x - 25)^2$. How does this form of the function allow you to find the dimensions that produce a maximum area?

e.   Discuss the strengths and weaknesses of the three strategies used in parts (b), (c), and (d).

Questions for Further Exploration

1. Suppose you were not restricted to a rectangular plot. Would you be able to use 100 meters of fencing to enclose a greater area? Explain.

2. In the project above, you found the maximum area that can be enclosed in a rectangular plot using 100 meters of fencing. If you doubled the amount of fencing, could you enclose twice as much area? Use numerical, graphical, and algebraic approaches and explain your reasoning.

3. Suppose the rectangular plot runs along a building, so that you need to fence only three sides. What dimensions will now yield a maximum area with 100 meters of fencing?

P Chapter Test

Take this test as you would take a test in class. After you are done, check your work against the answers in the back of the book.

1. Place the correct symbol ($<$ or $>$) between $-\frac{10}{3}$ and $-|-4|$.

2. Find the distance between the real numbers -17 and 39.

3. Evaluate $-x^2 + 6x + 7$ for (a) $x = -1$ and (b) $x = \frac{1}{2}$.

4. Plot the points $(-2, 5)$ and $(6, 0)$. Find the coordinates of the midpoint of the line segment joining the points and the distance between the points.

In Exercises 5–7, use the point-plotting method to graph the equation and identify any intercepts. Verify your results using a graphing utility.

5. $y = 4 - \frac{3}{4}|x|$
6. $y = 4 - (x - 2)^2$
7. $y = \sqrt{3 - x}$

8. Find the standard form of the equation of a circle with center $(4, -1)$ and a solution point $(1, 4)$. Then graph the equation.

In Exercises 9–11, find all solutions of the equation. Check your solution(s) algebraically and graphically.

9. $x^2 - 10x + 9 = 0$

10. $x + \sqrt{22 - 3x} = 6$

11. $|8x - 1| = 21$

12. Find an equation of the line that passes through the point $(0, 4)$ and is perpendicular to the line $5x + 2y = 3$.

13. The graph of $y^2(4 - x) = x^3$ is shown at the right. Does the graph represent y as a function of x? Explain.

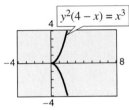

FIGURE FOR 13

In Exercises 14–16, use the function $f(x) = 10 - \sqrt{3 - x}$.

14. Evaluate: $f(-1)$
15. Simplify: $f(t - 3)$
16. Simplify: $\dfrac{f(x) - f(2)}{x - 2}$

In Exercises 17 and 18, find the open intervals for which the function is increasing, decreasing, or constant.

17. $h(x) = \frac{1}{4}x^4 - 2x^2$
18. $g(t) = |t + 2| - |t - 2|$

In Exercises 19 and 20, (a) identify the common function, (b) describe the transformation, and (c) sketch the graph of the function.

19. $f(x) = -2(x - 5)^3 + 3$
20. $f(x) = \sqrt{-x - 7} + 7$

21. Use the functions $f(x) = x^2$ and $g(x) = \sqrt{2 - x}$ to find the specified function and its domain.

    (a) $(f - g)(x)$ (b) $\left(\dfrac{f}{g}\right)(x)$ (c) $(f \circ g)(x)$ (d) $g^{-1}(x)$

22. Find the inverse of $f(x) = (x + 2)^2 - 1$, $x \geq -2$.

LIBRARY OF FUNCTIONS

In Chapter P, you were introduced to the concept of a *function*. As you proceed through the text, you will see that functions play a primary role in modeling real-life situations.

Over the past few hundred years, many different types of functions have been introduced and studied. Those that have proven to be most important in modeling real life have come to be known as *elementary functions*. There are three basic types of elementary functions: algebraic functions, exponential and logarithmic functions, and trigonometric and inverse trigonometric functions.

You will also encounter other types of functions in this text, such as functions defined by real-life data and piecewise-defined functions.

> ### Library of Functions
> Each time a new type of function is studied in detail in this text, it will be highlighted in a box like this one. For instance, a linear function is highlighted in the Library of Functions box in Section P.4 because lines are discussed in that section.
>
> In addition, there is a Library of Functions Summary inside the front cover which describes the functions listed below.

Algebraic Functions

Exponential and Logarithmic Functions

Trigonometric and Inverse Trigonometric Functions

Nonelementary Functions

Trigonometric Functions

Ron Watts/CORBIS

The Big Picture

In this chapter you will learn how to

❑ describe an angle and convert between degree and radian measures.

❑ identify a unit circle and its relationship to real numbers.

❑ evaluate trigonometric functions of any angle.

❑ use fundamental trigonometric identities.

❑ sketch graphs of trigonometric functions.

❑ evaluate inverse trigonometric functions.

❑ evaluate the composition of trigonometric functions.

❑ use trigonometric functions to model and solve real-life problems.

The Mauna Loa Observatory in Hawaii conducts research to understand the global carbon cycle. It is located far from pollution sources that would affect the gases being measured. (Source: NOAA/Climate Monitoring and Diagnostic Laboratory)

Important Vocabulary

As you encounter each new vocabulary term in this chapter, add the term and its definition to your notebook glossary.

- trigonometry (p. 124)
- angle (p. 124)
- initial side (p. 124)
- terminal side (p. 124)
- vertex (p. 124)
- positive angles (p. 124)
- negative angles (p. 124)
- coterminal angles (p. 124)
- central angle (p. 125)
- radian (p. 125)
- complementary angles (p. 127)
- supplementary angles (p. 127)

- degree (p. 127)
- linear speed (p. 129)
- angular speed (p. 129)
- unit circle (p. 135)
- sine (pp. 136, 143)
- cosecant (pp. 136, 143)
- cosine (pp. 136, 143)
- secant (pp. 136, 143)
- tangent (pp. 136, 143)
- cotangent (pp. 136, 143)
- periodic (p. 138)
- period (pp. 138, 166)

- hypotenuse (p. 143)
- opposite side (p. 143)
- adjacent side (p. 143)
- reference angle (p. 156)
- sine curve (p. 163)
- amplitude (p. 165)
- phase shift (p. 167)
- damping factor (p. 179)
- inverse sine function (p. 185)
- inverse cosine function (p. 187)
- inverse tangent function (p. 187)

Additional Resources   Text-specific additional resources are available to help you do well in this course. See page xvi for details.

1.1 Radian and Degree Measure

Angles

As derived from the Greek language, the word **trigonometry** means *"measurement of triangles."* Initially, trigonometry dealt with relationships among the sides and angles of triangles and was used in the development of astronomy, navigation, and surveying. With the development of calculus and the physical sciences in the 17th century, a different perspective arose—one that viewed the classic trigonometric relationships as *functions* with the set of real numbers as their domains. Consequently, the applications of trigonometry expanded to include a vast number of physical phenomena involving rotations and vibrations, including sound waves, light rays, planetary orbits, vibrating strings, pendulums, and orbits of atomic particles. The approach in this text incorporates *both* perspectives, starting with angles and their measure.

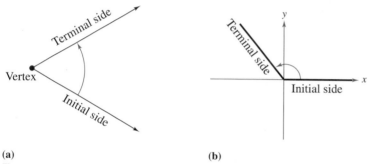

(a)

(b)

Figure 1.1

Bob Martin/Allsport

An **angle** is determined by rotating a ray (half-line) about its endpoint. The starting position of the ray is the **initial side** of the angle, and the position after rotation is the **terminal side,** as shown in Figure 1.1(a). The endpoint of the ray is the **vertex** of the angle. This perception of an angle fits a coordinate system in which the origin is the vertex and the initial side coincides with the positive *x*-axis. Such an angle is in **standard position,** as shown in Figure 1.1(b). **Positive angles** are generated by counterclockwise rotation, and **negative angles** by clockwise rotation, as shown in Figure 1.2. Angles are labeled with Greek letters α (alpha), β (beta), and θ (theta), as well as uppercase letters *A*, *B*, and *C*. In Figure 1.3, note that angles α and β have the same initial and terminal sides. Such angles are **coterminal.**

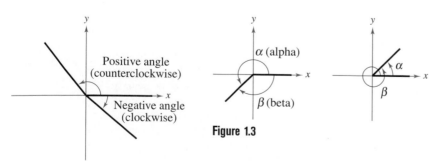

Figure 1.2

Figure 1.3

Radian Measure

One way to measure angles is in radians. This type of measure is especially useful in calculus. To define a radian, you can use a **central angle** of a circle, one whose vertex is the center of the circle, as shown in Figure 1.4.

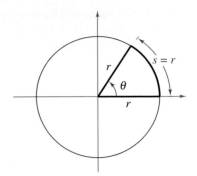

Arc Length = radius when $\theta = 1$ radian.
Figure 1.4

Definition of a Radian

One **radian** is the measure of a central angle θ that intercepts an arc s equal in length to the radius r of the circle. See Figure 1.4.

Because the circumference of a circle is $2\pi r$, it follows that a central angle of one full revolution (counterclockwise) corresponds to an arc length of $s = 2\pi r$. Therefore, 2π radians corresponds to 360°, π radians corresponds to 180°, and $\pi/2$ radians corresponds to 90°. Moreover, because $2\pi \approx 6.28$, there are just over six radius lengths in a full circle, as shown in Figure 1.5. In general, the radian measure of a central angle θ is obtained by dividing the arc length s by r. That is, $s/r = \theta$, where θ is *measured in radians*. Because the units of measure for s and r are the same, this ratio is unitless—it is simply a real number.

Because the radian measure of an angle of one full revolution is 2π, you can obtain the following.

$$\frac{1}{2} \text{ revolution} = \frac{2\pi}{2} = \pi \text{ radians}$$

$$\frac{1}{4} \text{ revolution} = \frac{2\pi}{4} = \frac{\pi}{2} \text{ radians}$$

$$\frac{1}{6} \text{ revolution} = \frac{2\pi}{6} = \frac{\pi}{3} \text{ radians}$$

These and other common angles are shown in Figure 1.6.

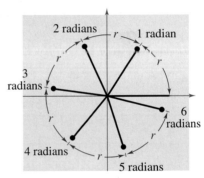

Figure 1.5

Figure 1.6

Recall that the four quadrants in a coordinate system are numbered I, II, III, and IV. Figure 1.7 shows which angles between 0 and 2π lie in each of the four quadrants. Note that angles between 0 and $\pi/2$ are acute and that angles between $\pi/2$ and π are obtuse.

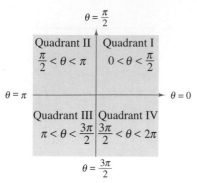

Figure 1.7

STUDY TIP

The phrase "the terminal side of θ lies in a quadrant" is often abbreviated by simply saying "θ lies in a quadrant." The terminal sides of the "quadrant angles" 0, $\pi/2$, π, and $3\pi/2$ do not lie within quadrants.

Two angles are coterminal if they have the same initial and terminal sides. For instance, the angles 0 and 2π are coterminal, as are the angles $\pi/6$ and $13\pi/6$. You can find an angle that is coterminal to a given angle θ by adding or subtracting 2π (one revolution), as demonstrated in Example 1. A given angle θ has infinitely many coterminal angles. For instance, $\theta = \pi/6$ is coterminal with

$$\frac{\pi}{6} + 2n\pi, \text{ where } n \text{ is an integer.}$$

EXAMPLE 1   Sketching and Finding Coterminal Angles

a. For the positive angle $13\pi/6$, subtract 2π to obtain a coterminal angle

$$\frac{13\pi}{6} - 2\pi = \frac{\pi}{6}. \qquad \text{See Figure 1.8(a).}$$

b. For the positive angle $3\pi/4$, subtract 2π to obtain a coterminal angle

$$\frac{3\pi}{4} - 2\pi = -\frac{5\pi}{4}. \qquad \text{See Figure 1.8(b).}$$

c. For the negative angle $-2\pi/3$, add 2π to obtain a coterminal angle

$$-\frac{2\pi}{3} + 2\pi = \frac{4\pi}{3}. \qquad \text{See Figure 1.8(c).}$$

The *Interactive* CD-ROM and *Internet* versions of this text show every example with its solution; clicking on the *Try It!* button brings up similar problems. Guided Examples and Integrated Examples show step-by-step solutions to additional examples. Integrated Examples are related to several concepts in the section.

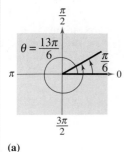

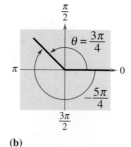

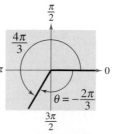

(a)   (b)   (c)

Figure 1.8

Two positive angles α and β are **complementary** (complements of each other) if their sum is $\pi/2$. Two positive angles are **supplementary** (supplements of each other) if their sum is π. See Figure 1.9.

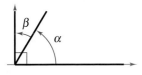

Complimentary angles

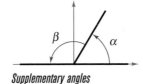

Supplementary angles

Figure 1.9

EXAMPLE 2   Complementary and Supplementary Angles

If possible, find the complement and the supplement of (a) $2\pi/5$ and (b) $4\pi/5$.

Solution

a. The complement of $2\pi/5$ is

$$\frac{\pi}{2} - \frac{2\pi}{5} = \frac{5\pi}{10} - \frac{4\pi}{10} = \frac{\pi}{10},$$

The supplement of $2\pi/5$ is

$$\pi - \frac{2\pi}{5} = \frac{3\pi}{5}.$$

b. Because $4\pi/5$ is greater than $\pi/2$, it has no complement. (Remember to use only *positive* angles for complements.) The supplement is

$$\pi - \frac{4\pi}{5} = \frac{\pi}{5}.$$

Degree Measure

A second way to measure angles is in terms of degrees. A measure of **one degree (1°)** is equivalent to a rotation of $1/360$ of a complete revolution about the vertex. To measure angles, it is convenient to mark degrees on the circumference of a circle, as shown in Figure 1.10. So, a full revolution (counterclockwise) corresponds to 360°, a half revolution to 180°, a quarter revolution to 90°, and so on.

Because 2π radians corresponds to one complete revolution, degrees and radians are related by the equations

$$360° = 2\pi \text{ rad} \qquad \text{and} \qquad 180° = \pi \text{ rad}.$$

From the latter equation, you obtain

$$1° = \frac{\pi}{180} \text{ rad} \qquad \text{and} \qquad 1 \text{ rad} = \frac{180°}{\pi}$$

which lead to the conversion rules at the top of the next page.

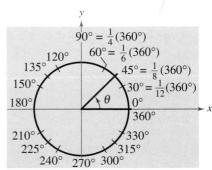

Figure 1.10

Conversions Between Degrees and Radians

1. To convert degrees to radians, multiply degrees by $\dfrac{\pi \text{ rad}}{180°}$.

2. To convert radians to degrees, multiply radians by $\dfrac{180°}{\pi \text{ rad}}$.

To apply these two conversion rules, use the basic relationship $\pi \text{ rad} = 180°$.
(See Figure 1.11.)

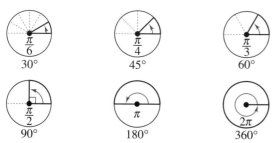

Figure 1.11

When no units of angle measure are specified, *radian measure is implied*. For instance, if you write $\theta = \pi$ or $\theta = 2$, you should mean $\theta = \pi$ radians or $\theta = 2$ radians.

EXAMPLE 3   Converting from Degrees to Radians

Express each angle in radian measure.

a. $135°$ **b.** $-270°$

Solution

a. $135° = (135 \text{ deg})\left(\dfrac{\pi \text{ rad}}{180 \text{ deg}}\right) = \dfrac{3\pi}{4} \text{ rad}$ Multiply by $\frac{\pi}{180}$.

b. $-270° = (-270 \text{ deg})\left(\dfrac{\pi \text{ rad}}{180 \text{ deg}}\right) = -\dfrac{3\pi}{2} \text{ rad}$ Multiply by $\frac{\pi}{180}$.

EXAMPLE 4   Converting from Radians to Degrees

Express each angle in degrees.

a. $-\dfrac{\pi}{2} \text{ rad}$ **b.** 2 rad

Solution

a. $-\dfrac{\pi}{2} \text{ rad} = \left(-\dfrac{\pi}{2} \text{ rad}\right)\left(\dfrac{180 \text{ deg}}{\pi \text{ rad}}\right) = -90°$ Multiply by $\frac{180}{\pi}$.

b. $2 \text{ rad} = (2 \text{ rad})\left(\dfrac{180 \text{ deg}}{\pi \text{ rad}}\right) = \dfrac{360}{\pi} \approx 114.59°$ Multiply by $\frac{180}{\pi}$.

Applications

The *radian measure* formula $\theta = s/r$ can be used to measure arc length along a circle. Specifically, for a circle of radius r, a central angle θ intercepts an arc of length s given by

$$s = r\theta \qquad\qquad \text{Length of circular arc}$$

where θ is measured in radians.

A computer simulation to accompany this concept appears in the *Interactive* CD-ROM and *Internet* versions of this text.

EXAMPLE 5   Finding Arc Length

A circle has a radius of 4 inches. Find the length of the arc intercepted by a central angle of 240°, as shown in Figure 1.12.

Solution
To use the formula $s = r\theta$, first convert 240° to radian measure.

$$240° = (240 \text{ deg})\left(\frac{\pi \text{ rad}}{180 \text{ deg}}\right) \qquad\qquad \text{Convert from degrees to radians.}$$

$$= \frac{4\pi}{3} \text{ radians} \qquad\qquad \text{Simplify.}$$

Then, using a radius of $r = 4$ inches, you can find the arc length to be

$$s = r\theta \qquad\qquad \text{Length of circular arc}$$

$$= 4\left(\frac{4\pi}{3}\right) \qquad\qquad \text{Substitute for } r \text{ and } \theta.$$

$$= \frac{16\pi}{3} \qquad\qquad \text{Simplify.}$$

$$\approx 16.76 \text{ inches.} \qquad\qquad \text{Use a calculator.}$$

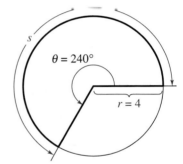

Figure 1.12

Note that the units for $r\theta$ are determined by the units for r because θ is given in radian measure and therefore has no units.

The formula for the length of a circular arc can be used to analyze the motion of a particle moving at a *constant speed* along a circular path.

Linear and Angular Speed

Consider a particle moving at a constant speed along a circular arc of radius r. If s is the length of the arc traveled in time t, then the **linear speed** of the particle is

$$\text{Linear speed} = \frac{\text{arc length}}{\text{time}} = \frac{s}{t}.$$

Moreover, if θ is the angle (in radian measure) corresponding to the arc length s, the **angular speed** of the particle is

$$\text{Angular speed} = \frac{\text{central angle}}{\text{time}} = \frac{\theta}{t}.$$

EXAMPLE 6   Finding Linear Speed

The second hand of a clock is 10.2 centimeters long, as shown in Figure 1.13. Find the linear speed of the tip of this second hand.

Solution

In one revolution, the arc length traveled is

$$s = 2\pi r$$

$$= 2\pi(10.2)$$

$$= 20.4\pi \text{ centimeters.}$$

The time required for the second hand to travel this distance is

$$t = 60 \text{ seconds} = 1 \text{ minute.}$$

So, the linear speed of the tip of the second hand is

$$\text{Linear speed} = \frac{s}{t}$$

$$= \frac{20.4\pi \text{ centimeters}}{60 \text{ seconds}}$$

$$\approx 1.068 \text{ cm/sec.}$$

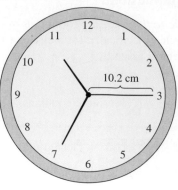

Figure 1.13

EXAMPLE 7   Finding Angular and Linear Speed

A 10 inch radius lawn roller makes 1.2 revolutions per second, as shown in Figure 1.14.

a. Find the angular speed of the roller in radians per second.

b. Find the speed of the tractor that is pulling the roller.

Solution

a. Because each revolution generates 2π radians, it follows that the roller turns $(1.2)(2\pi) = 2.4\pi$ radians per second. In other words, the angular speed is

$$\text{Angular speed} = \frac{\theta}{t}$$

$$= \frac{2.4\pi \text{ radians}}{1 \text{ second}}$$

$$= 2.4\pi \text{ rad/sec.}$$

b. The linear speed is

$$\text{Linear speed} = \frac{s}{t}$$

$$= \frac{r\theta}{t}$$

$$= \frac{10(2.4\pi) \text{ inches}}{1 \text{ second}} \approx 75.4 \text{ in./sec.}$$

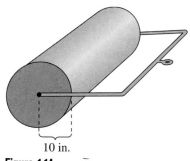

Figure 1.14

1.1 Exercises

In Exercises 1–4, estimate the angle to the nearest one-half radian.

1.

2.

3.

4.

In Exercises 5–10, determine the quadrant in which the angle lies. (The angle is given in radian measure.)

5. (a) $\dfrac{\pi}{5}$ (b) $\dfrac{7\pi}{5}$

6. (a) $\dfrac{7\pi}{4}$ (b) $\dfrac{11\pi}{4}$

7. (a) $-\dfrac{\pi}{12}$ (b) $-\dfrac{11\pi}{9}$

8. (a) -1 (b) -2

9. (a) 3.5 (b) 2.25

10. (a) 5.63 (b) -2.25

In Exercises 11–14, sketch the angle in standard position.

11. (a) $\dfrac{3\pi}{4}$ (b) $\dfrac{5\pi}{3}$

12. (a) $-\dfrac{7\pi}{4}$ (b) $-\dfrac{7\pi}{2}$

13. (a) $\dfrac{11\pi}{6}$ (b) 9π

14. (a) 4 (b) -3

In Exercises 15–18, determine two coterminal angles in radian measure (one positive and one negative) for the given angle.

15. (a) 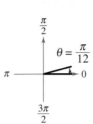   (b)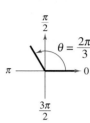

16. (a) 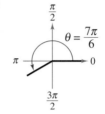   (b)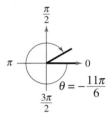

17. (a) $-\dfrac{11\pi}{4}$ (b) $-\dfrac{2\pi}{15}$

18. (a) $\dfrac{7\pi}{8}$ (b) $\dfrac{8\pi}{35}$

In Exercises 19–22, find (if possible) the complement and supplement of the angle.

19. (a) $\dfrac{\pi}{3}$ (b) $\dfrac{3\pi}{4}$

20. (a) $\dfrac{\pi}{12}$ (b) $\dfrac{11\pi}{12}$

21. (a) 1 (b) 2

22. (a) 3 (b) 1.5

In Exercises 23–26, estimate the number of degrees in the angle.

23.

24.

25.

26.

In Exercises 27–30, determine the quadrant in which the angle lies.

27. (a) $150°$ (b) $282°$

28. (a) $7.9°$ (b) $257.5°$

29. (a) $-132° \, 50'$ (b) $-336° \, 30'$

30. (a) $-260.25°$ (b) $-2.4°$

In Exercises 31–34, sketch the angle in standard position.

31. (a) $60°$ (b) $150°$

32. (a) $-270°$ (b) $-120°$

33. (a) $405°$ (b) $780°$

34. (a) $-450°$ (b) $-570°$

In Exercises 35–38, determine two coterminal angles in degree measure (one positive and one negative) for the given angle.

35. (a)   (b)

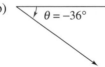

36. (a)   (b)

37. (a) $300°$ (b) $230°$

38. (a) $-445°$ (b) $-740°$

In Exercises 39–42, find (if possible) the complement and supplement of the angle.

39. (a) $24°$ (b) $126°$

40. (a) $87°$ (b) $167°$

41. (a) $79°$ (b) $150°$

42. (a) $130°$ (b) $170°$

In Exercises 43–46, express the angle in radian measure as a multiple of π. (Do not use a calculator.)

43. (a) $30°$ (b) $150°$

44. (a) $315°$ (b) $120°$

45. (a) $-20°$ (b) $-240°$

46. (a) $-270°$ (b) $144°$

In Exercises 47–54, convert the angle measure from degrees to radians. Round your answer to three decimal places.

47. $115°$ **48.** $83.7°$

49. $-216.35°$ **50.** $-46.52°$

51. $642°$ **52.** $0.54°$

53. $-0.78°$ **54.** $395°$

In Exercises 55–58, express the angle in degree measure. (Do not use a calculator.)

55. (a) $\dfrac{3\pi}{2}$ (b) $-\dfrac{7\pi}{6}$

56. (a) $-\dfrac{7\pi}{12}$ (b) $\dfrac{\pi}{9}$

57. (a) $\dfrac{7\pi}{3}$ (b) $-\dfrac{13\pi}{60}$

58. (a) $\dfrac{15\pi}{6}$ (b) $\dfrac{28\pi}{15}$

In Exercises 59–66, convert the angle measure from radians to degrees. Round your answer to three decimal places.

59. $\dfrac{\pi}{7}$ **60.** $\dfrac{8\pi}{13}$ **61.** $\dfrac{25\pi}{8}$

62. 6.5π **63.** -4.2π **64.** 4.8

65. -2 **66.** -0.48

In Exercises 67–70, use the angle-conversion capabilities of a graphing utility to convert the angle measure to decimal degree form. Round your answer to three decimal places if necessary.

67. (a) $64° \, 45'$ (b) $-124° \, 30'$

68. (a) $275° \, 10'$ (b) $9° \, 12'$

69. (a) $85° \, 18' \, 30''$ (b) $-408° \, 16' \, 25''$

70. (a) $-125° \, 36''$ (b) $330° \, 25''$

In Exercises 71–74, use the angle-conversion capabilities of a graphing utility to convert the angle measure to $D° \, M' \, S''$ form.

71. (a) $280.6°$ (b) $-115.8°$

72. (a) $-345.12°$ (b) $310.75°$

73. (a) 4.5 (b) -3.58

74. (a) -0.355 (b) 0.7865

In Exercises 75–78, find the angle in radians.

75.

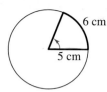
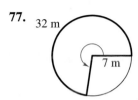

6 cm

5 cm

76.

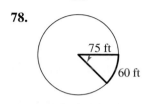

31 in.

12 in.

77.

32 m

7 m

78.

75 ft

60 ft

In Exercises 79–82, find the radian measure of the central angle of a circle of radius *r* that intercepts an arc of length *s*.

Radius r	Arc Length s
79. 15 inches	8 inches
80. 22 feet	10 feet
81. 14.5 centimeters	35 centimeters
82. 80 kilometers	160 kilometers

In Exercises 83–86, find the length of the arc on a circle of radius *r* intercepted by a central angle *θ*.

Radius r	Central Angle θ
83. 14 inches	180°
84. 9 feet	60°
85. 6 meters	$\dfrac{2\pi}{3}$ radians
86. 40 centimeters	$\dfrac{3\pi}{4}$ radians

Distance Between Cities **In Exercises 87 and 88, find the distance between the cities. Assume that earth is a sphere of radius 4000 miles and the cities are on the same meridian (one city is due north of the other).**

City	Latitude
87. Miami	25° 46′ 37″ N
Erie	42° 7′ 15″ N
88. Johannesburg, South Africa	26° 10′ S
Jerusalem, Israel	31° 47′ N

89. *Difference in Latitudes* Assuming that earth is a sphere of radius 6378 kilometers, what is the differ-

ence in latitude of two cities, one of which is 600 kilometers due north of the other?

90. *Difference in Latitudes* Assuming that earth is a sphere of radius 6378 kilometers, what is the difference in latitude of two cities, one of which is 800 kilometers due north of the other?

91. *Instrumentation* The pointer on a voltmeter is 6 centimeters in length. Find the angle through which the pointer rotates when it moves 2.5 centimeters on the scale.

6 cm

92. *Electric Hoist* An electric hoist is used to lift a piece of equipment 1 foot. The diameter of the drum on the hoist is 10 inches. Find the number of degrees through which the drum must rotate.

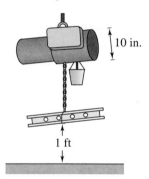

10 in.

1 ft

93. *Figure Skating* The number of revolutions made by a figure skater for each type of axel jump is given. Determine the measure of the angle generated as the skater performs each jump. Give the answer in both degrees and radians.

(a) single axel: $1\frac{1}{2}$

(b) double axel: $2\frac{1}{2}$

(c) triple axel: $3\frac{1}{2}$

94. *Linear Speed of an Earth Satellite* An earth satellite in circular orbit 1250 kilometers high makes one complete revolution every 90 minutes. What is its linear speed? Use 6400 kilometers for the radius of the earth.

95. *Angular Speed* A car is moving at a rate of 40 miles per hour, and the diameter of its wheels is 2.5 feet.

(a) Find the rotational speed of the wheels in revolutions per minute.

(b) Find the angular speed of the wheels in radians per minute.

96. *Circular Saw Speed* The circular blade on a saw has a diameter of 7.5 inches and rotates at 2400 revolutions per minute.

(a) Find the angular speed in radians per second.

(b) Find the linear speed of the saw teeth (in feet per second) as they contact the wood being cut.

|← 7.5 in. →|

97. *Floppy Disk* The radius of the magnetic disk in a 3.5-inch diskette is 1.68 inches. Find the linear speed of a point on the circumference of the disk if it is rotating at a speed of 360 revolutions per minute.

98. *Speed of a Bicycle* The radii of the sprocket assemblies and the wheel of a bicycle are 4 inches, 2 inches, and 14 inches, respectively. If the cyclist is pedaling at a rate of 1 revolution per second, find the speed of the bicycle in (a) feet per second and (b) miles per hour.

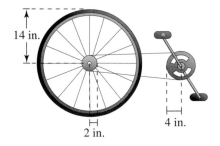

Synthesis

True or False? In Exercises 99–101, determine whether the statement is true or false. Justify your answer.

99. A degree is a larger unit of measure than a radian.

100. An angle that measures $-1260°$ lies in Quadrant III.

101. The angles of a triangle can have radian measures $2\pi/3$, $\pi/4$, and $\pi/12$.

102. *Writing* If the radius of a circle is increasing and the magnitude of a central angle is held constant, how is the length of the intercepted arc changing? Explain your reasoning.

103. *Writing* Write a short paragraph defining and giving examples of coterminal angles.

104. *Geometry* Show that the area of a circular sector of radius r with central angle θ is $A = \frac{1}{2}r^2\theta$, where θ is measured in radians.

Area of a Circular Sector In Exercises 105 and 106, use the result of Exercise 104 to find the area of the sector.

105.

$\frac{\pi}{3}$ 10 m

106.

12 ft

15 ft

107. *Graphical Reasoning* The formulas for the area of a circular sector and arc length are $A = \frac{1}{2}r^2\theta$ and $s = r\theta$, respectively. (r is the radius and θ is the angle measured in radians.)

(a) If $\theta = 0.8$, express the area and arc length as functions of r. What is the domain of each function? Graph the functions using a graphing utility. Use the graphs to determine which function changes more rapidly for changes in r when $r > 1$. Explain.

(b) If $r = 10$ centimeters, express the area and arc length as functions of θ. What is the domain of each function? Use a graphing utility to graph and identify the functions.

Review

In Exercises 108 and 109, determine the open intervals over which the function is increasing, decreasing, or constant.

108. $f(x) = -x^2 - 8$ **109.** $f(x) = 2x^4 - 3x^2$

In Exercises 110–113, sketch the graph of $y = x^5$ and the specified transformations.

110. $f(x) = (x - 2)^5$ **111.** $f(x) = x^5 - 4$

112. $f(x) = 2 - x^5$ **113.** $f(x) = -(x + 3)^5$

1.2 Trigonometric Functions: The Unit Circle

The Unit Circle

The two historical perspectives of trigonometry incorporate different methods for introducing the trigonometric functions. Our first introduction to these functions is based on the unit circle.

Consider the **unit circle** given by

$$x^2 + y^2 = 1 \qquad \text{Unit circle}$$

as shown in Figure 1.15.

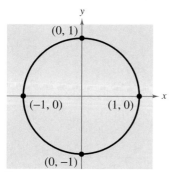

Figure 1.15

Imagine that the real number line is wrapped around this circle, with positive numbers corresponding to a counterclockwise wrapping and negative numbers corresponding to a clockwise wrapping, as shown in Figure 1.16.

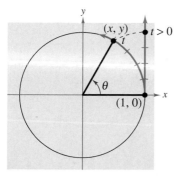 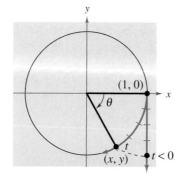

Figure 1.16

As the real number line is wrapped around the unit circle, each real number t corresponds to a point (x, y) on the circle. For example, the real number 0 corresponds to the point $(1, 0)$. Moreover, because the unit circle has a circumference of 2π, the real number 2π also corresponds to the point $(1, 0)$.

In general, each real number t also corresponds to a central angle θ (in standard position) whose radian measure is t. With this interpretation of t, the arc length formula $s = r\theta$ (with $r = 1$) indicates that the real number t is the length of the arc intercepted by the angle θ, given in radians.

What **You Should Learn:**

- How to identify a unit circle and its relationship to real numbers
- How to evaluate trigonometric functions using the unit circle
- How to use the domain and period to evaluate sine and cosine functions
- How to use a calculator to evaluate trigonometric functions

Why **You Should Learn It:**

Trigonometric functions are used to model the movement of an oscillating weight. For instance, in Exercise 59 on page 141, the displacement from equilibrium of an oscillating weight suspended by a spring is modeled as a function of time.

Richard Megna/Fundamental Photographs

A computer animation of this concept appears in the *Interactive* CD-ROM and *Internet* versions of this text.

The Trigonometric Functions

From the preceding discussion, it follows that the coordinates x and y are two functions of the real variable t. You can use these coordinates to define the six trigonometric functions of t.

sine	**cosecant**
cosine	**secant**
tangent	**cotangent**

These six functions are normally abbreviated sin, csc, cos, sec, tan, and cot, respectively.

Definitions of Trigonometric Functions

Let t be a real number and let (x, y) be the point on the unit circle corresponding to t.

$$\sin t = y \qquad\qquad \csc t = \frac{1}{y}, \quad y \neq 0$$

$$\cos t = x \qquad\qquad \sec t = \frac{1}{x}, \quad x \neq 0$$

$$\tan t = \frac{y}{x}, \quad x \neq 0 \qquad \cot t = \frac{x}{y}, \quad y \neq 0$$

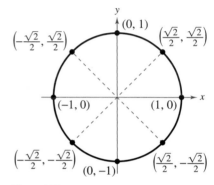

Figure 1.17

Note that the functions in the second column are the *reciprocals* of the corresponding functions in the first column.

In the definitions of the trigonometric functions, note that the tangent and secant are not defined when $x = 0$. For instance, because $t = \pi/2$ corresponds to $(x, y) = (0, 1)$, it follows that $\tan(\pi/2)$ and $\sec(\pi/2)$ are *undefined*. Similarly, the cotangent and cosecant are not defined when $y = 0$. For instance, because $t = 0$ corresponds to $(x, y) = (1, 0)$, cot 0 and csc 0 are *undefined*.

In Figure 1.17, the unit circle has been divided into eight equal arcs, corresponding to t-values of

$$0, \frac{\pi}{4}, \frac{\pi}{2}, \frac{3\pi}{4}, \pi, \frac{5\pi}{4}, \frac{3\pi}{2}, \frac{7\pi}{4}, \text{ and } 2\pi.$$

Similarly, in Figure 1.18, the unit circle has been divided into twelve equal arcs, corresponding to t-values of

$$0, \frac{\pi}{6}, \frac{\pi}{3}, \frac{\pi}{2}, \frac{2\pi}{3}, \frac{5\pi}{6}, \pi, \frac{7\pi}{6}, \frac{4\pi}{3}, \frac{3\pi}{2}, \frac{5\pi}{3}, \frac{11\pi}{6}, \text{ and } 2\pi.$$

Using the (x, y) coordinates in Figures 1.17 and 1.18, you can easily evaluate the trigonometric functions for common t-values. This procedure is demonstrated in Examples 1 and 2.

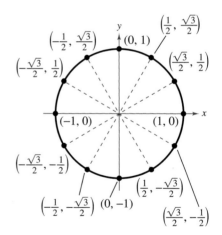

Figure 1.18

A computer animation of this concept appears in the *Interactive* CD-ROM and *Internet* versions of this text.

EXAMPLE 1   Evaluating Trigonometric Functions

Evaluate the six trigonometric functions at each real number.

a. $t = \dfrac{\pi}{6}$ **b.** $t = \dfrac{5\pi}{4}$ **c.** $t = 0$ **d.** $t = \pi$

Solution

For each t-value, begin by finding the corresponding point (x, y) on the unit circle.
Then use the definitions of trigonometric functions listed on page 136.

a. $t = \pi/6$ corresponds to the point $(x, y) = \left(\sqrt{3}/2,\, 1/2\right)$.

$$\sin \frac{\pi}{6} = y = \frac{1}{2} \qquad\qquad \csc \frac{\pi}{6} = \frac{1}{y} = 2$$

$$\cos \frac{\pi}{6} = x = \frac{\sqrt{3}}{2} \qquad\qquad \sec \frac{\pi}{6} = \frac{1}{x} = \frac{2}{\sqrt{3}} = \frac{2\sqrt{3}}{3}$$

$$\tan \frac{\pi}{6} = \frac{y}{x} = \frac{1/2}{\sqrt{3}/2} = \frac{1}{\sqrt{3}} \qquad\qquad \cot \frac{\pi}{6} = \frac{x}{y} = \sqrt{3}$$

b. $t = 5\pi/4$ corresponds to the point $(x, y) = \left(-\sqrt{2}/2,\, -\sqrt{2}/2\right)$.

$$\sin \frac{5\pi}{4} = y = -\frac{\sqrt{2}}{2} \qquad\qquad \csc \frac{5\pi}{4} = \frac{1}{y} = -\frac{2}{\sqrt{2}} = -\sqrt{2}$$

$$\cos \frac{5\pi}{4} = x = -\frac{\sqrt{2}}{2} \qquad\qquad \sec \frac{5\pi}{4} = \frac{1}{x} = -\frac{2}{\sqrt{2}} = -\sqrt{2}$$

$$\tan \frac{5\pi}{4} = \frac{y}{x} = \frac{-\sqrt{2}/2}{-\sqrt{2}/2} = 1 \qquad\qquad \cot \frac{5\pi}{4} = \frac{x}{y} = 1$$

c. $t = 0$ corresponds to the point $(x, y) = (1, 0)$.

$$\sin 0 = y = 0 \qquad\qquad \csc 0 = \frac{1}{y} \text{ is undefined.}$$

$$\cos 0 = x = 1 \qquad\qquad \sec 0 = \frac{1}{x} = 1$$

$$\tan 0 = \frac{y}{x} = \frac{0}{1} = 0 \qquad\qquad \cot 0 = \frac{x}{y} \text{ is undefined.}$$

d. $t = \pi$ corresponds to the point $(x, y) = (-1, 0)$.

$$\sin \pi = y = 0 \qquad\qquad \csc \pi = \frac{1}{y} \text{ is undefined.}$$

$$\cos \pi = x = -1 \qquad\qquad \sec \pi = \frac{1}{x} = -1$$

$$\tan \pi = \frac{y}{x} = \frac{0}{-1} = 0 \qquad\qquad \cot \pi = \frac{x}{y} \text{ is undefined.}$$

EXAMPLE 2   Evaluating Trigonometric Functions

Evaluate the six trigonometric functions at $t = -\dfrac{\pi}{3}$.

Solution

Moving *clockwise* around the unit circle, it follows that $t = -\pi/3$ corresponds to the point $(x, y) = \left(1/2, -\sqrt{3}/2\right)$.

$$\sin\left(-\frac{\pi}{3}\right) = -\frac{\sqrt{3}}{2} \qquad \csc\left(-\frac{\pi}{3}\right) = -\frac{2}{\sqrt{3}}$$

$$\cos\left(-\frac{\pi}{3}\right) = \frac{1}{2} \qquad \sec\left(-\frac{\pi}{3}\right) = 2$$

$$\tan\left(-\frac{\pi}{3}\right) = -\sqrt{3} \qquad \cot\left(-\frac{\pi}{3}\right) = -\frac{1}{\sqrt{3}}$$

Domain and Period of Sine and Cosine

The *domain* of the sine and cosine functions is the set of all real numbers. To determine the *range* of these two functions, consider the unit circle shown in Figure 1.19. Because $r = 1$, it follows that $\sin t = y$ and $\cos t = x$. Moreover, because (x, y) is on the unit circle, you know that $-1 \le y \le 1$ and $-1 \le x \le 1$. So, the values of sine and cosine also range between -1 and 1.

$$\begin{array}{ccc} -1 \le\ y\ \le 1 & & -1 \le\ x\ \le 1 \\ & \text{and} & \\ -1 \le \sin t \le 1 & & -1 \le \cos t \le 1 \end{array}$$

Adding 2π to each value of t in the interval $[0, 2\pi]$ completes a second revolution around the unit circle, as shown in Figure 1.20. The values of $\sin(t + 2\pi)$ and $\cos(t + 2\pi)$ correspond to those of $\sin t$ and $\cos t$. Similar results can be obtained for repeated revolutions (positive or negative) on the unit circle. This leads to the general result

$$\sin(t + 2\pi n) = \sin t$$

and

$$\cos(t + 2\pi n) = \cos t$$

for any integer n and real number t. Functions that behave in such a repetitive (or cyclic) manner are called **periodic.**

Definition of a Periodic Function

A function f is **periodic** if there exists a positive real number c such that

$$f(t + c) = f(t)$$

for all t in the domain of f. The smallest number c for which f is periodic is called the **period** of f.

Exploration

With your graphing utility in *radian* and *parametric* modes, enter

$X_{1T} = \cos T$ and $Y_{1T} = \sin T$

and use the following settings.

Tmin = 0, Tmax = 6.3,
Tstep = 0.1
Xmin = -1.5, Xmax = 1.5,
Xscl = 1
Ymin = -1, Ymax = 1,
Yscl = 1

1. Graph the entered equations and describe the graph.
2. Use the *trace* feature to move the cursor around the graph. What do the *t*-values represent? What do the *x*- and *y*-values represent?
3. What are the least and greatest values for *x* and *y*?

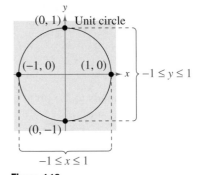

Figure 1.19

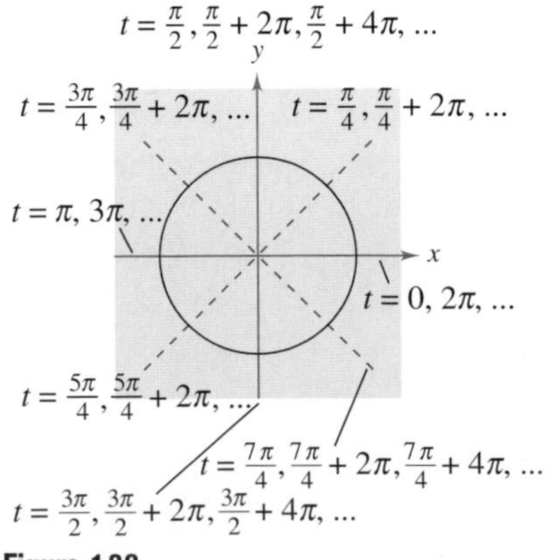

$$t = \tfrac{\pi}{2}, \tfrac{\pi}{2} + 2\pi, \tfrac{\pi}{2} + 4\pi, \ldots$$

$$t = \tfrac{3\pi}{4}, \tfrac{3\pi}{4} + 2\pi, \ldots \qquad t = \tfrac{\pi}{4}, \tfrac{\pi}{4} + 2\pi, \ldots$$

$$t = \pi, 3\pi, \ldots$$

$$t = 0, 2\pi, \ldots$$

$$t = \tfrac{5\pi}{4}, \tfrac{5\pi}{4} + 2\pi, \ldots$$

$$t = \tfrac{7\pi}{4}, \tfrac{7\pi}{4} + 2\pi, \tfrac{7\pi}{4} + 4\pi, \ldots$$

$$t = \tfrac{3\pi}{2}, \tfrac{3\pi}{2} + 2\pi, \tfrac{3\pi}{2} + 4\pi, \ldots$$

Figure 1.20

EXAMPLE 3   Using the Period to Evaluate the Sine and Cosine

a. Because $\dfrac{13\pi}{6} = 2\pi + \dfrac{\pi}{6}$, you have

$$\sin \frac{13\pi}{6} = \sin\left(2\pi + \frac{\pi}{6}\right) = \sin \frac{\pi}{6} = \frac{1}{2}.$$

b. Because $-\dfrac{7\pi}{2} = -4\pi + \dfrac{\pi}{2}$, you have

$$\cos\left(-\frac{7\pi}{2}\right) = \cos\left(-4\pi + \frac{\pi}{2}\right) = \cos \frac{\pi}{2} = 0.$$

Recall from Section P.6 that a function f is *even* if $f(-t) = f(t)$, and it is *odd* if $f(-t) = -f(t)$.

Even and Odd Trigonometric Functions

The cosine and secant functions are *even*.

$$\cos(-t) = \cos t \qquad \sec(-t) = \sec t$$

The sine, cosecant, tangent, and cotangent functions are *odd*.

$$\sin(-t) = -\sin t \qquad \csc(-t) = -\csc t$$
$$\tan(-t) = -\tan t \qquad \cot(-t) = -\cot t$$

Evaluating Trigonometric Functions with a Calculator

When evaluating a trigonometric function with a calculator, you need to set the calculator to the desired *mode* of measurement (degrees or radians).

Most calculators do not have keys for the cosecant, secant, and cotangent functions. To evaluate these functions, you can use the $\boxed{x^{-1}}$ key with their respective reciprocal functions sine, cosine, and tangent. For example, to evaluate $\csc(\pi/8)$, use the fact that

$$\csc \frac{\pi}{8} = \frac{1}{\sin(\pi/8)}$$

and enter the following keystroke sequence in radian mode.

$\boxed{(}$ $\boxed{\text{SIN}}$ $\boxed{(}$ π $\boxed{\div}$ 8 $\boxed{)}$ $\boxed{)}$ $\boxed{x^{-1}}$ $\boxed{\text{ENTER}}$ Display 2.6131259

EXAMPLE 4   Using a Calculator

Function	Mode	Calculator Keystrokes	Display
a. $\sin 2\pi/3$	Radian	$\boxed{\text{SIN}}$ $\boxed{(}$ 2 $\boxed{\pi}$ $\boxed{\div}$ 3 $\boxed{)}$ $\boxed{\text{ENTER}}$	0.8660254
b. $\cot 1.5$	Radian	$\boxed{(}$ $\boxed{\text{TAN}}$ 1.5 $\boxed{)}$ $\boxed{x^{-1}}$ $\boxed{\text{ENTER}}$	0.0709148

1.2   Exercises

In Exercises 1–2, determine the exact values of the six trigonometric functions of the angle θ.

1.

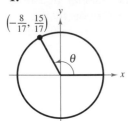

$\left(-\frac{8}{17}, \frac{15}{17}\right)$

2.

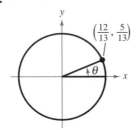

$\left(\frac{12}{13}, \frac{5}{13}\right)$

3.

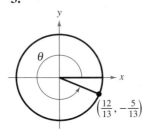

$\left(\frac{12}{13}, -\frac{5}{13}\right)$

4.

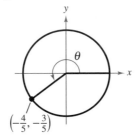

$\left(-\frac{4}{5}, -\frac{3}{5}\right)$

In Exercises 5–12, find the point (x, y) on the unit circle that corresponds to the real number t.

5. $t = \dfrac{\pi}{4}$ **6.** $t = \dfrac{\pi}{3}$ **7.** $t = \dfrac{7\pi}{6}$

8. $t = \dfrac{5\pi}{4}$ **9.** $t = \dfrac{4\pi}{3}$ **10.** $t = \dfrac{5\pi}{3}$

11. $t = \dfrac{3\pi}{2}$ **12.** $t = \pi$

In Exercises 13–22, evaluate (if possible) the sine, cosine, and tangent of the real number.

13. $t = \dfrac{\pi}{4}$ **14.** $t = \dfrac{\pi}{3}$

15. $t = -\dfrac{\pi}{6}$ **16.** $t = -\dfrac{\pi}{4}$

17. $t = -\dfrac{7\pi}{4}$ **18.** $t = -\dfrac{4\pi}{3}$

19. $t = \dfrac{11\pi}{6}$ **20.** $t = \dfrac{5\pi}{3}$

21. $t = -\dfrac{3\pi}{2}$ **22.** $t = -2\pi$

In Exercises 23–28, evaluate (if possible) the six trigonometric functions of the real number.

23. $t = \dfrac{3\pi}{4}$ **24.** $t = \dfrac{5\pi}{6}$ **25.** $t = \dfrac{\pi}{2}$

26. $t = \dfrac{3\pi}{2}$ **27.** $t = -\dfrac{\pi}{3}$ **28.** $t = -\dfrac{3\pi}{2}$

In Exercises 29–36, evaluate the trigonometric function using its period as an aid.

29. $\sin 5\pi$ **30.** $\cos 5\pi$

31. $\cos \dfrac{8\pi}{3}$ **32.** $\sin \dfrac{9\pi}{4}$

33. $\cos(-3\pi)$ **34.** $\sin(-3\pi)$

35. $\sin\left(-\dfrac{9\pi}{4}\right)$ **36.** $\cos\left(-\dfrac{8\pi}{3}\right)$

In Exercises 37–42, use the value of the trigonometric function to evaluate the indicated functions.

37. $\sin t = \frac{1}{3}$ **38.** $\sin(-t) = \frac{3}{8}$
  (a) $\sin(-t)$ (a) $\sin t$
  (b) $\csc(-t)$ (b) $\csc t$

39. $\cos(-t) = -\frac{1}{5}$ **40.** $\cos t = -\frac{3}{4}$
  (a) $\cos t$ (a) $\cos(-t)$
  (b) $\sec(-t)$ (b) $\sec(-t)$

41. $\sin t = \frac{4}{5}$ **42.** $\cos t = \frac{4}{5}$
  (a) $\sin(\pi - t)$ (a) $\cos(\pi - t)$
  (b) $\sin(t + \pi)$ (b) $\cos(t + \pi)$

In Exercises 43–52, use a calculator to evaluate the expression. Round to four decimal places.

43. $\sin \dfrac{\pi}{4}$ **44.** $\tan \dfrac{\pi}{3}$

45. $\csc 1.3$ **46.** $\cot 1$

47. $\cos(-1.7)$ **48.** $\cos(-2.5)$

49. $\csc 0.8$ **50.** $\sec 1.8$

51. $\sec 22.8$ **52.** $\sin(-0.9)$

Estimation **In Exercises 53 and 54, use the figure and a straightedge to approximate the value of the trigonometric function.**

53. (a) sin 5   (b) cos 2

54. (a) sin 0.75   (b) cos 2.5

Estimation **In Exercises 55 and 56, use the figure and a straightedge to approximate the solution of the equation, where $0 \leq t < 2\pi$.**

55. (a) $\sin t = 0.25$ (b) $\cos t = -0.25$

56. (a) $\sin t = -0.75$ (b) $\cos t = 0.75$

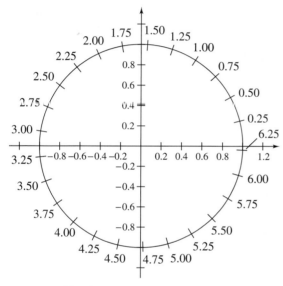

FIGURE FOR 53–56

Path of a Projectile **In Exercises 57 and 58, use the following information. The horizontal distance traveled by a projectile (in feet) with an initial speed of v feet per second is**

$$d = \frac{v^2}{32} \sin \theta$$

where θ is the angle at which the projectile is launched.

57. Find the horizontal distance traveled by a golf ball that is hit with an initial speed of 100 feet per second when the ball is hit at an angle of (a) $\theta = 30°$, (b) $\theta = 50°$, and (c) $\theta = 60°$.

58. Find the horizontal distance traveled by a model rocket launched with an initial speed of 120 feet per second when the rocket is launched at an angle of (a) $\theta = 60°$, (b) $\theta = 70°$, and (c) $\theta = 80°$.

59. *Harmonic Motion* The displacement from equilibrium of an oscillating weight suspended by a spring is

$$y(t) = \frac{1}{4} \cos 6t$$

where y is the displacement in feet and t is the time in seconds (see figure). Find the displacement when (a) $t = 0$, (b) $t = \frac{1}{4}$, and (c) $t = \frac{1}{2}$.

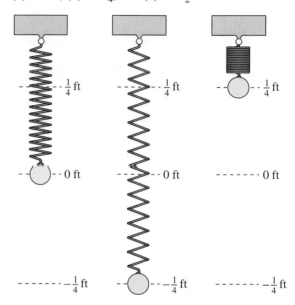

Equilibrium   Maximum negative   Maximum positive
               displacement   displacement

Simple Harmonic Motion

60. *Water Wave* The cross section of a water wave is

$$y(x) = 3 \sin \frac{\pi x}{4}$$

where y is the vertical height of the water wave (in feet) and x is the distance to the wave (in feet). Find the height when (a) $x = 1$, (b) $x = 2$, and (c) $x = 4$.

61. Verify that $\cos 2t \neq 2 \cos t$ by approximating $\cos 1.5$ and $2 \cos 0.75$.

62. Verify that $\sin(t_1 + t_2) \neq \sin t_1 + \sin t_2$ by approximating $\sin 0.25$, $\sin 0.75$, and $\sin 1$.

Synthesis

True or False?   In Exercises 63 and 64, determine whether the statement is true or false. Justify your answer.

63. Because $\sin(-t) = -\sin t$, it can be said that the sine of a negative angle is a negative number.

64. $\tan a = \tan(a - 6\pi)$

65. *Exploration*   Let (x_1, y_1) and (x_2, y_2) be points on the unit circle corresponding to $t = t_1$ and $t = \pi - t_1$, respectively.
    (a) Identify the symmetry of the points (x_1, y_1) and (x_2, y_2).
    (b) Make a conjecture about any relationship between $\sin t_1$ and $\sin(\pi - t_1)$.
    (c) Make a conjecture about any relationship between $\cos t_1$ and $\cos(\pi - t_1)$.

66. *Exploration*   Let (x_1, y_1) and (x_2, y_2) be points on the unit circle corresponding to $t = t_1$ and $t = t_1 + \pi$, respectively.
    (a) Identify the symmetry of the points (x_1, y_1) and (x_2, y_2).
    (b) Make a conjecture about any relationship between $\sin t_1$ and $\sin(t_1 + \pi)$.
    (c) Make a conjecture about any relationship between $\cos t_1$ and $\cos(t_1 + \pi)$.

67. Use the unit circle to verify that the cosine and secant functions are even and that the sine, cosecant, tangent, and cotangent functions are odd.

68. *Think About It*   Because $f(t) = \sin t$ is an odd function and $g(t) = \cos t$ is an even function, what can be said about the function $h(t) = f(t)g(t)$?

69. *Think About It*   Because $f(t) = \sin t$ and $g(t) = \tan t$ are odd functions, what can be said about the function $h(t) = f(t)g(t)$?

Review

In Exercises 70–73, find the inverse of the one-to-one function f. Use a graphing utility to graph both f and f^{-1} in the same viewing window.

70. $f(x) = \frac{1}{2}(3x - 2)$ 71. $f(x) = \frac{1}{4}x^3 + 1$

72. $f(x) = \sqrt{x^2 - 4}$, $x \geq 2$

73. $f(x) = \dfrac{2x}{x + 1}$, $x > -1$

In Exercises 74–77, find (a) the distance between the two points and (b) the midpoint of the segment joining the points.

74. $(-2, 7), (6, 3)$ 75. $(-5, 0), (3, 6)$
76. $\left(\frac{5}{2}, -1\right), \left(-\frac{3}{2}, 4\right)$ 77. $\left(-6, \frac{2}{3}\right), \left(\frac{3}{4}, \frac{1}{6}\right)$

In Exercises 78–83, sketch a graph of the equation. Use a graphing utility to verify your graph.

78. $y = -x - 7$ 79. $y = 9 - 4x$
80. $y = x^2 - 4x$ 81. $y = -x^3 - 3$
82. $y = \sqrt{9 - x}$ 83. $y = 5 - 2|3x|$

In Exercises 84–87, evaluate the function at each specified value of the independent variable and simplify.

84. $f(x) = 5x - 1$
    (a) $f(6)$ (b) $f(-1)$ (c) $f(x - 3)$
85. $f(x) = -x^2 - x + 3$
    (a) $f(4)$ (b) $f(-2)$ (c) $f(x - 2)$
86. $f(x) = x\sqrt{x - 3}$
    (a) $f(3)$ (b) $f(12)$ (c) $f(6)$
87. $f(x) = -\frac{1}{2}x|x + 1|$
    (a) $f(-4)$ (b) $f(10)$ (c) $f\left(-\frac{2}{3}\right)$

In Exercises 88–91, find the domain of the function.

88. $f(x) = \dfrac{4}{9 - x}$ 89. $f(x) = \dfrac{\sqrt{x - 5}}{x - 7}$
90. $f(x) = \sqrt{100 - x^2}$ 91. $f(x) = \sqrt[3]{16 - x^2}$

In Exercises 92 and 93, find the difference quotient.

92. $f(x) = x^2 - 2x + 9, \dfrac{f(3 + h) - f(3)}{h}, h \neq 0$

93. $f(x) = 5 + 6x - x^2, \dfrac{f(6 + h) - f(6)}{h}, h \neq 0$

The symbol ⊕ indicates an example or exercise that highlights algebraic techniques specifically used in calculus.

1.3 Right Triangle Trigonometry

The Six Trigonometric Functions

Our second look at the trigonometric functions is from a *right triangle* perspective. Consider a right triangle, one of whose acute angles is labeled θ, as shown in Figure 1.21. Relative to the angle θ, the three sides of the triangle are the **hypotenuse**, the **opposite side** (the side opposite the angle θ), and the **adjacent side** (the side adjacent to the angle θ).

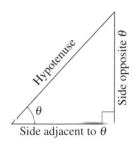

Figure 1.21

Using the lengths of these three sides, you can form six ratios that define the six trigonometric functions of the acute angle θ.

sine   cosecant   cosine   secant   tangent   cotangent

In the following definitions it is important to see that $0° < \theta < 90°$ and that for such angles the value of each trigonometric function is *positive*.

Right Triangle Definitions of Trigonometric Functions

Let θ be an *acute* angle of a right triangle. Then the six trigonometric functions of the angle θ are defined as follows.

$$\sin \theta = \frac{\text{opp}}{\text{hyp}} \qquad \cos \theta = \frac{\text{adj}}{\text{hyp}} \qquad \tan \theta = \frac{\text{opp}}{\text{adj}}$$

$$\csc \theta = \frac{\text{hyp}}{\text{opp}} \qquad \sec \theta = \frac{\text{hyp}}{\text{adj}} \qquad \cot \theta = \frac{\text{adj}}{\text{opp}}$$

The abbreviations "opp", "adj", and "hyp" represent the lengths of the three sides of a right triangle.

opp = the length of the side *opposite* θ

adj = the length of the side *adjacent* to θ

hyp = the length of the *hypotenuse*

Note that the functions in the second row above are the *reciprocals* of the corresponding functions in the first row.

What You Should Learn:

- How to evaluate trigonometric functions of acute angles
- How to use the fundamental trigonometric identities
- How to use a calculator to evaluate trigonometric functions
- How to use trigonometric functions to model and solve real-life problems

Why You Should Learn It:

You can use trigonometric functions to model and solve real-life problems. For instance, Exercise 69 on page 152 shows you how trigonometric functions can be used to approximate the height of a buidling.

Chen Yixin/China Stock

A computer animation of this concept appears in the *Interactive* CD-ROM and *Internet* versions of this text.

Trigonometric Identities

In trigonometry, a great deal of time is spent studying relationships between trigonometric functions (identities).

Exploration

Select a number t and use your graphing utility to calculate $(\cos t)^2 + (\sin t)^2$. Repeat this experiment for other values of t and explain why the answer is always the same. Is the result true in both *radian* and *degree* modes?

Fundamental Trigonometric Identities

Reciprocal Identities

$$\sin \theta = \frac{1}{\csc \theta} \qquad \cos \theta = \frac{1}{\sec \theta} \qquad \tan \theta = \frac{1}{\cot \theta}$$

$$\csc \theta = \frac{1}{\sin \theta} \qquad \sec \theta = \frac{1}{\cos \theta} \qquad \cot \theta = \frac{1}{\tan \theta}$$

Quotient Identities

$$\tan \theta = \frac{\sin \theta}{\cos \theta} \qquad \cot \theta = \frac{\cos \theta}{\sin \theta}$$

Pythagorean Identities

$$\sin^2 \theta + \cos^2 \theta = 1$$

$$1 + \tan^2 \theta = \sec^2 \theta$$

$$1 + \cot^2 \theta = \csc^2 \theta$$

Note that $\sin^2 \theta$ represents $(\sin \theta)^2$ not $\sin(\theta^2)$; $\cos^2 \theta$ represents $(\cos \theta)^2$, not $\cos(\theta^2)$; and so on.

EXAMPLE 4   Applying Trigonometric Identities

Let θ be an acute angle such that $\sin \theta = 0.6$. Find the values of (a) $\cos \theta$ and (b) $\tan \theta$ using trigonometric identities.

Solution

a. To find the value of $\cos \theta$, use the Pythagorean identity

$$\sin^2 \theta + \cos^2 \theta = 1.$$

So, you have

$$(0.6)^2 + \cos^2 \theta = 1$$

$$\cos^2 \theta = 1 - (0.6)^2 = 0.64$$

$$\cos \theta = \sqrt{0.64} = 0.8.$$

b. Now, knowing the sine and cosine of θ, you can find the tangent of θ to be

$$\tan \theta = \frac{\sin \theta}{\cos \theta} = \frac{0.6}{0.8} = 0.75.$$

Try using the definitions of $\cos \theta$ and $\tan \theta$, and the triangle shown in Figure 1.25, to check these results.

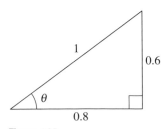

Figure 1.25

EXAMPLE 5 Using Trigonometric Identities

Use trigonometric identities to transform one side of the equation into the other.

a. $\cos \theta \sec \theta = 1$ **b.** $(\sec \theta + \tan \theta)(\sec \theta - \tan \theta) = 1$

Solution

a. $\cos \theta \sec \theta = 1$ Write original equation.

  $\left(\dfrac{1}{\sec \theta}\right)\sec \theta = 1$ Reciprocal identity

          $1 = 1$ Divide out common factor.

b. $(\sec \theta + \tan \theta)(\sec \theta - \tan \theta) = 1$ Write original equation.

  $\sec^2 \theta - \sec\theta \tan \theta + \sec \theta \tan \theta - \tan^2 \theta = 1$ Distributive Property

                $\sec^2\theta - \tan^2\theta = 1$ Simplify.

                      $1 = 1$ Pythagorean identity

Evaluating Trigonometric Functions with a Calculator

To use a calculator to evaluate trigonometric functions of angles measured in degrees, first set the calculator to degree mode and then proceed as demonstrated in Section 1.2.

EXAMPLE 6 Using a Calculator

Use a calculator to evaluate $\sec(5° \, 40' \, 12'')$.

Solution

Begin by converting to decimal form.

$$5° \, 40' \, 12'' = 5° + \left(\dfrac{40}{60}\right)° + \left(\dfrac{12}{3600}\right)° = 5.67°$$

Then use a calculator to evaluate $\sec 5.67°$.

$$\sec(5° \, 40' \, 12'') = \sec 5.67° = \dfrac{1}{\cos 5.67°} \approx 1.00492$$

Applications Involving Right Triangles

Many applications of trigonometry involve a process called **solving right triangles.** In this type of application, you are usually given one side of a right triangle and one of the acute angles and asked to find one of the other sides, *or* you are given two sides and asked to find one of the acute angles. In Example 7, the angle you are given is the **angle of elevation**, which denotes the angle from the horizontal upward to the object. In other applications you may be given the **angle of depression**, which denotes the angle from the horizontal downward to the object.

Reference Angles

The values of the trigonometric functions of angles greater than 90° (or less than 0°) can be determined from their values at corresponding acute angles called **reference angles.**

Definition of Reference Angle
Let θ be an angle in standard position. Its **reference angle** is the acute angle θ' formed by the terminal side of θ and the horizontal axis.

A computer simulation to accompany this concept appears in the *Interactive* CD-ROM and *Internet* versions of this text.

Figure 1.33 shows the reference angles for θ in Quadrants II, III, and IV.

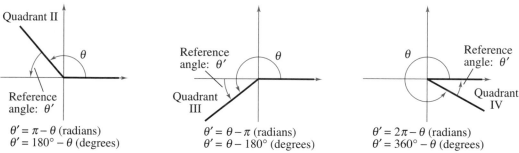

$\theta' = \pi - \theta$ (radians)
$\theta' = 180° - \theta$ (degrees)

$\theta' = \theta - \pi$ (radians)
$\theta' = \theta - 180°$ (degrees)

$\theta' = 2\pi - \theta$ (radians)
$\theta' = 360° - \theta$ (degrees)

Figure 1.33

EXAMPLE 4   Finding Reference Angles

Find the reference angle θ'.

a. $\theta = 300°$ **b.** $\theta = 2.3$ **c.** $\theta = -135°$

Solution

a. Because 300° lies in Quadrant IV, the angle it makes with the x-axis is

$\theta' = 360° - 300° = 60°$.   Degrees

b. Because 2.3 lies between $\pi/2 \approx 1.5708$ and $\pi \approx 3.1416$, it follows that it is in Quadrant II and its reference angle is

$\theta' = \pi - 2.3 \approx 0.8416$.   Radians

c. First, determine that $-135°$ is coterminal with 225°, which lies in Quadrant III. So, the reference angle is

$\theta' = 225° - 180° = 45°$.   Degrees

Figure 1.34 shows each angle θ and its reference angle θ'.

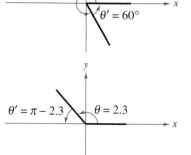

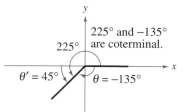

Figure 1.34

Trigonometric Functions of Real Numbers

To see how a reference angle is used to evaluate a trigonometric function, consider the point (x, y) on the terminal side of θ as shown in Figure 1.35. By definition, you know that

$$\sin \theta = \frac{y}{r} \quad \text{and} \quad \tan \theta = \frac{y}{x}.$$

For the right triangle with acute angle θ' and sides of lengths $|x|$ and $|y|$, you have

$$\sin \theta' = \frac{\text{opp}}{\text{hyp}} = \frac{|y|}{r}$$

and

$$\tan \theta' = \frac{\text{opp}}{\text{adj}} = \frac{|y|}{|x|}.$$

So, it follows that $\sin \theta$ and $\sin \theta'$ are equal, *except possibly in sign*. The same is true for $\tan \theta$ and $\tan \theta'$ *and* for the other four trigonometric functions. In all cases, the sign of the function value can be determined by the quadrant in which θ lies.

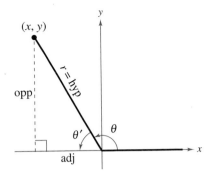

Figure 1.35

Evaluating Trigonometric Functions of Any Angle

To find the value of a trigonometric function of any angle θ:

1. Determine the function value for the associated reference angle θ'.
2. Depending on the quadrant in which θ lies, prefix the appropriate sign to the function value.

By using reference angles and the special angles discussed in the previous section, you can greatly extend the scope of *exact* trigonometric values. For instance, knowing the function values of 30° means that you know the function values of all angles for which 30° is a reference angle. For convenience, the following table gives the exact values of the trigonometric functions of special angles and quadrant angles.

STUDY TIP

Learning the table of values at the left is worth the effort. Doing so will increase both your efficiency and your confidence. Here is a pattern for the sine function that may help you remember the values.

θ	0°	30°	45°	60°	90°
$\sin \theta$	$\frac{\sqrt{0}}{2}$	$\frac{\sqrt{1}}{2}$	$\frac{\sqrt{2}}{2}$	$\frac{\sqrt{3}}{2}$	$\frac{\sqrt{4}}{2}$

Reverse the order to get cosine values of the same angles.

Trigonometric Values of Common Angles

θ (degrees)	0°	30°	45°	60°	90°	180°	270°
θ (radians)	0	$\frac{\pi}{6}$	$\frac{\pi}{4}$	$\frac{\pi}{3}$	$\frac{\pi}{2}$	π	$\frac{3\pi}{2}$
$\sin \theta$	0	$\frac{1}{2}$	$\frac{\sqrt{2}}{2}$	$\frac{\sqrt{3}}{2}$	1	0	-1
$\cos \theta$	1	$\frac{\sqrt{3}}{2}$	$\frac{\sqrt{2}}{2}$	$\frac{1}{2}$	0	-1	0
$\tan \theta$	0	$\frac{\sqrt{3}}{3}$	1	$\sqrt{3}$	Undef.	0	Undef.

1.4   E x e r c i s e s

In Exercises 1–4, determine the exact values of the six trigonometric functions of the angle θ.

1. (a)
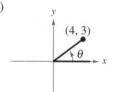
(b)
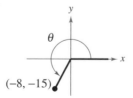

2. (a)
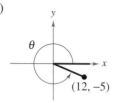
(b)
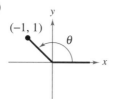

3. (a)
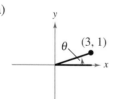
(b)
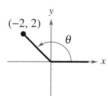

4. (a)
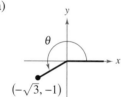
(b)
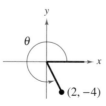

In Exercises 5–12, the given point is on the terminal side of an angle in standard position. Determine the exact values of the six trigonometric functions of the angle.

5. $(7, 24)$ **6.** $(8, 15)$ **7.** $(5, -12)$

8. $(-24, 10)$ **9.** $(-4, 10)$ **10.** $(-5, -6)$

11. $(-2, 9)$ **12.** $(6, -14)$

In Exercises 13–18, state the quadrant in which θ lies.

13. $\sin \theta < 0$ and $\cos \theta < 0$

14. $\sin \theta > 0$ and $\cos \theta > 0$

15. $\sin \theta > 0$ and $\tan \theta < 0$

16. $\sec \theta > 0$ and $\cot \theta < 0$

17. $\cot \theta > 0$ and $\cos \theta > 0$

18. $\tan \theta > 0$ and $\csc \theta < 0$

In Exercises 19–26, find the values of the six trigonometric functions of θ.

Function Value	Constraint
19. $\sin \theta = \frac{3}{5}$	θ lies in Quadrant II.
20. $\cos \theta = -\frac{4}{5}$	θ lies in Quadrant III.
21. $\tan \theta = -\frac{15}{8}$	$\sin \theta < 0$
22. $\csc \theta = 4$	$\cot \theta < 0$
23. $\sec \theta = -2$	$0 \le \theta \le \pi$
24. $\cot \theta$ is undefined.	$\frac{\pi}{2} \le \theta \le \frac{3\pi}{2}$
25. $\sin \theta = 0$	$\sec \theta = -1$
26. $\tan \theta$ is undefined.	$\csc \theta = -1$

In Exercises 27–30, the terminal side of θ lies on the given line in the specified quadrant. Find the values of the six trigonometric functions of θ.

Line	Quadrant
27. $y = -x$	II
28. $y = \frac{1}{3}x$	III
29. $y = 2x$	III
30. $4x + 3y = 0$	IV

In Exercises 31–38, evaluate the trigonometric function of the quadrant angle.

31. $\sec \pi$ **32.** $\tan \frac{\pi}{2}$ **33.** $\cot \frac{\pi}{2}$

34. $\csc \pi$ **35.** $\sec 0$ **36.** $\csc \frac{3\pi}{2}$

37. $\cot \pi$ **38.** $\csc \frac{\pi}{2}$

In Exercises 39–50, find the reference angle θ' and sketch θ and θ' in standard position.

39. $\theta = 208°$ **40.** $\theta = 322°$

41. $\theta = -245°$ **42.** $\theta = -145°$

43. $\theta = -292°$ **44.** $\theta = -95°$

45. $\theta = \frac{11\pi}{3}$ **46.** $\theta = \frac{17\pi}{6}$

47. $\theta = 3.5$ **48.** $\theta = 4.8$

49. $\theta = -3.68$ **50.** $\theta = -1.72$

In Exercises 51–64, evaluate the sine, cosine, and tangent of the angle without using a calculator.

51. 225° **52.** 300°

53. −750° **54.** −495°

55. −240° **56.** −330°

57. $\dfrac{5\pi}{3}$ **58.** $\dfrac{\pi}{4}$

59. $-\dfrac{\pi}{6}$ **60.** $-\dfrac{\pi}{2}$

61. $\dfrac{11\pi}{4}$ **62.** $\dfrac{10\pi}{3}$

63. $-\dfrac{7\pi}{6}$ **64.** $-\dfrac{20\pi}{3}$

In Exercises 65–78, use a calculator to evaluate the trigonometric function. Round you answer to four decimal places.

65. sin 10° **66.** sec 225°

67. tan 240° **68.** csc 330°

69. cos(−110°) **70.** cot(−220°)

71. sec(−280°) **72.** sin(−195°)

73. sin 0.65 **74.** sin(−0.65)

75. $\tan \dfrac{\pi}{9}$ **76.** $\tan\left(-\dfrac{\pi}{9}\right)$

77. $\csc\left(-\dfrac{8\pi}{9}\right)$ **78.** $\cos\left(-\dfrac{15\pi}{14}\right)$

In Exercises 79–84 find two solutions of the equation. Give your answers in both degrees (0° ≤ θ ≤ 360°) and radians (0 ≤ θ ≤ 2π). Do not use a calculator.

79. (a) $\sin \theta = \dfrac{1}{2}$ (b) $\sin \theta = -\dfrac{1}{2}$

80. (a) $\cos \theta = \dfrac{\sqrt{2}}{2}$ (b) $\cos \theta = -\dfrac{\sqrt{2}}{2}$

81. (a) $\csc \theta = \dfrac{2\sqrt{3}}{3}$ (b) $\cot \theta = -1$

82. (a) $\csc \theta = -\sqrt{2}$ (b) $\csc \theta = 2$

83. (a) $\sec \theta = -\dfrac{2\sqrt{3}}{3}$ (b) $\cos \theta = -\dfrac{1}{2}$

84. (a) $\cot \theta = -\sqrt{3}$ (b) $\sec \theta = \sqrt{2}$

In Exercises 85–96, use a calculator to approximate two values of θ (0° ≤ θ < 360°) that satisfy the equation. Round your answers to two decimal places.

85. sin θ = 0.8191 **86.** cos θ = 0.8746

87. tan θ = 0.6524 **88.** cot θ = 0.7521

89. sec θ = −1.2241 **90.** csc θ = −1.0038

91. sin θ = −0.4793 **92.** tan θ = −2.1832

93. cos θ = 0.9848 **94.** sin θ = 0.0175

95. tan θ = 1.192 **96.** cot θ = 5.671

In Exercises 97–102, find the indicated trigonometric value in the specified quadrant.

Function	*Quadrant*	*Trigonometric Value*
97. $\sin \theta = -\frac{3}{5}$	IV	cos θ
98. $\cot \theta = -3$	II	sin θ
99. $\tan \theta = \frac{3}{2}$	III	sec θ
100. $\csc \theta = -2$	IV	cot θ
101. $\cos \theta = \frac{5}{8}$	I	sec θ
102. $\sec \theta = -\frac{9}{4}$	III	tan θ

103. *Average Temperature* The average daily temperature T (in degrees Fahrenheit) for a certain city is

$$T = 45 - 23 \cos\left[\frac{2\pi}{365}(t - 32)\right]$$

where t is the time in days, with $t = 1$ corresponding to January 1. Find the average daily temperatures on the following days.

(a) January 1

(b) July 4 ($t = 185$)

(c) October 18 ($t = 291$)

104. *Sales* A company that sells seasonal products forecasts monthly sales over a 2-year period to be

$$S = 23.1 + 0.442t + 4.3 \sin \frac{\pi t}{6}$$

where S is measured in thousands of units and t is the time (in months), with $t = 1$ representing January 2000. Estimate sales for the following months.

(a) February 2000 (b) February 2001

(c) September 2000 (d) September 2001

105. *Distance* An airplane, flying at an altitude of 6 miles, is on a flight path that passes directly over an observer. If θ is the angle of elevation from the observer to the plane, find the distance from the observer to the plane when (a) $\theta = 30°$, (b) $\theta = 90°$, and (c) $\theta = 120°$.

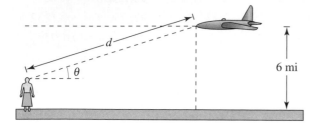

6 mi

Synthesis

True or False? **In Exercises 106–108, determine whether the statement is true or false. Justify your answer.**

106. $\sin 151° = \sin 29°$

107. $\csc\left(-\dfrac{7\pi}{6}\right) = \csc\left(-\dfrac{11\pi}{6}\right)$

108. $-\cot\left(\dfrac{3\pi}{4}\right) = \cot\left(-\dfrac{\pi}{4}\right)$

109. *Conjecture*

(a) Use a graphing utility to complete the table.

θ	0°	20°	40°	60°	80°
$\sin \theta$					
$\sin(180° - \theta)$					

(b) Make a conjecture about the relationship between $\sin \theta$ and $\sin(180° - \theta)$.

110. *Conjecture*

(a) Use a graphing utility to complete the table.

θ	0	0.3	0.6	0.9	1.2	1.5
$\cos\left(\dfrac{3\pi}{2} - \theta\right)$						
$-\sin \theta$						

(b) Make a conjecture about the relationship between $\cos\left(\dfrac{3\pi}{2} - \theta\right)$ and $-\sin \theta$.

111. Explain how reference angles are used to find the value of trigonometric functions of any angle θ.

112. *Writing* Consider an angle in standard position with $r = 12$ centimeters, as shown in the figure. Write a short paragraph describing the change in the magnitudes of x, y, $\sin \theta$, $\cos \theta$, and $\tan \theta$ as θ increases continually from 0° to 90°.

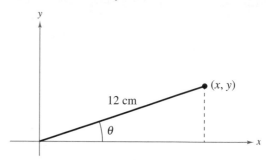

Review

In Exercises 113–116, find the equation of the line that passes through the two points.

113. $(-4, -2), (-3, 8)$ **114.** $(1, 5), (-8, 2)$

115. $\left(\frac{3}{2}, -1\right), \left(-\frac{1}{3}, 4\right)$ **116.** $(0, 1.1), (-4, 3.1)$

In Exercises 117–122, use the graph of f to sketch the graph of the specified function.

117. $f(x - 4)$

118. $f(x + 2)$

119. $f(x) + 4$

120. $f(x) - 1$

121. $2f(x)$

122. $\frac{1}{2}f(x)$

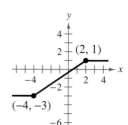

In Exercises 123–126, sketch a right triangle corresponding to the trigonometric function of the acute angle θ. Use the Pythagorean Theorem to determine the third side and then find the other five trigonometric functions of θ.

123. $\cot \theta = \dfrac{8}{3}$ **124.** $\sec \theta = 8$

125. $\tan \theta = \dfrac{7}{24}$ **126.** $\cos \theta = \dfrac{2}{9}$

1.5 Graphs of Sine and Cosine Functions

Basic Sine and Cosine Curves

In this section you will study techniques for sketching the graphs of the sine and cosine functions. The graph of the sine function is a **sine curve.** In Figure 1.37, the black portion of the graph represents one period of the function and is called **one cycle** of the sine curve. The gray portion of the graph indicates that the basic sine wave repeats indefinitely to the right and left. The graph of the cosine function is shown in Figure 1.38. To produce these graphs with a graphing utility, make sure you have set the graphing utility to *radian* mode.

  Recall from Section 1.2 that the domain of the sine and cosine functions is the set of all real numbers. Moreover, the range of each function is the interval $[-1, 1]$, and each function has a period of 2π. Do you see how this information is consistent with the basic graphs given in Figures 1.37 and 1.38?

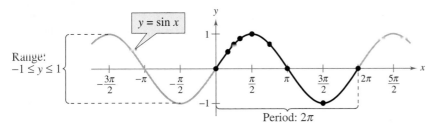

Figure 1.37

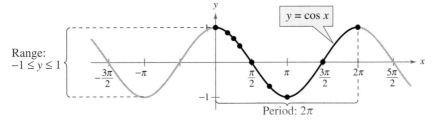

Figure 1.38

Jim Corwin/Tony Stone Images

The table below lists key points on the graphs of $y = \sin x$ and $y = \cos x$.

x	0	$\dfrac{\pi}{6}$	$\dfrac{\pi}{4}$	$\dfrac{\pi}{3}$	$\dfrac{\pi}{2}$	$\dfrac{3\pi}{4}$	π	$\dfrac{3\pi}{2}$	2π
$\sin x$	0	$\dfrac{1}{2}$	$\dfrac{\sqrt{2}}{2}$	$\dfrac{\sqrt{3}}{2}$	1	$\dfrac{\sqrt{2}}{2}$	0	-1	0
$\cos x$	1	$\dfrac{\sqrt{3}}{2}$	$\dfrac{\sqrt{2}}{2}$	$\dfrac{1}{2}$	0	$-\dfrac{\sqrt{2}}{2}$	-1	0	1

Note from Figure 1.37 and 1.38 that the sine graph is symmetric with respect to the *origin,* whereas the cosine graph is symmetric with respect to the *y-axis.* These properties of symmetry follow from the fact that the sine function is odd whereas the cosine function is even.

To sketch the graphs of the basic sine and cosine functions by hand, it helps to note five *key points* in one period of each graph: the *intercepts*, *maximum points*, and *minimum points*. See Figure 1.39

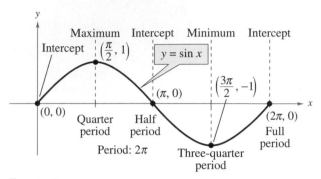

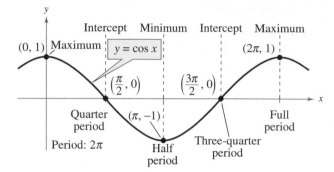

Figure 1.39

EXAMPLE 1   Using Key Points to Sketch a Sine Curve

Sketch the graph of $y = 2 \sin x$ on the interval $[-\pi, 4\pi]$.

Solution
Note that

$$y = 2 \sin x = 2(\sin x)$$

indicates that the y-values for the key points will have twice the magnitude of the graph of $y = \sin x$. Divide the period 2π into four equal parts to get the key points for $y = 2 \sin x$.

$$(0, 0), \qquad \left(\frac{\pi}{2}, 2\right), \qquad (\pi, 0), \qquad \left(\frac{3\pi}{2}, -2\right), \qquad \text{and} \qquad (2\pi, 0)$$

By connecting these key points with a smooth curve and extending the curve in both directions over the interval $[-\pi, 4\pi]$, you obtain the graph shown in Figure 1.40. Use a graphing utility to confirm this graph. Be sure to set the graphing utility to *radian* mode.

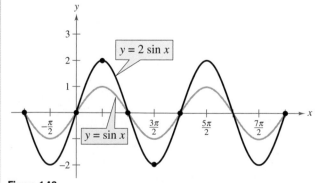

Figure 1.40

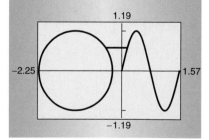

Amplitude and Period of Sine and Cosine Curves

In the rest of this section you will study the graphic effect of each of the constants a, b, c, and d in equations of the forms

$$y = d + a \sin(bx - c)$$

and

$$y = d + a \cos(bx - c).$$

A quick review of the transformations studied in Section P.7 should help in this investigation.

The constant factor a in $y = a \sin x$ acts as a *scaling factor*—a *vertical stretch* or *vertical shrink* of the basic sine curve. If $|a| > 1$, the basic sine curve is stretched, and if $|a| < 1$, the basic sine curve is shrunk. The result is that the graph of $y = a \sin x$ ranges between $-a$ and a instead of between -1 and 1. The absolute value of a is the **amplitude** of the function $y = a \sin x$. The range of the function $y = a \sin x$ is $-a \le y \le a$.

> ### Definition of Amplitude of Sine and Cosine Curves
>
> The **amplitude** of $y = a \sin x$ and $y = a \cos x$ represents half the distance between the maximum and minimum values of the function and is given by
>
> $$\text{Amplitude} = |a|.$$

EXAMPLE 2   Scaling: Vertical Shrinking and Stretching

On the same coordinate axes, sketch the graph of each function.

a. $y = \dfrac{1}{2} \cos x$ **b.** $y = 3 \cos x$

Solution

a. Because the amplitude of $y = \frac{1}{2} \cos x$ is $\frac{1}{2}$, the maximum value is $\frac{1}{2}$ and the minimum value is $-\frac{1}{2}$. Divide one cycle, $0 \le x \le 2\pi$, into four equal parts to get the key points

$$\left(0, \frac{1}{2}\right), \quad \left(\frac{\pi}{2}, 0\right), \quad \left(\pi, -\frac{1}{2}\right), \quad \left(\frac{3\pi}{2}, 0\right), \quad \text{and} \quad \left(2\pi, \frac{1}{2}\right).$$

b. A similar analysis shows that the amplitude of $y = 3 \cos x$ is 3, and the key points are

$$(0, 3), \quad \left(\frac{\pi}{2}, 0\right), \quad (\pi, -3), \quad \left(\frac{3\pi}{2}, 0\right), \quad \text{and} \quad (2\pi, 3).$$

The graphs of these two functions are shown in Figure 1.41. Use a graphing utility to confirm these graphs.

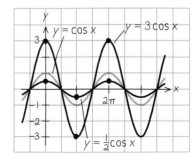

Figure 1.41

You know from Section P.7 that the graph of $y = -f(x)$ is a *reflection* in the *x*-axis of the graph of $y = f(x)$. For instance, the graph of

$$y = -3 \cos x$$

is a reflection of the graph of

$$y = 3 \cos x,$$

as shown in Figure 1.42.

Because $y = a \sin x$ completes one cycle from $x = 0$ to $x = 2\pi$, it follows that $y = a \sin bx$ completes one cycle from $x = 0$ to $x = 2\pi/b$.

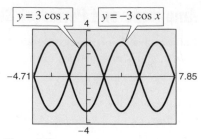

Figure 1.42

Period of Sine and Cosine Functions

Let b be a positive real number. The **period** of $y = a \sin bx$ and $y = a \cos bx$ is given by

$$\text{Period} = \frac{2\pi}{b}.$$

Note that if $0 < b < 1$, the period of $y = a \sin bx$ is greater than 2π and represents a *horizontal stretching* of the graph of $y = a \sin x$. Similarly, if $b > 1$, the period of $y = a \sin bx$ is less than 2π and represents a *horizontal shrinking* of the graph of $y = a \sin x$. If b is negative, the identities $\sin(-x) = -\sin x$ and $\cos(-x) = \cos x$ are used to rewrite the function.

Exploration

Use a graphing utility to graph $y = \sin bx$, where $b = 0.5$, 1, and 2. How does the value of b affect the graph?

EXAMPLE 3   Scaling: Horizontal Stretching

Sketch the graph of $y = \sin \dfrac{x}{2}$.

Solution
The amplitude is 1. Moreover, because $b = \frac{1}{2}$, the period is

$$\frac{2\pi}{b} = \frac{2\pi}{\frac{1}{2}} = 4\pi.$$

Now, divide the period-interval $[0, 4\pi]$ into four equal parts with the values π, 2π, and 3π, to obtain the key points on the graph

$$(0, 0), \quad (\pi, 1), \quad (2\pi, 0), \quad (3\pi, -1), \quad \text{and} \quad (4\pi, 0).$$

The graph is shown in Figure 1.43. Use a graphing utility to confirm this graph.

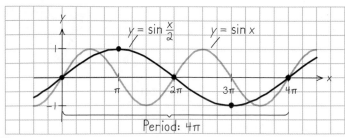

Figure 1.43

STUDY TIP

In general, to divide a period-interval into four equal parts, successively add "period/4," starting with the left endpoint of the interval. For instance, for the period-interval $[-\pi/6, \pi/2]$ of length $2\pi/3$, you would successively add

$$\frac{2\pi/3}{4} = \frac{\pi}{6}$$

to get $-\pi/6, 0, \pi/6, \pi/3,$ and $\pi/2$.

Translations of Sine and Cosine Curves

The constant c in the general equations

$$y = a \sin(bx - c) \qquad \text{and} \qquad y = a \cos(bx - c)$$

creates *horizontal translations* (shifts) of the basic sine and cosine curves. Comparing $y = a \sin bx$ with $y = a \sin(bx - c)$, you find that the graph of $y = a \sin(bx - c)$ completes one cycle from $bx - c = 0$ to $bx - c = 2\pi$. By solving for x, you can find the interval for one cycle to be

$$\overset{\text{Left endpoint}}{\overbrace{\frac{c}{b}}} \le x \le \underset{\underbrace{\text{Period}}}{\overset{\text{Right endpoint}}{\overbrace{\frac{c}{b}}} + \frac{2\pi}{b}}.$$

This implies that the period of $y = a \sin(bx - c)$ is $2\pi/b$, and the graph of $y = a \sin bx$ is shifted by an amount c/b. The number c/b is the **phase shift.**

Graphs of Sine and Cosine Functions

The graphs of $y = a \sin(bx - c)$ and $y = a \cos(bx - c)$ have the following characteristics. (Assume $b > 0$.)

$$\text{Amplitude} = |a| \qquad \text{Period} = 2\pi/b$$

The left and right endpoints of a one-cycle interval can be determined by solving the equations $bx - c = 0$ and $bx - c = 2\pi$.

EXAMPLE 4   Horizontal Translation

Analyze the graph of $y = \dfrac{1}{2} \sin\left(x - \dfrac{\pi}{3}\right)$.

Algebraic Solution

The amplitude is $\frac{1}{2}$ and the period is 2π. By solving the equations

$$x - \frac{\pi}{3} = 0 \qquad \text{and} \qquad x - \frac{\pi}{3} = 2\pi$$

$$x = \frac{\pi}{3} \qquad\qquad x = \frac{7\pi}{3}$$

you see that the interval $[\pi/3, 7\pi/3]$ corresponds to one cycle of the graph. Dividing this interval into four equal parts produces the following key points.

$$\left(\frac{\pi}{3}, 0\right), \quad \left(\frac{5\pi}{6}, \frac{1}{2}\right), \quad \left(\frac{4\pi}{3}, 0\right), \quad \left(\frac{11\pi}{6}, -\frac{1}{2}\right), \quad \left(\frac{7\pi}{3}, 0\right)$$

Graphical Solution

Use a graphing utility set in *radian* mode to graph $y = (1/2) \sin(x - \pi/3)$, as shown in Figure 1.44. Use the *minimum*, *maximum*, and *zero* or *root* features of the graphing utility to approximate the key points $(1.047, 0)$, $(2.618, 0.5)$, $(4.189, 0)$, $(5.760, -0.5)$, and $(7.330, 0)$.

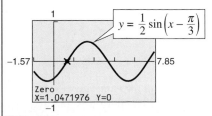

Figure 1.44

EXAMPLE 5   Horizontal Translation

Use a graphing utility to analyze the graph of $y = -3 \cos(2\pi x + 4\pi)$.

Solution

The amplitude is 3 and the period is $2\pi/2\pi = 1$. By solving the equations

$$2\pi x + 4\pi = 0 \qquad \text{and} \qquad 2\pi x + 4\pi = 2\pi$$

$$2\pi x = -4\pi \qquad\qquad\qquad 2\pi x = -2\pi$$

$$x = -2 \qquad\qquad\qquad\qquad x = -1$$

you see that the interval $[-2, -1]$ corresponds to one cycle of the graph. Dividing this interval into four equal parts produces the key points

$$(-2, -3), \quad (-7/4, 0), \quad (-3/2, 3), \quad (-5/4, 0), \quad \text{and} \quad (-1, -3).$$

The graph is shown in Figure 1.45.

The final type of transformation is the *vertical translation* caused by the constant *d* in the equations

$$y = d + a \sin(bx - c) \qquad \text{and} \qquad y = d + a \cos(bx - c).$$

The shift is *d* units upward for $d > 0$ and *d* units downward for $d < 0$. In other words, the graph oscillates about the horizontal line $y = d$ instead of about the *x*-axis.

EXAMPLE 6   Vertical Translation

Use a graphing utility to analyze the graph of $y = 2 + 3 \cos 2x$.

Solution

The amplitude is 3 and the period is π. The key points over the interval $[0, \pi]$ are

$$(0, 5), \qquad (\pi/4, 2), \qquad (\pi/2, -1), \qquad (3\pi/4, 2), \qquad \text{and} \qquad (\pi, 5).$$

The graph is shown in Figure 1.46. Compared with the graph of $f(x) = 3 \cos 2x$, the graph of $y = 2 + 3 \cos 2x$ is shifted upward two units.

EXAMPLE 7   Finding an Equation for a Graph

Find the amplitude, period, and phase shift for the sine function whose graph is shown in Figure 1.47. Write an equation for this graph.

Solution

The amplitude for this sine curve is 2. The period is 2π, and there is a right phase shift of $\pi/2$. So, you can write

$$y = 2 \sin\left(x - \frac{\pi}{2}\right).$$

In Example 7, you can find a cosine function with the same graph. The amplitude for this cosine curve is 2. The period is 2π, and there is a right phase shift of π. So, you can write

$$y = 2 \cos(x - \pi).$$

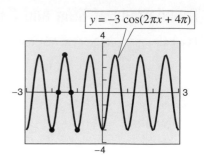

$y = -3 \cos(2\pi x + 4\pi)$

Figure 1.45

A computer animation of this example appears in the *Interactive* CD-ROM and *Internet* versions of this text.

Exploration

Use a graphing utility to graph $y = d + \sin x$, where $d = -2, 1,$ and 3. How does the value of *d* affect the graph?

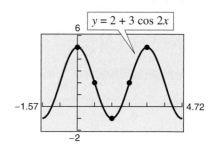

$y = 2 + 3 \cos 2x$

Figure 1.46

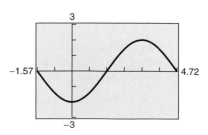

Figure 1.47

Mathematical Modeling

Sine and cosine functions can be used to model many real-life situations, including electric currents, musical tones, radio waves, tides, and weather patterns.

EXAMPLE 8   Finding a Trigonometric Model

Throughout the day, the depth of water at the end of a dock varies with the tides. The table shows the depths (in meters) at various times during the morning.

t (time)	Midnight	2 A.M.	4 A.M.	6 A.M.	8 A.M.	10 A.M.	Noon
y (depth)	2.55	3.80	4.40	3.80	2.55	1.80	2.27

a. Use a trigonometric function to model this data.

b. Find the depths at 9 A.M. and 3 P.M.

c. A boat needs at least 3 meters of water to moor at the dock. During what times in the afternoon can it safely dock?

Solution

a. Begin by graphing the data, as shown in Figure 1.48. You can use either a sine or cosine model. Suppose you use a cosine model of the form

$$y = a \cos(bt - c) + d.$$

The difference between the maximum height and minimum height of the graph is twice the amplitude of the function. So, the amplitude is

$$a = \tfrac{1}{2}[(\text{maximum depth}) - (\text{minimum depth})] = \tfrac{1}{2}(4.4 - 1.8) = 1.3.$$

The cosine function completes one half of a cycle between when the maximum and minimum depths occur. So, the period is

$$p = 2[(\text{time of min. depth}) - (\text{time of max. depth})] = 2(10 - 4) = 12$$

which implies that $b = 2\pi/p \approx 0.524$. Because high tide occurs 4 hours after midnight, consider the left endpoint to be $c/b = 4$, so $c \approx 2.094$. Moreover, because the average depth is $\tfrac{1}{2}(4.4 + 1.8) = 3.1$, it follows that $d = 3.1$. So, you can model the depth with the function

$$y = 1.3 \cos(0.524t - 2.094) + 3.1.$$

b. The depths at 9 A.M. and 3 P.M. are as follows:

$$y = 1.3 \cos(0.524 \cdot 9 - 2.094) + 3.1 \approx 1.97 \text{ meters} \qquad \text{9 A.M.}$$

$$y = 1.3 \cos(0.524 \cdot 15 - 2.094) + 3.1 \approx 4.23 \text{ meters.} \qquad \text{3 P.M.}$$

c. Using a graphing utility, graph the model with the line $y = 3$, as shown in Figure 1.49. From the graph, it follows that the depth is at least 3 meters between 12:54 P.M. ($t \approx 12.9$) and 7:06 P.M. ($t \approx 19.1$).

Note that a sine model for the data in Example 8 is

$y = 1.3 \sin(0.524t - 0.524) + 3.1$. Can you see how to find this?

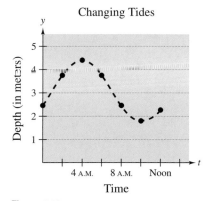

Changing Tides

Figure 1.48

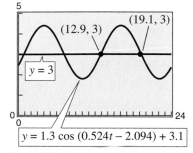

$y = 1.3 \cos(0.524t - 2.094) + 3.1$

Figure 1.49

1.5 Exercises

In Exercises 1–14, find the period and amplitude.

1. $y = 3 \sin 2x$

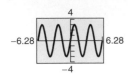

2. $y = 2 \cos 3x$

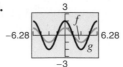

3. $y = \dfrac{5}{2} \cos \dfrac{x}{2}$

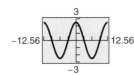

4. $y = -3 \sin \dfrac{x}{3}$

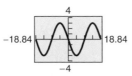

5. $y = \dfrac{2}{3} \sin \pi x$

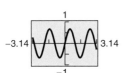

6. $y = \dfrac{3}{2} \cos \dfrac{\pi x}{2}$

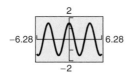

7. $y = -2 \sin x$

8. $y = -\cos \dfrac{2x}{5}$

9. $y = 3 \sin 6x$

10. $y = \dfrac{1}{3} \sin 10x$

11. $y = \dfrac{1}{4} \cos \dfrac{2x}{3}$

12. $y = \dfrac{5}{2} \cos \dfrac{x}{4}$

13. $y = 3 \sin 4\pi x$

14. $y = \dfrac{3}{4} \cos \dfrac{\pi x}{12}$

In Exercises 15–22, describe the relationship between the graphs of f and g.

15. $f(x) = \sin x$
$g(x) = \sin(x - \pi)$

16. $f(x) = \cos x$
$g(x) = \cos(x + \pi)$

17. $f(x) = \cos 2x$
$g(x) = -\cos 2x$

18. $f(x) = \sin 3x$
$g(x) = \sin(-3x)$

19. $f(x) = \cos x$
$g(x) = -5 \cos x$

20. $f(x) = \sin x$
$g(x) = \sin 3x$

21. $f(x) = \sin x$
$g(x) = 4 + \sin x$

22. $f(x) = \cos 4x$
$g(x) = -6 + \cos 4x$

In Exercises 23–26, describe the relationship between the graphs of f and g.

23.

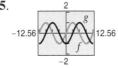

24.

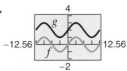

25.

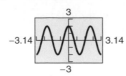

26.

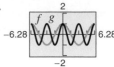

In Exercises 27–34, sketch the graphs of the two functions in the same coordinate plane. (Include two full periods.)

27. $f(x) = -2 \sin x$
$g(x) = 4 \sin x$

28. $f(x) = \sin x$
$g(x) = \sin \dfrac{x}{3}$

29. $f(x) = \cos x$
$g(x) = 4 + \cos x$

30. $f(x) = 2 \cos 2x$
$g(x) = -\cos 4x$

31. $f(x) = -\dfrac{1}{2} \sin \dfrac{x}{2}$
$g(x) = 5 - \dfrac{1}{2} \sin \dfrac{x}{2}$

32. $f(x) = 4 \sin \pi x$
$g(x) = 4 \sin \pi x - 1$

33. $f(x) = 2 \cos x$
$g(x) = 2 \cos(x + \pi)$

34. $f(x) = -\cos x$
$g(x) = -\cos\left(x - \dfrac{\pi}{2}\right)$

Conjecture   **In Exercises 35–38, use a graphing utility to graph f and g in the same viewing window. (Include two full periods.) Make a conjecture about the functions.**

35. $f(x) = \sin x$
$g(x) = \cos\left(x - \dfrac{\pi}{2}\right)$

36. $f(x) = \sin x$
$g(x) = -\cos\left(x + \dfrac{\pi}{2}\right)$

37. $f(x) = \cos x$
$g(x) = -\sin\left(x - \dfrac{\pi}{2}\right)$

38. $f(x) = \cos x$
$g(x) = -\cos(x - \pi)$

In Exercises 39–54, sketch the graph of the function by hand. Use a graphing utility to verify your sketch. (Include two full periods.)

39. $y = -2 \sin 4x$ **40.** $y = -3 \cos 2x$

41. $y = \cos 2\pi x$ **42.** $y = \dfrac{5}{2} \sin \dfrac{\pi x}{4}$

43. $y = -2 \sin \dfrac{2\pi x}{3}$ **44.** $y = -10 \cos \dfrac{\pi x}{6}$

45. $y = \sin\left(x - \dfrac{\pi}{4}\right)$ **46.** $y = \dfrac{1}{2} \sin(x - \pi)$

47. $y = 8 \cos(x + \pi)$ **48.** $y = 6 \cos\left(x + \dfrac{\pi}{6}\right)$

49. $y = \dfrac{1}{10} \cos 60\pi x$ **50.** $y = -4 + 5 \cos \dfrac{\pi t}{12}$

51. $y = 2 - 2 \sin \dfrac{2\pi x}{3}$ **52.** $y = 2 \cos x - 3$

53. $y = \dfrac{2}{3} \cos\left(\dfrac{x}{2} - \dfrac{\pi}{4}\right)$ **54.** $y = -3 \cos(6x + \pi)$

In Exercises 55–62, use a graphing utility to graph the function. (Include two full periods.) Identify the amplitude and period of the graph.

55. $y = -2 \sin(4x + \pi)$ **56.** $y = -4 \sin\left(\dfrac{2}{3}x - \dfrac{\pi}{3}\right)$

57. $y = \cos\left(2\pi x - \dfrac{\pi}{2}\right) + 1$

58. $y = 3 \cos\left(\dfrac{\pi x}{2} + \dfrac{\pi}{2}\right) - 3$

59. $y = 5 \sin(\pi - 2x) + 10$

60. $y = 5 \cos(\pi - 2x) + 6$

61. $y = \frac{1}{100} \sin 120\pi t$ **62.** $y = -\frac{97}{100} \cos 50\pi t$

Graphical Reasoning **In Exercises 63–66, find a and d for the function $f(x) = a \cos x + d$ so that the graph of f matches the figure.**

63.

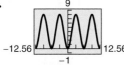

64.

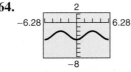

65.

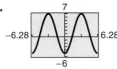

66.

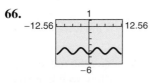

Graphical Reasoning **In Exercises 67–70, find a, b, and c for the function $f(x) = a \sin(bx - c)$ so that the graph of f matches the figure.**

67.

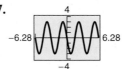

68.

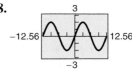

69.

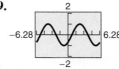

70.

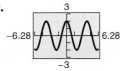

In Exercises 71–74, use a graphing utility to graph y_1 and y_2 for all real numbers x in the interval $[-2\pi, 2\pi]$. Use the graphs to find the real numbers x such that $y_1 = y_2$.

71. $y_1 = \sin x$ **72.** $y_1 = \cos x$
     $y_2 = -\dfrac{1}{2}$ $y_2 = 1$

73. $y_1 = \cos x$ **74.** $y_1 = \sin x$
     $y_2 = \dfrac{\sqrt{2}}{2}$ $y_2 = \dfrac{\sqrt{3}}{2}$

75. ***Respiratory Cycle*** For a person at rest, the velocity v (in liters per second) of air flow during a respiratory cycle is

$$v = 0.85 \sin \dfrac{\pi t}{3}$$

where t is the time (in seconds). (Inhalation occurs when $v > 0$, and exhalation occurs when $v < 0$.)

(a) Use a graphing utility to graph v.

(b) Find the time for one full respiratory cycle.

(c) Find the number of cycles per minute.

76. ***Respiratory Cycle*** The model in Exercise 75 is for a person at rest. How might the model change for a person who is exercising? Explain.

Sales **In Exercises 77 and 78, use a graphing utility to graph the sales function over 1 year, where S is the sales in thousands of units and t is the time in months, with $t = 1$ corresponding to January. Determine the months of maximum and minimum sales.**

77. $S = 22.3 - 3.4 \cos \dfrac{\pi t}{6}$

78. $S = 74.50 + 43.75 \sin \dfrac{\pi t}{6}$

79. Ferris wheel You are riding a Ferris wheel. Your height h (in feet) above the ground at any time t (in seconds) can be modeled by

$$h = 25 \sin \frac{\pi}{15}(t - 75) + 30.$$

The ferris wheel turns for 135 seconds before it stops to let the first passengers off.

(a) Graph the model.

(b) What are your minimum and maximum heights above the ground?

80. Blood Pressure The pressure P (in millimeters of mercury) against the walls of the blood vessels of a certain person is modeled by

$$P = 100 - 20 \cos \frac{8\pi}{3}t$$

where t is the time (in seconds). Graph the model. If one cycle is equivalent to one heartbeat, what is the person's pulse rate in heartbeats per minute?

81. Fuel Consumption The daily consumption C (in gallons) of diesel fuel on a farm is modeled by

$$C = 30.3 + 21.6 \sin\left(\frac{2\pi t}{365} + 10.9\right)$$

where t is the time in days, with $t = 1$ corresponding to January 1.

(a) What is the period of the model? Is it what you expected? Explain.

(b) What is the average daily fuel consumption? Which term of the model did you use? Explain.

(c) Use a graphing utility to graph the model. Use the graph to approximate the time of the year when consumption exceeds 40 gallons per day.

82. Data Analysis The motion of an oscillating weight suspended by a spring was measured by a motion detector. The data was collected, and the approximate maximum displacements from equilibrium ($y = 2$) are labeled in the figure. The distance y from the motion detector is measured in centimeters and the time t is measured in seconds.

(a) Is y a function of t? Explain.

(b) Approximate the amplitude and period.

(c) Find a model for the data.

(d) Use a graphing utility to graph the model in part (c). Compare the result with the data in the figure.

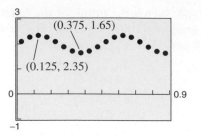

FIGURE FOR 82

83. Data Analysis The normal monthly precipitation p (in inches) for Seattle, Washington, for month t is shown in the table, with $t = 1$ corresponding to January. (Source: National Climatic Data Center)

t	1	2	3	4	5	6
p	5.4	4.0	3.8	2.5	1.8	1.6

t	7	8	9	10	11	12
p	0.9	1.2	1.9	3.3	5.7	6.0

(a) Find a trigonometric model for the normal monthly precipitation in Seattle.

(b) Use a graphing utility to graph the data and the model found in part (a). How well does the model fit the data?

84. Data Analysis The table shows the normal daily high temperatures for Honolulu H and Chicago C (in degrees Fahrenheit) for month t, with $t = 1$ corresponding to January. (Source: NOAA)

t	1	2	3	4	5	6
H	80.1	80.5	81.6	82.8	84.7	86.5
C	29.0	33.5	45.8	58.6	70.1	79.6

t	7	8	9	10	11	12
H	87.5	88.7	88.5	86.9	84.1	81.2
C	83.7	81.8	74.8	63.3	48.4	34.0

(a) A model for Honolulu is

$$H(t) = 84.40 + 4.28 \sin\left(\frac{\pi t}{6} + 3.86\right).$$

Find a trigonometric model for Chicago.

(b) Use a graphing utility to graph the data points and the model for the temperatures in Honolulu. How well does the model fit the data?

(c) Use a graphing utility to graph the data points and the model for the temperatures in Chicago. How well does the model fit the data?

(d) Use the models to estimate the average annual temperature in each city. Which term of the models did you use? Explain.

(e) What are the periods of the two models? Are they what you expected? Explain.

(f) Which city has the greater variability in temperature throughout the year? Which factor of the models determines this variability? Explain.

Synthesis

True or False? **In Exercises 85 and 86, determine whether the statement is true or false. Justify your answer.**

85. The graph of $y = 6 - \dfrac{3}{4}\sin\dfrac{3x}{10}$ has a period of $\dfrac{20\pi}{3}$.

86. The graph of $y = -\cos x$ is a reflection of the graph of $y = \sin(x + \pi/2)$ in the x-axis.

87. ***Writing*** Use a graphing utility to graph the function $y = a \sin x$ for $a = \dfrac{1}{2}$, $a = \dfrac{3}{2}$, and $a = -3$. Write a paragraph describing the changes in the graph for the specified changes in a.

88. ***Writing*** Use a graphing utility to graph the function $y = d + \sin x$ for $d = 2$, $d = 3.5$, and $d = -2$. Write a paragraph describing the changes in the graph for the specified changes in d.

89. ***Writing*** Use a graphing utility to graph the function $y = \sin bx$ for $b = \dfrac{1}{2}$, $b = \dfrac{3}{2}$, and $b = 4$. Write a paragraph describing the changes in the graph for the specified changes in b.

90. ***Writing*** Use a graphing utility to graph the function $y = \sin(x - c)$ for $c = 1$, $c = 3$, and $c = -2$. Write a paragraph describing the changes in the graph for the specified changes in c.

91. ***Exploration*** In Section 1.2 it was shown that $f(x) = \cos x$ is an even function and $g(x) = \sin x$ is an odd function. Use a graphing utility to graph h and use the graph to determine whether h is even, odd, or neither.

(a) $h(x) = \cos^2 x$

(b) $h(x) = \sin^2 x$

(c) $h(x) = \sin x \cos x$

92. ***Conjecture*** If f is an even function and g is an odd function, use the results of Exercise 91 to make a conjecture about the following.

(a) $h(x) = [f(x)]^2$

(b) $h(x) = [g(x)]^2$

(c) $h(x) = f(x)g(x)$

93. ***Graphical Reasoning*** The figure shows the graphs of the functions $f(x) = 1 - \dfrac{1}{2}x^2$ and $g(x) = \cos x$. Identify the graphs and explain your reasoning.

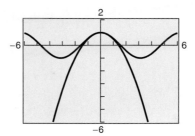

Review

In Exercises 94–97, evaluate the combination of functions when $f(x) = 2x^2 - 5$ and $g(x) = x - 3$ at the indicated value of x.

94. $(f + g)(-x)$ when $x = -2$

95. $(f - g)(x)$ when $x = 4$

96. $(fg)(-x)$ when $x = -3$

97. $\left(\dfrac{f}{g}\right)(x)$ when $x = \dfrac{3}{2}$

In Exercises 98–101, convert the angle measure from radians to degrees. Round your answer to three decimal places.

98. $\dfrac{13\pi}{2}$

99. $-\dfrac{\pi}{9}$

100. 8.5π

101. -0.48

1.6 Graphs of Other Trigonometric Functions

Graph of the Tangent Function

Recall that the tangent function is odd. That is, $\tan(-x) = -\tan x$. Consequently, the graph of $y = \tan x$ is symmetric with respect to the origin. You also know from the identity $\tan x = \sin x/\cos x$ that the tangent is undefined when $\cos x = 0$. Two such values are $x = \pm\pi/2 \approx \pm 1.5708$.

x	$-\dfrac{\pi}{2}$	-1.57	-1.5	-1	0	1	1.5	1.57	$\dfrac{\pi}{2}$
$\tan x$	Undef.	-1255.8	-14.1	-1.56	0	1.56	14.1	1255.8	Undef.

$\tan x$ approaches $-\infty$ as x approaches $-\pi/2$ from the right.

$\tan x$ approaches ∞ as x approaches $\pi/2$ from the left.

As indicated in the table, $\tan x$ increases without bound as x approaches $\pi/2$ from the left, and it decreases without bound as x approaches $-\pi/2$ from the right. So, the graph of $y = \tan x$ has *vertical asymptotes* at $x = \pi/2$ and $-\pi/2$, as shown in Figure 1.50. Moreover, because the period of the tangent function is π, vertical asymptotes also occur when $x = \pi/2 + n\pi$, where n is an integer. The domain of the tangent function is the set of all real numbers other than $x = \pi/2 + n\pi$, and the range is the set of all real numbers.

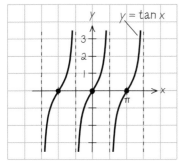

Period: π

Domain: all $x \neq \dfrac{\pi}{2} + n\pi$

Range: $(-\infty, \infty)$

Vertical asymptotes: $x = \dfrac{\pi}{2} + n\pi$

Figure 1.50

Sketching the graph of a function of the form $y = a\tan(bx - c)$ is similar to sketching the graph of $y = a\sin(bx - c)$ in that you locate key points that identify the intercepts and asymptotes. Two consecutive asymptotes can be found by solving the equations $bx - c = -\pi/2$ and $bx - c = \pi/2$. The midpoint between two consecutive asymptotes is an x-intercept of the graph. The period of the function $y = a\tan(bx - c)$ is the distance between two consecutive asymptotes. The amplitude of a tangent function is not defined. After plotting the asymptotes and the x-intercept, plot a few additional points between the two asymptotes and sketch one cycle. Finally, sketch one or two additional cycles to the left and right.

What You Should Learn:

- How to sketch the graphs of tangent functions
- How to sketch the graphs of cotangent functions
- How to sketch the graphs of secant and cosecant functions
- How to sketch the graphs of damped trigonometric functions

Why You Should Learn It:

You can use tangent, cotangent, secant, and cosecant functions to model real-life data. For instance, Exercise 70 on page 182 shows you how a tangent function can be used to model and analyze the distance between a television camera and a parade unit.

A. Ramey/PhotoEdit

EXAMPLE 1   Sketching the Graph of a Tangent Function

Sketch the graph of $y = \tan \dfrac{x}{2}$.

Solution

By solving the equations

$$\frac{x}{2} = -\frac{\pi}{2} \quad \text{and} \quad \frac{x}{2} = \frac{\pi}{2}$$

$$x = -\pi \qquad\qquad\quad x = \pi$$

you can see that two consecutive asymptotes occur at $x = -\pi$ and $x = \pi$. Between these two asymptotes, plot a few points, including the x-intercept, as shown in the table. Three cycles of the graph are shown in Figure 1.51. Use a graphing utility to confirm this graph.

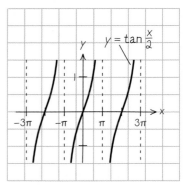

Figure 1.51

x	$-\pi$	$-\dfrac{\pi}{2}$	0	$\dfrac{\pi}{2}$	π
$\tan \dfrac{x}{2}$	Undef.	-1	0	1	Undef.

EXAMPLE 2   Sketching the Graph of a Tangent Function

Sketch the graph of $y = -3 \tan 2x$.

A computer animation of this example appears in the *Interactive* CD-ROM and Internet versions of this text.

Algebraic Solution

By solving the equations $2x = -\pi/2$ and $2x = \pi/2$, you can see that two consecutive asymptotes occur at $x = -\pi/4$ and $x = \pi/4$. Between these two asymptotes, plot a few points, including the x-intercept, as shown in the table. Three complete cycles of the graph are shown in Figure 1.52.

x	$-\dfrac{\pi}{4}$	$-\dfrac{\pi}{8}$	0	$\dfrac{\pi}{8}$	$\dfrac{\pi}{4}$
$-3 \tan 2x$	Undef.	3	0	-3	Undef.

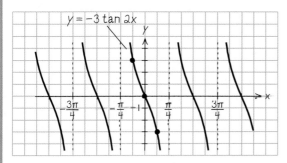

Figure 1.52

Graphical Solution

Use a graphing utility set in *radian* mode and *dot* mode to graph $y = -3 \tan 2x$, as shown in Figure 1.53. Use the *zero* or *root* feature or the *zoom* and *trace* features of the graphing utility to approximate the key points as shown in the table.

x	0.7854	-0.3927	0	0.3927	0.7854
$-3 \tan 2x$	Undef.	3	0	-3	Undef.

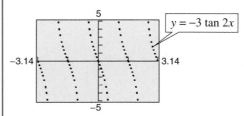

Figure 1.53

By comparing the graphs in Examples 1 and 2, you can see that the graph of $y = a \tan(bx - c)$ is increasing between consecutive vertical asymptotes if $a > 0$ and decreasing between consecutive vertical asymptotes if $a < 0$. In other words, the graph for $a < 0$ is a reflection in the x-axis of the graph for $a > 0$.

Graph of the Cotangent Function

The graph of the cotangent function is similar to the graph of the tangent function. It also has a period of π. However, from the identity

$$y = \cot x = \frac{\cos x}{\sin x}$$

you can see that the cotangent function has vertical asymptotes $x = n\pi$, where n is an integer, because $\sin x$ is zero at these x-values. The graph of the cotangent function is shown in Figure 1.54.

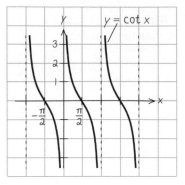

Period: π
Domain: all $x \neq n\pi$
Range: $(-\infty, \infty)$
Vertical asymptotes: $x = n\pi$
Figure 1.54

EXAMPLE 3   Sketching the Graph of a Cotangent Function

Sketch the graph of $y = 2 \cot \dfrac{x}{3}$.

Solution

To locate two consecutive vertical asymptotes of the graph, solve the equations $x/3 = 0$ and $x/3 = \pi$ to see that two consecutive asymptotes occur at $x = 0$ and $x = 3\pi$.

Then, between these two asymptotes, plot a few points, including the x-intercept, as shown in the table. Three cycles of the graph are shown in Figure 1.55. Use a graphing utility to confirm this graph. [Enter the function as $y = 2/\tan(x/3)$.] Note that the period is 3π, the distance between consecutive asymptotes.

x	0	$\dfrac{3\pi}{4}$	$\dfrac{3\pi}{2}$	$\dfrac{9\pi}{4}$	3π
$2 \cot \dfrac{x}{3}$	Undef.	2	0	-2	Undef.

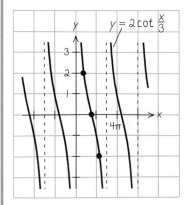

Figure 1.55

Graphs of the Reciprocal Functions

The graphs of the two remaining trigonometric functions can be obtained from the graphs of the sine and cosine functions using the reciprocal identities

$$\csc x = \frac{1}{\sin x} \quad \text{and} \quad \sec x = \frac{1}{\cos x}.$$

For instance, at a given value of x, the y-coordinate for $\sec x$ is the reciprocal of the y-coordinate for $\cos x$. Of course, when $\cos x = 0$, the reciprocal does not exist. Near such values of x, the behavior of the secant function is similar to that of the tangent function. In other words, the graphs of

$$\tan x = \frac{\sin x}{\cos x} \quad \text{and} \quad \sec x = \frac{1}{\cos x}$$

have vertical asymptotes at $x = \pi/2 + n\pi$, where n is an integer and the cosine is zero at these x-values. Similarly,

$$\cot x = \frac{\cos x}{\sin x} \quad \text{and} \quad \csc x = \frac{1}{\sin x}$$

have vertical asymptotes where $\sin x = 0$, that is, at $x = n\pi$.

To sketch the graph of a secant or cosecant function, you should first make a sketch of its reciprocal function. For instance, to sketch the graph of $y = \csc x$, first sketch the graph of $y = \sin x$. Then take reciprocals of the y-coordinates to obtain points on the graph of $y = \csc x$. You can use this procedure to obtain the graphs shown in Figure 1.56.

Exploration

Use a graphing utility to graph the functions $y_1 = \sin x$ and $y_2 = \csc x = 1/\sin x$ in the same viewing window. How are the graphs related? What happens to the graph of the cosecant function as x approaches the zeros of the sine function?

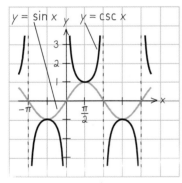

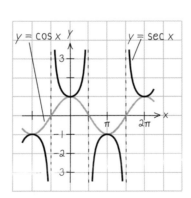

Period: 2π

Domain: all $x \neq n\pi$

Range: all y not in $(-1, 1)$

Vertical asymptotes: $x = n\pi$

Symmetry: origin

Period: 2π

Domain: all $x \neq \dfrac{\pi}{2} + n\pi$

Range: all y not in $(-1, 1)$

Vertical asymptotes: $x = \dfrac{\pi}{2} + n\pi$

Symmetry: y-axis

Figure 1.56

In comparing the graphs of the secant and cosecant functions with those of the sine and cosine functions, note that the "hills" and "valleys" are interchanged. For example, a hill (or maximum point) on the sine curve corresponds to a valley (a local minimum) on the cosecant curve. Similarly, a valley (or minimum point) on the sine curve corresponds to a hill (a local maximum) on the cosecant curve, as shown in Figure 1.57.

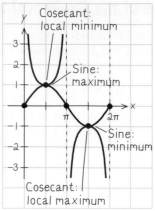

Figure 1.57

EXAMPLE 4   Comparing Trigonometric Graphs

Use a graphing utility to compare the graphs of

$$y = 2 \sin\left(x + \frac{\pi}{4}\right) \qquad \text{and} \qquad y = 2 \csc\left(x + \frac{\pi}{4}\right).$$

Solution

The two graphs are shown in Figure 1.58. Note how the hills and valleys of the graphs are related. For the function $y = 2 \sin[x + (\pi/4)]$, the amplitude is 2 and the period is 2π. By solving the double inequality

$$0 < x + \frac{\pi}{4} < 2\pi \qquad \Longrightarrow \qquad -\frac{\pi}{4} < x < \frac{7\pi}{4}$$

you can see that one cycle of the sine function corresponds to the interval from $x = -\pi/4$ to $x = 7\pi/4$. The graph of this sine function is represented by the gray curve in Figure 1.58. Because the sine function is zero at the endpoints of this interval, the corresponding cosecant function

$$y = 2 \csc\left(x + \frac{\pi}{4}\right) = 2\left(\frac{1}{\sin[x + (\pi/4)]}\right)$$

has vertical asymptotes at $x = -\pi/4, 3\pi/4, 7\pi/4$, etc. The graph of the cosecant function is represented by the black curve.

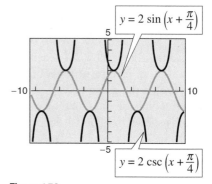

Figure 1.58

EXAMPLE 5   Comparing Trigonometric Graphs

Use a graphing utility to compare the graphs of

$$y = \cos 2x \qquad \text{and} \qquad y = \sec 2x.$$

Solution

Begin by graphing $y_1 = \cos 2x$ and $y_2 = \sec 2x = 1/\cos 2x$ in the same viewing window, as shown in Figure 1.59. Note that the x-intercepts of $y = \cos 2x$

$$\left(-\frac{\pi}{4}, 0\right), \qquad \left(\frac{\pi}{4}, 0\right), \qquad \left(\frac{3\pi}{4}, 0\right), \dots$$

correspond to the vertical asymptotes

$$x = -\frac{\pi}{4}, \qquad x = \frac{\pi}{4}, \qquad x = \frac{3\pi}{4}, \dots$$

of the graph of $y = \sec 2x$.

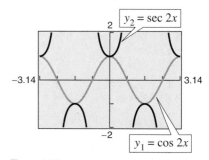

Figure 1.59

Damped Trigonometric Graphs

A *product* of two functions can be graphed using properties of the individual functions. For instance, consider the function

$$f(x) = x \sin x$$

as the product of the functions $y = x$ and $y = \sin x$. Using properties of absolute value and the fact that $|\sin x| \le 1$, you have $0 \le |x| \, |\sin x| \le |x|$. Consequently,

$$-|x| \le x \sin x \le |x|$$

which means that the graph of $f(x) = x \sin x$ lies between the lines $y = -x$ and $y = x$. Furthermore, because

$$f(x) = x \sin x = \pm x \qquad \text{at} \qquad x = \frac{\pi}{2} + n\pi$$

$$f(x) = x \sin x = 0 \qquad \text{at} \qquad x = n\pi$$

the graph of f touches the line $y = -x$ or the line $y = x$ at $x = \pi/2 + n\pi$ and has x-intercepts at $x = n\pi$. A sketch of f is shown in Figure 1.60. In the function $f(x) = x \sin x$, the factor x is called the **damping factor.**

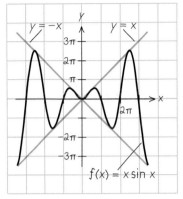

Figure 1.60

EXAMPLE 6   Analyzing a Damped Sine Curve

Analyze the graph of

$$f(x) = x^2 \sin 3x.$$

Solution

Consider $f(x)$ as the product of the two functions

$$y = x^2 \qquad \text{and} \qquad y = \sin 3x$$

each of which has the set of real numbers as its domain. For any real number x, you know that $x^2 \ge 0$ and $|\sin 3x| \le 1$. Therefore, $|x^2| \, |\sin 3x| \le x^2$, which means that

$$-x^2 \le x^2 \sin 3x \le x^2.$$

Furthermore, because

$$f(x) = x^2 \sin 3x = \pm x^2 \qquad \text{at} \qquad x = \frac{\pi}{6} + \frac{n\pi}{3}$$

and

$$f(x) = x^2 \sin 3x = 0 \qquad \text{at} \qquad x = \frac{n\pi}{3}$$

the graph of f touches the curves $y = -x^2$ and $y = x^2$ at $x = \frac{\pi}{6} + \frac{n\pi}{3}$ and has intercepts at $x = \frac{n\pi}{3}$. The graph is shown in Figure 1.61.

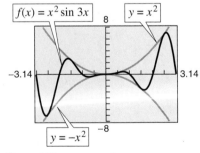

Figure 1.61

Figure 1.62 summarizes the graphs, domains, ranges, and periods of the six basic trigonometric functions.

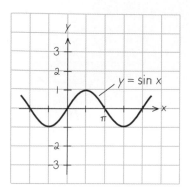

Domain: all reals

Range: $[-1, 1]$
Period: 2π

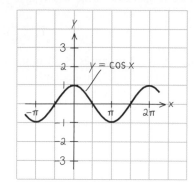

Domain: all reals

Range: $[-1, 1]$
Period: 2π

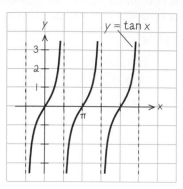

Domain: all $x \neq \dfrac{\pi}{2} + n\pi$

Range: $(-\infty, \infty)$
Period: π

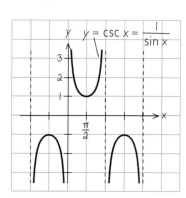

Domain: all $x \neq n\pi$

Range: $(-\infty, -1]$ and $[1, \infty)$
Period: 2π

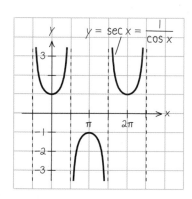

Domain: all $x \neq \dfrac{\pi}{2} + n\pi$

Range: $(-\infty, -1]$ and $[1, \infty)$
Period: 2π

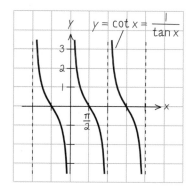

Domain: all $x \neq n\pi$

Range: $(-\infty, \infty)$
Period: π

Figure 1.62

Writing About Math   *Combining Trigonometric Functions*

Recall from Section P.8 that functions can be combined arithmetically. This also applies to trigonometric functions. For each of the functions $h(x) = x + \sin x$ and $h(x) = \cos x - \sin 3x$, (a) identify two simpler functions f and g that comprise the combination, (b) use a table to show how to obtain the numerical values of $h(x)$ from the numerical values of $f(x)$ and $g(x)$, and (c) use graphing utility graphs of f and g to show how h may be formed.

Can you find functions

$$f(x) = d + a\sin(bx - c) \qquad \text{and} \qquad g(x) = d + a\cos(bx - c)$$

such that $f(x) + g(x) = 0$ for all x? If so, write a short paragraph explaining how you found the functions.

1.6 E X E R C I S E S

In Exercises 1–8, match the function with its graph. State the period of the function. [The graphs are labeled (a), (b), (c), (d), (e), (f), (g), and (h).]

(a)
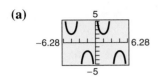

(b)
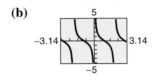

(c)
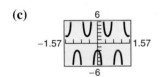

(d)
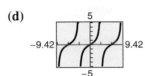

(e)
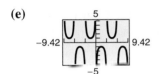

(f)
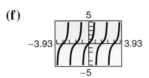

(g)
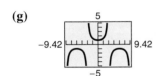

(h)
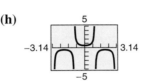

1. $y = \sec \dfrac{x}{2}$

2. $y = \tan \dfrac{x}{2}$

3. $y = \tan 2x$

4. $y = 2 \csc x$

5. $y = \cot \dfrac{\pi x}{2}$

6. $y = \dfrac{1}{2} \sec \dfrac{\pi x}{2}$

7. $y = -\csc x$

8. $y = -2 \sec 2\pi x$

In Exercises 9–30, sketch the graph of the function. (Include two full periods.) Use a graphing utility to verify your result.

9. $y = \frac{1}{3} \tan x$

10. $y = \frac{1}{4} \tan 2x$

11. $y = -2 \tan 2x$

12. $y = -3 \tan \pi x$

13. $y = -\frac{1}{2} \sec x$

14. $y = \frac{1}{4} \sec 2x$

15. $y = -\sec \pi x$

16. $y = 2 \sec 4x$

17. $y = \sec \pi x - 3$

18. $y = -2 \sec 4x + 2$

19. $y = \csc \dfrac{x}{2}$

20. $y = -\csc \dfrac{x}{3}$

21. $y = \dfrac{1}{2} \cot \dfrac{x}{2}$

22. $y = 3 \cot \dfrac{\pi x}{2}$

23. $y = \frac{1}{2} \sec 2x$

24. $y = -\frac{1}{2} \tan \pi x$

25. $y = 2 \tan \dfrac{\pi x}{4}$

26. $y = \sec(x + \pi)$

27. $y = \csc(\pi - x)$

28. $y = \sec(\pi - x)$

29. $y = 2 \cot\left(x - \dfrac{\pi}{2}\right)$

30. $y = \dfrac{1}{4} \csc\left(x + \dfrac{\pi}{4}\right)$

In Exercises 31–40, use a graphing utility to graph the function. (Include two full periods.) Describe your viewing window.

31. $y = \dfrac{1}{3} \tan \dfrac{x}{3}$

32. $y = -2 \tan 2\pi x$

33. $y = -2 \sec 4x$

34. $y = \dfrac{1}{4} \sec \pi x$

35. $y = \tan\left(x - \dfrac{\pi}{4}\right)$

36. $y = -\csc(4x - \pi)$

37. $y = \dfrac{1}{4} \cot\left(x + \dfrac{\pi}{2}\right)$

38. $y = 0.1 \tan\left(\dfrac{\pi x}{4} + \dfrac{\pi}{4}\right)$

39. $y = \dfrac{1}{2} \sec(2x - \pi)$

40. $y = \dfrac{1}{3} \sec\left(\dfrac{\pi x}{2} + \dfrac{\pi}{2}\right)$

In Exercises 41–44, use a graph to solve the equation in the interval $[-2\pi, 2\pi]$.

41. $\tan x = 1$

42. $\cot x = -\sqrt{3}$

43. $\sec x = -2$

44. $\csc x = \sqrt{2}$

In Exercises 45 and 46, use the graph of the function to determine whether the function is even, odd, or neither.

45. $f(x) = \sec x$

46. $f(x) = \tan x$

In Exercises 47–50, use a graphing utility to graph the two equations in the same viewing window. Use the graphs to lend evidence that the expressions are equivalent. Verify the results algebraically.

47. $y_1 = \sin x \csc x, \quad y_2 = 1$

48. $y_1 = \sin x \sec x, \quad y_2 = \tan x$

49. $y_1 = \dfrac{\cos x}{\sin x}, \quad y_2 = \cot x$

50. $y_1 = \sec^2 x - 1, \quad y_2 = \tan^2 x$

In Exercises 51–54, match the function with its graph. Describe the behavior of the function as *x* approaches zero. [The graphs are labeled (a), (b), (c), and (d).]

(a)
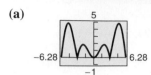

(b)
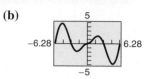

(c)
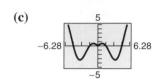

(d)
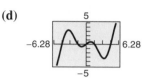

51. $f(x) = x \cos x$

52. $f(x) = |x \sin x|$

53. $g(x) = |x| \sin x$

54. $g(x) = |x| \cos x$

Conjecture In Exercises 55–58, use a graphing utility to graph the functions *f* and *g*. Use the graphs to make a conjecture about the relationship between the functions.

55. $f(x) = \sin x + \cos\left(x + \dfrac{\pi}{2}\right)$, $g(x) = 0$

56. $f(x) = \sin x - \cos\left(x + \dfrac{\pi}{2}\right)$, $g(x) = 2 \sin x$

57. $f(x) = \sin^2 x$, $g(x) = \frac{1}{2}(1 - \cos 2x)$

58. $f(x) = \cos^2 \dfrac{\pi x}{2}$, $g(x) = \dfrac{1}{2}(1 + \cos \pi x)$

In Exercises 59–62, use a graphing utility to graph the function and the damping factor of the function in the same viewing window. Describe the behavior of the function as *x* increases without bound.

59. $f(x) = x \cos x$

60. $g(x) = x^2 \cos x$

61. $f(x) = x^3 \cos x$

62. $h(x) = x^3 \sin x$

Exploration In Exercises 63–68, use a graphing utility to graph the function. Describe the behavior of the function as *x* approaches zero.

63. $f(x) = \dfrac{6}{x} + \cos x$

64. $f(x) = \sin x - \dfrac{4}{x}$

65. $f(x) = \dfrac{\sin x}{x}$

66. $f(x) = \dfrac{1 - \cos x}{x}$

67. $f(x) = \dfrac{\tan x}{x}$

68. $f(x) = \dfrac{x}{\cot x}$

69. *Distance* A plane flying at an altitude of 5 miles over level ground will pass directly over a radar antenna. Let *d* be the ground distance from the antenna to the point directly under the plane and let *x* be the angle of elevation to the plane from the antenna. Write *d* as a function of *x* and graph the function over the interval $0 < x < \pi$.

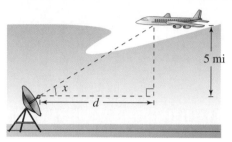

70. *Television Coverage* A television camera is on a reviewing platform 36 meters from the street on which a parade will be passing from left to right. Express the distance *d* from the camera to a particular unit in the parade as a function of the angle *x*, and graph the function over the interval $-\pi/2 < x < \pi/2$. (Consider *x* as negative when a unit in the parade approaches from the left.)

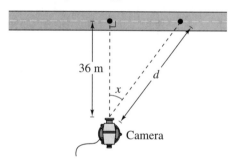

71. *Predator-Prey Model* Suppose the population of a certain predator at time *t* (in months) in a region is estimated to be $P = 10{,}000 + 3000 \sin(2\pi t/24)$ and the population of its primary food source (its prey) is estimated to be $p = 15{,}000 + 5000 \cos(2\pi t/24)$. Use the graph of the models to explain the oscillations in the size of each population.

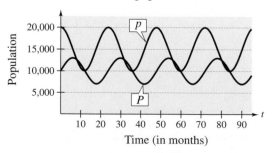

72. *Normal Temperatures* The normal monthly high temperatures in degrees Fahrenheit for Erie, Pennsylvania, are approximated by

$$H(t) = 54.33 - 20.38 \cos \frac{\pi t}{6} - 15.69 \sin \frac{\pi t}{6}$$

and the normal monthly low temperatures are approximated by

$$L(t) = 39.36 - 15.70 \cos \frac{\pi t}{6} - 14.16 \sin \frac{\pi t}{6}$$

where t is the time (in months), with $t = 1$ corresponding to January. (Source: National Oceanic and Atmospheric Association)

(a) Use a graphing utility to graph each function. What is the period of each function?

(b) During what part of the year is the difference between the normal high and low temperatures greatest? When is it smallest?

(c) The sun is the farthest north in the sky around June 21, but the graph shows the warmest temperatures at a later date. Approximate the lag time of the temperatures relative to the position of the sun.

73. *Harmonic Motion* An object weighing W pounds is suspended from a ceiling by a steel spring. The weight is pulled downward (positive direction) from its equilibrium position and released. The resulting motion of the weight is described by the function

$$y = \frac{1}{t^2 + 1} \cos 4t, \quad t > 0$$

where y is the distance (in feet) and t is the time (in seconds).

(a) Use a graphing utility to graph the function.

(b) Describe the behavior of the displacement function for increasing values of time t.

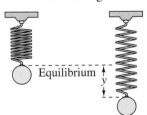

74. *Sales* The projected monthly sales S (in thousands of units) of a seasonal product is modeled by

$$S = 74 + 3t + 40 \sin \frac{\pi t}{6}$$

where t is the time (in months), with $t = 1$ corresponding to January.

(a) Graph the sales function over 1 year. Use a graphing utility to verify your graph.

(b) Which month should have the greatest number of sales? Which month should have the least number of sales? Justify your answers.

75. *Sales* The projected monthly sales S (in thousands of units) of a seasonal product is modeled by

$$S = 52 + 5t - 28 \cos \frac{\pi t}{6}$$

where t is the time (in months), with $t = 1$ corresponding to January.

(a) Graph the sales function over 1 year. Use a graphing utility to verify your graph.

(b) Which month should have the greatest number of sales? Which month should have the least number of sales? Justify your answers.

76. *Numerical and Graphical Reasoning* A crossed belt connects a 10-centimeter pulley on an electric motor with a 20-centimeter pulley on a saw arbor. The electric motor runs at 1700 revolutions per minute.

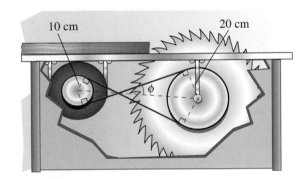

(a) Determine the number of revolutions per minute of the saw.

(b) How does crossing the belt affect the saw in relation to the motor?

(c) Let L be the total length of the belt. Write L as a function of ϕ, where ϕ is measured in radians. What is the domain of the function? (*Hint:* Add the lengths of the straight sections of the belt and the length of belt around each pulley.)

(d) Use a graphing utility to complete the table.

ϕ	0.3	0.6	0.9	1.2	1.5
L					

(e) As ϕ increases, do the lengths of the straight sections of the belt change faster or slower than the lengths of the belt around each pulley?

(f) Use a graphing utility to graph the function over the appropriate domain.

Synthesis

True or False? **In Exercises 77–79, determine whether the statement is true or false. Justify your answer.**

77. The graph of $y = -\dfrac{1}{8} \tan\left(\dfrac{x}{2} + \pi\right)$ has an asymptote at $x = -3\pi$.

78. The graph of $y = -2 \csc\left(x + \dfrac{\pi}{3}\right)$ has an asymptote at $x = -\dfrac{\pi}{3}$.

79. In the graph of $y = 2^x \sin x$, as x approaches $-\infty$, y approaches 0.

80. *Writing* Describe the behavior of $f(x) = \tan x$ as x approaches $\pi/2$ from the left and from the right.

81. *Writing* Describe the behavior of $f(x) = \csc x$ as x approaches π from the left and from the right.

82. *Graphical Reasoning* Consider the two functions $f(x) = 2 \sin x$ and $g(x) = \frac{1}{2} \csc x$ on the interval $(0, \pi)$.

(a) Use a graphing utility to graph f and g in the same viewing window.

(b) Approximate the interval where $f > g$.

(c) Describe the behavior of each of the functions as x approaches π. How is the behavior of g related to the behavior of f as x approaches π?

83. *Pattern Recognition*

(a) Use a graphing utility to graph each function.

$$y_1 = \frac{4}{\pi}\left(\sin \pi x + \frac{1}{3} \sin 3\pi x\right)$$

$$y_2 = \frac{4}{\pi}\left(\sin \pi x + \frac{1}{3} \sin 3\pi x + \frac{1}{5} \sin 5\pi x\right)$$

(b) Identify the pattern in part (a) and find a function y_3 that continues the pattern one more term. Use a graphing utility to graph y_3.

(c) The graphs of parts (a) and (b) approximate the periodic function in the figure. Find a function y_4 that is a better approximation.

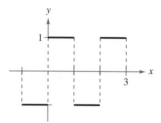

Review

In Exercises 84–87, determine whether the function is one-to-one. If it is, find its inverse.

84. $f(x) = -10$

85. $f(x) = (x - 7)^2 + 3$

86. $f(x) = \sqrt{3x - 14}$

87. $f(x) = \sqrt[3]{x - 5}$

In Exercises 88 and 89, find the exact values of the six trigonometric functions of the angle θ.

88.

89.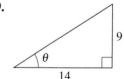

1.7 Inverse Trigonometric Functions

Inverse Sine Function

Recall from Section P.9 that for a function to have an inverse, it must pass the Horizontal Line Test. From Figure 1.63 it is obvious that $y = \sin x$ does not pass the test because different values of x yield the same y-value.

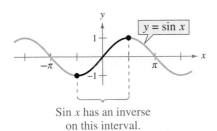

Sin x has an inverse
on this interval.

Figure 1.63

However, if you restrict the domain to the interval $-\pi/2 \le x \le \pi/2$ (corresponding to the black portion of the graph in Figure 1.63), the following properties hold.

1. On the interval $[-\pi/2, \pi/2]$, the function $y = \sin x$ is increasing.
2. On the interval $[-\pi/2, \pi/2]$, $y = \sin x$ takes on its full range of values, $-1 \le \sin x \le 1$.
3. On the interval $[-\pi/2, \pi/2]$, $y = \sin x$ passes the Horizontal Line Test.

So, on the restricted domain $-\pi/2 \le x \le \pi/2$, $y = \sin x$ has a unique inverse called the **inverse sine function.** It is denoted by

$$y = \arcsin x \quad \text{or} \quad y = \sin^{-1} x.$$

The notation $\sin^{-1} x$ is consistent with the inverse function notation $f^{-1}(x)$. The arcsin x notation (read as "the arcsine of x") comes from the association of a central angle with its intercepted *arc length* on a unit circle. So, arcsin x means the angle (or arc) whose sine is x. Both notations, arcsin x and $\sin^{-1} x$, are commonly used in mathematics, so remember that $\sin^{-1} x$ denotes the *inverse* sine function rather than $1/\sin x$. The values of arcsin x lie in the interval $-\pi/2 \le \arcsin x \le \pi/2$. The graph of $y = \arcsin x$ is shown in Example 2.

Definition of Inverse Sine Function

The **inverse sine function** is defined by

$$y = \arcsin x \quad \text{if and only if} \quad \sin y = x$$

where $-1 \le x \le 1$ and $-\pi/2 \le y \le \pi/2$. The domain of $y = \arcsin x$ is $[-1, 1]$ and the range is $[-\pi/2, \pi/2]$.

When evaluating the inverse sine function, it helps to remember the phrase "the arcsine of x is the angle (or number) whose sine is x."

EXAMPLE 1   Evaluating the Inverse Sine Function

If possible, find the exact value.

a. $\arcsin\left(-\dfrac{1}{2}\right)$ **b.** $\sin^{-1}\dfrac{\sqrt{3}}{2}$ **c.** $\sin^{-1} 2$

Solution

a. Because $\sin(-\pi/6) = -\dfrac{1}{2}$ for $-\pi/2 \le y \le \pi/2$, it follows that

$$\arcsin\left(-\frac{1}{2}\right) = -\frac{\pi}{6}.$$ Angle whose sine is $-\frac{1}{2}$

b. Because $\sin(\pi/3) = \sqrt{3}/2$ for $-\pi/2 \le y \le \pi/2$, it follows that

$$\sin^{-1}\frac{\sqrt{3}}{2} = \frac{\pi}{3}.$$ Angle whose sine is $\frac{\sqrt{3}}{2}$

c. It is not possible to evaluate $y = \sin^{-1} x$ when $x = 2$ because there is no angle whose sine is 2. Remember that the domain of the inverse sine function is $[-1, 1]$.

EXAMPLE 2   Graphing the Arcsine Function

Sketch a graph of $y = \arcsin x$.

Solution

By definition, the equations

$$y = \arcsin x \qquad \text{and} \qquad \sin y = x$$

are equivalent for $-\pi/2 \le y \le \pi/2$. So, their graphs are the same. From the interval $[-\pi/2, \pi/2]$, you can assign values to y in the second equation to make a table of values.

y	$-\dfrac{\pi}{2}$	$-\dfrac{\pi}{4}$	$-\dfrac{\pi}{6}$	0	$\dfrac{\pi}{6}$	$\dfrac{\pi}{4}$	$\dfrac{\pi}{2}$
$x = \sin y$	-1	$-\dfrac{\sqrt{2}}{2}$	$-\dfrac{1}{2}$	0	$\dfrac{1}{2}$	$\dfrac{\sqrt{2}}{2}$	1

The resulting graph for

$$y = \arcsin x$$

is shown in Figure 1.64. Note that it is the reflection (in the line $y = x$) of the black portion of the graph in Figure 1.63. Use a graphing utility to confirm this graph. Be sure you see that Figure 1.64 shows the *entire* graph of the inverse sine function. Remember that the range of $y = \arcsin x$ is the closed interval $[-\pi/2, \pi/2]$.

Library of Functions

The inverse trigonometric functions are obtained from the trigonometric functions in much the same way that the logarithmic function was developed from the exponential function. However, unlike the exponential function, the trigonometric functions are not one-to-one, and so it is necessary to restrict their domains to regions that pass the Horizontal Line Test. Consult the Library of Functions Summary inside the front cover for a description of the inverse trigonometric functions.

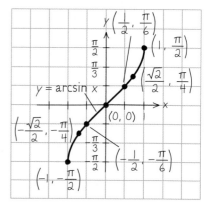

Figure 1.64

Other Inverse Trigonometric Functions

The cosine function is decreasing on the interval $0 \leq x \leq \pi$, as shown in Figure 1.65.

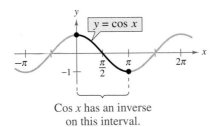

Figure 1.65

Consequently, on this interval the cosine function has an inverse function—the **inverse cosine function**—denoted by

$$y = \arccos x \qquad \text{or} \qquad y = \cos^{-1} x.$$

Because these equations are equivalent for $0 \leq y \leq \pi$, their values are the same, as shown in the table below.

y	0	$\dfrac{\pi}{6}$	$\dfrac{\pi}{3}$	$\dfrac{\pi}{2}$	$\dfrac{2\pi}{3}$	$\dfrac{5\pi}{6}$	π
$x = \cos y$	1	$\dfrac{\sqrt{3}}{2}$	$\dfrac{1}{2}$	0	$-\dfrac{1}{2}$	$-\dfrac{\sqrt{3}}{2}$	-1

Similarly, you can define an **inverse tangent function** by restricting the domain of $y = \tan x$ to the interval $(-\pi/2, \pi/2)$. The following list summarizes the definitions of the three most common inverse trigonometric functions. The remaining three are discussed in Exercises 80–82 in this section.

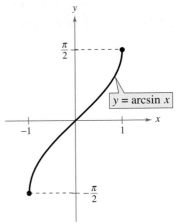

Domain: $[-1, 1]$; *Range:* $\left[-\dfrac{\pi}{2}, \dfrac{\pi}{2}\right]$

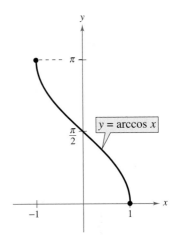

Domain: $[-1, 1]$; *Range:* $[0, \pi]$

Definition of Inverse Trigonometric Functions

Function	Domain	Range
$y = \arcsin x$ if and only if $\sin y = x$	$-1 \leq x \leq 1$	$-\dfrac{\pi}{2} \leq y \leq \dfrac{\pi}{2}$
$y = \arccos x$ if and only if $\cos y = x$	$-1 \leq x \leq 1$	$0 \leq y \leq \pi$
$y = \arctan x$ if and only if $\tan y = x$	$-\infty < x < \infty$	$-\dfrac{\pi}{2} < y < \dfrac{\pi}{2}$

The graphs of these three inverse trigonometric functions are shown in Figure 1.66.

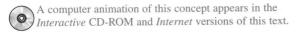

A computer animation of this concept appears in the *Interactive* CD-ROM and *Internet* versions of this text.

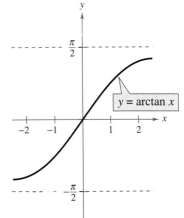

Domain: $(-\infty, \infty)$; *Range:* $\left(-\dfrac{\pi}{2}, \dfrac{\pi}{2}\right)$

Figure 1.66

EXAMPLE 3   Evaluating Inverse Trigonometric Functions

Find the exact value.

a. $\arccos \dfrac{\sqrt{2}}{2}$ **b.** $\arccos(-1)$ **c.** $\arctan 0$ **d.** $\arctan(-1)$

Solution

a. Because $\cos(\pi/4) = \sqrt{2}/2$, and $\pi/4$ lies in $[0, \pi]$, it follows that

$$\arccos \dfrac{\sqrt{2}}{2} = \dfrac{\pi}{4}. \qquad \text{Angle whose cosine is } \dfrac{\sqrt{2}}{2}$$

b. Because $\cos \pi = -1$, and π lies in $[0, \pi]$, it follows that

$$\arccos(-1) = \pi. \qquad \text{Angle whose cosine is } -1$$

c. Because $\tan 0 = 0$, and 0 lies in $(-\pi/2, \pi/2)$, it follows that

$$\arctan 0 = 0. \qquad \text{Angle whose tangent is } 0$$

d. Because $\tan(-\pi/4) = -1$ and $-\pi/4$ lies in $(-\pi/2, \pi/2)$, it follows that

$$\arctan(-1) = -\dfrac{\pi}{4}. \qquad \text{Angle whose tangent is } -1$$

You can use the $\boxed{\text{SIN}^{-1}}$, $\boxed{\text{COS}^{-1}}$, and $\boxed{\text{TAN}^{-1}}$ keys on your calculator to approximate values of trigonometric functions.

EXAMPLE 4   Calculators and Inverse Trigonometric Functions

Use a calculator to approximate the value (if possible).

a. $\arctan(-8.45)$ **b.** $\arcsin 0.2447$ **c.** $\arccos 2$

Solution

Function	Mode	Graphing Calculator Keystrokes
a. $\arctan(-8.45)$	Radian	$\boxed{\text{TAN}^{-1}}$ $\boxed{(}$ $\boxed{(-)}$ 8.45 $\boxed{)}$ $\boxed{\text{ENTER}}$

From the display, it follows that $\arctan(-8.45) \approx -1.453001$.

b. $\arcsin 0.2447$	Radian	$\boxed{\text{SIN}^{-1}}$ 0.2447 $\boxed{\text{ENTER}}$

From the display, it follows that $\arcsin 0.2447 \approx 0.2472103$.

c. $\arccos 2$	Radian	$\boxed{\text{COS}^{-1}}$ 2 $\boxed{\text{ENTER}}$

In real number mode, the calculator should display an error message because the domain of the inverse cosine function is $[-1, 1]$.

In Example 4, if you had set the calculator to *degree* mode, the display would have been in degrees rather than in radians. This convention is peculiar to calculators. By definition, the values of inverse trigonometric functions are always in *radians*.

Compositions of Functions

Recall from Section P.9 that inverse functions possess the properties

$$f(f^{-1}(x)) = x \qquad \text{and} \qquad f^{-1}(f(x)) = x.$$

The inverse trigonometric versions of these properties are given below.

Inverse Properties

If $-1 \le x \le 1$ and $-\pi/2 \le y \le \pi/2$, then

$$\sin(\arcsin x) = x \qquad \text{and} \qquad \arcsin(\sin y) = y.$$

If $-1 \le x \le 1$ and $0 \le y \le \pi$, then

$$\cos(\arccos x) = x \qquad \text{and} \qquad \arccos(\cos y) = y.$$

If x is a real number and $-\pi/2 < y < \pi/2$, then

$$\tan(\arctan x) = x \qquad \text{and} \qquad \arctan(\tan y) = y.$$

Keep in mind that these inverse properties do not apply for arbitrary values of x and y. For instance,

$$\arcsin\left(\sin\frac{3\pi}{2}\right) = \arcsin(-1) = -\frac{\pi}{2} \ne \frac{3\pi}{2}.$$

In other words, the property

$$\arcsin(\sin y) = y$$

is not valid for values of y outside the interval $[-\pi/2, \pi/2]$.

Exploration

Use a graphing utility to graph $y = \arcsin(\sin x)$. What are the domain and range of this function? Explain why $\arcsin(\sin 4)$ does not equal 4.

Now graph $y = \sin(\arcsin x)$ and determine the domain and range. Explain why $\sin(\arcsin 4)$ is not defined.

EXAMPLE 5   Using Inverse Properties

If possible, find the exact value.

a. $\tan[\arctan(-5)]$ **b.** $\arcsin\left(\sin\dfrac{5\pi}{3}\right)$ **c.** $\cos(\cos^{-1}\pi)$

Solution

a. Because -5 lies in the domain of $\arctan x$, the inverse property applies, and you have

$$\tan[\arctan(-5)] = -5.$$

b. In this case, $5\pi/3$ does not lie within the range of the arcsine function, $-\pi/2 \le y \le \pi/2$. However, $5\pi/3$ is coterminal with

$$\frac{5\pi}{3} - 2\pi = -\frac{\pi}{3}$$

which does lie in the range of the arcsine function, and you have

$$\arcsin\left(\sin\frac{5\pi}{3}\right) = \arcsin\left[\sin\left(-\frac{\pi}{3}\right)\right] = -\frac{\pi}{3}.$$

c. The expression $\cos(\cos^{-1}\pi)$ is not defined because $\cos^{-1}\pi$ is not defined. Remember that the domain of the inverse cosine function is $[-1, 1]$.

Example 6 shows how to use right triangles to find exact values of compositions of inverse functions. Then, Example 7 shows how to use triangles to convert a trigonometric expression into an algebraic expression. This conversion technique is used frequently in calculus.

EXAMPLE 6   Evaluating Compositions of Functions

Find the exact value of (a) $\tan\left(\arccos\dfrac{2}{3}\right)$ and (b) $\cos\left[\arcsin\left(-\dfrac{3}{5}\right)\right]$.

Algebraic Solution

a. If you let $u = \arccos\frac{2}{3}$, then $\cos u = \frac{2}{3}$. Because $\cos u$ is positive, u is a first-quadrant angle. You can sketch and label angle u as shown in Figure 1.67(a). Consequently,

$$\tan\left(\arccos\frac{2}{3}\right) = \tan u = \frac{\text{opp}}{\text{adj}} = \frac{\sqrt{5}}{2}.$$

b. If you let $u = \arcsin\left(-\frac{3}{5}\right)$, then $\sin u = -\frac{3}{5}$. Because $\sin u$ is negative, u is a fourth-quadrant angle. You can sketch and label angle u as shown in Figure 1.67(b). Consequently,

$$\cos\left[\arcsin\left(-\frac{3}{5}\right)\right] = \cos u = \frac{\text{adj}}{\text{hyp}} = \frac{4}{5}.$$

Graphical Solution

a. Use a graphing utility set in *radian* mode to graph $y = \tan(\arccos x)$, as shown in Figure 1.68(a). Use the *value* feature or *zoom* and *trace* features of the graphing utility to find that the value of the composition of functions when $x = \frac{2}{3} \approx 0.66667$ is $y = 1.118 \approx \sqrt{5}/2$.

b. Use a graphing utility set in *radian* mode to graph $y = \cos(\arcsin x)$, as shown in Figure 1.68(b). Use the *value* feature or *zoom* and *trace* features of the graphing utility to find that the value of the composition of functions when $x = -\frac{3}{5} = -0.6$ is $y = 0.8 = \frac{4}{5}$.

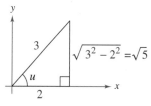

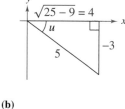

(a)   **(b)**

Figure 1.67

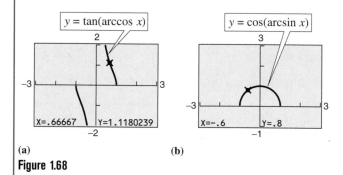

(a)   **(b)**

Figure 1.68

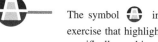 The symbol indicates an example or exercise that highlights algebraic techniques specifically used in calculus.

EXAMPLE 7   Some Problems from Calculus

Write each of the following as an algebraic expression in x.

a. $\sin(\arccos 3x), \quad 0 \leq x \leq \dfrac{1}{3}$ **b.** $\cot(\arccos 3x), \quad 0 \leq x \leq \dfrac{1}{3}$

Solution

If you let $u = \arccos 3x$, then $\cos u = 3x$. Because $\cos u = 3x/1 = \text{adj/hyp}$ you can sketch a right triangle with acute angle u, as shown in Figure 1.69. From this triangle, you can easily convert each expression to algebraic form.

a. $\sin(\arccos 3x) = \sin u = \dfrac{\text{opp}}{\text{hyp}} = \sqrt{1 - 9x^2}, \qquad 0 \leq x < \dfrac{1}{3}$

b. $\cot(\arccos 3x) = \cot u = \dfrac{\text{adj}}{\text{opp}} = \dfrac{3x}{\sqrt{1 - 9x^2}}, \qquad 0 \leq x < \dfrac{1}{3}$

A similar argument can be made here for x-values lying in the interval $\left[-\frac{1}{3}, 0\right]$.

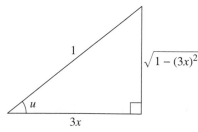

Figure 1.69

1.7 Exercises

1. *Numerical and Graphical Analysis* Consider the function $y = \arcsin x$.

   (a) Use a graphing utility to complete the table.

x	-1	-0.8	-0.6	-0.4	-0.2
y					

x	0	0.2	0.4	0.6	0.8	1
y						

   (b) Plot the points from the table in part (a) and graph the function. (Do not use a graphing utility.)

   (c) Use a graphing utility to graph the inverse sine function and compare the result with your hand drawn graph in part (b).

   (d) Determine any intercepts and symmetry of the graph.

2. *Numerical and Graphical Analysis* Consider the function $y = \arccos x$.

   (a) Use a graphing utility to complete the table.

x	-1	-0.8	-0.6	-0.4	-0.2
y					

x	0	0.2	0.4	0.6	0.8	1
y						

   (b) Plot the points from the table in part (a) and graph the function. (Do not use a graphing utility.)

   (c) Use a graphing utility to graph the inverse cosine function and compare the result with your hand-drawn graph in part (b).

   (d) Determine any intercepts and symmetry of the graph.

3. *Numerical and Graphical Analysis* Consider the function $y = \arctan x$.

   (a) Use a graphing utility to complete the table.

x	-10	-8	-6	-4	-2
y					

x	0	2	4	6	8	10
y						

   (b) Plot the points from the table in part (a) and graph the function. (Do not use a graphing utility.)

   (c) Use a graphing utility to graph the inverse tangent function and compare the result with your hand-drawn graph in part (b).

   (d) Determine the horizontal asymptotes of the graph.

In Exercises 4–7, write each trigonometric expression in inverse function form, or vice versa. For example,

$$\sin \frac{\pi}{2} = 1 \quad \Longrightarrow \quad \arcsin 1 = \frac{\pi}{2}.$$

4. $\sin\left(\dfrac{\pi}{6}\right) = \dfrac{1}{2}$ $\Longrightarrow$

5. $\tan\left(-\dfrac{\pi}{4}\right) = -1$ $\Longrightarrow$

6. $\arccos \dfrac{\sqrt{2}}{2} = \dfrac{\pi}{4}$ $\Longrightarrow$

7. $\arcsin(-1) = -\dfrac{\pi}{2}$ $\Longrightarrow$

In Exercises 8–15, evaluate the expression using a calculator.

8. (a) $\arcsin \dfrac{1}{2}$ (b) $\arcsin 0$

9. (a) $\arccos \dfrac{1}{2}$ (b) $\arccos 0$

10. (a) $\arctan \dfrac{\sqrt{3}}{3}$ (b) $\arctan 1$

11. (a) $\arccos\left(-\dfrac{\sqrt{2}}{2}\right)$ (b) $\arcsin\left(-\dfrac{\sqrt{2}}{2}\right)$

12. (a) $\arctan\left(-\sqrt{3}\right)$ (b) $\arctan \sqrt{3}$

13. (a) $\arccos\left(-\dfrac{1}{2}\right)$ (b) $\arcsin \dfrac{\sqrt{2}}{2}$

14. (a) $\arcsin\left(-\dfrac{\sqrt{3}}{2}\right)$ (b) $\arctan\left(-\dfrac{\sqrt{3}}{3}\right)$

15. (a) $\arcsin(-1)$ (b) $\arccos 1$

In Exercises 16 and 17, determine the missing coordinates of the points on the graph of the function.

16.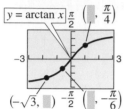

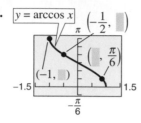

17.

In Exercises 18–23, use a calculator to approximate the value of the expression. (Round your answer to two decimal places.)

18. (a) $\arccos 0.22$ (b) $\arcsin 0.45$

19. (a) $\arcsin(-0.75)$ (b) $\arccos(-0.7)$

20. (a) $\arctan(-6)$ (b) $\arctan 18$

21. (a) $\arcsin 0.41$ (b) $\arccos 0.36$

22. (a) $\arccos(-0.51)$ (b) $\arcsin(-0.125)$

23. (a) $\arctan 0.98$ (b) $\arctan 4.7$

In Exercises 24 and 25, use a graphing utility to graph f, g, and $y = x$ in the same viewing window to verify geometrically that g is the inverse of f. (Be sure to properly restrict the domain of f.)

24. $f(x) = \tan x, \quad g(x) = \arctan x$

25. $f(x) = \sin x, \quad g(x) = \arcsin x$

In Exercises 26–29, use an inverse trigonometric function to write θ as a function of x.

26.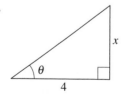
     27.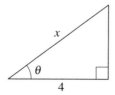

28.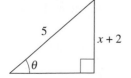
     29.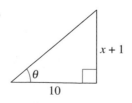

In Exercises 30–37, use the properties of inverse functions to evaluate the expression.

30. $\sin(\arcsin 0.7)$ **31.** $\tan(\arctan 35)$

32. $\cos[\arccos(-0.3)]$ **33.** $\sin[\arcsin(-0.1)]$

34. $\arcsin(\sin 3\pi)$ **35.** $\arccos\left(\cos\dfrac{7\pi}{2}\right)$

36. $\arctan\left(\tan\dfrac{11\pi}{6}\right)$ **37.** $\arcsin\left(\sin\dfrac{7\pi}{4}\right)$

In Exercises 38–45, find the exact value of the expression. Use a graphing utility to verify your result. (*Hint:* Make a sketch of a right triangle.)

38. $\sin\left(\arctan\dfrac{4}{3}\right)$ **39.** $\sec\left(\arcsin\dfrac{3}{5}\right)$

40. $\cos\left(\arcsin\dfrac{24}{25}\right)$ **41.** $\csc\left[\arctan\left(-\dfrac{12}{5}\right)\right]$

42. $\sec\left[\arctan\left(-\dfrac{3}{5}\right)\right]$ **43.** $\tan\left[\arcsin\left(-\dfrac{3}{8}\right)\right]$

44. $\sin\left[\arccos\left(-\dfrac{2}{3}\right)\right]$ **45.** $\cot\left(\arctan\dfrac{6}{11}\right)$

 In Exercises 46–53, write an algebraic expression that is equivalent to the expression. (*Hint:* Sketch a right triangle, as demonstrated in Example 7.)

46. $\cot(\arctan x)$ **47.** $\sin(\arctan x)$

48. $\sin(\arccos x)$ **49.** $\sec[\arcsin(x - 1)]$

50. $\tan\left(\arccos\dfrac{x}{5}\right)$ **51.** $\cot\left(\arctan\dfrac{1}{x}\right)$

52. $\csc\left(\arctan\dfrac{x}{\sqrt{7}}\right)$ **53.** $\cos\left(\arcsin\dfrac{x - h}{r}\right)$

In Exercises 54 and 55, use a graphing utility to graph f and g in the same viewing window to verify that the two are equal. Explain why they are equal. Identify any asymptotes of the graphs.

54. $f(x) = \sin(\arctan 2x), \quad g(x) = \dfrac{2x}{\sqrt{1 + 4x^2}}$

55. $f(x) = \tan\left(\arccos\dfrac{x}{2}\right), \quad g(x) = \dfrac{\sqrt{4 - x^2}}{x}$

In Exercises 56–59, complete the equation.

56. $\arctan\dfrac{14}{x} = \arcsin(\rule{1.5em}{0.8em}), \quad x \neq 0$

57. $\arcsin\dfrac{\sqrt{36 - x^2}}{6} = \arccos(\rule{1.5em}{0.8em}), \quad 0 \leq x \leq 6$

58. $\arccos \dfrac{3}{\sqrt{x^2 - 2x + 10}} = \arcsin(\;\rule{1cm}{0.5pt}\;)$

59. $\arccos \dfrac{x - 2}{2} = \arctan(\;\rule{1cm}{0.5pt}\;), \quad |x - 2| < 2$

In Exercises 60–67, use a graphing utility to graph the function.

60. $y = 2 \arccos x$

61. $y = \arcsin \dfrac{x}{2}$

62. $f(x) = \arcsin(x - 2)$

63. $g(t) = \arccos(t + 2)$

64. $f(x) = \arctan 2x$

65. $f(x) = \pi + \arctan x$

66. $h(v) = \tan(\arccos v)$

67. $f(x) = \arccos \dfrac{x}{4}$

In Exercises 68 and 69, write the given function in terms of the sine function by using the identity

$$A \cos \omega t + B \sin \omega t = \sqrt{A^2 + B^2} \sin\left(\omega t + \arctan \dfrac{A}{B}\right).$$

Use a graphing utility to graph both forms of the function. What does the graph imply?

68. $f(t) = 3 \cos 2t + 3 \sin 2t$

69. $f(t) = 4 \cos \pi t + 3 \sin \pi t$

70. *Docking a Boat*   A boat is pulled in by means of a winch located on a dock 10 feet above the deck of the boat. Let θ be the angle of elevation from the boat to the winch and let s be the length of the rope from the winch to the boat.

(a) Write θ as a function of s.

(b) Find θ when $s = 52$ feet and when $s = 26$ feet.

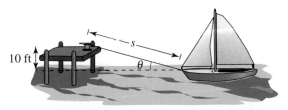

71. *Photography*   A television camera at ground level is filming the lift-off of a space shuttle at a point 750 meters from the launch pad. Let θ be the angle of elevation to the shuttle and let s be the height of the shuttle.

(a) Write θ as a function of s.

(b) Find θ when $s = 400$ meters and when $s = 1600$ meters.

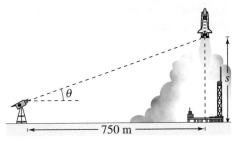

FIGURE FOR 71

72. *Rock Salt*   Different types of granular substances naturally settle at different angles when stored in cone-shaped piles. This angle θ is called the *angle of repose*. When rock salt is stored in a cone-shaped pile 11 feet high, the diameter of the pile's base is about 34 feet.   (Source: Bulk-Store Structures, Inc.)

(a) Find the angle of repose for rock salt.

(b) How tall is a pile of rock salt that has a base diameter of 40 feet?

73. *Granular Angle of Repose*   When whole corn is stored in a cone-shaped pile 20 feet high, the diameter of the pile's base is about 82 feet.

(a) Find the angle of repose for whole corn.

(b) How tall is a pile of corn that has a base diameter of 100 feet?

74. *Photography*   A photographer is taking a picture of a 3-foot painting hung in an art gallery. The camera lens is 1 foot below the lower edge of the painting. The angle β subtended by the camera lens x feet from the painting is

$$\beta = \arctan \dfrac{3x}{x^2 + 4}, \quad x > 0.$$

(a) Use a graphing utility to graph β as a function of x.

(b) Move the cursor to approximate the distance from the picture when β is maximum.

(c) Identify the asymptote of the graph and discuss its meaning in the context of the problem.

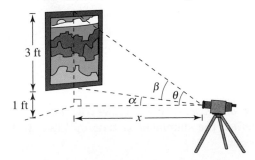

75. *Area* In calculus, it is shown that the area of the region bounded by the graphs of $y = 0$, $y = 1/(x^2 + 1)$, $x = a$, and $x = b$ is

Area $= \arctan b - \arctan a$.

Find the areas for the following values of a and b.

(a) $a = 0, b = 1$ (b) $a = -1, b = 1$
(c) $a = 0, b = 3$ (d) $a = -1, b = 3$

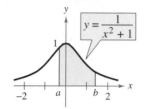

76. *Angle of Elevation* An airplane flies at an altitude of 5 miles toward a point directly over an observer. Consider θ and x as shown in the figure.

(a) Write θ as a function of x.

(b) Find θ when $x = 10$ miles and $x = 3$ miles.

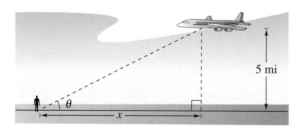

77. *Security Patrol* A security car with its spotlight on is parked 20 meters from a long warehouse. Consider θ and x as shown in the figure.

(a) Write θ as a function of x.

(b) Find θ when $x = 5$ meters and when $x = 12$ meters.

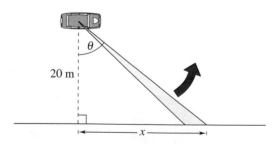

Synthesis

True or False? In Exercises 78 and 79, determine whether the statement is true or false. Justify your answer.

78. $\sin \dfrac{5\pi}{6} = \dfrac{1}{2}$ ⟹ $\arcsin \dfrac{1}{2} = \dfrac{5\pi}{6}$

79. $\arctan x = \dfrac{\arcsin x}{\arccos x}$

80. Define the inverse cotangent function by restricting the domain of the cotangent function to the interval $(0, \pi)$, and sketch the inverse function's graph.

81. Define the inverse secant function by restricting the domain of the secant function to the intervals $[0, \pi/2)$ and $(\pi/2, \pi]$, and sketch the inverse function's graph.

82. Define the inverse cosecant function by restricting the domain of the cosecant function to the intervals $[-\pi/2, 0)$ and $(0, \pi/2]$, and sketch the inverse function's graph.

83. Use the results of Exercises 80–82 to evaluate the following without using a calculator.

(a) $\operatorname{arcsec} \sqrt{2}$ (b) $\operatorname{arcsec} 1$
(c) $\operatorname{arccot}\left(-\sqrt{3}\right)$ (d) $\operatorname{arccsc} 2$

In Exercises 84–89, prove the identity.

84. $\arcsin(-x) = -\arcsin x$

85. $\arctan(-x) = -\arctan x$

86. $\arccos(-x) = \pi - \arccos x$

87. $\arctan x + \arctan \dfrac{1}{x} = \dfrac{\pi}{2}, \quad x > 0$

88. $\arcsin x + \arccos x = \dfrac{\pi}{2}$

89. $\arcsin x = \arctan \dfrac{x}{\sqrt{1 - x^2}}$

Review

In Exercises 90–93, evaluate the sine, cosine, and tangent of the angle without using a calculator.

90. $840°$

91. $-585°$

92. $\dfrac{17\pi}{3}$

93. $-\dfrac{19\pi}{4}$

In Exercises 94–97, sketch the graph of the function. (Include two full periods.) Use a graphing utility to verify your result.

94. $y = \dfrac{1}{2} \sin(x + \pi)$

95. $y = 2 - \cos(x + \pi)$

96. $y = \dfrac{1}{2} \tan\left(x + \dfrac{\pi}{2}\right)$

97. $y = 4 \cot \dfrac{1}{2} x$

1.8 Applications and Models

Applications Involving Right Triangles

In this section, the three angles of a right triangle are denoted by the letters A, B, and C (where C is the right angle), and the lengths of the sides opposite these angles by the letters a, b, and c (where c is the hypotenuse).

What You Should Learn:

- How to solve real-life problems involving right triangles
- How to solve real-life problems involving directional bearings
- How to solve real-life problems involving harmonic motion

Why You Should Learn It:

You can use trigonometric functions to model and solve real-life problems. For instance, Exercise 60 on page 205 shows you how a trigonometric function can be used to model the harmonic motion of a buoy.

EXAMPLE 1 Solving a Right Triangle

Solve the right triangle shown in Figure 1.70.

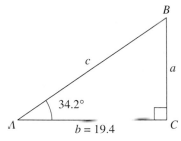

Figure 1.70

Solution

Because $C = 90°$, it follows that $A + B = 90°$ and $B = 90° - 34.2° = 55.8°$. To solve for a, use the fact that

$$\tan A = \frac{\text{opp}}{\text{adj}} = \frac{a}{b} \implies a = b \tan A.$$

So, $a = 19.4 \tan 34.2° \approx 13.18$. Similarly, to solve for c, use the fact that

$$\cos A = \frac{\text{adj}}{\text{hyp}} = \frac{b}{c} \implies c = \frac{b}{\cos A}.$$

So, $c = \dfrac{19.4}{\cos 34.2°} \approx 23.46$.

Mary Kate Denny/PhotoEdit

Recall from Section 1.3 that the term *angle of elevation* denotes the angle from the horizontal upward to an object and that the term *angle of depression* denotes the angle from the horizontal downward to an object. An angle of elevation and an angle of depression are shown in Figure 1.71.

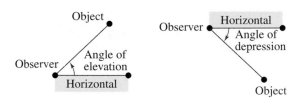

Figure 1.71

EXAMPLE 2   Finding a Side of a Right Triangle

A safety regulation states that the maximum angle of elevation for a rescue ladder is 72°. If a fire department's longest ladder is 110 feet, what is the maximum safe rescue height?

Solution

A sketch is shown in Figure 1.72. From the equation $\sin A = a/c$, it follows that

$$a = c \sin A = 110 \sin 72° \approx 104.6.$$

So, the maximum safe rescue height is about 104.6 feet above the height of the fire truck.

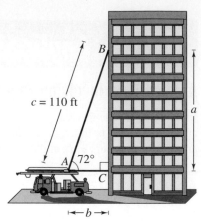

Figure 1.72

EXAMPLE 3   Finding a Side of a Right Triangle

At a point 200 feet from the base of a building, the angle of elevation to the *bottom* of a smokestack is 35°, and the angle of elevation to the *top* is 53°, as shown in Figure 1.73. Find the height s of the smokestack alone.

Solution

Note from Figure 1.73 that this problem involves two right triangles. In the smaller right triangle, use the fact that $\tan 35° = a/200$ to conclude that the height of the building is

$$a = 200 \tan 35°.$$

Now, in the larger right triangle, use the equation

$$\tan 53° = \frac{a + s}{200}$$

to conclude that $s = 200 \tan 53° - a$. So, the height of the smokestack is

$$s = 200 \tan 53° - a = 200 \tan 53° - 200 \tan 35° \approx 125.4 \text{ feet.}$$

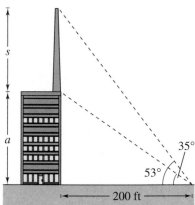

Figure 1.73

EXAMPLE 4   Finding an Angle of Depression

A swimming pool is 20 meters long and 12 meters wide. The bottom of the pool has a constant slant so that the water depth is 1.3 meters at the shallow end and 4 meters at the deep end, as shown in Figure 1.74. Find the angle of depression of the bottom of the pool.

Solution

Using the tangent function, you see that

$$\tan A = \frac{\text{opp}}{\text{adj}} = \frac{2.7}{20} = 0.135.$$

So, the angle of depression is

$$A = \arctan 0.135 \approx 0.13419 \text{ radian} \approx 7.69°.$$

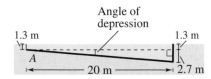

Figure 1.74

31.

Trigonometry and Bearings

32.

In surveying and navigation, directions are generally given in terms of **bearings.** A bearing measures the acute angle a path or line of sight makes with a fixed north–south line, as shown in Figure 1.75. For instance, the bearing of S 35° E in Figure 1.75(a) means 35 degrees east of south.

33.

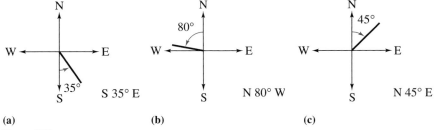

(a)　　　　　　　　**(b)**　　　　　　　　**(c)**

Figure 1.75

EXAMPLE 5　Finding Directions in Terms of Bearings

A ship leaves port at noon and heads due west at 20 knots (nautical miles per hour). At 2 P.M. the ship changes course to N 54° W, as shown in Figure 1.76. Find the ship's bearing and distance from the port of departure at 3 P.M.

34.

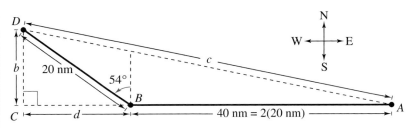

Figure 1.76

Solution

In triangle *BCD*, you have $B = 90° - 54° = 36°$. The two sides of this triangle can be determined to be

$$b = 20 \sin 36° \quad \text{and} \quad d = 20 \cos 36°.$$

In triangle *ACD*, you can find angle *A* as follows.

$$\tan A = \frac{b}{d + 40} = \frac{20 \sin 36°}{20 \cos 36° + 40} \approx 0.2092494$$

$$A \approx \arctan 0.2092494 \approx 0.2062732 \text{ radian} \approx 11.82°$$

35.

The angle with the north–south line is $90° - 11.82° = 78.18°$. Therefore, the bearing of the ship is

N 78.18° W.　　　　　　　　　　　　　　　　Bearing

Finally, from triangle *ACD*, you have $\sin A = b/c$, which yields

$$c = \frac{b}{\sin A} = \frac{20 \sin 36°}{\sin 11.82°} \approx 57.4 \text{ nautical miles.} \quad \text{Distance from port}$$

In Exercises 169–172, write an algebraic expression for the given expression.

169. $\sec[\arcsin(x - 1)]$ **170.** $\tan\left(\arccos \dfrac{x}{2}\right)$

171. $\sin\left(\arccos \dfrac{x^2}{4 - x^2}\right)$ **172.** $\csc(\arcsin 10x)$

1.8 173. *Railroad Grade*   A train travels 3.5 kilometers on a straight track with a grade of $1° \, 10'$. What is the vertical rise of the train in that distance?

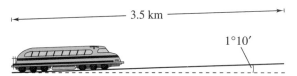

174. *Distance Between Towns*   A passenger in an airplane flying at an altitude of 37,000 feet sees two towns directly to the left of the airplane. The angles of depression to the towns are $32°$ and $76°$. How far apart are the towns?

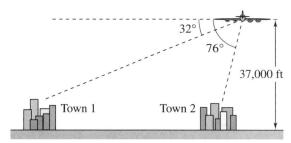

175. *Navigation*   An airplane flying at 490 miles per hour has a bearing of N 46° W. After flying 1.8 hours, how far north and how far west has the plane traveled from its point of departure?

176. *Distance*   From city A to city B, a plane flies 650 miles at a bearing of N 48° E. From city B to city C, the plane flies 810 miles at a bearing of S 65° E. Find the distance from A to C and the bearing from A to C.

177. *Wave Motion*   A buoy oscillates in simple harmonic motion as waves go past. At a given time it is noted that the buoy moves a total of 6 feet from its low point to its high point, returning to its high point every 15 seconds (see figure). Write an equation that describes the motion of the buoy if at $t = 0$ it is at its high point.

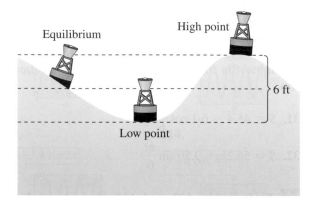

Synthesis

True or False?   **In Exercises 178–181, determine whether the statement is true or false. Justify your answer.**

178. $\dfrac{\sin 60°}{\sin 30°} = \sin 2°$ **179.** $\tan[(0.8)^2] = \tan^2(0.8)$

180. $y = \sin \theta$ is not a function because $\sin 30° = \sin 150°$.

181. $\tan \dfrac{3\pi}{4} = -1 \quad \Longrightarrow \quad \arctan(-1) = \dfrac{3\pi}{4}$

182. In your own words, explain the meaning of (a) an angle in standard position, (b) a negative angle, (c) coterminal angles, and (d) an obtuse angle.

183. A fan motor turns at a given angular speed. How does the speed of the tips of the blades change if a fan of greater diameter is installed on the motor? Explain.

Chapter Project   *Fitting a Model to Data*

In this project, you will find and use models relating to the carbon dioxide level of Earth's atmosphere.

Since 1958, the Mauna Loa Climate Observatory in Hawaii has been collecting data on the carbon dioxide level of Earth's atmosphere. The table shows the average monthly readings for January of each year from 1974 through 1999. The readings measure the carbon dioxide concentration in parts per million. (Source: National Oceanic Atmospheric Administration, Climate Monitoring and Diagnostic Laboratory, CarbonCycle-Greenhouse Gases)

1974	1975	1976	1977	1978	1979	1980	1981	1982	1983	1984	1985	1986
329.7	331.1	332.0	333.8	335.4	336.7	338.7	340.1	341.2	342.8	344.5	345.9	347.2

1987	1988	1989	1990	1991	1992	1993	1994	1995	1996	1997	1998	1999
349.0	351.5	352.9	354.2	355.5	356.3	357.0	358.6	360.6	362.4	363.5	366.5	368.6

(a) Enter the data in the table into a graphing utility. (Let $t = 0$ represent 1970.) Draw a scatter plot for the data. Does the data appear to be best modeled with a linear or quadratic model?

(b) Find the linear or quadratic model that you think best fits the data.

Questions for Further Exploration

1. The data in the table represents the carbon dioxide levels for January of each year. Throughout each year, the level oscillated as follows.

   - In April, the average reading was about 2.5 parts per million higher than the average reading given by the model in part (b) above.

   - In July, the average reading was the same as the average reading given by the model in part (b) above.

   - In October, the average reading was about 2.5 parts per million lower than the average reading given by the model in part (b) above.

   Use a sine function to rewrite the model found in part (b) above so that the model incorporates the described oscillations.

2. Use a graphing utility to graph the revised model.

3. Make a careful sketch of the model for 1 year. What physical factors on Earth would contribute to the oscillation in the carbon dioxide level during the year?

4. Is the model you found periodic? Explain your reasoning.

5. Use the model to estimate the level of carbon dioxide in Earth's atmosphere in the following years.

   (a) 2000

   (b) 2010

   (c) 2020

6. What significance does the trend in the carbon dioxide level in Earth's atmosphere have?

1 Chapter Test
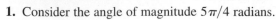

Take this test as you would take a test in class. After you are done, check your work against the answers in the back of the book.

1. Consider the angle of magnitude $5\pi/4$ radians.
   (a) Sketch the angle in standard position.
   (b) Determine two coterminal angles (one positive and one negative).
   (c) Convert the angle to degree measure.

2. A truck is moving at a rate of 90 kilometers per hour, and the diameter of its wheels is 1 meter. Find the angular speed of the wheels in radians per minute.

3. Find the exact values of the six trigonometric functions of the angle θ shown in the figure.

4. Given that $\tan \theta = \frac{11}{6}$, find the other five trigonometric functions of θ.

5. Determine the reference angle θ' of the angle $\theta = 290°$ and sketch θ and θ' in standard position.

6. Determine the quadrant in which θ lies if $\sec \theta < 0$ and $\tan \theta > 0$.

7. Find two values of θ in degrees $(0 \leq \theta < 360°)$ if $\cos \theta = -\sqrt{3}/2$. (Do not use a calculator.)

8. Use a calculator to approximate two values of θ in radians $(0 \leq \theta < 2\pi)$ if $\csc \theta = 1.030$. Round the result to two decimal places.

9. Find the five remaining trigonometric functions of θ, given that $\sec \theta = \frac{12}{10}$ and $\tan \theta < 0$.

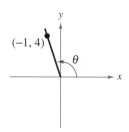

FIGURE FOR 3

In Exercises 10–15, graph the function through two full periods without using a graphing utility.

10. $g(x) = -2 \sin\left(x - \frac{\pi}{4}\right)$
11. $f(\alpha) = \frac{1}{2} \tan 2\alpha$
12. $f(x) = \frac{1}{2} \sec(x - \pi)$
13. $f(x) = 2 \cos(\pi - 2x) + 3$
14. $f(x) = 2 \csc\left(x + \frac{\pi}{2}\right)$
15. $f(x) = \frac{1}{4} \cot\left(x - \frac{\pi}{2}\right)$

In Exercises 16 and 17, use a graphing utility to graph the function. If the function is periodic, find its period.

16. $y = \sin 2\pi x + 2 \cos \pi x$
17. $y = 6t \cos(0.25t), \quad 0 \leq t \leq 32$

18. Find a, b, and c for the function $f(x) = a \sin(bx + c)$ so that the graph of f matches the graph at the right.

19. Find the exact value of $\tan\left(\arccos \frac{2}{3}\right)$ without using a calculator.

In Exercises 20–22, sketch the graph of the function.

20. $f(x) = 2 \arcsin\left(\frac{1}{2}x\right)$
21. $f(x) = 2 \arccos x$
22. $f(x) = \arctan \frac{x}{2}$

23. A plane is 160 miles north and 110 miles east of an airport. If the pilot wants to fly directly to the airport, what bearing should be taken?

24. From a 100-foot roof of a condominium on the coast, a tourist sights a cruise ship. The angle of depression is $2.5°$. How far is the ship from the shore line?

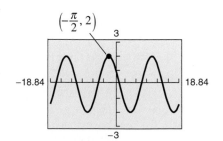

FIGURE FOR 18

Analytic Trigonometry

2.1 Using Fundamental Identities

2.2 Verifying Trigonometric Identities

2.3 Solving Trigonometric Equations

2.4 Sum and Difference Formulas

2.5 Multiple-Angle and Product-Sum Formulas

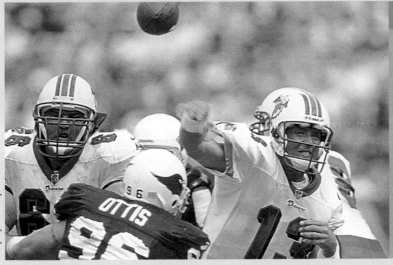

Andy Lyons/Allsport

A football thrown by a quarterback follows a parabolic path. The horizontal distance the football travels depends not only on the speed of the throw, but on the angle at which the ball is thrown.

The Big Picture

In this chapter you will learn how to

❏ use fundamental trigonometric identities to evaluate trigonometric functions and simplify trigonometric expressions.

❏ verify trigonometric identities.

❏ use standard algebraic techniques and inverse trigonometric functions to solve trigonometric equations.

❏ use sum and difference formulas, multiple-angle formulas, power-reducing formulas, half-angle formulas, and product-sum formulas to rewrite and evaluate trigonometric functions.

Important Vocabulary

As you encounter each new vocabulary term in this chapter, add the term and its definition to your notebook glossary.

- reduction formulas (p. 246)
- double-angle formulas (p. 251)
- power-reducing formulas (p. 253)
- half-angle formulas (p. 254)
- product-to-sum formulas (p. 255)
- sum-to-product formulas (p. 256)

Additional Resources Text-specific additional resources are available to help you do well in this course. See page xvi for details.

2.1 Using Fundamental Identities

Introduction

In Chapter 1, you studied the basic definitions, properties, graphs, and applications of the individual trigonometric functions. In this chapter, you will learn how to use the fundamental identities to

1. evaluate trigonometric functions,
2. simplify trigonometric expressions,
3. develop additional trigonometric identities, and
4. solve trigonometric equations.

What You Should Learn:

- How to recognize and write fundamental trigonometric identities
- How to use fundamental trigonometric identities to evaluate trigonometric functions, simplify trigonometric expressions, and rewrite trigonometric expressions

Why You Should Learn It:

Fundamental trigonometric identities can be used to simplify trigonometric expressions. For instance, Exercise 103 on page 223 shows you how trigonometric identities can be used to simplify an expression for the rate of change of a function, a concept used in calculus.

Fundamental Trigonometric Identities

Reciprocal Identities

$$\sin u = \frac{1}{\csc u} \qquad \cos u = \frac{1}{\sec u} \qquad \tan u = \frac{1}{\cot u}$$

$$\csc u = \frac{1}{\sin u} \qquad \sec u = \frac{1}{\cos u} \qquad \cot u = \frac{1}{\tan u}$$

Quotient Identities

$$\tan u = \frac{\sin u}{\cos u} \qquad \cot u = \frac{\cos u}{\sin u}$$

Pythagorean Identities

$$\sin^2 u + \cos^2 u = 1 \qquad 1 + \tan^2 u = \sec^2 u \qquad 1 + \cot^2 u = \csc^2 u$$

Cofunction Identities

$$\sin\left(\frac{\pi}{2} - u\right) = \cos u \qquad \cos\left(\frac{\pi}{2} - u\right) = \sin u$$

$$\tan\left(\frac{\pi}{2} - u\right) = \cot u \qquad \cot\left(\frac{\pi}{2} - u\right) = \tan u$$

$$\sec\left(\frac{\pi}{2} - u\right) = \csc u \qquad \csc\left(\frac{\pi}{2} - u\right) = \sec u$$

Even/Odd Identities

$$\sin(-u) = -\sin u \qquad \sec(-u) = \sec u \qquad \tan(-u) = -\tan u$$

$$\csc(-u) = -\csc u \qquad \cos(-u) = \cos u \qquad \cot(-u) = -\cot u$$

Pythagorean identities are sometimes used in radical form such as

$$\sin u = \pm\sqrt{1 - \cos^2 u} \quad \text{or} \quad \tan u = \pm\sqrt{\sec^2 u - 1}$$

where the sign depends on the choice of u.

Using the Fundamental Identities

One common use of trigonometric identities is to use given values of trigonometric functions to evaluate other trigonometric functions.

EXAMPLE 1   Using Identities to Evaluate a Function

Use the values $\sec u = -\frac{3}{2}$ and $\tan u > 0$ to find the values of all six trigonometric functions.

Solution
Using a reciprocal identity, you have

$$\cos u = \frac{1}{\sec u} = \frac{1}{-3/2} = -\frac{2}{3}.$$

Using a Pythagorean identity, you have

$$\sin^2 u = 1 - \cos^2 u \qquad \text{Pythagorean identity}$$

$$= 1 - \left(-\frac{2}{3}\right)^2 \qquad \text{Substitute } -\frac{2}{3} \text{ for } \cos u.$$

$$= 1 - \frac{4}{9} \qquad \text{Simplify.}$$

$$= \frac{5}{9}. \qquad \text{Simplify.}$$

Because $\sec u < 0$ and $\tan u > 0$, it follows that u lies in Quadrant III. Moreover, because $\sin u$ is negative when u is in Quadrant III, you can choose the negative root and obtain $\sin u = -\sqrt{5}/3$. Now, knowing the values of the sine and cosine, you can find the values of all six trigonometric functions.

$$\sin u = -\frac{\sqrt{5}}{3} \qquad\qquad \csc u = \frac{1}{\sin u} = -\frac{3}{\sqrt{5}}$$

$$\cos u = -\frac{2}{3} \qquad\qquad \sec u = \frac{1}{\cos u} = -\frac{3}{2}$$

$$\tan u = \frac{\sin u}{\cos u} = \frac{-\sqrt{5}/3}{-2/3} = \frac{\sqrt{5}}{2} \qquad \cot u = \frac{1}{\tan u} = \frac{2}{\sqrt{5}}$$

EXAMPLE 2   Simplifying a Trigonometric Expression

Simplify $\sin x \cos^2 x - \sin x$.

Solution
First factor out a common monomial factor and then use a fundamental identity.

$$\sin x \cos^2 x - \sin x = \sin x(\cos^2 x - 1) \qquad \text{Factor out monomial factor.}$$

$$= -\sin x(1 - \cos^2 x) \qquad \text{Distributive Property}$$

$$= -\sin x(\sin^2 x) \qquad \text{Pythagorean identity}$$

$$= -\sin^3 x \qquad \text{Multiply.}$$

STUDY TIP

You can use a graphing utility to check the result of Example 2. To do this, graph

$$y_1 = \sin x \cos^2 x - \sin x$$

and

$$y_2 = -\sin^3 x$$

in the same viewing window, as shown below. The two graphs *appear* to coincide, so the expressions *appear* to be equivalent. Remember that in order to be certain that two expressions are equivalent, you need to show their equivalence algebraically, as in Example 2.

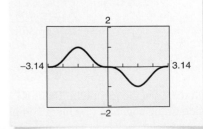

EXAMPLE 3   Verifying Trigonometric Identities

Determine which of the following appears to be an identity.

a. $\cos 3x \overset{?}{=} 4 \cos^3 x - 3 \cos x$ **b.** $\cos 3x \overset{?}{=} \sin\left(3x - \dfrac{\pi}{2}\right)$

Numerical Solution

a. Use the *table* feature of a graphing utility set in *radian* mode to create a table that shows the values of $y_1 = \cos 3x$ and $y_2 = 4 \cos^3 x - 3 \cos x$ for different values of x, as shown in Figure 2.1(a). The values of y_1 and y_2 appear to be identical, so $\cos 3x = 4 \cos^3 x - 3 \cos x$ appears to be an identity.

b. Create a table that shows the values of $y_1 = \cos 3x$ and $y_2 = \sin(3x - \pi/2)$ for different values of x, as shown in Figure 2.1(b). The values of y_1 and y_2 are not identical, so $\cos 3x = \sin(3x - \pi/2)$ is not an identity.

X	Y₁	Y₂
-.5	.07074	.07074
-.25	.73169	.73169
0	1	1
.25	.73169	.73169
.5	.07074	.07074
.75	-.6282	-.6282
1	-.99	-.99

X=1

X	Y₁	Y₂
-.5	.07074	-.0707
-.25	.73169	-.7317
0	1	-1
.25	.73169	-.7317
.5	.07074	-.0707
.75	-.6282	.62817
1	-.99	.98999

X=1

(a)   (b)

Figure 2.1

Graphical Solution

a. Use a graphing utility set in *radian* mode to graph $y_1 = \cos 3x$ and $y_2 = 4 \cos^3 x - 3 \cos x$ in the same viewing window, as shown in Figure 2.2(a). Because the graphs appear to coincide, $\cos 3x = 4 \cos^3 x - 3 \cos x$ appears to be an identity.

b. Graph $y_1 = \cos 3x$ and $y_2 = \sin(3x - \pi/2)$ in the same viewing window, as shown in Figure 2.2(b). Because the graphs do not coincide, $\cos 3x = \sin(3x - \pi/2)$ is not an identity.

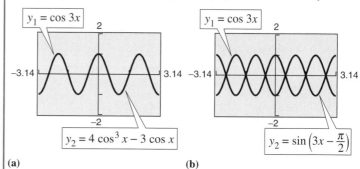

(a)   (b)

Figure 2.2

EXAMPLE 4   Verifying a Trigonometric Identity

Verify the identity $\dfrac{\sin \theta}{1 + \cos \theta} + \dfrac{\cos \theta}{\sin \theta} = \csc \theta.$

Algebraic Solution

$$\frac{\sin \theta}{1 + \cos \theta} + \frac{\cos \theta}{\sin \theta} = \frac{(\sin \theta)(\sin \theta) + (\cos \theta)(1 + \cos \theta)}{(1 + \cos \theta)(\sin \theta)}$$

$$= \frac{\sin^2 \theta + \cos^2 \theta + \cos \theta}{(1 + \cos \theta)(\sin \theta)} \qquad \text{Multiply.}$$

$$= \frac{1 + \cos \theta}{(1 + \cos \theta)(\sin \theta)} \qquad \text{Pythagorean identity}$$

$$= \csc \theta \qquad \text{Divide out common factor and use reciprocal identity.}$$

Notice how the identity in Example 4 is verified. You start with the left side of the equation (the more complicated side) and use the fundamental trigonometric identities to simplify it until you obtain the right side.

Graphical Solution

Use a graphing utility set in *radian* and *dot* modes to graph y_1 and y_2 in the same viewing window, as shown in Figure 2.3. Because the graphs appear to coincide, this equation appears to be an identity.

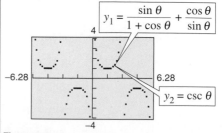

Figure 2.3

When factoring trigonometric expressions, it is helpful to find a polynomial form that fits the expression, as shown in Example 5.

EXAMPLE 5   Factoring Trigonometric Expressions

Factor each expression.

a. $\sec^2 \theta - 1$

b. $4 \tan^2 \theta + \tan \theta - 3$

Solution

a. Here the expression has the polynomial form $u^2 - v^2$ (the difference of two squares), which factors as

$$\sec^2 \theta - 1 = (\sec \theta - 1)(\sec \theta + 1).$$

b. This expression has the polynomial form, $ax^2 + bx + c$, and it factors as

$$4 \tan^2 \theta + \tan \theta - 3 = (4 \tan \theta - 3)(\tan \theta + 1).$$

On occasion, factoring or simplifying can best be done by first rewriting the expression in terms of just *one* trigonometric function or in terms of *sine* or *cosine* *alone*. These strategies are illustrated in Examples 6 and 7.

EXAMPLE 6   Factoring a Trigonometric Expression

Factor $\csc^2 x - \cot x - 3$.

Solution

You can use the identity $\csc^2 x = 1 + \cot^2 x$ to rewrite the expression in terms of the cotangent alone.

$$\csc^2 x - \cot x - 3 = (1 + \cot^2 x) - \cot x - 3 \qquad \text{Pythagorean identity}$$

$$= \cot^2 x - \cot x - 2 \qquad \text{Combine like terms.}$$

$$= (\cot x - 2)(\cot x + 1) \qquad \text{Factor.}$$

EXAMPLE 7   Simplifying a Trigonometric Expression

Simplify $\sin t + \cot t \cos t$.

Solution

Begin by rewriting the expression in terms of sine and cosine.

$$\sin t + \cot t \cos t = \sin t + \left(\frac{\cos t}{\sin t}\right)\cos t \qquad \text{Quotient identity}$$

$$= \frac{\sin^2 t + \cos^2 t}{\sin t} \qquad \text{Add fractions.}$$

$$= \frac{1}{\sin t} \qquad \text{Pythagorean identity}$$

$$= \csc t \qquad \text{Reciprocal identity}$$

The last two examples in this section involve techniques for rewriting expressions into forms that are used in calculus.

EXAMPLE 8 Rewriting a Trigonometric Expression

Rewrite $\dfrac{1}{1 + \sin x}$ so that it is *not* in fractional form.

Solution

From the Pythagorean identity $\cos^2 x = 1 - \sin^2 x = (1 - \sin x)(1 + \sin x)$, you can see that by multiplying both the numerator and the denominator by $(1 - \sin x)$ you will produce a monomial denominator.

$$\frac{1}{1 + \sin x} = \frac{1}{1 + \sin x} \cdot \frac{1 - \sin x}{1 - \sin x} \qquad \text{Multiply numerator and denominator by } (1 - \sin x).$$

$$= \frac{1 - \sin x}{1 - \sin^2 x} \qquad \text{Multiply.}$$

$$= \frac{1 - \sin x}{\cos^2 x} \qquad \text{Pythagorean identity}$$

$$= \frac{1}{\cos^2 x} - \frac{\sin x}{\cos^2 x} \qquad \text{Write as separate fractions.}$$

$$= \frac{1}{\cos^2 x} - \frac{\sin x}{\cos x} \cdot \frac{1}{\cos x} \qquad \text{Write as separate fractions.}$$

$$= \sec^2 x - \tan x \sec x \qquad \text{Reciprocal and quotient identities}$$

EXAMPLE 9 Trigonometric Substitution

Use the substitution $x = 2 \tan \theta$, $0 < \theta < \pi/2$, to express $\sqrt{4 + x^2}$ as a trigonometric function of θ.

Solution

Begin by letting $x = 2 \tan \theta$. Then you can obtain

$$\sqrt{4 + x^2} = \sqrt{4 + (2 \tan \theta)^2} \qquad \text{Substitute } 2 \tan \theta \text{ for } x.$$

$$= \sqrt{4(1 + \tan^2 \theta)} \qquad \text{Distributive Property}$$

$$= \sqrt{4 \sec^2 \theta} \qquad \text{Pythagorean identity}$$

$$= 2 \sec \theta. \qquad \sec \theta > 0 \text{ for } 0 < \theta < \tfrac{\pi}{2}$$

Figure 2.4 shows the right triangle illustration of this substitution. For $0 < \theta < \pi/2$, you have

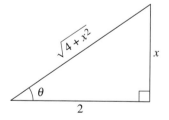

Figure 2.4

opposite $= x$, adjacent $= 2$, and hypotenuse $= \sqrt{4 + x^2}$.

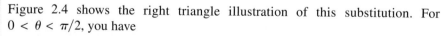

Try using these expressions to obtain the result shown above.

2.1   E x e r c i s e s

In Exercises 1–14, use the given values to evaluate (if possible) the remaining trigonometric functions.

1. $\sin x = \dfrac{\sqrt{3}}{2},\quad \cos x = \dfrac{1}{2}$

2. $\csc \theta = \dfrac{5}{3},\quad \tan \theta = \dfrac{3}{4}$

3. $\sec \theta = \sqrt{2},\quad \sin \theta = -\dfrac{\sqrt{2}}{2}$

4. $\tan x = \dfrac{\sqrt{3}}{3},\quad \cos x = -\dfrac{\sqrt{3}}{2}$

5. $\tan x = \dfrac{7}{24},\quad \sec x = -\dfrac{25}{24}$

6. $\cot \phi = -5,\quad \sin \phi = \dfrac{\sqrt{26}}{26}$

7. $\sec \phi = -1,\quad \sin \phi = 0$

8. $\cos\left(\dfrac{\pi}{2} - x\right) = \dfrac{3}{5},\quad \cos x = \dfrac{4}{5}$

9. $\sin(-x) = -\dfrac{2}{3},\quad \tan x = -\dfrac{2\sqrt{5}}{5}$

10. $\csc x = 5,\quad \cos x > 0$

11. $\tan \theta = 4,\quad \sin \theta < 0$

12. $\sec \theta = -3,\quad \tan \theta < 0$

13. $\sin \theta = -1,\quad \cot \theta = 0$

14. $\tan \theta$ is undefined, $\sin \theta > 0$

In Exercises 15–18, fill in the blanks. (*Note:* $x \to c^{+}$ indicates that x approaches c from the right, and $x \to c^{-}$ indicates that x approaches c from the left.)

15. As $x \to \dfrac{\pi}{2}^{-}$, $\sin x \to$ ____ and $\csc x \to$ ____ .

16. As $x \to 0^{+}$, $\cos x \to$ ____ and $\sec x \to$ ____ .

17. As $x \to \dfrac{\pi}{2}^{-}$, $\tan x \to$ ____ and $\cot x \to$ ____ .

18. As $x \to \pi^{+}$, $\sin x \to$ ____ and $\csc x \to$ ____ .

In Exercises 19–24, match the trigonometric expression with one of the following.

(a) -1 (b) $\cos x$ (c) $\cot x$

(d) 1 (e) $-\tan x$ (f) $\sin x$

19. $\csc x \sin x$

20. $\tan x \cos x$

21. $\tan^2 x - \sec^2 x$

22. $(1 - \sin^2 x)(\sec x)$

23. $\dfrac{\sin(-x)}{\cos(-x)}$

24. $\dfrac{\sin[(\pi/2) - x]}{\cos[(\pi/2) - x]}$

In Exercises 25–30, match the trigonometric expression with one of the following.

(a) $\csc x$ (b) $\cot x$ (c) $\cos^2 x$

(d) $\sin x \tan x$ (e) $\sec^2 x$ (f) $\sec^2 x + \tan^2 x$

25. $\cos x \csc x$

26. $\sin^2 x(\csc^2 x - 1)$

27. $\sec^4 x - \tan^4 x$

28. $\cot x \sec x$

29. $\dfrac{\sec^2 x - 1}{\sin^2 x}$

30. $\dfrac{\cos^2[(\pi/2) - x]}{\cos x}$

In Exercises 31–44, use the fundamental identities to simplify the expression. Use the *table* feature of a graphing utility to check your result numerically.

31. $\cot x \sin x$

32. $\cos \beta \tan \beta$

33. $\sin \phi(\csc \phi - \sin \phi)$

34. $\sec^2 x(1 - \sin^2 x)$

35. $\dfrac{\cot x}{\csc x}$

36. $\dfrac{\sec \theta}{\csc \theta}$

37. $\sec \alpha \cdot \dfrac{\sin \alpha}{\tan \alpha}$

38. $\dfrac{1}{\tan^2 x + 1}$

39. $\dfrac{\sin(-x)}{\cos x}$

40. $\dfrac{\tan^2 \theta}{\sec^2 \theta}$

41. $\sin\left(\dfrac{\pi}{2} - x\right)\csc x$

42. $\cot\left(\dfrac{\pi}{2} - x\right)\cos x$

43. $\dfrac{\cos^2 y}{1 - \sin y}$

44. $\cos t(1 + \tan^2 t)$

In Exercises 45–52, verify the identity algebraically. Then use a graphing utility to check your result graphically.

45. $\csc \theta \tan \theta = \sec \theta$

46. $\cos \theta \sec \theta - \cos^2 \theta = \sin^2 \theta$

47. $\dfrac{\csc \theta}{\sec \theta} + \dfrac{\cos \theta}{\sin \theta} = 2 \cot \theta$

48. $\dfrac{\sec^2 \theta - \tan^2 \theta + \tan \theta}{\sec \theta} = \cos \theta + \sin \theta$

The *Interactive* CD-ROM and *Internet* versions of this text contain step-by-step solutions to all odd-numbered Section and Review Exercises. They also provide Tutorial Exercises, which link to Guided Examples for additional help.

49. $1 - \dfrac{\sin^2 \theta}{1 - \cos \theta} = -\cos \theta$

50. $\dfrac{\tan \theta}{1 + \sec \theta} + \dfrac{1 + \sec \theta}{\tan \theta} = 2 \csc \theta$

51. $\dfrac{\cot(-\theta)}{\csc \theta} = -\cos \theta$ **52.** $\dfrac{\csc\left(\dfrac{\pi}{2} - \theta\right)}{\tan(-\theta)} = -\csc \theta$

In Exercises 53–60, verify the identity algebraically. Then use the *table* feature of a graphing utility to check your result numerically.

53. $\sin \theta + \cos \theta \cot \theta = \csc \theta$

54. $(\sec \theta - \tan \theta)(\csc \theta + 1) = \cot \theta$

55. $\dfrac{\cos \theta}{1 - \sin \theta} = \sec \theta + \tan \theta$

56. $\dfrac{1 + \csc \theta}{\cot \theta + \cos \theta} = \sec \theta$

57. $\dfrac{\sin \theta}{\csc \theta} + \dfrac{\cos \theta}{\sec \theta} = 1$

58. $\dfrac{1 + \csc \theta}{\sec \theta} - \cot \theta = \cos \theta$

59. $\dfrac{1 + \cos \theta}{\sin \theta} + \dfrac{\sin \theta}{1 + \cos \theta} = 2 \csc \theta$

60. $\dfrac{\sin \theta + \cos \theta}{\sin \theta} - \dfrac{\cos \theta - \sin \theta}{\cos \theta} = \sec \theta \csc \theta$

In Exercises 61–68, factor the expression and use the fundamental identities to simplify. Use a graphing utility to check your result graphically.

61. $\cot^2 x - \cot^2 x \cos^2 x$

62. $\sec^2 x \tan^2 x + \sec^2 x$

63. $\sin^2 x \sec^2 x - \sin^2 x$

64. $\dfrac{\csc^2 x - 1}{\csc x - 1}$

65. $\tan^4 x + 2 \tan^2 x + 1$

66. $1 - 2 \sin^2 x + \sin^4 x$

67. $\sin^4 x - \cos^4 x$

68. $\sec^3 x - \sec^2 x - \sec x + 1$

In Exercises 69–72, perform the multiplication and use the fundamental identities to simplify.

69. $(\sin x + \cos x)^2$

70. $(\cot x + \csc x)(\cot x - \csc x)$

71. $(\sec x + 1)(\sec x - 1)$

72. $(3 - 3 \sin x)(3 + 3 \sin x)$

In Exercises 73–76, perform the addition or subtraction and use the fundamental identities to simplify.

73. $\dfrac{1}{1 + \cos x} + \dfrac{1}{1 - \cos x}$

74. $\dfrac{1}{\sec x + 1} - \dfrac{1}{\sec x - 1}$

75. $\dfrac{\cos x}{1 + \sin x} + \dfrac{1 + \sin x}{\cos x}$

76. $\tan x - \dfrac{\sec^2 x}{\tan x}$

In Exercises 77–80, rewrite the expression so that it is *not* in fractional form.

77. $\dfrac{\sin^2 y}{1 - \cos y}$ **78.** $\dfrac{5}{\tan x + \sec x}$

79. $\dfrac{3}{\sec x - \tan x}$ **80.** $\dfrac{\tan^2 x}{\csc x + 1}$

Numerical and Graphical Analysis   **In Exercises 81–84, use a graphing utility to complete the table and graph the functions. Make a conjecture about y_1 and y_2.**

x	0.2	0.4	0.6	0.8	1.0	1.2	1.4
y_1							
y_2							

81. $y_1 = \cos\left(\dfrac{\pi}{2} - x\right), \quad y_2 = \sin x$

82. $y_1 = \cos x + \sin x \tan x, \quad y_2 = \sec x$

83. $y_1 = \dfrac{\cos x}{1 - \sin x}, \quad y_2 = \dfrac{1 + \sin x}{\cos x}$

84. $y_1 = \sec^4 x - \sec^2 x, \quad y_2 = \tan^2 x + \tan^4 x$

In Exercises 85–88, use a graphing utility to determine which of the six trigonometric functions is equal to the expression.

85. $\cos x \cot x + \sin x$ **86.** $\sin x(\cot x + \tan x)$

87. $\sec x - \dfrac{\cos x}{1 + \sin x}$

88. $\dfrac{1}{2}\left(\dfrac{1 + \sin \theta}{\cos \theta} + \dfrac{\cos \theta}{1 + \sin \theta}\right)$

In Exercises 89–94, use the trigonometric substitution to write the algebraic expression as a trigonometric function of θ, where $0 < \theta < \pi/2$.

89. $\sqrt{25 - x^2}, \quad x = 5 \sin \theta$

90. $\sqrt{16 - 4x^2}, \quad x = 2 \sin \theta$

91. $\sqrt{x^2 - 9}, \quad x = 3 \sec \theta$

92. $\sqrt{x^2 - 4}, \quad x = 2 \sec \theta$

93. $\sqrt{x^2 + 25}, \quad x = 5 \tan \theta$

94. $\sqrt{x^2 + 100}, \quad x = 10 \tan \theta$

In Exercises 95–98, use a graphing utility to solve the equation for θ, where $0 \le \theta < 2\pi$.

95. $\sin \theta = \sqrt{1 - \cos^2 \theta}$ **96.** $\cos \theta = -\sqrt{1 - \sin^2 \theta}$

97. $\sec \theta = \sqrt{1 + \tan^2 \theta}$ **98.** $\tan \theta = \sqrt{\sec^2 \theta - 1}$

In Exercises 99–102, use the *table* feature of a graphing utility to demonstrate the identity for the given values of θ.

99. $\csc^2 \theta - \cot^2 \theta = 1$, (a) $\theta = 132°$, (b) $\theta = \dfrac{2\pi}{7}$

100. $\tan^2 \theta + 1 = \sec^2 \theta$, (a) $\theta = 346°$, (b) $\theta = 3.1$

101. $\cos\left(\dfrac{\pi}{2} - \theta\right) = \sin \theta$, (a) $\theta = 80°$, (b) $\theta = 0.8$

102. $\sin(-\theta) = -\sin \theta$, (a) $\theta = 250°$, (b) $\theta = \tfrac{1}{2}$

103. *Rate of Change* The rate of change of the function $f(x) = -\csc x - \sin x$ is given by the expression $\csc x \cot x - \cos x$. Show that this expression can also be written as $\cos x \cot^2 x$.

Synthesis

True or False? In Exercises 104–107, determine whether the statement is true or false. Justify your answer.

104. $\dfrac{\sin k\theta}{\cos k\theta} = \tan \theta, \quad k$ is a constant.

105. $\dfrac{1}{5 \cos \theta} = 5 \sec \theta$

106. $\sin \theta \csc \theta = 1$ **107.** $\sin \theta \csc \phi = 1$

108. Express each of the other trigonometric functions of θ in terms of $\sin \theta$.

109. Express each of the other trigonometric functions of θ in terms of $\cos \theta$.

Review

In Exercises 110–115, find the reference angle θ' and sketch θ and θ' in standard position.

110. $\theta = 254°$ **111.** $\theta = 341°$

112. $\theta = -178°$ **113.** $\theta = -212°$

114. $\theta = \dfrac{13\pi}{15}$ **115.** $\theta = \dfrac{35\pi}{6}$

In Exercises 116–119, sketch the graph of the function.

116. $f(x) = \dfrac{1}{2} \sin \pi x$ **117.** $f(x) = -2 \tan \dfrac{\pi x}{2}$

118. $f(x) = \dfrac{1}{2} \cot\left(x + \dfrac{\pi}{4}\right)$

119. $f(x) = \dfrac{3}{2} \cos(x - \pi) + 3$

In Exercises 120–123, solve the right triangle shown in the figure. (Round your answers to two decimal places.)

120. $B = 80°, \ a = 16$ **121.** $A = 28°, \ c = 20$

122. $a = 14, \ b = 8$ **123.** $b = 6.2, \ c = 12.54$

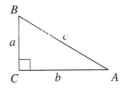

2.2 Verifying Trigonometric Identities

Introduction

In this section, you will study techniques for verifying trigonometric identities. In the next section, you will study techniques for solving trigonometric equations. The key to both verifying identities *and* solving equations is the ability to use the fundamental identities and the rules of algebra to rewrite trigonometric expressions.

Remember that a *conditional equation* is an equation that is true for only some of the values in its domain. For example, the conditional equation

$$\sin x = 0 \qquad \text{Conditional equation}$$

is true only for $x = n\pi$, where n is an integer. When you find these values, you are *solving* the equation.

On the other hand, an equation that is true for all real values in the domain of the variable is an *identity*. For example, the familiar equation

$$\sin^2 x = 1 - \cos^2 x \qquad \text{Identity}$$

is true for all real numbers x. So, it is an identity.

Verifying Trigonometric Identities

Although there are similarities, proving that a trigonometric equation is an identity is quite different from solving an equation. There is no well-defined set of rules to follow in verifying trigonometric identities, and the process is best learned by practice.

Guidelines for Verifying Trigonometric Identities

1. Work with one side of the equation at a time. It is often better to work with the more complicated side first.

2. Look for opportunities to factor an expression, add fractions, square a binomial, or create a monomial denominator.

3. Look for opportunities to use the fundamental identities. Note which functions are in the final expression you want. Sines and cosines pair up well, as do secants and tangents, and cosecants and cotangents.

4. If the preceding guidelines do not help, try converting all terms to sines and cosines.

5. Try something! Even making an attempt that leads to a dead end gives insight.

Nancy Dudley/Stock Boston

A computer animation of this concept appears in the *Interactive* CD-ROM and *Internet* versions of this text.

EXAMPLE 1   Verifying a Trigonometric Identity

Verify the identity $\dfrac{\sec^2 \theta - 1}{\sec^2 \theta} = \sin^2 \theta$.

Solution

Because the left side is more complicated, start with it.

$$\frac{\sec^2 \theta - 1}{\sec^2 \theta} = \frac{(\tan^2 \theta + 1) - 1}{\sec^2 \theta} \qquad \text{Pythagorean identity}$$

$$= \frac{\tan^2 \theta}{\sec^2 \theta} \qquad \text{Simplify.}$$

$$= \tan^2 \theta (\cos^2 \theta) \qquad \text{Reciprocal identity}$$

$$= \frac{\sin^2 \theta}{\cos^2 \theta}(\cos^2 \theta) \qquad \text{Quotient identity}$$

$$= \sin^2 \theta \qquad \text{Simplify.}$$

The *Interactive* CD-ROM and *Internet* versions of this text offer a built-in graphing calculator, which can be used with the Examples, Explorations, and Exercises.

Here is another way to verify the identity in Example 1.

$$\frac{\sec^2 \theta - 1}{\sec^2 \theta} = \frac{\sec^2 \theta}{\sec^2 \theta} - \frac{1}{\sec^2 \theta}$$

$$= 1 - \cos^2 \theta$$

$$= \sin^2 \theta$$

There can be more than one way to verify an identity. Your method may differ from that used by your instructor or fellow students. This is a good chance to be creative and establish your own style, but try to be as efficient as possible.

EXAMPLE 2   Combining Fractions Before Using Identities

Verify the identity $\dfrac{1}{1 - \sin \alpha} + \dfrac{1}{1 + \sin \alpha} = 2 \sec^2 \alpha$.

Algebraic Solution

$$\frac{1}{1 - \sin \alpha} + \frac{1}{1 + \sin \alpha} = \frac{1 + \sin \alpha + 1 - \sin \alpha}{(1 - \sin \alpha)(1 + \sin \alpha)} \qquad \text{Add fractions.}$$

$$= \frac{2}{1 - \sin^2 \alpha} \qquad \text{Simplify.}$$

$$= \frac{2}{\cos^2 \alpha} \qquad \text{Pythagorean identity}$$

$$= 2 \sec^2 \alpha \qquad \text{Reciprocal identity}$$

Numerical Solution

Use the *table* feature of a graphing utility set in *radian* mode to create a table that shows the values of $y_1 = 1/(1 - \sin x) + 1/(1 + \sin x)$ and $y_2 = 2/\cos^2 x$ for different values of x, as shown in Figure 2.5. From the table, you can see that the values appear to be identical, so $1/(1 - \sin x) + 1/(1 + \sin x) = 2 \sec^2$ appears to be an identity.

X	Y1	Y2
-.5	2.5969	2.5969
-.25	2.1304	2.1304
0	2	2
.25	2.1304	2.1304
.5	2.5969	2.5969
.75	3.7357	3.7357
1	6.851	6.851
X=-.5		

Figure 2.5

EXAMPLE 3   Verifying a Trigonometric Identity

Verify the identity $(\tan^2 x + 1)(\cos^2 x - 1) = -\tan^2 x$.

Solution

By applying identities before multiplying, you obtain the following.

$$(\tan^2 x + 1)(\cos^2 x - 1) = (\sec^2 x)(-\sin^2 x) \qquad \text{Pythagorean identities}$$

$$= -\frac{\sin^2 x}{\cos^2 x} \qquad \text{Reciprocal identity}$$

$$= -\left(\frac{\sin x}{\cos x}\right)^2 \qquad \text{Rule of exponents}$$

$$= -\tan^2 x \qquad \text{Quotient identity}$$

You can use a graphing utility to verify your results by graphing both sides of the original identity in the same viewing window, as shown in Figure 2.6. The graphs appear to coincide.

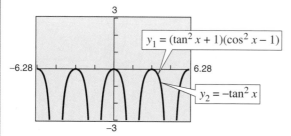

$y_1 = (\tan^2 x + 1)(\cos^2 x - 1)$

$y_2 = -\tan^2 x$

Figure 2.6

EXAMPLE 4   Converting to Sines and Cosines

Verify the identity $\tan x + \cot x = \sec x \csc x$.

Solution

In this case there appear to be no fractions to add, no products to find, and no opportunity to use one of the Pythagorean identities. So, try converting the left side into sines and cosines to see what happens.

$$\tan x + \cot x = \frac{\sin x}{\cos x} + \frac{\cos x}{\sin x} \qquad \text{Quotient identities}$$

$$= \frac{\sin^2 x + \cos^2 x}{\cos x \sin x} \qquad \text{Add fractions.}$$

$$= \frac{1}{\cos x \sin x} \qquad \text{Pythagorean identity}$$

$$= \frac{1}{\cos x} \cdot \frac{1}{\sin x} \qquad \text{Product of fractions}$$

$$= \sec x \csc x \qquad \text{Reciprocal identities}$$

Recall from algebra that *rationalizing the denominator* is, on occasion, a powerful simplification technique. A related form of this technique works for simplifying trigonometric expressions as well.

EXAMPLE 5   Verifying a Trigonometric Identity

Verify the identity $\sec x + \tan x = \dfrac{\cos x}{1 - \sin x}$.

Algebraic Solution

Work with the *right* side because you can create a monomial denominator by multiplying the numerator and denominator by $(1 + \sin x)$.

$$\frac{\cos x}{1 - \sin x} = \frac{\cos x}{1 - \sin x} \left(\frac{1 + \sin x}{1 + \sin x} \right) \qquad \begin{array}{l}\text{Multiply numerator} \\ \text{and denominator by} \\ (1 + \sin x).\end{array}$$

$$= \frac{\cos x + \cos x \sin x}{1 - \sin^2 x} \qquad \text{Multiply.}$$

$$= \frac{\cos x + \cos x \sin x}{\cos^2 x} \qquad \text{Pythagorean identity}$$

$$= \frac{\cos x}{\cos^2 x} + \frac{\cos x \sin x}{\cos^2 x} \qquad \text{Separate fractions.}$$

$$- \frac{1}{\cos x} + \frac{\sin x}{\cos x} \qquad \text{Simplify.}$$

$$= \sec x + \tan x \qquad \text{Identities}$$

Graphical Solution

Use a graphing utility set in *radian* and *dot* modes to graph $y_1 = \sec x + \tan x = 1/\cos x + \tan x$ and $y_2 = \cos x/(1 - \sin x)$ in the same viewing window, as shown in Figure 2.7. Because the graphs appear to coincide, $\sec x + \tan x = \cos x/(1 - \sin x)$ appears to be an identity.

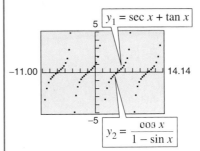

Figure 2.7

In Examples 1 through 5, you have been verifying trigonometric identities by working with one side of the equation and converting to the form given on the other side. On occasion it is practical to work with each side *separately* to obtain one common form equivalent to both sides. This is illustrated in Example 6.

EXAMPLE 6   Working with Each Side Separately

Verify the identity $\dfrac{\cot^2 \theta}{1 + \csc \theta} = \dfrac{1 - \sin \theta}{\sin \theta}$.

Solution

Working with the left side, you have

$$\frac{\cot^2 \theta}{1 + \csc \theta} = \frac{\csc^2 \theta - 1}{1 + \csc \theta} \qquad \text{Pythagorean identity}$$

$$= \frac{(\csc \theta - 1)(\csc \theta + 1)}{1 + \csc \theta} \qquad \text{Factor.}$$

$$= \csc \theta - 1. \qquad \text{Simplify.}$$

Now, simplifying the right side, you have

$$\frac{1 - \sin \theta}{\sin \theta} = \frac{1}{\sin \theta} - \frac{\sin \theta}{\sin \theta} \qquad \text{Separate fractions.}$$

$$= \csc \theta - 1. \qquad \text{Reciprocal identity}$$

The identity is verified because both sides are equal to $\csc \theta - 1$.

In Example 7, powers of trigonometric functions are rewritten as more complicated sums of products of trigonometric functions. This is a common procedure used in calculus.

EXAMPLE 7   Three Examples from Calculus

Verify each identity.

a. $\tan^4 x = \tan^2 x \sec^2 x - \tan^2 x$

b. $\sin^3 x \cos^4 x = (\cos^4 x - \cos^6 x)\sin x$

c. $\sec^4 x \, \tan^2 x = (\tan^2 x + \tan^4 x)\sec^2 x$

Solution

a. $\tan^4 x = (\tan^2 x)(\tan^2 x)$ Write as separate factors.

$\qquad = \tan^2 x(\sec^2 x - 1)$ Pythagorean identity

$\qquad = \tan^2 x \sec^2 x - \tan^2 x$ Multiply.

b. $\sin^3 x \cos^4 x = \sin^2 x \cos^4 x \sin x$ Write as separate factors.

$\qquad = (1 - \cos^2 x)\cos^4 x \sin x$ Pythagorean identity

$\qquad = (\cos^4 x - \cos^6 x)\sin x$ Multiply.

c. $\sec^4 x \, \tan^2 x = \sec^2 x \tan^2 x \sec^2 x$ Write as separate facors.

$\qquad = (1 + \tan^2 x)\tan^2 x \sec^2 x$ Pythagorean identity

$\qquad = (\tan^2 x + \tan^4 x)\sec^2 x$ Multiply.

Writing About Math   *Error Analysis*

Suppose you are tutoring a student in trigonometry. One of the homework problems your student encounters asks whether the following statement is an identity.

$$\tan^2 x \sin^2 x \stackrel{?}{=} \frac{5}{6} \tan^2 x$$

Your student does not attempt to verify the equivalence algebraically, but mistakenly uses only a graphical approach. Using window settings of $-3\pi \leq x \leq 3\pi$ with an x-scale of $\pi/2$, and $-20 \leq y \leq 20$ with a y-scale of 1, your student graphs both sides of the expression on a graphing utility and concludes that the statement is an identity.

Write a short paragraph explaining what is wrong with your student's reasoning.

2.2 Exercises

In Exercises 1–10, verify the identity.

1. $\sin t \csc t = 1$

2. $\tan y \cot y = 1$

3. $\dfrac{\csc^2 x}{\cot x} = \csc x \sec x$

4. $\cot^2 y(\sec^2 y - 1) = 1$

5. $\cos^2 \beta - \sin^2 \beta = 1 - 2\sin^2 \beta$

6. $\cos^2 \beta - \sin^2 \beta = 2\cos^2 \beta - 1$

7. $\tan^2 \theta + 6 = \sec^2 \theta + 5$

8. $2 - \csc^2 z = 1 - \cot^2 z$

9. $\cos x + \sin x \tan x = \sec x$

10. $\dfrac{\cot^3 t}{\csc t} = \cos t(\csc^2 t - 1)$

Numerical, Graphical, and Algebraic Analysis **In Exercises 11–18, use a graphing utility to complete the table and graph the functions. Use both as evidence that $y_1 = y_2$. Then verify the identity algebraically.**

x	0.2	0.4	0.6	0.8	1.0	1.2	1.4
y_1							
y_2							

11. $y_1 = \dfrac{1}{\sec x \tan x}, \quad y_2 = \csc x - \sin x$

12. $y_1 = \dfrac{\csc x - 1}{1 - \sin x}, \quad y_2 = \csc x$

13. $y_1 = \csc x - \sin x, \quad y_2 = \cos x \cot x$

14. $y_1 = \sec x - \cos x, \quad y_2 = \sin x \tan x$

15. $y_1 = \sin x + \cos x \cot x, \quad y_2 = \csc x$

16. $y_1 = \dfrac{\sec x + \tan x}{\sec x - \tan x}, \quad y_2 = (\sec x + \tan x)^2$

17. $y_1 = \dfrac{1}{\tan x} + \dfrac{1}{\cot x}, \quad y_2 = \tan x + \cot x$

18. $y_1 = \dfrac{1}{\cos x} - \dfrac{1}{\sec x}, \quad y_2 = \sec x - \cos x$

In Exercises 19 and 20, find the error.

19. $(1 + \tan x)[1 + \cot(-x)]$

$= (1 + \tan x)(1 + \cot x)$

$= 1 + \cot x + \tan x + \tan x \cot x$

$= 1 + \cot x + \tan x + 1$

$= 2 + \cot x + \tan x$

20. $\dfrac{1 + \sec(-\theta)}{\sin(-\theta) + \tan(-\theta)} = \dfrac{1 - \sec \theta}{\sin \theta - \tan \theta}$

$= \dfrac{1 - \sec \theta}{(\sin \theta)\left[1 - \left(\dfrac{1}{\cos \theta}\right)\right]}$

$= \dfrac{1 - \sec \theta}{\sin \theta(1 - \sec \theta)}$

$= \dfrac{1}{\sin \theta}$

$= \csc \theta$

In Exercises 21 and 22, fill in the missing step(s).

21. $\sec^4 x - 2\sec^2 x + 1 = (\sec^2 x - 1)^2$

$=$ ▨

$= \tan^4 x$

22. $\dfrac{\tan x - \cot x}{\tan x + \cot x} = \dfrac{\dfrac{\sin x}{\cos x} - \dfrac{\cos x}{\sin x}}{\dfrac{\sin x}{\cos x} + \dfrac{\cos x}{\sin x}}$

$=$ ▨

$= \dfrac{\sin^2 x - \cos^2 x}{1}$

$= \sin^2 x - \cos^2 x$

$=$ ▨

$= 1 - 2\cos^2 x$

In Exercises 23–32, verify the identity.

23. $\sin^{1/2} x \cos x - \sin^{5/2} x \cos x = \cos^3 x \sqrt{\sin x}$

24. $\sec^6 x(\sec x \tan x) - \sec^4 x(\sec x \tan x) = \sec^5 x \tan^3 x$

25. $\tan\left(\dfrac{\pi}{2} - x\right)\sec x = \csc x$

26. $\dfrac{\sin[(\pi/2) - x]}{\cos[(\pi/2) - x]} = \cot x$

27. $\dfrac{\sec(-x)}{\csc(-x)} = -\tan x$

28. $(1 + \sin y)[1 + \sin(-y)] = \cos^2 y$

29. $\dfrac{\cos(-\theta)}{1 + \sin(-\theta)} = \sec\theta + \tan\theta$

30. $\dfrac{1 + \csc(-\theta)}{\cos(-\theta) + \cot(-\theta)} = \sec\theta$

31. $\dfrac{\sin x \cos y + \cos x \sin y}{\cos x \cos y - \sin x \sin y} = \dfrac{\tan x + \tan y}{1 - \tan x \tan y}$

32. $\dfrac{\tan x + \tan y}{1 - \tan x \tan y} = \dfrac{\cot x + \cot y}{\cot x \cot y - 1}$

In Exercises 33–40, verify the identity algebraically, and use the *table* feature of a graphing utility to confirm it numerically.

33. $\dfrac{\tan x + \cot y}{\tan x \cot y} = \tan y + \cot x$

34. $\dfrac{\cos x - \cos y}{\sin x + \sin y} + \dfrac{\sin x - \sin y}{\cos x + \cos y} = 0$

35. $\sqrt{\dfrac{1 + \sin\theta}{1 - \sin\theta}} = \dfrac{1 + \sin\theta}{|\cos\theta|}$

36. $\sqrt{\dfrac{1 - \cos\theta}{1 + \cos\theta}} = \dfrac{1 - \cos\theta}{|\sin\theta|}$

37. $\cos^2 x + \cos^2\left(\dfrac{\pi}{2} - x\right) = 1$

38. $\sec^2 y - \cot^2\left(\dfrac{\pi}{2} - y\right) = 1$

39. $\sec x \sin\left(\dfrac{\pi}{2} - x\right) = 1$

40. $\csc^2\left(\dfrac{\pi}{2} - x\right) - 1 = \tan^2 x$

In Exercises 41–52, verify the identity algebraically, and use a graphing utility to confirm it graphically.

41. $2\sec^2 x - 2\sec^2 x \sin^2 x - \sin^2 x - \cos^2 x = 1$

42. $\csc x(\csc x - \sin x) + \dfrac{\sin x - \cos x}{\sin x} + \cot x = \csc^2 x$

43. $2 + \cos^2 x - 3\cos^4 x = \sin^2 x(2 + 3\cos^2 x)$

44. $4\tan^4 x + \tan^2 x - 3 = \sec^2 x(4\tan^2 x - 3)$

45. $\csc^4 x - 2\csc^2 x + 1 = \cot^4 x$

46. $\sin x(1 - 2\cos^2 x + \cos^4 x) = \sin^5 x$

47. $\sec^4\theta - \tan^4\theta = 1 + 2\tan^2\theta$

48. $\csc^4\theta - \cot^4\theta = 2\csc^2\theta - 1$

49. $\dfrac{\sin\beta}{1 - \cos\beta} = \dfrac{1 + \cos\beta}{\sin\beta}$

50. $\dfrac{\cot\alpha}{\csc\alpha - 1} = \dfrac{\csc\alpha + 1}{\cot\alpha}$

51. $\dfrac{\tan^3\alpha - 1}{\tan\alpha - 1} = \tan^2\alpha + \tan\alpha + 1$

52. $\dfrac{\sin^3\beta + \cos^3\beta}{\sin\beta + \cos\beta} = 1 - \sin\beta\cos\beta$

Conjecture **In Exercises 53–56, use a graphing utility to graph the trigonometric function. Use the graph to make a conjecture about a simplification of the expression. Verify the resulting identity algebraically.**

53. $y = \dfrac{1}{\cot x + 1} + \dfrac{1}{\tan x + 1}$

54. $y = \dfrac{\cos x}{1 - \tan x} + \dfrac{\sin x \cos x}{\sin x - \cos x}$

55. $y = \dfrac{1}{\sin x} - \dfrac{\cos^2 x}{\sin x}$

56. $y = \sin t + \dfrac{\cot^2 t}{\csc t}$

In Exercises 57–60, use the cofunction identities to evaluate the expression without using a calculator.

57. $\sin^2 25° + \sin^2 65°$

58. $\cos^2 14° + \cos^2 76°$

59. $\cos^2 20° + \cos^2 52° + \cos^2 38° + \cos^2 70°$

60. $\sin^2 12° + \sin^2 40° + \sin^2 50° + \sin^2 78°$

🔴 **In Exercises 61–64, the powers of trigonometric functions are rewritten to be useful in calculus. Verify the identity.**

61. $\tan^5 x = \tan^3 x \sec^2 x - \tan^3 x$

62. $\tan^3 x \sec^3 x = \sec^4 x(\sec x \tan x) - \sec^2 x(\sec x \tan x)$

63. $\cos^3 x \sin^2 x = (\sin^2 x - \sin^4 x)\cos x$

64. $\sin^4 x + \cos^4 x = 1 - 2\cos^2 x + 2\cos^4 x$

65. *Friction* The forces acting on an object weighing W units on an inclined plane positioned at an angle of θ with the horizontal are modeled by

$$\mu W \cos \theta = W \sin \theta$$

where μ is the coefficient of friction. Solve the equation for μ and simplify the result.

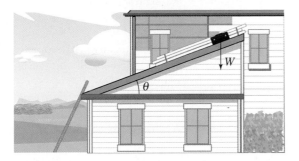

66. *Shadow of a Sundial* The length s of a shadow cast by a vertical *gnomon* (column or shaft on a sundial) of height h when the angle of the sun above the horizon is θ can be modeled by the equation

$$s = \frac{h \sin(90° - \theta)}{\sin \theta}.$$

Show that the equation is equivalent to $s = h \cot \theta$.

67. *Rate of Change* The rate of change of the function $f(x) = \sin x + \csc x$ is given by $\cos x - \csc x \cot x$. Show that the expression for the rate of change can also be given by $-\cos x \cot^2 x$.

Synthesis

True or False? **In Exercises 68–71, determine whether the statement is true or false. Justify your answer.**

68. There can be more than one way to verify a trigonometric identity.

69. Of the six trigonometric functions, two are even.

70. $\tan^2(x) = \tan(x^2)$

71. $\sin(x^2) = \sin^2(x)$

Think About It **In Exercises 72–75, explain why the equation is *not* an identity and find one value of the variable for which the equation is not true.**

72. $\sqrt{\tan^2 x} = \tan x$ **73.** $\sin \theta = \sqrt{1 - \cos^2 \theta}$

74. $\tan \theta = \sqrt{\sec^2 \theta - 1}$

75. $\sqrt{\sin^2 x + \cos^2 x} = \sin x + \cos x$

76. Verify that for all integers n

$$\cos\left[\frac{(2n + 1)\pi}{2}\right] = 0.$$

77. Verify that for all integers n

$$\sin\left[\frac{(12n + 1)\pi}{6}\right] = \frac{1}{2}.$$

Review

In Exercises 78–83, solve the equation by factoring.

78. $x^2 - 5x - 14 = 0$ **79.** $x^2 + 5x - 36 = 0$

80. $x^2 - 19x + 88 = 0$ **81.** $2x^2 - 7x - 15 = 0$

82. $36x^2 - 16 = 0$ **83.** $9x^2 - 1 = 0$

In Exercises 84–91, use the Quadratic Formula to solve the quadratic equation.

84. $x^2 + 4x - 10 = 0$ **85.** $x^2 - 8x + 4 = 0$

86. $3x^2 + 7x + 3 = 0$ **87.** $2x^2 - 8x + 5 = 0$

88. $6x^2 + 20x + 8 = 0$ **89.** $9x^2 + 4x - 12 = 0$

90. $-4x^2 - 7x + 18 = 0$ **91.** $11x^2 + 24x + 6 = 0$

In Exercises 92 and 93, find the measure of the angle in radians.

92.   **93.**

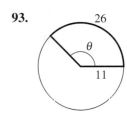

In Exercises 94–97, state the quadrant in which θ lies.

94. $\csc \theta > 0$ and $\tan \theta < 0$

95. $\cot \theta > 0$ and $\cos \theta < 0$

96. $\sec \theta > 0$ and $\sin \theta < 0$

97. $\cot \theta > 0$ and $\sec \theta < 0$

2.3 Solving Trigonometric Equations

Introduction

To solve a trigonometric equation, use standard algebraic techniques such as collecting like terms and factoring. Your preliminary goal is to isolate the trigonometric function involved in the equation.

EXAMPLE 1 Solving a Trigonometric Equation

$2 \sin x - 1 = 0$	Original equation
$2 \sin x = 1$	Add 1 to each side.
$\sin x = \frac{1}{2}$	Divide each side by 2.

To solve for x, note that the equation $\sin x = \frac{1}{2}$ has solutions $x = \pi/6$ and $x = 5\pi/6$ in the interval $[0, 2\pi)$. Moreover, because $\sin x$ has a period of 2π, there are infinitely many other solutions, which can be written as

$$x = \frac{\pi}{6} + 2n\pi \quad \text{and} \quad x = \frac{5\pi}{6} + 2n\pi \qquad \text{General solution}$$

where n is an integer, as shown in Figure 2.8.

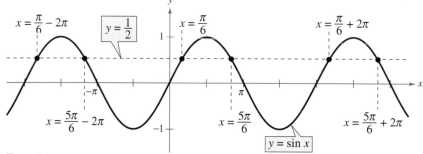

Figure 2.8

Another way to see that the equation $\sin x = \frac{1}{2}$ has infinitely many solutions is indicated in Figure 2.9. Any angles that are coterminal with $\pi/6$ or $5\pi/6$ are also solutions of the equation.

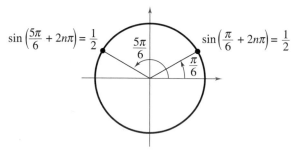

Figure 2.9

What You Should Learn:

- How to use standard algebraic techniques to solve trigonometric equations
- How to solve trigonometric equations of quadratic type
- How to solve trigonometric equations involving multiple angles
- How to use inverse trigonometric functions to solve trigonometric equations

Why You Should Learn It:

You can use trigonometric equations to solve a variety of real-life problems. For instance, Exercise 75 on page 242 shows you how solving a trigonometric equation can help answer questions about the position of the sun in Cheyenne, Wyoming.

EXAMPLE 2 Collecting Like Terms

Find all solutions of $\sin x + \sqrt{2} = -\sin x$ in the interval $[0, 2\pi)$.

Algebraic Solution

Rewrite the equation so that $\sin x$ is isolated on one side of the equation.

$$\sin x + \sqrt{2} = -\sin x \qquad \text{Write original equation.}$$

$$\sin x + \sin x = -\sqrt{2} \qquad \begin{array}{l}\text{Add } \sin x \text{ and subtract}\\ \sqrt{2} \text{ from each side.}\end{array}$$

$$2\sin x = -\sqrt{2} \qquad \text{Combine like terms.}$$

$$\sin x = -\frac{\sqrt{2}}{2} \qquad \text{Divide each side by 2.}$$

The solutions in the interval $[0, 2\pi)$ are

$$x = \frac{5\pi}{4} \qquad \text{and} \qquad x = \frac{7\pi}{4}.$$

Numerical Solution

Use the *table* feature of a graphing utility set in *radian* mode to create a table that shows the values of $y_1 = \sin x + \sqrt{2}$ and $y_2 = -\sin x$ for different values of x. Your table should go from $x = 0$ to $x = 2\pi$ using increments in the table of $\pi/8$, as shown in Figure 2.10. From the table, you can see that the values of y_1 and y_2 appear to be identical when $x \approx 3.927 \approx 5\pi/4$ and $x \approx 5.4978 \approx 7\pi/4$. These values are the approximate solutions of $\sin x + \sqrt{2} = -\sin x$.

X	Y1	Y2
3.1416	1.4142	1E-13
3.5343	1.0315	.38268
3.927	.70711	.70711
4.3197	.49033	.92388
4.7124	.41421	1
5.1051	.49033	.92388
5.4978	.70711	.70711

X=5.497787143782

Figure 2.10

EXAMPLE 3 Extracting Square Roots

Solve $3 \tan^2 x - 1 = 0$.

Solution

Rewrite the equation so that $\tan x$ is isolated on one side of the equation.

$$3 \tan^2 x - 1 = 0 \qquad \text{Write original equation.}$$

$$3 \tan^2 x = 1 \qquad \text{Add 1 to each side.}$$

$$\tan^2 x = \frac{1}{3} \qquad \text{Divide each side by 3.}$$

$$\tan x = \pm\frac{1}{\sqrt{3}} \qquad \text{Extract square roots.}$$

Because $\tan x$ has a period of π, first find all solutions in the interval $[0, \pi)$. These are $x = \pi/6$ and $x = 5\pi/6$. Finally, add $n\pi$ to each of these solutions to get the general form

$$x = \frac{\pi}{6} + n\pi \qquad \text{and} \qquad x = \frac{5\pi}{6} + n\pi \qquad \text{General solution}$$

where n is an integer. The graph of $y = 3 \tan^2 x - 1$, shown in Figure 2.11, confirms this result.

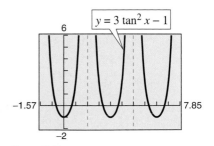

Figure 2.11

The equations in Examples 1, 2, and 3 involved only one trigonometric function. When two or more functions occur in the same equation, collect all terms on one side and try to separate the functions by factoring or by using appropriate identities. This may produce factors that yield no solutions, as illustrated in Example 4.

EXAMPLE 4   Factoring

Solve $\cot x \cos^2 x = 2 \cot x$.

Solution

$\cot x \cos^2 x = 2 \cot x$	Write original equation.
$\cot x \cos^2 x - 2 \cot x = 0$	Subtract 2 cot x from each side.
$\cot x(\cos^2 x - 2) = 0$	Factor.

By setting each of these factors equal to zero, you obtain the following.

$$\cot x = 0 \qquad \text{and} \qquad \cos^2 x - 2 = 0$$

$$x = \frac{\pi}{2} \qquad\qquad\qquad \cos^2 x = 2$$

$$\cos x = \pm\sqrt{2}$$

The equation $\cot x = 0$ has the solution $x = \pi/2$. No solution is obtained from $\cos x = \pm\sqrt{2}$ because $\pm\sqrt{2}$ are outside the range of the cosine function. Therefore, the general form of the solution is obtained by adding multiples of π to $x = \pi/2$, to get

$$x = \frac{\pi}{2} + n\pi \qquad\qquad \text{General solution}$$

where n is an integer. The graph of $y = \cot x \cos^2 x - 2 \cot x$, shown in Figure 2.12, confirms this result.

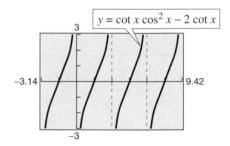

Figure 2.12

In Example 4, don't make the mistake of dividing each side of the equation by $\cot x$. Doing this would lose the solutions. Can you see why?

Equations of Quadratic Type

Many trigonometric equations are of quadratic type. Here are a few examples.

Quadratic in sin x	*Quadratic in sec x*
$2 \sin^2 x - \sin x - 1 = 0$	$\sec^2 x - 3 \sec x - 2 = 0$
$2(\sin x)^2 - \sin x - 1 = 0$	$(\sec x)^2 - 3 \sec x - 2 = 0$

To solve equations of this type, factor the quadratic or, if factoring is not possible, use the Quadratic Formula.

EXAMPLE 5   Factoring an Equation of Quadratic Type

Find all solutions of $2 \sin^2 x - \sin x - 1 = 0$ in the interval $[0, 2\pi)$.

Algebraic Solution

Treating the equation as a quadratic in $\sin x$ and factoring produces the following.

$$2 \sin^2 x - \sin x - 1 = 0 \qquad \text{Write original equation.}$$

$$(2 \sin x + 1)(\sin x - 1) = 0 \qquad \text{Factor.}$$

Setting each factor equal to zero, you obtain the following solutions in the interval $[0, 2\pi)$.

$$2 \sin x + 1 = 0 \qquad \text{and} \qquad \sin x - 1 = 0$$

$$\sin x = -\frac{1}{2} \qquad\qquad\qquad \sin x = 1$$

$$x = \frac{7\pi}{6}, \frac{11\pi}{6} \qquad\qquad\qquad x = \frac{\pi}{2}$$

So, the solutions in the interval $[0, 2\pi)$ are $\pi/2$, $7\pi/6$, and $11\pi/6$.

Graphical Solution

Use a graphing utility set in *radian* mode to graph $y = 2 \sin^2 x - \sin x - 1$ for $0 \le x < 2\pi$, as shown in Figure 2.13. Use the *zero* or *root* feature or the *zoom* and *trace* features to approximate the x-intercepts to be

$$x \approx 1.571 \approx \frac{\pi}{2}, \quad x \approx 3.665 \approx \frac{7\pi}{6}, \quad x \approx 5.760 \approx \frac{11\pi}{6}.$$

These values are the approximate solutions of $2 \sin^2 x - \sin x - 1 = 0$.

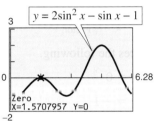

Figure 2.13

In Example 5, the general solution would be

$$x = \frac{\pi}{2} + 2n\pi, \quad x = \frac{7\pi}{6} + 2n\pi, \quad x = \frac{11\pi}{6} + 2n\pi \qquad \text{General solution}$$

where n is an integer.

When working with an equation of quadratic type, be sure that the equation involves a *single* trigonometric function, as shown in the next example.

EXAMPLE 6   Rewriting with a Single Trigonometric Function

Solve $2 \sin^2 x + 3 \cos x - 3 = 0$.

Solution

Begin by rewriting the equation so that it has only cosine functions.

$$2 \sin^2 x + 3 \cos x - 3 = 0 \qquad \text{Write original equation.}$$

$$2(1 - \cos^2 x) + 3 \cos x - 3 = 0 \qquad \text{Pythagorean identity}$$

$$2 \cos^2 x - 3 \cos x + 1 = 0 \qquad \text{Combine like terms and multiply each side by } -1.$$

$$(2 \cos x - 1)(\cos x - 1) = 0 \qquad \text{Factor.}$$

By setting each factor equal to zero, you can find the solutions in the interval $[0, 2\pi)$ to be $x = 0$, $x = \pi/3$, and $x = 5\pi/3$. The general solution is therefore

$$x = 2n\pi, \quad x = \frac{\pi}{3} + 2n\pi, \quad x = \frac{5\pi}{3} + 2n\pi \qquad \text{General solution}$$

where n is an integer. The graph of $y = 2 \sin^2 x + 3 \cos x - 3$, shown in Figure 2.14, confirms this result.

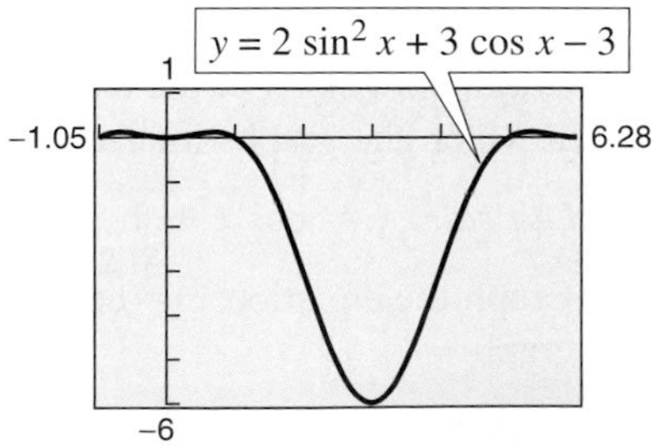

Figure 2.14

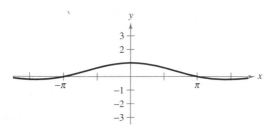

FIGURE FOR 72

73. Harmonic Motion   A weight is oscillating on the end of a spring. The position of the weight relative to the point of equilibrium is

$$y = \tfrac{1}{12}(\cos 8t - 3 \sin 8t)$$

where y is the displacement (in meters) and t is the time (in seconds). Find the times when the weight is at the point of equilibrium $(y = 0)$ for $0 \le t \le 1$.

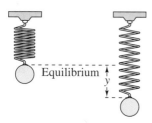

74. Sales   The monthly sales (in thousands of units) of a seasonal product are approximated by

$$S = 74.50 + 43.75 \sin \frac{\pi t}{6}$$

where t is the time (in months), with $t = 1$ corresponding to January. Determine the months when sales exceed 100,000 units.

75. Position of the Sun   Cheyenne, Wyoming, has a latitude of $41° $ N. At this latitude, the position of the sun at sunrise can be modeled by

$$D = 31 \sin\left(\frac{2\pi}{365}t - 1.4\right)$$

where t is the time (in days) and $t = 1$ represents January 1. In this model, D represents the number of degrees north or south of due east that the sun rises. Use a graphing calculator to determine the days that the sun is more than $20°$ north of due east at sunrise.

76. Projectile Motion   A batted baseball leaves the bat at an angle of θ with the horizontal and an initial velocity of $v_0 = 100$ feet per second. The ball is caught by an outfielder 300 feet from home plate. Find θ if the range r of a projectile is

$$r = \tfrac{1}{32}v_0^2 \sin 2\theta.$$

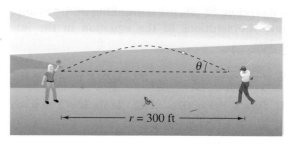

FIGURE FOR 76

77. Projectile Motion   A sharpshooter intends to hit a target 1000 yards away with a gun that has a muzzle velocity of 1200 feet per second. Neglecting air resistance, determine the minimum angle of elevation of the gun if the range is $r = \tfrac{1}{32}v_0^2 \sin 2\theta$.

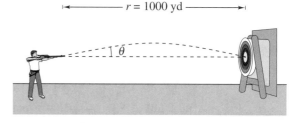

78. Area   The area of a rectangle inscribed in one arch of the graph of $y = \cos x$ is

$$A = 2x \cos x, \quad -\frac{\pi}{2} < x < \frac{\pi}{2}.$$

(a) Use a graphing utility to graph the area function, and approximate the area of the largest inscribed rectangle.

(b) Determine the values of x for which $A \ge 1$.

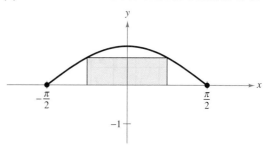

79. Damped Harmonic Motion   The displacement from equilibrium of a weight oscillating on the end of a spring is $y = (1/t) \cos 1.9t$ where y is the displacement (in feet) and t is the time (in seconds). Use a graphing utility to graph the displacement function for $0 \le t \le 10$. Find the time beyond which the displacement does not exceed 1 foot from equilibrium.

80. *Data Analysis* The table gives the unemployment rate r for the years 1989 through 1996 in the United States. The time t is measured in years, with $t = 0$ corresponding to 1990. (Source: U.S. Bureau of Labor Statistics)

t	-1	0	1	2
r	5.3	5.6	6.8	7.5

t	3	4	5	6
r	6.9	6.1	5.6	5.4

(a) Create a scatter plot of the data.

(b) Which of the following models best represents the data? Explain your reasoning.

   (i) $r = 1.10 \cos(t + 0.93) + 6.35$

   (ii) $r = 1.02 \sin(0.98t + 5.12) + 6.26$

   (iii) $r = 1.04 \sin[0.98(t + 5.81)] + 6.34$

   (iv) $r = 1.20 \sin[0.95(t + 5.48)] + 6.20$

(c) What term in the model gives the average unemployment rate? What is the rate?

(d) Economists study the lengths of business cycles, such as cycles in the unemployment rate. Based on this short span of time, use the model to give the length of this cycle.

(e) Use the model to estimate the next time the unemployment rate will be 5.5% or less.

81. *Quadratic Approximation* Consider the function

$$f(x) = 3 \sin(0.6x - 2).$$

(a) Approximate the zero of the function in the interval $[0, 6]$.

(b) A quadratic approximation agreeing with f at $x = 5$ is

$$g(x) = -0.45x^2 + 5.52x - 13.70.$$

Use a graphing utility to graph f and g in the same viewing window. Describe the result.

(c) Use the Quadratic Formula to find the zeros of g. Compare the zero in the interval $[0, 6]$ with the result of part (a).

Synthesis

True or False? **In Exercises 82 and 83, determine whether the statement is true or false. Justify your answer.**

82. All trigonometric equations have either an infinite number of solutions or no solution.

83. The solutions of any trigonometric equation can always be found from its solutions in the interval $[0, 2\pi)$.

84. *Writing* Describe the difference between verifying an identity and solving an equation.

Review

In Exercises 85–88, convert the angle measure from degrees to radians. Round your result to three decimal places.

85. $124°$ **86** $486°$

87. $-0.41°$ **88.** $-210.55°$

In Exercises 89–92, solve for x.

89.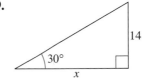

90.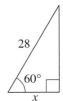

91.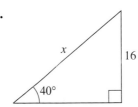

92.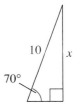

In Exercises 93–96, sketch the graph of the function.

93. $f(x) = \dfrac{1}{4} \sin\left(x - \dfrac{\pi}{2}\right)$ **94.** $f(x) = -6 \cos \dfrac{\pi x}{4}$

95. $f(x) = \dfrac{1}{2} \cot\left(x - \dfrac{\pi}{4}\right)$ **96.** $f(x) = \dfrac{1}{2} \sec(2x + \pi)$

2.4   Sum and Difference Formulas

Using Sum and Difference Formulas

In this and the following section, you will study the uses of several trigonometric identities and formulas. For proofs of these formulas see Appendix A.

Sum and Difference Formulas

$$\sin(u + v) = \sin u \cos v + \cos u \sin v$$

$$\sin(u - v) = \sin u \cos v - \cos u \sin v$$

$$\cos(u + v) = \cos u \cos v - \sin u \sin v$$

$$\cos(u - v) = \cos u \cos v + \sin u \sin v$$

$$\tan(u + v) = \frac{\tan u + \tan v}{1 - \tan u \tan v}$$

$$\tan(u - v) = \frac{\tan u - \tan v}{1 + \tan u \tan v}$$

Exploration

Use a graphing utility to graph $y = \cos(x + 2)$ and $y = \cos x + \cos 2$ in the same viewing window. What can you conclude about the graphs? Is it true that $\cos(x + 2) = \cos x + \cos 2$?

Use a graphing utility to graph $y = \sin(x + 4)$ and $y = \sin x + \sin 4$ in the same viewing window. What can you conclude about the graphs? Is it true that $\sin(x + 4) = \sin x + \sin 4$?

Examples 1 and 2 show how sum and difference formulas can be used to find exact values of trigonometric functions involving sums or differences of special angles.

EXAMPLE 1   Evaluating a Trigonometric Function

Find the exact value of $\cos 75°$.

Solution

To find the exact value of $\cos 75°$, use the fact that $75° = 30° + 45°$. Consequently, the formula for $\cos(u + v)$ yields

$$\cos 75° = \cos(30° + 45°)$$

$$= \cos 30° \cos 45° - \sin 30° \sin 45°$$

$$= \frac{\sqrt{3}}{2}\left(\frac{\sqrt{2}}{2}\right) - \frac{1}{2}\left(\frac{\sqrt{2}}{2}\right)$$

$$= \frac{\sqrt{6} - \sqrt{2}}{4}.$$

Try checking this result on your calculator. You will find that $\cos 75° \approx 0.259$.

What You Should Learn:

- How to use sum and difference formulas to evaluate trigonometric functions
- How to use sum and difference formulas to verify identities and solve trigonometric equations

Why You Should Learn It:

You can use sum and difference formulas to rewrite trigonometric expressions. For instance, Exercise 65 on page 249 shows how to use sum and difference formulas to rewrite a trigonometric expression in a form that helps you find the equation of a standing wave.

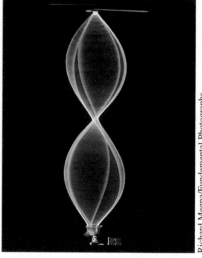

Richard Megna/Fundamental Photographs

EXAMPLE 2   Evaluating a Trigonometric Function

Find the exact value of $\sin \dfrac{\pi}{12}$.

Solution
Using the fact that $\pi/12 = \pi/3 - \pi/4$ together with the formula for $\sin(u - v)$, you obtain

$$\sin \frac{\pi}{12} = \sin\left(\frac{\pi}{3} - \frac{\pi}{4}\right) = \sin \frac{\pi}{3} \cos \frac{\pi}{4} - \cos \frac{\pi}{3} \sin \frac{\pi}{4}$$

$$= \frac{\sqrt{3}}{2}\left(\frac{\sqrt{2}}{2}\right) - \frac{1}{2}\left(\frac{\sqrt{2}}{2}\right) = \frac{\sqrt{6} - \sqrt{2}}{4}.$$

EXAMPLE 3   Evaluating a Trigonometric Expression

Find the exact value of $\sin(u + v)$ given that

$$\sin u = \frac{4}{5}, \text{ where } 0 < u < \frac{\pi}{2} \quad \text{and} \quad \cos v = -\frac{12}{13}, \text{ where } \frac{\pi}{2} < v < \pi.$$

Solution
Because $\sin u = 4/5$ and u is in Quadrant I, $\cos u = 3/5$, as shown in Figure 2.23(a). Because $\cos v = -12/13$ and v is in Quadrant II, $\sin v = 5/13$, as shown in Figure 2.23(b). You can find $\sin(u + v)$ as follows.

$$\sin(u + v) = \sin u \cos v + \cos u \sin v$$

$$= \left(\frac{4}{5}\right)\left(-\frac{12}{13}\right) + \left(\frac{3}{5}\right)\left(\frac{5}{13}\right) = -\frac{48}{65} + \frac{15}{65} = -\frac{33}{65}$$

EXAMPLE 4   An Application of a Sum Formula

Evaluate $\cos(\arctan 1 + \arccos x)$.

Solution
This expression fits the formula for $\cos(u + v)$. Angles $u = \arctan 1$ and $v = \arccos x$ are shown in Figure 2.24.

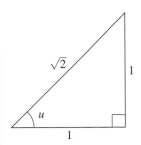
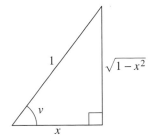

Figure 2.24

$$\cos(u + v) = \cos(\arctan 1)\cos(\arccos x) - \sin(\arctan 1)\sin(\arccos x)$$

$$= \frac{1}{\sqrt{2}} \cdot x - \frac{1}{\sqrt{2}} \cdot \sqrt{1 - x^2} = \frac{x - \sqrt{1 - x^2}}{\sqrt{2}}.$$

Hipparchus, considered the most eminent of Greek astronomers, was born about 160 B.C. in Nicea. He is credited with the invention of trigonometry. He also derived the sum and difference formulas for $\sin(A \pm B)$ and $\cos(A \pm B)$.

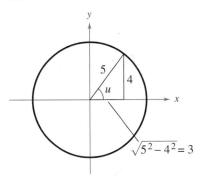

(a)

(b)
Figure 2.23

EXAMPLE 5   Proving a Cofunction Identity

Prove the cofunction identity $\cos\left(\dfrac{\pi}{2} - x\right) = \sin x$.

Solution

Using the formula for $\cos(u - v)$, you have

$$\cos\left(\frac{\pi}{2} - x\right) = \cos\frac{\pi}{2}\cos x + \sin\frac{\pi}{2}\sin x$$

$$= (0)(\cos x) + (1)(\sin x)$$

$$= \sin x.$$

Sum and difference formulas can be used to derive **reduction formulas** involving expressions such as

$$\sin\left(\theta + \frac{n\pi}{2}\right) \quad \text{and} \quad \cos\left(\theta + \frac{n\pi}{2}\right)$$

where n is an integer, as shown in the following example.

EXAMPLE 6   Deriving Reduction Formulas

Simplify each expression.

a. $\cos\left(\theta - \dfrac{3\pi}{2}\right)$ **b.** $\tan(\theta + 3\pi)$

Solution

a. Using the formula for $\cos(u - v)$, you have

$$\cos\left(\theta - \frac{3\pi}{2}\right) = \cos\theta\cos\frac{3\pi}{2} + \sin\theta\sin\frac{3\pi}{2}$$

$$= (\cos\theta)(0) + (\sin\theta)(-1)$$

$$= -\sin\theta.$$

b. Using the formula for $\tan(u + v)$, you have

$$\tan(\theta + 3\pi) = \frac{\tan\theta + \tan 3\pi}{1 - \tan\theta\tan 3\pi}$$

$$= \frac{\tan\theta + 0}{1 - (\tan\theta)(0)}$$

$$= \tan\theta.$$

Note that the period of $\tan\theta$ is π, so the period of $\tan(\theta + 3\pi)$ is the same as the period of $\tan\theta$.

The next example was taken from calculus. It is used to derive the formula for the derivative of the sine function.

EXAMPLE 7   An Application from Calculus

Verify that

$$\frac{\sin(x + h) - \sin x}{h} = (\cos x)\left(\frac{\sin h}{h}\right) - (\sin x)\left(\frac{1 - \cos h}{h}\right)$$

where $h \neq 0$.

Solution

Using the formula for $\sin(u + v)$, you have

$$\frac{\sin(x + h) - \sin x}{h} = \frac{\sin x \cos h + \cos x \sin h - \sin x}{h}$$

$$= \frac{\cos x \sin h - \sin x(1 - \cos h)}{h}$$

$$= (\cos x)\left(\frac{\sin h}{h}\right) - (\sin x)\left(\frac{1 - \cos h}{h}\right).$$

EXAMPLE 8   Solving a Trigonometric Equation

Find all solutions of $\sin\left(x + \dfrac{\pi}{4}\right) + \sin\left(x - \dfrac{\pi}{4}\right) = -1$

in the interval $[0, 2\pi)$.

Algebraic Solution

Using sum and difference formulas, rewrite the equation as

$$\sin x \cos \frac{\pi}{4} + \cos x \sin \frac{\pi}{4} + \sin x \cos \frac{\pi}{4} - \cos x \sin \frac{\pi}{4} = -1$$

$$2 \sin x \cos \frac{\pi}{4} = -1$$

$$2(\sin x)\left(\frac{\sqrt{2}}{2}\right) = -1$$

$$\sin x = -\frac{1}{\sqrt{2}}$$

$$\sin x = -\frac{\sqrt{2}}{2}.$$

Therefore, the only solutions in the interval $[0, 2\pi)$ are

$$x = \frac{5\pi}{4} \qquad \text{and} \qquad x = \frac{7\pi}{4}.$$

Graphical Solution

Use a graphing utility set in *radian* mode to graph $y = \sin\left(x + \dfrac{\pi}{4}\right) + \sin\left(x - \dfrac{\pi}{4}\right) + 1,$

as shown in Figure 2.25. Use the *zero* or *root* feature or the *zoom* and *trace* features to approximate the x-intercepts in the interval $[0, 2\pi)$ to be

$$x \approx 3.927 \approx \frac{5\pi}{4} \text{ and } x \approx 5.498 \approx \frac{7\pi}{4}.$$

These values are the approximate solutions in the interval $[0, 2\pi)$ of

$$\sin\left(x + \frac{\pi}{4}\right) + \sin\left(x - \frac{\pi}{4}\right) = -1.$$

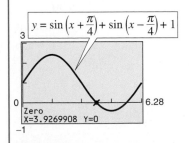

Figure 2.25

2.4   E x e r c i s e s

In Exercises 1–8, find the exact value of each expression.

1. (a) $\cos\left(\dfrac{\pi}{6} + \dfrac{\pi}{3}\right)$ (b) $\cos\dfrac{\pi}{6} + \cos\dfrac{\pi}{3}$

2. (a) $\sin\left(\dfrac{2\pi}{3} + \dfrac{3\pi}{4}\right)$ (b) $\sin\dfrac{2\pi}{3} + \sin\dfrac{3\pi}{4}$

3. (a) $\sin\left(\dfrac{7\pi}{6} - \dfrac{\pi}{3}\right)$ (b) $\sin\dfrac{7\pi}{6} - \sin\dfrac{\pi}{3}$

4. (a) $\cos\left(\dfrac{5\pi}{4} - \dfrac{\pi}{6}\right)$ (b) $\cos\dfrac{5\pi}{4} - \cos\dfrac{\pi}{6}$

5. (a) $\cos(0° + 135°)$ (b) $\cos 0° + \cos 135°$

6. (a) $\cos(240° - 0°)$ (b) $\cos 240° - \cos 0°$

7. (a) $\sin(315° - 60°)$ (b) $\sin 315° - \sin 60°$

8. (a) $\sin(390° + 120°)$ (b) $\sin 390° + \sin 120°$

In Exercises 9–18, use the sum and difference formulas to find the exact values of the sine, cosine, and tangent of the angle.

9. $75° = 30° + 45°$ 10. $15° = 45° - 30°$

11. $105° = 60° + 45°$ 12. $165° = 135° + 30°$

13. $195° = 225° - 30°$ 14. $285° = 330° - 45°$

15. $\dfrac{11\pi}{12} = \dfrac{3\pi}{4} + \dfrac{\pi}{6}$ 16. $\dfrac{17\pi}{12} = \dfrac{7\pi}{6} + \dfrac{\pi}{4}$

17. $-\dfrac{\pi}{12} = \dfrac{\pi}{6} - \dfrac{\pi}{4}$ 18. $-\dfrac{19\pi}{12} = \dfrac{2\pi}{3} - \dfrac{9\pi}{4}$

In Exercises 19–28, use the sum and difference formulas to write the expression as the sine, cosine, or tangent of an angle.

19. $\cos 40° \cos 15° - \sin 40° \sin 15°$

20. $\sin 110° \cos 80° + \cos 110° \sin 80°$

21. $\sin 340° \cos 50° - \cos 340° \sin 50°$

22. $\cos 20° \cos 30° + \sin 20° \sin 30°$

23. $\dfrac{\tan 325° - \tan 86°}{1 + \tan 325° \tan 86°}$ 24. $\dfrac{\tan 140° - \tan 60°}{1 + \tan 140° \tan 60°}$

25. $\sin 3 \cos 1.2 - \cos 3 \sin 1.2$

26. $\cos 0.88 \cos 0.34 + \sin 0.88 \sin 0.34$

27. $\cos\dfrac{\pi}{7} \cos\dfrac{\pi}{5} - \sin\dfrac{\pi}{7} \sin\dfrac{\pi}{5}$

28. $\sin\dfrac{2\pi}{9} \cos\dfrac{\pi}{10} + \cos\dfrac{2\pi}{9} \sin\dfrac{\pi}{10}$

Numerical, Graphical, and Algebraic Analysis **In Exercises 29–34, use a graphing utility to complete the table and graph the two functions. Use both table and graph as evidence that $y_1 = y_2$. Then verify the identity algebraically.**

x	0.2	0.4	0.6	0.8	1.0	1.2	1.4
y_1							
y_2							

29. $y_1 = \sin\left(\dfrac{\pi}{2} + x\right), \quad y_2 = \cos x$

30. $y_1 = \sin(3\pi - x), \quad y_2 = \sin x$

31. $y_1 = \sin\left(\dfrac{\pi}{6} + x\right), \quad y_2 = \dfrac{1}{2}\left(\cos x + \sqrt{3}\sin x\right)$

32. $y_1 = \cos\left(\dfrac{5\pi}{4} - x\right), \quad y_2 = -\dfrac{\sqrt{2}}{2}(\cos x + \sin x)$

33. $y_1 = \cos(x + \pi)\cos(x - \pi), \quad y_2 = \cos^2 x$

34. $y_1 = \sin(x + \pi)\sin(x - \pi), \quad y_2 = \sin^2 x$

In Exercises 35–38, find the exact value of the trigonometric function given that

$\sin u = 5/13,$ where $0 < u < \pi/2$

$\cos v = -3/5,$ where $\pi/2 < v < \pi.$

35. $\sin(u + v)$ 36. $\cos(v - u)$

37. $\cos(u + v)$ 38. $\sin(u - v)$

In Exercises 39–42, find the exact value of the trigonometric function given that

$\sin u = 7/25,$ where $\pi/2 < u < \pi$

$\cos v = 4/5,$ where $3\pi/2 < v < 2\pi.$

39. $\cos(u + v)$ 40. $\sin(u + v)$

41. $\sin(v - u)$ 42. $\cos(u - v)$

In Exercises 43–50, verify the identity.

43. $\cos(\pi - \theta) + \sin\left(\dfrac{\pi}{2} + \theta\right) = 0$

44. $\sin(\theta + \pi) + \cos\left(\theta - \dfrac{\pi}{2}\right) = 0$

45. $\tan(x + \pi) - \tan(\pi - x) = 2 \tan x$

46. $\tan\left(\dfrac{\pi}{4} - \theta\right) = \dfrac{1 - \tan \theta}{1 + \tan \theta}$

47. $\sin(x + y) + \sin(x - y) = 2 \sin x \cos y$

48. $\cos(x + y) + \cos(x - y) = 2 \cos x \cos y$

49. $\cos(x + y) \cos(x - y) = \cos^2 x - \sin^2 y$

50. $\sin(x + y) \sin(x - y) = \sin^2 x - \sin^2 y$

In Exercises 51–54, write the trigonometric expression as an algebraic expression.

51. $\sin(\arcsin x + \arccos x)$

52. $\cos(\arccos x - \arcsin x)$

53. $\sin(\arctan 2x - \arccos x)$

54. $\cos(\arcsin x - \arctan 2x)$

In Exercises 55–60, find all solutions of the equation in the interval $[0, 2\pi)$. Use a graphing utility to verify your results.

55. $\sin\left(x + \dfrac{\pi}{3}\right) + \sin\left(x - \dfrac{\pi}{3}\right) = 1$

56. $\sin\left(x + \dfrac{\pi}{6}\right) - \sin\left(x - \dfrac{\pi}{6}\right) = \dfrac{1}{2}$

57. $\cos\left(x + \dfrac{\pi}{4}\right) - \cos\left(x - \dfrac{\pi}{4}\right) = 1$

58. $\cos\left(x + \dfrac{\pi}{6}\right) - \cos\left(x - \dfrac{\pi}{6}\right) = 1$

59. $\tan(x + \pi) + 2 \sin(x + \pi) = 0$

60. $2 \sin\left(x + \dfrac{\pi}{2}\right) + 3 \tan(\pi - x) = 0$

In Exercises 61–64, use a graphing utility to approximate all solutions of the equation in the interval $[0, 2\pi)$.

61. $\cos\left(x + \dfrac{\pi}{4}\right) + \cos\left(x - \dfrac{\pi}{4}\right) = 1$

62. $\sin\left(x + \dfrac{\pi}{2}\right) - \cos\left(x + \dfrac{3\pi}{2}\right) = 0$

63. $\tan(x + \pi) - \cos\left(x + \dfrac{\pi}{2}\right) = 0$

64. $\tan(\pi - x) + 2 \cos\left(x + \dfrac{3\pi}{2}\right) = 0$

65. *Standing Waves*   The equation of a standing wave is obtained by adding the displacements of two waves traveling in opposite directions. Assume that each of the waves has amplitude A, period T, and wavelength λ. If the models for these waves are y_1 and y_2 given below, show that

$$y_1 + y_2 = 2A \cos \dfrac{2\pi t}{T} \cos \dfrac{2\pi x}{\lambda}.$$

$$y_1 = A \cos 2\pi\left(\dfrac{t}{T} - \dfrac{x}{\lambda}\right), \quad y_2 = A \cos 2\pi\left(\dfrac{t}{T} + \dfrac{x}{\lambda}\right)$$

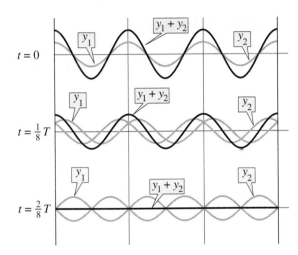

66. *Harmonic Motion*   A weight is attached to a spring suspended vertically from a ceiling. When a driving force is applied to the system, the weight moves vertically from its equilibrium position, and this motion is modeled by

$$y = \tfrac{1}{3} \sin 2t + \tfrac{1}{4} \cos 2t$$

where y is the distance from equilibrium (in feet) and t is the time (in seconds).

(a) Use a graphing utility to graph the model.

(b) Write the model in the form

$$y = \sqrt{a^2 + b^2} \sin(Bt + C).$$

Use a graphing utility to verify your result.

(c) Find the amplitude of the oscillations of the weight.

(d) Find the frequency of the oscillations of the weight.

Synthesis

True or False? In Exercises 67–70, determine whether the statement is true or false. Justify your answer.

67. $\sin(u \pm v) = \sin u \pm \sin v$

68. $\cos(u \pm v) = \cos u \pm \cos v$

69. It is not possible to find the exact value of $\sin 75°$.

70. $\sin\left(x - \dfrac{11\pi}{2}\right) = \cos x$

In Exercises 71–74, verify the identity.

71. $\cos(n\pi + \theta) = (-1)^n \cos \theta$, n is an integer.

72. $\sin(n\pi + \theta) = (-1)^n \sin \theta$, n is an integer.

73. $a \sin B\theta + b \cos B\theta = \sqrt{a^2 + b^2} \sin(B\theta + C)$
where $C = \arctan(b/a)$ and $a > 0$.

74. $a \sin B\theta + b \cos B\theta = \sqrt{a^2 + b^2} \cos(B\theta - C)$
where $C = \arctan(a/b)$ and $b > 0$.

In Exercises 75–78, use the formulas given in Exercises 73 and 74 to write the expression in the following forms. Use a graphing utility to verify your results.

(a) $\sqrt{a^2 + b^2} \sin(B\theta + C)$

(b) $\sqrt{a^2 + b^2} \cos(B\theta - C)$

75. $\sin \theta + \cos \theta$

76. $3 \sin 2\theta + 4 \cos 2\theta$

77. $12 \sin 3\theta + 5 \cos 3\theta$

78. $\sin 2\theta - \cos 2\theta$

In Exercises 79 and 80, use the formulas given in Exercises 73 and 74 to write the trigonometric expression in the form $a \sin B\theta + b \cos B\theta$.

79. $2 \sin\left(\theta + \dfrac{\pi}{2}\right)$

80. $5 \cos\left(\theta + \dfrac{\pi}{4}\right)$

81. ***Conjecture*** Three squares of side s are placed side by side. Make a conjecture about the relationship between the sum $u + v$ and w. Prove your conjecture by using the identity for the tangent of the sum of two angles.

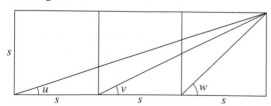

Conjecture In Exercises 82 and 83, use a graphing utility to graph the trigonometric function. Use the graph to make a conjecture about a simplification of the expression. Verify the resulting identity algebraically.

82. $g(x) = \cos(\pi + x)$

83. $h(\theta) = \tan(\pi + \theta)$

84. Verify the identity used in calculus.

$$\dfrac{\cos(x + h) - \cos x}{h} = \dfrac{\cos x(\cos h - 1)}{h} - \dfrac{\sin x \sin h}{h}$$

85. ***Exploration*** Let $x = \pi/6$ in the identity in Exercise 84 and define the functions f and g as follows.

$$f(h) = \dfrac{\cos(\pi/6 + h) - \cos(\pi/6)}{h}$$

$$g(h) = \cos\dfrac{\pi}{6}\left(\dfrac{\cos h - 1}{h}\right) - \sin\dfrac{\pi}{6}\left(\dfrac{\sin h}{h}\right)$$

(a) What are the domains of these functions?

(b) Use a graphing utility to complete the table.

h	0.01	0.02	0.05	0.1	0.2	0.5
$f(h)$						
$g(h)$						

(c) Use a graphing utility to graph the functions.

(d) Use the table and graph to make a conjecture about the values of the functions as $h \to 0$.

86. Use the sum formulas for sine and cosine to derive the formula

$$\tan(u + v) = \dfrac{\tan u + \tan v}{1 - \tan u \tan v}.$$

Review

In Exercises 87–90, find the x- and y-intercepts of the graph of the equation. Use a graphing utility to verify your results.

87. $y = -\frac{1}{2}(x - 10) + 14$

88. $y = x^2 - 3x - 40$

89. $y = |2x - 9| - 5$

90. $y = 2x\sqrt{x + 7}$

In Exercises 91–94, evaluate the expression without using a calculator.

91. $\arccos\left(\dfrac{\sqrt{3}}{2}\right)$

92. $\arctan(-\sqrt{3})$

93. $\arcsin 1$

94. $\arctan 0$

2.5 Multiple-Angle and Product-Sum Formulas

Multiple-Angle Formulas

In this section you will study four other categories of trigonometric identities.

1. The first category involves functions of multiple angles such as $\sin ku$ and $\cos ku$.
2. The second category involves squares of trigonometric functions such as $\sin^2 u$.
3. The third category involves functions of half-angles such as $\sin(u/2)$.
4. The fourth category involves products of trigonometric functions such as $\sin u \cos v$.

The most commonly used multiple-angle formulas are the **double-angle formulas.** They are used often, so you should learn them. (Proofs of the double-angle formulas are given in Appendix A.)

Double-Angle Formulas

$$\sin 2u = 2 \sin u \cos u \qquad \tan 2u = \frac{2 \tan u}{1 - \tan^2 u}$$

$$\cos 2u = \cos^2 u - \sin^2 u$$
$$= 2 \cos^2 u - 1$$
$$= 1 - 2 \sin^2 u$$

Note that $\sin 2u \neq 2 \sin u$, $\cos 2u \neq 2 \cos u$, and $\tan 2u \neq 2 \tan u$.

EXAMPLE 1   Solving a Multiple-Angle Equation

Find all solutions of $2 \cos x + \sin 2x = 0$.

Solution
Begin by rewriting the equation so that it involves functions of x (rather than $2x$). Then factor and solve as usual.

$$2 \cos x + \sin 2x = 0 \qquad \text{Write original equation.}$$

$$2 \cos x + 2 \sin x \cos x = 0 \qquad \text{Double-angle formula}$$

$$2 \cos x(1 + \sin x) = 0 \qquad \text{Factor.}$$

$$\cos x = 0, \quad 1 + \sin x = 0 \qquad \text{Set factors equal to zero.}$$

$$x = \frac{\pi}{2}, \frac{3\pi}{2} \qquad x = \frac{3\pi}{2} \qquad \text{Solutions in } [0, 2\pi)$$

Therefore, the general solution is

$$x = \frac{\pi}{2} + 2n\pi \quad \text{and} \quad x = \frac{3\pi}{2} + 2n\pi \qquad \text{General solution}$$

where n is an integer. Try verifying these solutions graphically.

What You Should Learn:
- How to use multiple-angle formulas to rewrite and evaluate trigonometric functions
- How to use power-reducing formulas to rewrite and evaluate trigonometric functions
- How to use half-angle formulas to rewrite and evaluate trigonometric functions
- How to use product-sum formulas to rewrite and evaluate trigonometric functions

Why You Should Learn It:

You can use a variety of trigonometric formulas to rewrite trigonometric functions in more convenient forms. For instance, Exercise 105 on page 260 shows you how to use a half-angle formula to determine the apex angle of a sound wave cone from the speed of an airplane.

NASA-Liaison Agency

A computer animation of this concept appears in the *Interactive* CD-ROM and *Internet* versions of this text.

EXAMPLE 2   Using Double-Angle Formulas in Sketching Graphs

Analyze the graph of $y = 4 \cos^2 x - 2$ over the interval $[0, 2\pi]$.

Solution

Using a double-angle formula, you can rewrite the given function as

$$y = 4 \cos^2 x - 2$$
$$= 2(2 \cos^2 x - 1)$$
$$= 2 \cos 2x.$$

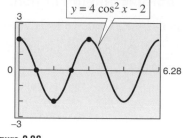

Using the techniques discussed in Section 1.5, you can recognize that the graph of this function has an amplitude of 2 and a period of π. The key points in the interval $[0, \pi]$ are as follows.

Maximum	*Intercept*	*Minimum*	*Intercept*	*Maximum*
$(0, 2)$	$\left(\dfrac{\pi}{4}, 0\right)$	$\left(\dfrac{\pi}{2}, -2\right)$	$\left(\dfrac{3\pi}{4}, 0\right)$	$(\pi, 2)$

Figure 2.26

Two cycles of the graph are shown in Figure 2.26.

EXAMPLE 3   Evaluating Functions Involving Double Angles

Use the following to find $\sin 2\theta$, $\cos 2\theta$, and $\tan 2\theta$.

$$\cos \theta = \frac{5}{13}, \qquad \frac{3\pi}{2} < \theta < 2\pi$$

Solution

From Figure 2.27, you can see that

$$\sin \theta = \frac{y}{r} = -\frac{12}{13}.$$

Consequently, you can write

$$\sin 2\theta = 2 \sin \theta \cos \theta = 2\left(-\frac{12}{13}\right)\left(\frac{5}{13}\right) = -\frac{120}{169}$$

$$\cos 2\theta = 2 \cos^2 \theta - 1 = 2\left(\frac{25}{169}\right) - 1 = -\frac{119}{169}$$

$$\tan 2\theta = \frac{\sin 2\theta}{\cos 2\theta} = \frac{120}{119}.$$

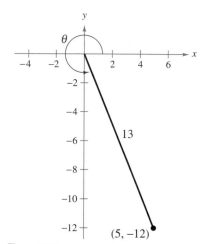

Figure 2.27

The double-angle formulas are not restricted to angles 2θ and θ. Other *double* combinations, such as 4θ and 2θ or 6θ and 3θ, are also valid. Here are two examples.

$$\sin 4\theta = 2 \sin 2\theta \cos 2\theta \qquad \text{and} \qquad \cos 6\theta = \cos^2 3\theta - \sin^2 3\theta$$

By using double-angle formulas together with the sum formulas derived in the previous section, you can form other multiple-angle formulas.

STUDY T!P

Notice that you cannot solve Example 3 by simply evaluating $\theta = \cos^{-1}(5/13)$ and then calculating $\sin 2\theta$, $\cos 2\theta$, and $\tan 2\theta$. Do you see why?

EXAMPLE 4   Deriving a Triple-Angle Formula

Express $\sin 3x$ in terms of $\sin x$.

Solution

$$\sin 3x = \sin(2x + x)$$

$$= \sin 2x \cos x + \cos 2x \sin x \qquad \text{Sum formula}$$

$$= 2 \sin x \cos x \cos x + (1 - 2 \sin^2 x)\sin x \qquad \text{Double-angle formula}$$

$$= 2 \sin x \cos^2 x + \sin x - 2 \sin^3 x \qquad \text{Multiply.}$$

$$= 2 \sin x(1 - \sin^2 x) + \sin x - 2 \sin^3 x \qquad \text{Pythagorean identity}$$

$$= 2 \sin x - 2 \sin^3 x + \sin x - 2 \sin^3 x \qquad \text{Multiply.}$$

$$= 3 \sin x - 4 \sin^3 x \qquad \text{Simplify.}$$

Power-Reducing Formulas

The double-angle formulas can be used to obtain the following **power-reducing formulas.** (Proofs of the power-reducing formulas are given in Appendix A.)

Power-Reducing Formulas

$$\sin^2 u = \frac{1 - \cos 2u}{2} \qquad \cos^2 u = \frac{1 + \cos 2u}{2} \qquad \tan^2 u = \frac{1 - \cos 2u}{1 + \cos 2u}$$

Example 5 shows a typical power reduction that is used in calculus.

EXAMPLE 5   Reducing a Power

Rewrite $\sin^4 x$ as a sum of first powers of the cosines of multiple angles.

Solution

Note the repeated use of power-reducing formulas.

$$\sin^4 x = (\sin^2 x)^2 \qquad \text{Property of exponents}$$

$$= \left(\frac{1 - \cos 2x}{2}\right)^2 \qquad \text{Power-reducing formula}$$

$$= \frac{1}{4}(1 - 2 \cos 2x + \cos^2 2x) \qquad \text{Expand binomial.}$$

$$= \frac{1}{4}\left(1 - 2 \cos 2x + \frac{1 + \cos 4x}{2}\right) \qquad \text{Power-reducing formula}$$

$$= \frac{1}{4} - \frac{1}{2}\cos 2x + \frac{1}{8} + \frac{1}{8}\cos 4x \qquad \text{Distributive Property}$$

$$= \frac{3}{8} - \frac{1}{2}\cos 2x + \frac{1}{8}\cos 4x \qquad \text{Simplify.}$$

$$= \frac{1}{8}(3 - 4 \cos 2x + \cos 4x) \qquad \text{Factor.}$$

Half-Angle Formulas

You can derive some useful alternative forms of the power-reducing formulas by replacing u with $u/2$. The results are called **half-angle formulas.**

Half-Angle Formulas

$$\sin \frac{u}{2} = \pm \sqrt{\frac{1 - \cos u}{2}}$$

$$\cos \frac{u}{2} = \pm \sqrt{\frac{1 + \cos u}{2}}$$

$$\tan \frac{u}{2} = \frac{1 - \cos u}{\sin u} = \frac{\sin u}{1 + \cos u}$$

The signs of $\sin(u/2)$ and $\cos(u/2)$ depend on the quadrant in which $u/2$ lies.

A computer animation of this concept appears in the *Interactive* CD-ROM and *Internet* versions of this text.

EXAMPLE 6   Using a Half-Angle Formula

Find the exact value of sin 105°.

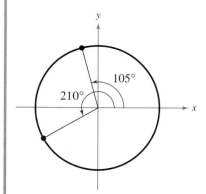

Figure 2.28

Solution

Begin by noting that 105° is half of 210°. Then, using the half-angle formula for $\sin(u/2)$ and the fact that 105° lies in Quadrant II (see Figure 2.28), you have

$$\sin 105° = \sqrt{\frac{1 - \cos 210°}{2}}$$

$$= \sqrt{\frac{1 - (-\cos 30°)}{2}}$$

$$= \sqrt{\frac{1 + (\sqrt{3}/2)}{2}} = \frac{\sqrt{2 + \sqrt{3}}}{2}.$$

The positive square root is chosen because $\sin \theta$ is positive in Quadrant II.

Use your calculator to verify the result obtained in Example 6. That is, evaluate sin 105° and $(\sqrt{2 + \sqrt{3}})/2$.

EXAMPLE 7 Solving a Trigonometric Equation

Find all solutions of

$$2 - \sin^2 x = 2\cos^2 \frac{x}{2}$$

in the interval $[0, 2\pi)$.

Algebraic Solution

$$2 - \sin^2 x = 2\cos^2 \frac{x}{2} \qquad \text{Write original equation.}$$

$$2 - \sin^2 x = 2\left(\pm\sqrt{\frac{1+\cos x}{2}}\right)^2 \qquad \text{Half-angle formula}$$

$$2 - \sin^2 x = 2\left(\frac{1+\cos x}{2}\right) \qquad \text{Simplify.}$$

$$2 - \sin^2 x = 1 + \cos x \qquad \text{Simplify.}$$

$$2 - (1 - \cos^2 x) = 1 + \cos x \qquad \text{Pythagorean identity}$$

$$\cos^2 x - \cos x = 0 \qquad \text{Simplify.}$$

$$\cos x(\cos x - 1) = 0 \qquad \text{Factor.}$$

By setting the factors $\cos x$ and $(\cos x - 1)$ equal to zero, you find that the solutions in the interval $[0, 2\pi)$ are

$$x = \frac{\pi}{2}, \quad x = \frac{3\pi}{2}, \quad \text{and} \quad x = 0.$$

Graphical Solution

Use a graphing utility set in *radian* mode to graph $y = 2 - \sin^2 x - 2\cos^2(x/2)$, as shown in Figure 2.29. Use the *zero* or *root* feature or the *zoom* and *trace* features to approximate the x-intercepts in the interval $[0, 2\pi)$ to be

$$x = 0, \quad x \approx 1.5708 \approx \frac{\pi}{2}, \quad \text{and} \quad x \approx 4.7124 \approx \frac{3\pi}{2}.$$

These values are the approximate solutions in the interval $[0, 2\pi)$ of $2 - \sin^2 x = \cos^2 \frac{x}{2}$.

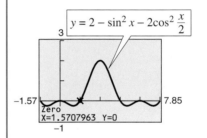

$$y = 2 - \sin^2 x - 2\cos^2 \frac{x}{2}$$

Figure 2.29

Product-to-Sum Formulas

Each of the following **product-to-sum formulas** is easily verified using the sum and difference formulas discussed in the preceding section.

Product-to-Sum Formulas

$$\sin u \sin v = \frac{1}{2}\left[\cos(u - v) - \cos(u + v)\right]$$

$$\cos u \cos v = \frac{1}{2}\left[\cos(u - v) + \cos(u + v)\right]$$

$$\sin u \cos v = \frac{1}{2}\left[\sin(u + v) + \sin(u - v)\right]$$

$$\cos u \sin v = \frac{1}{2}\left[\sin(u + v) - \sin(u - v)\right]$$

EXAMPLE 8   Writing Products as Sums

Rewrite the product as a sum or difference.

$$\cos 5x \sin 4x$$

Solution

$$\cos 5x \sin 4x = \frac{1}{2}[\sin(5x + 4x) - \sin(5x - 4x)]$$

$$= \frac{1}{2} \sin 9x - \frac{1}{2} \sin x$$

Occasionally, it is useful to reverse the procedure and write a sum of trigonometric functions as a product. This can be accomplished with the following **sum-to-product formulas.** (A proof of the first formula is given in Appendix A.)

Sum-to-Product Formulas

$$\sin x + \sin y = 2 \sin\left(\frac{x + y}{2}\right) \cos\left(\frac{x - y}{2}\right)$$

$$\sin x - \sin y = 2 \cos\left(\frac{x + y}{2}\right) \sin\left(\frac{x - y}{2}\right)$$

$$\cos x + \cos y = 2 \cos\left(\frac{x + y}{2}\right) \cos\left(\frac{x - y}{2}\right)$$

$$\cos x - \cos y = -2 \sin\left(\frac{x + y}{2}\right) \sin\left(\frac{x - y}{2}\right)$$

EXAMPLE 9   Using a Sum-to-Product Formula

Find the exact value of

$$\cos 195° + \cos 105°.$$

Solution

Using the appropriate sum-to-product formula, you obtain

$$\cos 195° + \cos 105° = 2 \cos\left(\frac{195° + 105°}{2}\right) \cos\left(\frac{195° - 105°}{2}\right)$$

$$= 2 \cos 150° \cos 45°$$

$$= 2\left(-\frac{\sqrt{3}}{2}\right)\left(\frac{\sqrt{2}}{2}\right)$$

$$= -\frac{\sqrt{6}}{2}.$$

EXAMPLE 10   Solving a Trigonometric Equation

Find all solutions of $\sin 5x + \sin 3x = 0$ in the interval $[0, 2\pi)$.

Algebraic Solution

$$\sin 5x + \sin 3x = 0 \qquad \text{Write original equation.}$$

$$2 \sin\left(\frac{5x + 3x}{2}\right) \cos\left(\frac{5x - 3x}{2}\right) = 0 \qquad \text{Sum-to-product formula}$$

$$2 \sin 4x \cos x = 0 \qquad \text{Simplify.}$$

By setting the factor $\sin 4x$ equal to zero, you can find that the solutions in the interval $[0, 2\pi)$ are

$$x = 0, \frac{\pi}{4}, \frac{\pi}{2}, \frac{3\pi}{4}, \pi, \frac{5\pi}{4}, \frac{3\pi}{2}, \frac{7\pi}{4}.$$

Moreover, the equation $\cos x = 0$ yields no additional solutions.

Graphical Solution

Use a graphing utility set in *radian* mode to graph $y = \sin 5x + \sin 3x$, as shown in Figure 2.30. Use the *zero* or *root* feature or the *zoom* and *trace* features to approximate the *x*-intercepts in the interval $[0, 2\pi)$ to be

$$x \approx 0, \; x \approx 0.7854 \approx \frac{\pi}{4}, \; x \approx 1.5708 \approx \frac{\pi}{2},$$

$$x \approx 2.3562 \approx \frac{3\pi}{4}, \; x \approx 3.1416 \approx \pi, \; x \approx 3.9270 \approx \frac{5\pi}{4},$$

$$x \approx 4.7124 \approx \frac{3\pi}{2}, \; x \approx 5.4978 \approx \frac{7\pi}{4}.$$

These values are the approximate solutions in the interval $[0, 2\pi)$ of $\sin 5x + \sin 3x = 0$.

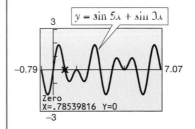

Figure 2.30

Note in Example 10 that the general solution would be $x = \dfrac{n\pi}{4}$, where n is an integer.

EXAMPLE 11   Verifying a Trigonometric Identity

Verify the identity $\dfrac{\sin t + \sin 3t}{\cos t + \cos 3t} = \tan 2t$.

Solution

Using appropriate sum-to-product formulas, you have

$$\frac{\sin t + \sin 3t}{\cos t + \cos 3t} = \frac{2 \sin 2t \cos(-t)}{2 \cos 2t \cos(-t)} = \frac{\sin 2t}{\cos 2t} = \tan 2t.$$

W r i t i n g   A b o u t   M a t h   *Deriving an Area Formula*

Describe how you can use a double-angle formula or a half-angle formula to derive a formula for the area of an isosceles triangle. Use a labeled sketch to illustrate your derivation. Then write two examples that show how your formula can be used.

2.5　E X E R C I S E S

In Exercises 1–8, use the figure to find the exact value of the trigonometric function.

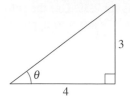

1. $\sin \theta$　　　　　　　　　**2.** $\tan \theta$

3. $\cos 2\theta$　　　　　　　　**4.** $\sin 2\theta$

5. $\tan 2\theta$　　　　　　　　**6.** $\sec 2\theta$

7. $\csc 2\theta$　　　　　　　　**8.** $\cot 2\theta$

In Exercises 9–16, use a graphing utility to approximate the equation's solutions in the interval $[0, 2\pi)$. If possible, find the exact solutions algebraically.

9. $\sin 2x - \sin x = 0$　　　**10.** $\sin 2x + \cos x = 0$

11. $8 \sin x \cos x = 1$　　　　**12.** $\sin 2x \sin x = \cos x$

13. $\cos 2x + \cos x = 0$　　**14.** $\tan 2x - \cot x = 0$

15. $\sin 4x = -2 \sin 2x$　　**16.** $(\sin 2x + \cos 2x)^2 = 1$

In Exercises 17–20, use a double-angle formula to rewrite the expression. Use a graphing utility to graph both expressions to verify that both forms are the same.

17. $8 \sin x \cos x$　　　　　　**18.** $6 \sin x \cos x + 4$

19. $5 - 10 \sin^2 x$

20. $(\cos x + \sin x)(\cos x - \sin x)$

In Exercises 21–24, find the exact values of $\sin 2u$, $\cos 2u$, and $\tan 2u$ using the double-angle formulas.

21. $\sin u = \frac{3}{5},\quad 0 < u < \pi/2$

22. $\cos u = -\frac{2}{7},\quad \pi/2 < u < \pi$

23. $\tan u = \frac{1}{2},\quad \pi < u < 3\pi/2$

24. $\cot u = -6,\quad 3\pi/2 < u < 2\pi$

In Exercises 25–30, rewrite the expression in terms of the first power of the cosine. Use a graphing utility to graph both expressions and verify that both forms are the same.

25. $\cos^4 x$　　　　　　　　**26.** $\sin^4 x$

27. $\sin^2 x \cos^2 x$　　　　　**28.** $\cos^6 x$

29. $\sin^2 x \cos^4 x$　　　　　**30.** $\sin^4 x \cos^2 x$

In Exercises 31–38, use the figure to find the exact value of the trigonometric function.

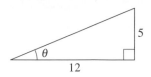

31. $\cos \dfrac{\theta}{2}$　　　　　　　**32.** $\sin \dfrac{\theta}{2}$

33. $\tan \dfrac{\theta}{2}$　　　　　　　**34.** $\sec \dfrac{\theta}{2}$

35. $\csc \dfrac{\theta}{2}$　　　　　　　**36.** $\cot \dfrac{\theta}{2}$

37. $2 \sin \dfrac{\theta}{2} \cos \dfrac{\theta}{2}$　　　**38.** $2 \cos \dfrac{\theta}{2} \tan \dfrac{\theta}{2}$

In Exercises 39–46, use the half-angle formulas to determine the exact values of the sine, cosine, and tangent of the angle.

39. $15°$　　　　　　　　　**40.** $165°$

41. $112° \, 30'$　　　　　　**42.** $157° \, 30'$

43. $\dfrac{\pi}{8}$　　　　　　　　　**44.** $\dfrac{\pi}{12}$

45. $\dfrac{3\pi}{8}$　　　　　　　　**46.** $\dfrac{7\pi}{12}$

In Exercises 47–50, find the exact values of $\sin(u/2)$, $\cos(u/2)$ and $\tan(u/2)$ using the half-angle formulas.

47. $\sin u = \frac{5}{13},\quad \pi/2 < u < \pi$

48. $\cos u = \frac{7}{25},\quad 0 < u < \pi/2$

49. $\tan u = -\frac{8}{5},\quad 3\pi/2 < u < 2\pi$

50. $\cot u = 7,\quad \pi < u < 3\pi/2$

In Exercises 51–54, use the half-angle formulas to simplify the expression.

51. $\sqrt{\dfrac{1 - \cos 6x}{2}}$　　　**52.** $\sqrt{\dfrac{1 + \cos 4x}{2}}$

53. $-\sqrt{\dfrac{1 - \cos 8x}{1 + \cos 8x}}$　　**54.** $-\sqrt{\dfrac{1 - \cos(x - 1)}{2}}$

In Exercises 55–58, find the exact zeros of the function in the interval $[0, 2\pi)$. Use a graphing utility to graph the function and verify your answers.

55. $f(x) = \sin \dfrac{x}{2} - \cos x$

56. $h(x) = \sin \dfrac{x}{2} + \cos x - 1$

57. $h(x) = \cos \dfrac{x}{2} - \sin x$ **58.** $g(x) = \tan \dfrac{x}{2} - \sin x$

In Exercises 59–64, use the product-to-sum formulas to write the product as a sum or difference.

59. $6 \sin \dfrac{\pi}{3} \cos \dfrac{\pi}{3}$ **60.** $4 \sin \dfrac{\pi}{3} \cos \dfrac{5\pi}{6}$

61. $\sin 5\theta \cos 3\theta$ **62.** $5 \sin 3\alpha \sin 4\alpha$

63. $5 \cos(-5\beta) \cos 3\beta$ **64.** $\cos 2\theta \cos 4\theta$

In Exercises 65–72, use the sum-to-product formulas to write the sum or difference as a product.

65. $\sin 60° + \sin 30°$ **66.** $\cos 120° + \cos 30°$

67. $\sin 5\theta - \sin \theta$ **68.** $\sin x + \sin 7x$

69. $\sin(\alpha + \beta) - \sin(\alpha - \beta)$

70. $\cos(\phi + 2\pi) + \cos \phi$

71. $\cos\left(\theta + \dfrac{\pi}{2}\right) - \cos\left(\theta - \dfrac{\pi}{2}\right)$

72. $\sin\left(x + \dfrac{\pi}{2}\right) + \sin\left(x - \dfrac{\pi}{2}\right)$

In Exercises 73–76, find the exact zeros of the function in the interval $[0, 2\pi)$. Use a graphing utility to graph the function and verify your answers.

73. $g(x) = \sin 6x + \sin 2x$

74. $h(x) = \cos 2x - \cos 6x$

75. $f(x) = \dfrac{\cos 2x}{\sin 3x - \sin x} - 1$

76. $f(x) = \sin^2 3x - \sin^2 x$

In Exercises 77–80, use the figure to find the exact value of the trigonometric function in two ways.

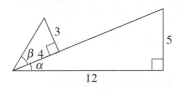

77. $\sin^2 \alpha$ **78.** $\cos^2 \alpha$

79. $\sin \alpha \cos \beta$ **80.** $\cos \alpha \sin \beta$

In Exercises 81–92, verify the identity algebraically. Use a graphing utility to confirm the identity graphically.

81. $\csc 2\theta = \dfrac{\csc \theta}{2 \cos \theta}$ **82.** $\sec 2\theta = \dfrac{\sec^2 \theta}{2 - \sec^2 \theta}$

83. $\cos^2 2\alpha - \sin^2 2\alpha = \cos 4\alpha$

84. $\cos^4 x - \sin^4 x = \cos 2x$

85. $(\sin x + \cos x)^2 = 1 + \sin 2x$

86. $1 + \cos 10y = 2 \cos^2 5y$

87. $\sec \dfrac{u}{2} = \pm \sqrt{\dfrac{2 \tan u}{\tan u + \sin u}}$

88. $\tan \dfrac{u}{2} = \csc u - \cot u$

89. $\cos 3\beta = \cos^3 \beta - 3 \sin^2 \beta \cos \beta$

90. $\sin 4\beta = 4 \sin \beta \cos \beta(1 - 2 \sin^2 \beta)$

91. $\dfrac{\cos 4x - \cos 2x}{2 \sin 3x} = -\sin x$

92. $\dfrac{\cos 3x - \cos x}{\sin 3x - \sin x} = -\tan 2x$

In Exercises 93 and 94, graph the function by using the power-reducing formulas.

93. $f(x) = \sin^2 x$ **94.** $f(x) = \cos^2 x$

In Exercises 95–98, (a) use a graphing utility to graph the function and approximate the maximum and minimum points on the graph in the interval $[0, 2\pi]$, and (b) solve the trigonometric equation and verify that the x-coordinates of the maximum and minimum points of f are among its solutions (calculus is required to find the trigonometric equation).

Function	*Trigonometric Equation*
95. $f(x) = 4 \sin \dfrac{x}{2} + \cos x$	$2 \cos \dfrac{x}{2} - \sin x = 0$
96. $f(x) = \cos 2x - 2 \sin x$	$-2 \cos x(2 \sin x + 1) = 0$
97. $f(x) = 2 \cos \dfrac{x}{2} + \sin 2x$	$2 \cos 2x - \sin \dfrac{x}{2} = 0$
98. $f(x) = 2 \sin \dfrac{x}{2}$ $- 5 \cos\left(2x - \dfrac{\pi}{4}\right)$	$10 \sin\left(2x - \dfrac{\pi}{4}\right)$ $+ \cos \dfrac{x}{2} = 0$

In Exercises 99–102, write the trigonometric expression as an algebraic expression.

99. $\sin(2 \arcsin x)$ **100.** $\cos(2 \arccos x)$

101. $\cos(2 \arcsin x)$ **102.** $\sin(2 \arctan x)$

103. *Projectile Motion*   The range of a projectile fired at an angle θ with the horizontal and with an initial velocity of v_0 feet per second is

$$r = \tfrac{1}{32} v_0 \sin 2\theta$$

where r is measured in feet. Determine the expression for the range in terms of θ.

104. *Area*   The length of each of the two equal sides of an isosceles triangle is 10 meters. The angle between the equal sides is θ.

(a) Express the area of the triangle as a function of $\theta/2$.

(b) Express the area of the triangle as a function of θ and determine the value of θ such that the area is a maximum.

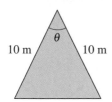

105. *Mach Number*   The mach number M of an airplane is the ratio of its speed to the speed of sound. When an airplane travels faster than the speed of sound, the sound waves form a cone behind the airplane. The mach number is related to the apex angle θ of the cone by

$$\sin \frac{\theta}{2} = \frac{1}{M}.$$

Find the angle θ that corresponds to a mach number of 4.5.

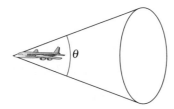

106. *Railroad Track*   When two railroad tracks merge, the overlapping portions of the tracks are in the shape of a circular arcs. The radius of the arcs r (in

feet) and the angle θ are related by

$$\frac{x}{2} = 2r \sin^2 \frac{\theta}{2}.$$

Write a formula for x in terms of $\cos \theta$.

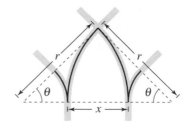

FIGURE FOR 106

Synthesis

True or False?   **In Exercises 107 and 108, determine whether the statement is true or false. Justify your answer.**

107. $\sin \dfrac{x}{2} = -\sqrt{\dfrac{1 - \cos x}{2}}, \qquad \pi \le x \le 2\pi.$

108. The graph of $y = 4 - 8 \sin^2 x$ has a maximum at $(\pi, 4)$.

109. *Conjecture*   Consider the function

$$f(x) = 2 \sin x \left[2 \cos^2 \left(\frac{x}{2} \right) - 1 \right].$$

(a) Use a graphing utility to graph the function.

(b) Make a conjecture about the function that is an identity with f.

(c) Verify your conjecture algebraically.

Review

In Exercises 110 and 111, find (if possible) the complement and supplement of each angle.

110. (a) $55°$ (b) $162°$

111. (a) $\dfrac{\pi}{18}$ (b) $\dfrac{9\pi}{20}$

In Exercises 112–115, sketch the graph of the function. Use a graphing utility to verify your graph.

112. $f(x) = \dfrac{3}{2} \cos 2x$ **113.** $f(x) = \dfrac{5}{2} \sin \dfrac{1}{2} x$

114. $f(x) = \dfrac{1}{2} \tan 2\pi x$ **115.** $f(x) = \dfrac{1}{4} \sec \dfrac{\pi x}{2}$

2 Chapter Summary

What did you learn?

Section 2.1	Review Exercises
☐ How to recognize and write the fundamental trigonometric identities	1–10
☐ How to use fundamental trigonometric identities to evaluate trigonometric functions, simplify trigonometric expressions, and rewrite trigonometric expressions	11–24

Section 2.2

☐ How to verify trigonometric identities	25–36

Section 2.3

☐ How to use standard algebraic techniques to solve trigonometric equations	37–50
☐ How to solve trigonometric equations of quadratic type	51–54
☐ How to solve trigonometric equations involving multiple angles	55–60
☐ How to use inverse trigonometric functions to solve trigonometric equations	61–64

Section 2.4

☐ How to use sum and difference formulas to evaluate trigonometric functions	65–78
☐ How to use sum and difference formulas to verify identities and solve trigonometric equations	79–86

Section 2.5

☐ How to use multiple-angle formulas to rewrite and evaluate trigonometric formulas	87–95
☐ How to use power-reducing formulas to rewrite and evaluate trigonometric formulas	96–99
☐ How to use half-angle formulas to rewrite and evaluate trigonometric functions	100–108
☐ How to use product-sum formulas to rewrite and evaluate trigonometric functions	109–117

2   Review Exercises

2.1 In Exercises 1–10, name the trigonometric function equivalent to the expression.

1. $\dfrac{1}{\cos x}$

2. $\dfrac{1}{\sin x}$

3. $\dfrac{1}{\sec x}$

4. $\dfrac{1}{\tan x}$

5. $\dfrac{\cos x}{\sin x}$

6. $\sqrt{1 + \tan^2 x}$

7. $\csc\left(\dfrac{\pi}{2} - x\right)$

8. $\cot\left(\dfrac{\pi}{2} - x\right)$

9. $\sec(-x)$

10. $\tan(-x)$

In Exercises 11–14, use the given values and trigonometric identities to evaluate (if possible) the other trigonometric functions of the angle.

11. $\sin x = \dfrac{3}{5}, \qquad \cos x = \dfrac{4}{5}$

12. $\tan \theta = \dfrac{3}{2}, \qquad \sec \theta = \dfrac{\sqrt{13}}{2}$

13. $\sin\left(\dfrac{\pi}{2} - x\right) = \dfrac{1}{\sqrt{2}}, \qquad \sin x = -\dfrac{1}{\sqrt{2}}$

14. $\csc\left(\dfrac{\pi}{2} - \theta\right) = 3, \qquad \sin \theta = \dfrac{2\sqrt{2}}{3}$

In Exercises 15–22, use the fundamental identities to simplify the trigonometric expression. Use the *table* feature of a graphing utility to check your result numerically.

15. $\dfrac{1}{\cot^2 x + 1}$

16. $\dfrac{\sec^2 x - 1}{\sec x - 1}$

17. $\dfrac{\sin^2 \alpha - \cos^2 \alpha}{\sin^2 \alpha - \sin \alpha \cos \alpha}$

18. $\dfrac{\sin^3 \beta + \cos^3 \beta}{\sin \beta + \cos \beta}$

19. $\tan^2 \theta (\csc^2 \theta - 1)$

20. $\csc^2 x(1 - \cos^2 x)$

21. $\tan\left(\dfrac{\pi}{2} - x\right)\sec x$

22. $\dfrac{\sin(-x)\cot x}{\sin\left(\dfrac{\pi}{2} - x\right)}$

23. *Rate of Change*   The rate of change of the function $f(x) = 2\sqrt{\sin x}$ is given by the expression $\sin^{-1/2} x \cos x$. Show that this expression can also be written as $\cot x \sqrt{\sin x}$.

24. *Rate of Change*   The rate of change of the function $f(x) = \csc x - \cot x$ is the expression $\csc^2 x - \csc x \cot x$. Show that this expression can also be written as

$$\dfrac{(1 - \cos x)}{\sin^2 x}.$$

2.2 In Exercises 25–36, verify the identity.

25. $\cos x(\tan^2 x + 1) = \sec x$

26. $\sec^2 x \cot x - \cot x = \tan x$

27. $\sin^3 \theta + \sin \theta \cos^2 \theta = \sin \theta$

28. $\cot^2 x - \cos^2 x = \cot^2 x \cos^2 x$

29. $\sin^5 x \cos^2 x = (\cos^2 x - 2\cos^4 x + \cos^6 x)\sin x$

30. $\cos^3 x \sin^2 x = (\sin^2 x - \sin^4 x)\cos x$

31. $\sqrt{\dfrac{1 - \sin \theta}{1 + \sin \theta}} = \dfrac{1 - \sin \theta}{|\cos \theta|}$

32. $\sqrt{1 - \cos x} = \dfrac{|\sin x|}{\sqrt{1 + \cos x}}$

33. $\dfrac{\csc(-x)}{\sec(-x)} = -\cot x$

34. $\dfrac{1 + \sec(-x)}{\sin(-x) + \tan(-x)} = -\csc x$

35. $\sin^2 x + \sin^2\left(\dfrac{\pi}{2} - x\right) = 1$

36. $\csc x \sin\left(\dfrac{\pi}{2} - x\right) = \cot x$

2.3 In Exercises 37 and 38, verify that the *x*-values are solutions of the equation.

37. $2\cos^2 4x - 1 = 0$

   (a) $x = \dfrac{\pi}{16}$ (b) $x = \dfrac{3\pi}{16}$

38. $2\sin^2 x - \sin x - 1 = 0$

   (a) $x = \dfrac{\pi}{2}$ (b) $x = \dfrac{7\pi}{6}$

In Exercises 39–50, solve the equation for $0 \le x < 2\pi$.

39. $2\sin x - 1 = 0$

40. $\tan x + 1 = 0$

41. $\sin x = \sqrt{3} - \sin x$

42. $4\cos x = 1 + 2\cos x$

43. $3\sqrt{3}\tan x = 3$

44. $\dfrac{1}{2}\sec x - 1 = 0$

45. $3 \csc^2 x = 4$

46. $4 \tan^2 x - 1 = \tan^2 x$

47. $4 \cos^2 x - 3 = 0$

48. $\sin x(\sin x + 1) = 0$

49. $\sin x - \tan x = 0$

50. $\csc x - 2 \cot x = 0$

In Exercises 51–60, find all solutions of the equation in the interval $[0, 2\pi)$. Use a graphing utility to verify your answers.

51. $2 \cos^2 x - \cos x = 1$

52. $2 \sin^2 x - 3 \sin x = -1$

53. $\cos^2 x + \sin x = 1$

54. $\sin^2 x + 2 \cos x = 2$

55. $2 \sin 2x - \sqrt{2} = 0$

56. $\sqrt{3} \tan 3x = 0$

57. $\cos 4x(\cos x - 1) = 0$

58. $3 \csc^2 5x = -4$

59. $\cos 4x - 7 \cos 2x = 8$

60. $\sin 4x - \sin 2x = 0$

In Exercises 61–64, use the inverse functions where necessary to find all solutions of the equation in the interval $[0, 2\pi)$. Use a graphing utility to verify your answers.

61. $\sin^2 x - 2 \sin x = 0$

62. $2 \cos^2 x + 3 \cos x = 0$

63. $\tan^2 \theta + \tan \theta - 12 = 0$

64. $\sec^2 x + 6 \tan x + 4 = 0$

2.4 **In Exercises 65–68, find the exact value of the sine, cosine, and tangent of the angle by using the sum or difference formula.**

65. $285° = 225° + 60°$

66. $345° = 300° + 45°$

67. $\dfrac{5\pi}{12} = \dfrac{2\pi}{3} - \dfrac{\pi}{4}$

68. $\dfrac{19\pi}{12} = \dfrac{11\pi}{6} - \dfrac{\pi}{4}$

In Exercises 69–72, write the expression as the sine, cosine, or tangent of an angle.

69. $\sin 140° \cos 50° + \cos 140° \sin 50°$

70. $\cos 25° \cos 15° - \sin 25° \sin 15°$

71. $\dfrac{\tan 25° + \tan 10°}{1 - \tan 25° \tan 10°}$

72. $\dfrac{\tan 68° - \tan 115°}{1 + \tan 68° \tan 115°}$

In Exercises 73–78, find the exact value of the trigonometric function given that $\sin u = \frac{3}{4}$, $\cos v = -\frac{5}{13}$, and u and v are in Quadrant II.

73. $\sin(u + v)$

74. $\tan(u + v)$

75. $\cos(u - v)$

76. $\sin(u - v)$

77. $\cos(u + v)$

78. $\tan(u - v)$

In Exercises 79–84, verify the identity.

79. $\cos\left(x + \dfrac{\pi}{2}\right) = -\sin x$

80. $\sin\left(x - \dfrac{3\pi}{2}\right) = \cos x$

81. $\cot\left(\dfrac{\pi}{2} - x\right) = \tan x$

82. $\sin(\pi - x) = \sin x$

83. $\cos 3x = 4 \cos^3 x - 3 \cos x$

84. $\dfrac{\sin(\alpha + \beta)}{\cos \alpha \cos \beta} = \tan \alpha + \tan \beta$

In Exercises 85 and 86, find all solutions of the equation in the interval $[0, 2\pi)$.

85. $\sin\left(x + \dfrac{\pi}{2}\right) - \sin\left(x - \dfrac{\pi}{2}\right) = \sqrt{3}$

86. $\cos\left(x + \dfrac{3\pi}{4}\right) - \cos\left(x - \dfrac{3\pi}{4}\right) = 0$

2.5 **In Exercises 87–92, use double-angle formulas to verify the identity algebraically and use a graphing utility to confirm it graphically.**

87. $6 \sin x \cos x = 3 \sin 2x$

88. $4 \sin x \cos x + 2 = 2 \sin 2x + 2$

89. $1 - 4 \sin^2 x \cos^2 x = \cos^2 2x$

90. $\sin 4x = 8 \cos^3 x \sin x - 4 \cos x \sin x$

91. $\dfrac{\sin 2\alpha}{\cos^2 \alpha - \sin^2 \alpha} = \tan 2\alpha$

92. $\tan^2 x = \dfrac{1 - \cos 2x}{1 + \cos 2x}$

In Exercises 93 and 94, find the exact values of $\sin 2u$, $\cos 2u$, and $\tan 2u$ using the double-angle formulas.

93. $\sin u = -\dfrac{5}{7}, \qquad \pi < u < \dfrac{3\pi}{2}$

94. $\cos u = -\dfrac{2}{\sqrt{5}}, \qquad \dfrac{\pi}{2} < u < \pi$

95. *Projectile Motion* A baseball leaves the hand of the first baseman at an angle of θ with the horizontal and an initial velocity of $v_0 = 80$ feet per second. The ball is caught by the second baseman 100 feet away. Find θ if the range r of a projectile is

$$r = \dfrac{1}{32} v_0^2 \sin 2\theta.$$

In Exercises 96–99, use the power-reducing formulas to rewrite the expression in terms of the first power of the cosine of the angle.

96. $\sin^6 x$

97. $\cos^4 x \sin^4 x$

98. $\cos^4 2x$

99. $\sin^4 2x$

In Exercises 100–103, use the half-angle formulas to determine the exact value of the sine, cosine, and tangent of the angle.

100. $105°$

101. $67° \, 30'$

102. $\dfrac{7\pi}{8}$

103. $\dfrac{11\pi}{12}$

In Exercises 104 and 105, use the half-angle formulas to simplify the expression.

104. $-\sqrt{\dfrac{1 + \cos 10x}{2}}$

105. $\dfrac{\sin 6x}{1 + \cos 6x}$

In Exercises 106 and 107, use the half-angle formulas to verify the identity.

106. $\sqrt{\dfrac{1 - \cos^2 x}{1 + \cos x}} = \sqrt{2}\left|\sin \dfrac{x}{2}\right|$

107. $\dfrac{\sec x - 1}{\tan x} = \tan \dfrac{x}{2}$

108. *Volume* A trough for feeding cattle is 4 meters long and its cross sections are isosceles triangles with the two equal sides being $\frac{1}{2}$ meter. The angle between the equal sides is θ.

(a) Express the trough's volume as a function of $\theta/2$.

(b) Express the volume of the trough as a function of θ and determine the value of θ such that the volume is maximum.

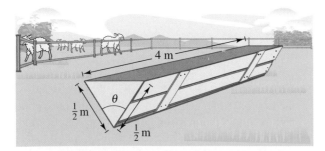

In Exercises 109–112, write the trigonometric expression as a product.

109. $\cos 3\theta + \cos 2\theta$ **110.** $\sin 5\theta - \sin 3\theta$

111. $\cos \dfrac{3\pi}{4} - \cos \dfrac{\pi}{4}$ **112.** $\sin\left(x + \dfrac{\pi}{4}\right) - \sin\left(x - \dfrac{\pi}{4}\right)$

In Exercises 113–116, write the trigonometric expression as a sum or difference.

113. $\sin 3\alpha \sin 2\alpha$ **114.** $3 \sin 2x \sin 3x$

115. $6 \sin \dfrac{\pi}{4} \cos \dfrac{\pi}{4}$ **116.** $\cos \dfrac{x}{2} \cos \dfrac{x}{4}$

117. *Harmonic Motion* A weight is attached to a spring suspended vertically from a ceiling. When a driving force is applied to the system, the weight moves vertically from its equilibrium position, and this motion is described by the model

$$y = 1.5 \sin 8t - 0.5 \cos 8t$$

where y is the distance from equilibrium measured in feet and t is the time in seconds.

(a) Write the model in the form
$y = \sqrt{a^2 + b^2} \sin(Bt + C)$.

(b) Use a graphing utility to graph the model.

(c) Find the amplitude of the weight's oscillations.

(d) Find the frequency of the weight's oscillations.

Synthesis

True or False? **In Exercises 118–121, determine whether the statement is true or false. Justify your answer.**

118. If $\dfrac{\pi}{2} < \theta < \pi$, then $\cos \dfrac{\theta}{2} < 0$.

119. $\sin(x + y) = \sin x + \sin y$

120. $4 \sin(-x) \cos(-x) = -2 \sin 2x$

121. $4 \sin 45° \cos 15° = 1 + \sqrt{3}$

122. List the reciprocal identities, quotient identities, and Pythagorean identities from memory.

123. Is $\cos \theta = \sqrt{1 - \sin^2 \theta}$ an identity? Explain.

In Exercises 124 and 125, use the graphs of y_1 and y_2 to determine how to change y_2 to a new function y_3 so that $y_1 = y_3$.

124. $y_1 = \sec^2\left(\dfrac{\pi}{2} - x\right)$ **125.** $y_1 = \dfrac{\cos 3x}{\cos x}$

$\quad\quad y_2 = \cot^2 x$ $\quad\quad y_2 = (2 \sin x)^2$

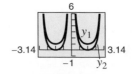

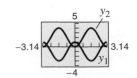

Chapter Project *Projectile Motion*

Parametric equations are equations that express two variables in terms of a third variable, called the parameter. In this project, you will use parametric equations to model the path of a projectile. These parametric equations express position x and position y in terms of time t. For any time t, the horizontal position $x(t)$ and vertical position $y(t)$ of a projectile (ignoring air resistance) launched at ground level is given by the parameter equations

$$x(t) = (v_0 \cos \theta)t$$

$$y(t) = (v_0 \sin \theta)t - 16t^2.$$

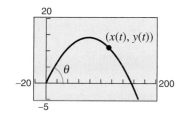

In these equations, θ is the angle with the horizontal and v_0 is the initial velocity in feet per second, as indicated in the figure at the right.

Set your graphing utility to parametric and degree modes, and use the viewing window

$$0 \le t \le 5$$

$$-20 \le x \le 200$$

$$-5 \le y \le 20.$$

Letting $v_0 = 88$ feet per second and $\theta = 20°$, graph the parametric equations

$$x(t) = (88 \cos 20°)t$$

$$y(t) = (88 \sin 20°)t - 16t^2.$$

Use the *zoom* and *trace* features to (a) find the maximum height attained by the projectile, (b) find the time at which the maximum height occurs, (c) determine the length of time that the projectile is in the air, and (d) determine the range of the projectile.

Questions for Further Exploration

1. Verify algebraically the range of the projectile by solving the equation $(88 \sin 20°)t - 16t^2 = 0$ for t and then evaluating $x(t)$ at this t-value.

2. Use a graphing utility to find the maximum height and range of the projectile when $\theta = 30°$ and $v_0 = 132$ feet per second.

3. Let $v_0 = 60$ feet per second. Find the maximum range for the angles $\theta = 20°, 30°, 40°, 50°,$ and $60°$. In general, what angle should you use to produce the maximum range?

4. What is the relationship between the time the projectile reaches its maximum height and the time it takes for the projectile to return to the ground? Explain.

5. Eliminate t from the parametric equations

$$x(t) = (v_0 \cos \theta)t \quad \text{and} \quad y(t) = (v_0 \sin \theta)t - 16t^2$$

by solving for t in the first equation and substituting this value into the equation for y. Show in this case that the height of the projectile is given by the equation

$$y = (\tan \theta)x - \frac{16 \sec^2 \theta}{v_0^2}x^2.$$

Use this equation to find the angle θ corresponding to a maximum range of 200 feet and initial velocity of $v_0 = 80$ feet per second.

2   Chapter Test

Take this test as you would take a test in class. After you are done, check your work against the answers given in the back of the book.

1. If $\tan \theta = \frac{6}{5}$ and $\cos \theta < 0$, use the fundamental identities to evaluate the other five trigonometric functions of θ.

2. Use the fundamental identities to simplify $\csc^2 \beta(1 - \cos^2 \beta)$.

3. Factor and simplify $\dfrac{\sec^4 x - \tan^4 x}{\sec^2 x + \tan^2 x}$.

4. Add and simplify $\dfrac{\cos \theta}{\sin \theta} + \dfrac{\sin \theta}{\cos \theta}$.

5. Determine the values of θ, $0 \le \theta < 2\pi$, for which $\tan \theta = -\sqrt{\sec^2 \theta - 1}$ is true.

6. Use a graphing utility to graph the functions $y_1 = \cos x + \sin x \tan x$ and $y_2 = \sec x$. Make a conjecture about y_1 and y_2. Verify the result algebraically.

The *Interactive* CD-ROM and *Internet* versions of this text provide answers to the Chapter Tests and Cumulative Tests. They also offer Chapter Pre-Tests (that test key skills and concepts covered in previous chapters) and Chapter Post-Tests, both of which have randomly generated exercises with diagnostic capabilities.

In Exercises 7–12, verify the identity.

7. $\sin \theta \sec \theta = \tan \theta$

8. $\sec^2 x \tan^2 x + \sec^2 x = \sec^4 x$

9. $\dfrac{\csc \alpha + \sec \alpha}{\sin \alpha + \cos \alpha} = \cot \alpha + \tan \alpha$

10. $\cos\left(x + \dfrac{\pi}{2}\right) = -\sin x$

11. $\sin(n\pi + \theta) = (-1)^n \sin \theta$, where n is an integer.

12. $(\sin x + \cos x)^2 = 1 + \sin 2x$

13. Find the exact values of the sine, cosine, and tangent of the angle $-\dfrac{7\pi}{12}$.

14. Use the fundamental identities to simplify the expression $\dfrac{\sin^4 x}{\tan^2 x}$.

15. Write $3 \sin 2\theta \sin 6\theta$ as a sum or difference.

16. Write $\cos 5\theta + \cos 3\theta$ as a product.

In Exercises 17–20, find all solutions of the equation in the interval $[0, 2\pi)$.

17. $\tan^2 x + \tan x = 0$

18. $\sin 2\alpha - \cos \alpha = 0$

19. $4 \cos^2 x - 3 = 0$

20. $\csc^2 x - \csc x - 2 = 0$

21. Use a graphing utility to approximate the solutions of the equation $5 \cos x - x = 0$ in the interval $[0, 2\pi)$ accurate to three decimal places.

22. Use the figure at the right to find the exact values of $\sin 2u$ and $\tan 2u$.

23. Explain why the equation $\cos^2 x + \cos x - 6 = 0$ has no solution.

24. Find the exact value of $\tan 105°$.

25. The index of refraction n of a transparent material is the ratio of the speed of light in a vacuum to the speed of light in the material. For the glass triangular prism in the figure at the right, $n = 1.5$ and $\alpha = 60°$. Find the angle θ for the glass prism if

$$n = \dfrac{\sin(\theta/2 + \alpha/2)}{\sin(\theta/2)}.$$

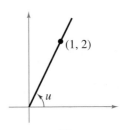

FIGURE FOR 22

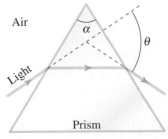

FIGURE FOR 25

Additional Topics in Trigonometry

3

The Big Picture

In this chapter you will learn how to

❏ use the Law of Sines and the Law of Cosines to solve oblique triangles.

❏ find areas of oblique triangles.

❏ represent vectors as directed line segments and perform mathematical operations on vectors.

❏ find direction angles of vectors.

❏ find dot products of two vectors and use properties of the dot product.

Superstock

In November of 1999, 81.4% of U.S. airline flights arrived on time. Factors such as severe weather, aircraft maintenance, and air traffic control decisions can cause flight delays. (Source: U.S. Department of Transportation)

Important Vocabulary

As you encounter each new vocabulary term in this chapter, add the term and its definition to your notebook glossary.

- oblique triangle (p. 268)
- Law of Sines (p. 268)
- Law of Cosines (p. 277)
- Heron's Area Formula (p. 280)
- directed line segment (p. 284)
- initial point (p. 284)
- terminal point (p. 284)
- vector **v** in the plane (p. 284)
- standard position (p. 285)
- component form of a vector **v** (p. 285)

- zero vector (p. 285)
- magnitude of **v** (p. 285)
- unit vector (p. 285)
- scalar multiplication (p. 286)
- vector addition (p. 286)
- resultant (p. 286)
- standard unit vectors (p. 289)
- horizontal and vertical components of **v** (p. 289)

- linear combination of vectors (p. 289)
- direction angle (p. 290)
- dot product (p. 298)
- angle between two nonzero vectors (p. 299)
- orthogonal vectors (p. 300)
- vector components (p. 301)
- projection (p. 301)

Additional Resources   Text-specific additional resources are available to help you do well in this course. See page xvi for details.

3.1   Law of Sines

Introduction

In Chapter 1 you looked at techniques for solving right triangles. In this section and the next, you will solve **oblique triangles**—triangles that have no right angles. As standard notation, the angles of a triangle are labeled A, B, and C, and their opposite sides are labeled a, b, and c, as shown in Figure 3.1.

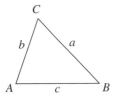

Figure 3.1

To solve an oblique triangle, you need to know the measure of at least one side and any two other parts of the triangle—two sides, two angles, or one angle and one side. This breaks down into the following four cases.

1. Two angles and any side (AAS or ASA)
2. Two sides and an angle opposite one of them (SSA)
3. Three sides (SSS)
4. Two sides and their included angle (SAS)

The first two cases can be solved using the **Law of Sines,** whereas the last two cases require the Law of Cosines (Section 3.2). A proof of the Law of Sines is given in Appendix A.

What You Should Learn:

• How to use the Law of Sines to solve oblique triangles (AAS, ASA, or SSA)
• How to find areas of oblique triangles
• How to use the Law of Sines to model and solve real-life problems

Why You Should Learn It:

You can use the Law of Sines to solve real-life problems involving oblique triangles. For instance, Exercise 35 on page 275 shows how the Law of Sines can be used to help determine the distance from a ranger station to a forest fire.

Superstock

Law of Sines

If ABC is a triangle with sides a, b, and c, then

$$\frac{a}{\sin A} = \frac{b}{\sin B} = \frac{c}{\sin C}.$$

Oblique Triangles

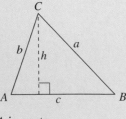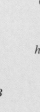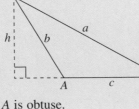

A is acute. A is obtuse.

The Law of Sines can also be written in the reciprocal form

$$\frac{\sin A}{a} = \frac{\sin B}{b} = \frac{\sin C}{c}.$$

EXAMPLE 1   Given Two Angles and One Side—AAS

For the triangle in Figure 3.2, $C = 102.3°$, $B = 28.7°$, and $b = 27.4$ feet. Find the remaining angle and sides.

Solution

The third angle of the triangle is

$$A = 180° - B - C$$
$$= 180° - 28.7° - 102.3°$$
$$= 49.0°.$$

By the Law of Sines, you have

$$\frac{a}{\sin 49°} = \frac{b}{\sin 28.7°} = \frac{c}{\sin 102.3°}. \qquad \frac{a}{\sin A} = \frac{b}{\sin B} = \frac{c}{\sin C}$$

Using $b = 27.4$ produces

$$a = \frac{27.4}{\sin 28.7°}(\sin 49°) \approx 43.06 \text{ feet}$$

and

$$c = \frac{27.4}{\sin 28.7°}(\sin 102.3°) \approx 55.75 \text{ feet}.$$

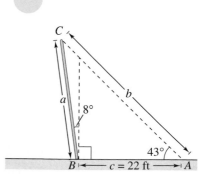

C
$b = 27.4$ ft
$102.3°$
a
$28.7°$
A c B

Figure 3.2

STUDY T!P

When you are solving triangles, a careful sketch is useful as a quick test for the feasibility of an answer. Remember that the longest side lies opposite the largest angle, and the shortest side lies opposite the smallest angle.

EXAMPLE 2   Given Two Angles and One Side—ASA

A pole tilts *toward* the sun at an 8° angle from the vertical, and it casts a 22-foot shadow. The angle of elevation from the tip of the shadow to the top of the pole is 43°. How tall is the pole?

Solution

From Figure 3.3, note that $A = 43°$ and $B = 90° + 8° = 98°$. So, the third angle is

$$C - 180° - A - B$$
$$= 180° - 43° - 98°$$
$$= 39°.$$

By the Law of Sines, you have

$$\frac{a}{\sin 43°} = \frac{c}{\sin 39°}. \qquad \frac{a}{\sin A} = \frac{c}{\sin C}$$

Because $c = 22$ feet, the length of the pole is

$$a = \frac{22}{\sin 39°}(\sin 43°) \approx 23.84 \text{ feet}.$$

For practice, try reworking Example 2 for a pole that tilts *away from* the sun under the same conditions.

Figure 3.3

The Ambiguous Case (SSA)

In Examples 1 and 2 you saw that two angles and one side determine a unique triangle. However, if two sides and one opposite angle are given, three possible situations can occur: (1) no such triangle exists, (2) one such triangle exists, or (3) two distinct triangles may satisfy the conditions.

A computer simulation to accompany this concept appears in the *Interactive* CD-ROM and *Internet* versions of this text.

The Ambiguous Case (SSA)

Consider a triangle in which you are given a, b, and A ($h = b \sin A$).

	A is acute.	A is acute.	A is acute.	A is acute.	A is obtuse.	A is obtuse.
Sketch			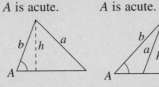			
Necessary condition	$a < h$	$a = h$	$a > b$	$h < a < b$	$a \le b$	$a > b$
Possible triangles	None	One	One	Two	None	One

EXAMPLE 3   Single-Solution Case—SSA

For the triangle in Figure 3.4, $a = 22$ inches, $b = 12$ inches, and $A = 42°$. Find the remaining side and angles.

Solution
By the Law of Sines, you have

$$\frac{22}{\sin 42°} = \frac{12}{\sin B} \qquad\qquad \frac{a}{\sin A} = \frac{b}{\sin B}$$

$$\sin B = 12\left(\frac{\sin 42°}{22}\right) \approx 0.3649803$$

$$B \approx 21.41°. \qquad\qquad B \text{ is acute.}$$

Now you can determine that

$$C \approx 180° - 42° - 21.41°$$

$$= 116.59°.$$

Then the remaining side is given by

$$\frac{c}{\sin 116.59°} = \frac{22}{\sin 42°} \qquad\qquad \frac{c}{\sin C} = \frac{a}{\sin A}$$

$$c = \sin 116.59°\left(\frac{22}{\sin 42°}\right) \approx 29.40 \text{ inches.}$$

Notice in Example 3 that A is acute and $a > b$, which results in one possible triangle.

b = 12 in.   *a* = 22 in.
$42°$
A   *c*   *B*

Figure 3.4   *One solution: $a > b$*

EXAMPLE 4   No-Solution Case—SSA

Show that there is no triangle for which $a = 15$, $b = 25$, and $A = 85°$.

Solution

Begin by making the sketch shown in Figure 3.5. From this figure it appears that no triangle is formed. You can verify this by using the Law of Sines.

$$\frac{a}{\sin A} = \frac{b}{\sin B}$$

$$\frac{15}{\sin 85°} = \frac{25}{\sin B}$$

$$\sin B = 25\left(\frac{\sin 85°}{15}\right) \approx 1.660 > 1$$

This contradicts the fact that $|\sin B| \leq 1$. So, no triangle can be formed having sides $a = 15$ and $b = 25$ and an angle of $A = 85°$.

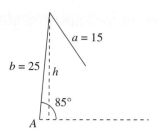

Figure 3.5   *No solution: $a < h$*

The *Interactive* CD-ROM and *Internet* versions of this text show every example with its solution; clicking on the *Try It!* button brings up similar problems. Guided Examples and Integrated Examples show step-by-step solutions to additional examples. Integrated Examples are related to several concepts in the section.

EXAMPLE 5   Two-Solution Case—SSA

Find two triangles for which $a = 12$ meters, $b = 31$ meters, and $A = 20.5°$.

Solution

Because $h = b \sin A = 31(\sin 20.5°) \approx 10.86$ meters, you can conclude that there are two possible triangles (because $h < a < b$). By the Law of Sines, you have

$$\frac{a}{\sin A} = \frac{b}{\sin B}$$

$$\sin B = b\left(\frac{\sin A}{a}\right) = 31\left(\frac{\sin 20.5°}{12}\right) \approx 0.9047.$$

There are two angles between $0°$ and $180°$ whose sine is 0.9047: $B_1 \approx 64.8°$ and $B_2 \approx 180° - 64.8° = 115.2°$. For $B_1 \approx 64.8°$, you obtain

$$C \approx 180° - 20.5° - 64.8° = 94.7°$$

$$c = \frac{a}{\sin A}(\sin C) = \frac{12}{\sin 20.5°}(\sin 94.7°) \approx 34.15 \text{ meters.}$$

For $B_2 \approx 115.2°$, you obtain

$$C \approx 180° - 20.5° - 115.2° = 44.3°$$

$$c = \frac{a}{\sin A}(\sin C) = \frac{12}{\sin 20.5°}(\sin 44.3°) \approx 23.93 \text{ meters.}$$

The resulting triangles are shown in Figure 3.6.

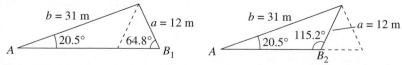

Figure 3.6

Area of an Oblique Triangle

The procedure used to prove the Law of Sines leads to a simple formula for the area of an oblique triangle. Referring to Figure 3.7, note that each triangle has a height of

$$h = b \sin A.$$

Consequently, the area of each triangle is

$$\text{Area} = \frac{1}{2}(\text{base})(\text{height}) = \frac{1}{2}(c)(b \sin A) = \frac{1}{2}bc \sin A.$$

By similar arguments, you can develop the formulas

$$\text{Area} = \frac{1}{2}ab \sin C = \frac{1}{2}ac \sin B.$$

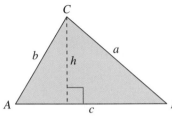

A is acute.

A is obtuse.

Figure 3.7

Area of an Oblique Triangle

The area of any triangle is one-half the product of the lengths of two sides times the sine of their included angle. That is,

$$\text{Area} = \frac{1}{2}bc \sin A = \frac{1}{2}ab \sin C = \frac{1}{2}ac \sin B.$$

Note that if angle A is 90°, the formula gives the area for a right triangle as

$$\text{Area} = \frac{1}{2}bc = \frac{1}{2}(\text{base})(\text{height}).$$

Similar results are obtained for angles C and B equal to 90°.

EXAMPLE 6   Finding the Area of an Oblique Triangle

Find the area of a triangular lot having two sides of lengths 90 meters and 52 meters and an included angle of 102°.

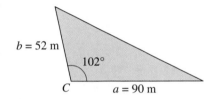

Figure 3.8

Solution

Consider $a = 90$ m, $b = 52$ m, and angle $C = 102°$, as shown in Figure 3.8. Then the area of the triangle is

$$\text{Area} = \frac{1}{2}ab \sin C = \frac{1}{2}(90)(52)(\sin 102°) \approx 2289 \text{ square meters.}$$

Application

EXAMPLE 7   An Application of the Law of Sines

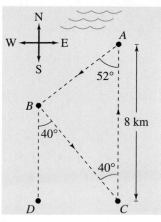

The course for a boat race starts at point A and proceeds in the direction S 52° W to point B, then in the direction S 40° E to point C, and finally back to A, as shown in Figure 3.9. The point C lies 8 kilometers directly south of point A. Approximate the total distance of the race course.

Solution

Because lines BD and AC are parallel, it follows that $\angle BCA \cong \angle DBC$. Consequently, triangle ABC has the measures shown in Figure 3.10. For angle B, you have $B = 180° - 52° - 40° = 88°$. Using the Law of Sines

$$\frac{a}{\sin 52°} = \frac{b}{\sin 88°} = \frac{c}{\sin 40°}$$

you can let $b = 8$ and obtain

$$a = \frac{8}{\sin 88°}(\sin 52°) \approx 6.308$$

and

$$c = \frac{8}{\sin 88°}(\sin 40°) \approx 5.145.$$

The total length of the course is approximately

Length $\approx 8 + 6.308 + 5.145 = 19.453$ kilometers.

Figure 3.9

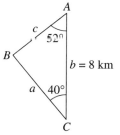

Figure 3.10

Writing About Math   *Using the Law of Sines*

In this section, you have been using the Law of Sines to solve *oblique* triangles. Can the Law of Sines also be used to solve a right triangle? If so, write a short paragraph explaining how to use the Law of Sines to solve the following two triangles. Is there an easier way to solve these triangles?

a. (ASA)   **b.** (SSA)

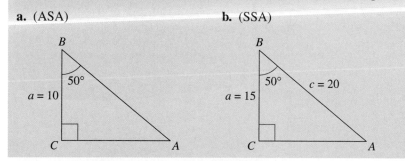

3.1 Exercises

In Exercises 1–14, use the given information to solve the triangle.

1. $A = 30°$, $a = 12$, $B = 45°$

2. $C = 120°$, $c = 15$, $B = 45°$

3. $A = 10°$, $a = 4.5$, $B = 60°$

4. $C = 135°$, $c = 45$, $B = 10°$

5. $A = 36°$, $a = 10$, $b = 4$

6. $A = 60°$, $a = 9$, $c = 10$

7. $A = 150°$, $C = 20°$, $a = 200$

8. $A = 24.3°$, $C = 54.6°$, $c = 2.68$

9. $A = 83°\ 20'$, $C = 54.6°$, $c = 18.1$

10. $A = 5°\ 40'$, $B = 8°\ 15'$, $b = 4.8$

11. $B = 15°\ 30'$, $a = 4.5$, $b = 6.8$

12. $C = 85°\ 20'$, $a = 35$, $c = 50$

13. $A = 110°\ 15'$, $a = 48$, $b = 16$

14. $B = 2°\ 45'$, $b = 6.2$, $c = 5.8$

In Exercises 15–20, use the given information to solve the triangle. If two solutions exist, find both.

15. $A = 58°$, $a = 4.5$, $b = 12.8$

16. $A = 94°$, $a = 14.6$, $b = 14.6$

17. $A = 58°$, $a = 11.4$, $b = 12.8$

18. $A = 58°$, $a = 4.5$, $b = 5$

19. $A = 110°$, $a = 125$, $b = 100$

20. $A = 140°$, $a = 48$, $b = 46$

In Exercises 21 and 22, find a value for b such that the triangle has (a) one, (b) two, and (c) no solution(s).

21. $A = 36°$, $a = 5$ 22. $A = 60°$, $a = 10$

In Exercises 23–28, find the area of the triangle having the indicated sides and angle.

23. $C = 110°$, $a = 6$, $b = 10$

24. $B = 74°\ 30'$, $a = 103$, $c = 58$

25. $A = 38°\ 45'$, $b = 67$, $c = 85$

26. $A = 5°\ 15'$, $b = 4.5$, $c = 22$

27. $B = 130°$, $a = 92$, $c = 30$

28. $C = 84°\ 30'$, $a = 16$, $b = 20$

29. **Height**   A flagpole is located on a slope that makes an angle of $14°$ with the horizontal. The pole casts a 16-meter shadow up the slope when the angle of elevation of the sun is $20°$.

   (a) Draw a triangle that represents the problem.

   (b) Write an equation involving the unknown quantity.

   (c) Find the height of the flagpole.

30. **Height**   You are standing 40 meters from the base of a tree that is leaning $8°$ from vertical away from you. The angle of elevation from your feet to the top of the tree is $20°\ 50'$.

   (a) Draw a triangle that represents the problem.

   (b) Write an equation involving the unknown slant height of the tree.

   (c) Find the slant height of the tree.

31. **Flight Path**   A plane flies 500 kilometers with a bearing of N $44°$ W from B to C. The plane then flies southwest 840 kilometers from C to A. Find the bearing of the flight from C to A.

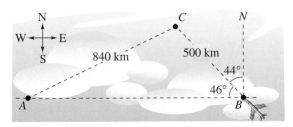

32. **Bridge Design**   A bridge is to be built across a small lake from B to C. The bearing from B to C is S $41°$ W. From a point A, 100 meters from B, the bearings to B and C are S $74°$ E and S $28°$ E, respectively. Find the distance from B to C.

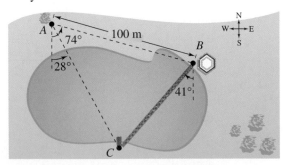

33. Railroad Track Design The circular arc of a railroad curve has a chord of length 3000 feet and a central angle of 40°.

(a) Draw a diagram to represent the problem. Show the known quantities on the diagram and use variables r and s to represent the radius of the arc and the length of the arc, respectively.

(b) Find the radius r of the circular arc.

(c) Find the length s of the circular arc.

34. Glide Path A pilot has just started on the glide path for landing at an airport where the length of the runway is 9000 feet. The angles of depression from the plane to the ends of the runway are 17.5° and 18.8°.

(a) Draw a diagram to represent the problem.

(b) Find the air distance the plane must travel until touching down on the near end of the runway.

(c) Find the ground distance the plane must travel until touching down.

(d) Find the altitude of the plane when the pilot begins the descent.

35. Locating a Fire Two fire towers A and B are 30 kilometers apart. The bearing from A to B is N 65° E. A fire is spotted by a ranger in each tower, and its bearings from A and B are N 28° E and N 16.5° W, respectively. Find the distance of the fire from each tower.

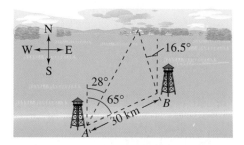

36. Distance A boat is sailing due east parallel to the shoreline at a speed of 10 miles per hour. At a given time the bearing to the lighthouse is S 70° E, and 15 minutes later the bearing is S 63° E. Find the distance from the boat to the shoreline if the lighthouse is at the shoreline.

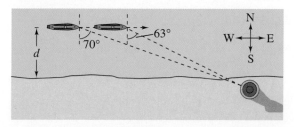

37. Distance A family is traveling due west on a road that passes a famous landmark. At a given time the bearing to the landmark is N 62° W, and after the family travels 5 miles farther the bearing is N 38° W. What is the closest the family will come to the landmark while on the road?

38. Altitude The angles of elevation to an airplane from two points A and B on level ground are 55° and 72°, respectively. The points A and B are 2.2 miles apart, and the airplane is east of both points in the same vertical plane. Find the altitude of the plane.

39. Engine Design The connecting rod in an engine is 6 inches in length and the radius of the crankshaft is $1\frac{1}{2}$ inches. Let d be the distance the piston is from the top of its stroke for the angle θ.

(a) Use a graphing utility to complete the table.

θ	0°	45°	90°	135°	180°
d					

(b) The spark plug fires at $\theta = 5°$ before top dead center. How far is the piston from the top of its stroke at this time?

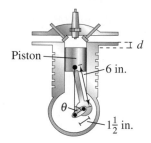

40. Angle of Elevation A 10-meter telephone pole casts a 17-meter shadow directly down a slope when the angle of elevation of the sun is 42°. Find θ, the angle of elevation of the ground.

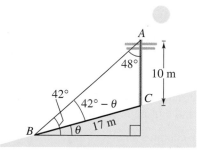

41. *Distance* The angles of elevation θ and ϕ to an airplane are being continuously monitored at two observation points A and B, respectively, which are 2 miles apart. Write an equation giving the distance d between the plane and point B in terms of θ and ϕ.

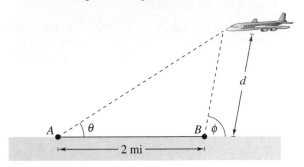

42. *Graphical and Numerical Analysis* In the figure, α and β are positive angles.

(a) Write α as a function of β.

(b) Use a graphing utility to graph the function. Determine its domain and range.

(c) Write c as a function of β.

(d) Use a graphing utility to graph the function in part (c). Determine its domain and range.

(e) Use a graphing utility to complete the table. What can you conclude?

β	0.4	0.8	1.2	1.6	2.0	2.4	2.8
α							
c							

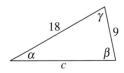

Synthesis

True or False? **In Exercises 43–45, determine whether the statement is true or false. Justify your answer.**

43. If any three sides or angles of an oblique triangle are known, then the triangle can be solved.

44. If a triangle contains an obtuse angle, then it must be oblique.

45. Two sides and an opposite angle determine a unique triangle.

46. *Graphical Analysis*

(a) Write the area A of the shaded region in the figure as a function of θ.

(b) Use a graphing utility to graph the area function.

(c) Determine the domain of the area function. Explain how the area of the region and the domain of the function would change if the 8-centimeter line segment were decreased in length.

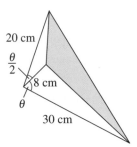

Review

In Exercises 47–50, use the given values to evaluate (if possible) the other four trigonometric functions.

47. $\cos \theta = \frac{5}{13}$, $\sin \theta = -\frac{12}{13}$

48. $\tan \theta = \frac{2}{9}$, $\csc \theta = -\frac{\sqrt{85}}{2}$

49. $\sec \theta = -\frac{\sqrt{122}}{11}$, $\cot \theta = -11$

50. $\cot \theta$ is undefined, $\cos \theta < 0$

In Exercises 51 and 52, verify the identity.

51. $\sec^2 x(\csc^2 x - 1) = \csc^2 x$

52. $\csc \theta \tan \theta - \dfrac{1 - \sin^2 \theta}{\cos \theta} = \dfrac{1 - \cos^2 \theta}{\cos \theta}$

In Exercises 53–56, write the product as a sum or difference.

53. $6 \sin 8\theta \cos 3\theta$

54. $2 \cos 2\theta \cos 5\theta$

55. $3 \cos \dfrac{\pi}{6} \sin \dfrac{5\pi}{3}$

56. $\dfrac{5}{2} \sin \dfrac{3\pi}{4} \sin \dfrac{5\pi}{6}$

3.2 Law of Cosines

Introduction

Two cases remain in the list of conditions needed to solve an oblique triangle—SSS and SAS. To use the Law of Sines, you must know at least one side and its opposite angle. If you are given three sides (SSS), or two sides and their included angle (SAS), none of the ratios in the Law of Sines would be complete. In such cases you can use the **Law of Cosines.** A proof of the Law of Cosines is given in Appendix A.

Law of Cosines

Standard Form	Alternative Form
$a^2 = b^2 + c^2 - 2bc \cos A$	$\cos A = \dfrac{b^2 + c^2 - a^2}{2bc}$
$b^2 = a^2 + c^2 - 2ac \cos B$	$\cos B = \dfrac{a^2 + c^2 - b^2}{2ac}$
$c^2 = a^2 + b^2 - 2ab \cos C$	$\cos C = \dfrac{a^2 + b^2 - c^2}{2ab}$

EXAMPLE 1 Three Sides of a Triangle—SSS

Find the three angles of the triangle in Figure 3.11.

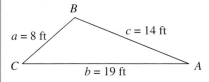

Figure 3.11

Solution

It is a good idea first to find the angle opposite the longest side—side b in this case. Using the Law of Cosines, you find that

$$\cos B = \frac{a^2 + c^2 - b^2}{2ac} = \frac{8^2 + 14^2 - 19^2}{2(8)(14)} \approx -0.45089.$$

Because $\cos B$ is negative, you know that B is an *obtuse* angle given by $B \approx 116.80°$. At this point knowing that $B \approx 116.80°$, it is simpler to use the Law of Sines to determine A.

$$\sin A = a\left(\frac{\sin B}{b}\right) \approx 8\left(\frac{\sin 116.80°}{19}\right) \approx 0.37582$$

Because B is obtuse, you know that A must be acute, because a triangle can have at most one obtuse angle. So, $A \approx 22.08°$ and

$$C \approx 180° - 22.08° - 116.80° = 41.12°.$$

Heron's Area Formula

The Law of Cosines can be used to establish the following formula for the area of a triangle. This formula is called **Heron's Area Formula** after the Greek mathematician Heron (ca. 100 B.C.). For a proof of Heron's Area Formula, see Appendix A.

> **Heron of Alexandria (ca. 100 B.C.)** was a Greek geometer and inventor. His works include the means of finding the areas of triangles, quadrilaterals, regular polygons having 3 to 12 sides, and circles, as well as finding the surface area and volume of three-dimensional objects.

> ### Heron's Area Formula
>
> Given any triangle with sides of lengths a, b, and c, the area of the triangle is
>
> $$\text{Area} = \sqrt{s(s - a)(s - b)(s - c)}$$
>
> where $s = (a + b + c)/2$.

EXAMPLE 5 Using Heron's Area Formula

Find the area of the triangular region having sides of lengths $a = 43$ meters, $b = 53$ meters, and $c = 72$ meters.

Solution

Because $s = (a + b + c)/2 = 168/2 = 84$, Heron's Area Formula yields

$$\text{Area} = \sqrt{s(s - a)(s - b)(s - c)}$$
$$= \sqrt{84(41)(31)(12)} \approx 1131.89 \text{ square meters.}$$

Writing About Math *The Area of a Triangle*

You have now studied three different formulas for the area of a triangle.

Standard Formula: $\qquad$ $\text{Area} = \frac{1}{2}bh$

Oblique Triangle: $\qquad$ $\text{Area} = \frac{1}{2}bc \sin A = \frac{1}{2}ab \sin C = \frac{1}{2}ac \sin B$

Heron's Area Formula: $\qquad$ $\text{Area} = \sqrt{s(s - a)(s - b)(s - c)}$

Use the most appropriate formula to find the area of each triangle. Show your work and write a short paragraph stating your reasons for choosing each formula.

a.

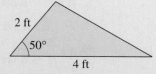

b.

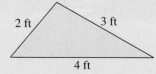

c.

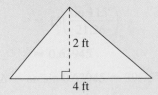

d.

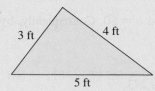

3.2   E x e r c i s e s

In Exercises 1–10, use the Law of Cosines to solve the triangle.

1. $a = 6$, $b = 8$, $c = 12$
2. $a = 9$, $b = 3$, $c = 11$
3. $A = 50°$, $b = 15$, $c = 30$
4. $C = 108°$, $a = 10$, $b = 6.5$
5. $a = 9$, $b = 12$, $c = 15$
6. $a = 45$, $b = 30$, $c = 72$
7. $a = 75.4$, $b = 48$, $c = 48$
8. $a = 1.42$, $b = 0.75$, $c = 1.25$
9. $B = 8° \, 15'$, $a = 26$, $c = 18$
10. $B = 75° \, 20'$, $a = 6.2$, $c = 9.5$

In Exercises 11–16, solve the parallelogram. (The lengths of the diagonals are given by c and d.)

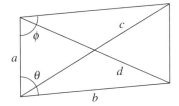

	a	b	c	d	θ	ϕ
11.	4	8			30°	
12.	25	36				110°
13.	10	14	20			
14.	40	60		80		
15.	10		18	12		
16.		25	50	35		

In Exercises 17–22, use Heron's Area Formula to find the area of the triangle.

17. $a = 5$, $b = 9$, $c = 10$
18. $a = 14$, $b = 17$, $c = 7$
19. $a = 3.5$, $b = 10.2$, $c = 9$
20. $a = 75.4$, $b = 52$, $c = 52$
21. $a = 20$, $b = 20$, $c = 10$
22. $a = 4.45$, $b = 1.85$, $c = 3.00$

23. **Navigation** A plane flies 810 miles from A to B with a bearing of N 75° E. Then it flies 648 miles from B to C with a bearing of N 32° E. Draw a diagram to represent the problem and find the straight-line distance and bearing from C to A.

24. **Navigation** A boat race runs along a triangular course marked by buoys A, B, and C. The race starts with the boats headed west for 2500 meters. The other two sides of the course lie to the north of the first side, and their lengths are 1100 meters and 2000 meters. Draw a diagram to represent the problem and find the bearings for the last two legs of the race.

25. **Surveying** To approximate the length of a marsh, a surveyor walks 380 meters from point A to point B, then turns 80° and walks 240 meters to point C. Approximate the length AC of the marsh.

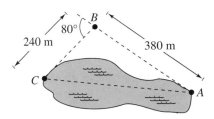

26. **Streetlight Design** Determine the angle θ in the design of the streetlight shown in the figure.

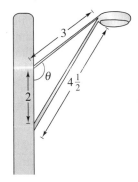

27. **Distance** Two ships leave a port at 9 A.M. One travels at a bearing of N 53° W at 12 miles per hour and the other travels at a bearing of S 67° W at 16 miles per hour. Approximate how far apart they are at noon that day.

3.3 Vectors in the Plane

Introduction

Many quantities in geometry and physics, such as area, time, and temperature, can be represented by a single real number. Other quantities, such as force and velocity, involve both *magnitude* and *direction* and cannot be completely characterized by a single real number. To represent such a quantity, you can use a **directed line segment,** as shown in Figure 3.15. The directed line segment $\overrightarrow{PQ}$ has **initial point** P and **terminal point** Q. Its **magnitude,** or **length,** is denoted by $\|PQ\|$.

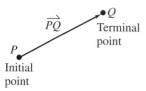

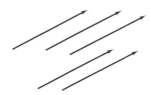

Figure 3.15

Figure 3.16

Two directed line segments that have the same magnitude and direction are equivalent. For example, the directed line segments in Figure 3.16 are all equivalent. The set of all directed line segments that are equivalent to a given directed line segment $\overrightarrow{PQ}$ is a **vector v in the plane,** written $\mathbf{v} = \overrightarrow{PQ}$. Vectors are denoted by lowercase, boldface letters such as **u v**, and **w.**

EXAMPLE 1   Vector Representation by Directed Line Segments

Let **u** be represented by the directed line segment from $P = (0, 0)$ to $Q = (3, 2)$, and let **v** be represented by the directed line segment from $R = (1, 2)$ to $S = (4, 4)$, as shown in Figure 3.17. Show that $\mathbf{u} = \mathbf{v}$.

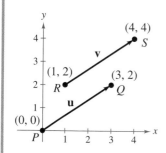

Figure 3.17

Solution

From the distance formula, it follows that $\overrightarrow{PQ}$ and $\overrightarrow{RS}$ have the *same magnitude.*

$$\|\overrightarrow{PQ}\| = \sqrt{(3 - 0)^2 + (2 - 0)^2} = \sqrt{13}$$

$$\|\overrightarrow{RS}\| = \sqrt{(4 - 1)^2 + (4 - 2)^2} = \sqrt{13}$$

Moreover, both line segments have the *same direction*, because they are both directed toward the upper right on lines having a slope of $\frac{2}{3}$. So, $\overrightarrow{PQ}$ and $\overrightarrow{RS}$ have the same magnitude and direction, and it follows that $\mathbf{u} = \mathbf{v}$.

Component Form of a Vector

The directed line segment whose initial point is the origin is often the most convenient representative of a set of equivalent directed line segments. This representative of the vector **v** is in **standard position.**

A vector whose initial point is at the origin $(0, 0)$ can be uniquely represented by the coordinates of its terminal point (v_1, v_2). This is the **component form of a vector v,** written

$$\mathbf{v} = \langle v_1, v_2 \rangle.$$

The coordinates v_1 and v_2 are the *components* of **v**. If both the initial point and the terminal point lie at the origin, **v** is the **zero vector** and is denoted by $\mathbf{0} = \langle 0, 0 \rangle$.

Component Form of a Vector

The component form of the vector with initial point $P = (p_1, p_2)$ and terminal point $Q = (q_1, q_2)$ is

$$\overrightarrow{PQ} = \langle q_1 - p_1, q_2 - p_2 \rangle = \langle v_1, v_2 \rangle = \mathbf{v}.$$

The **magnitude** (or length) of **v** is

$$\|\mathbf{v}\| = \sqrt{(q_1 - p_1)^2 + (q_2 - p_2)^2} = \sqrt{v_1^2 + v_2^2}.$$

If $\|\mathbf{v}\| = 1$, **v** is a **unit vector.** Moreover, $\|\mathbf{v}\| = 0$ if and only if **v** is the zero vector **0**.

Two vectors $\mathbf{u} = \langle u_1, u_2 \rangle$ and $\mathbf{v} = \langle v_1, v_2 \rangle$ are *equal* if and only if $u_1 = v_1$ and $u_2 = v_2$. For instance, in Example 1, the vector **u** from $P = (0, 0)$ to $Q = (3, 2)$ is

$$\mathbf{u} = \overrightarrow{PQ} = \langle 3 - 0, 2 - 0 \rangle = \langle 3, 2 \rangle$$

and the vector **v** from $R = (1, 2)$ to $S = (4, 4)$ is

$$\mathbf{v} = \overrightarrow{RS} = \langle 4 - 1, 4 - 2 \rangle = \langle 3, 2 \rangle.$$

EXAMPLE 2 Finding the Component Form of a Vector

Find the component form and length of the vector **v** that has initial point $(4, -7)$ and terminal point $(-1, 5)$.

Solution
Let $P = (4, -7) = (p_1, p_2)$ and $Q = (-1, 5) = (q_1, q_2)$ as shown in Figure 3.18. Then, the components of $\mathbf{v} = \langle v_1, v_2 \rangle$ are

$$v_1 = q_1 - p_1 = -1 - 4 = -5$$

$$v_2 = q_2 - p_2 = 5 - (-7) = 12.$$

So, $\mathbf{v} = \langle -5, 12 \rangle$ and the magnitude of **v** is

$$\|\mathbf{v}\| = \sqrt{(-5)^2 + 12^2} = \sqrt{169} = 13.$$

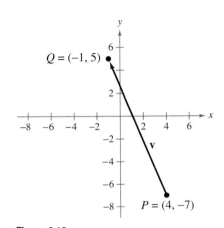

Figure 3.18

3.4   Vectors and Dot Products

The Dot Product of Two Vectors

So far you have studied two vector operations—vector addition and multiplication by a scalar—each of which yields another vector. In this section you will study a third vector operation, the **dot product.** This product yields a scalar, rather than a vector.

Definition of a Dot Product

The **dot product** of $\mathbf{u} = \langle u_1, u_2 \rangle$ and $\mathbf{v} = \langle v_1, v_2 \rangle$ is

$$\mathbf{u} \cdot \mathbf{v} = u_1 v_1 + u_2 v_2.$$

Properties of the Dot Product

Let $\mathbf{u}$, $\mathbf{v}$, and $\mathbf{w}$ be vectors in the plane or in space and let c be a scalar.

1. $\mathbf{u} \cdot \mathbf{v} = \mathbf{v} \cdot \mathbf{u}$
2. $\mathbf{0} \cdot \mathbf{v} = 0$
3. $\mathbf{u} \cdot (\mathbf{v} + \mathbf{w}) = \mathbf{u} \cdot \mathbf{v} + \mathbf{u} \cdot \mathbf{w}$
4. $\mathbf{v} \cdot \mathbf{v} = \|\mathbf{v}\|^2$
5. $c(\mathbf{u} \cdot \mathbf{v}) = c\mathbf{u} \cdot \mathbf{v} = \mathbf{u} \cdot c\mathbf{v}$

Proofs of properties 1, 4, and 5 are given in Appendix A.

EXAMPLE 1   Finding Dot Products

Find each dot product.

a. $\langle 4, 5 \rangle \cdot \langle 2, 3 \rangle$

b. $\langle 2, -1 \rangle \cdot \langle 1, 2 \rangle$

c. $\langle 0, 3 \rangle \cdot \langle 4, -2 \rangle$

Solution

a. $\langle 4, 5 \rangle \cdot \langle 2, 3 \rangle = 4(2) + 5(3) = 8 + 15 = 23$

b. $\langle 2, -1 \rangle \cdot \langle 1, 2 \rangle = 2(1) + (-1)(2) = 2 - 2 = 0$

c. $\langle 0, 3 \rangle \cdot \langle 4, -2 \rangle = 0(4) + 3(-2) = 0 - 6 = -6$

In Example 1, be sure you see that the dot product of two vectors is a scalar (a real number), not a vector. Moreover, notice that the dot product can be positive, zero, or negative.

What You Should Learn:

- How to find the dot product of two vectors and use properties of the dot product
- How to find angles between vectors
- How to determine whether two vectors are orthogonal
- How to write vectors as sums of two vector components
- How to use vectors to find the work done by a force

Why You Should Learn It:

You can use the dot product of two vectors to solve real-life problems involving two vector quantities. For instance, Exercise 55 on page 305 shows you how the dot product can be used to find the force necessary to keep a truck from rolling down a hill.

Alan Thornton/Tony Stone Images

EXAMPLE 2   Using Properties of Dot Products

Let $\mathbf{u} = \langle -1, 3 \rangle$, $\mathbf{v} = \langle 2, -4 \rangle$, and $\mathbf{w} = \langle 1, -2 \rangle$. Find each dot product.

a. $(\mathbf{u} \cdot \mathbf{v})\mathbf{w}$ **b.** $\mathbf{u} \cdot 2\mathbf{v}$

Solution

Begin by finding the dot product of $\mathbf{u}$ and $\mathbf{v}$.

$$\mathbf{u} \cdot \mathbf{v} = \langle -1, 3 \rangle \cdot \langle 2, -4 \rangle$$
$$= (-1)(2) + 3(-4)$$
$$= -14$$

a. $(\mathbf{u} \cdot \mathbf{v})\mathbf{w} = -14\langle 1, -2 \rangle$
$$= \langle -14, 28 \rangle$$

b. $\mathbf{u} \cdot 2\mathbf{v} = 2(\mathbf{u} \cdot \mathbf{v})$
$$= 2(-14)$$
$$= -28$$

Notice that the first product is a vector, whereas the second is a scalar. Can you see why?

EXAMPLE 3   Dot Product and Length

The dot product of $\mathbf{u}$ with itself is 5. What is the length of $\mathbf{u}$?

Solution

Because $\|\mathbf{u}\|^2 = \mathbf{u} \cdot \mathbf{u} = 5$, it follows that

$$\|\mathbf{u}\| = \sqrt{\mathbf{u} \cdot \mathbf{u}}$$
$$= \sqrt{5}.$$

The Angle Between Two Vectors

The **angle between two nonzero vectors** is the angle θ, $0 \le \theta \le \pi$, between its respective standard position vectors, as shown in Figure 3.30. This angle can be found using the dot product. (Note that the angle between the zero vector and another vector is not defined.) See Appendix A for a proof.

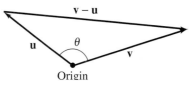

Figure 3.30

Angle   Between   Two   Vectors

If θ is the angle between two nonzero vectors $\mathbf{u}$ and $\mathbf{v}$, then

$$\cos \theta = \frac{\mathbf{u} \cdot \mathbf{v}}{\|\mathbf{u}\|\,\|\mathbf{v}\|}.$$

EXAMPLE 4   Finding the Angle Between Two Vectors

Find the angle between $\mathbf{u} = \langle 4, 3 \rangle$ and $\mathbf{v} = \langle 3, 5 \rangle$.

Solution

$$\cos \theta = \frac{\mathbf{u} \cdot \mathbf{v}}{\|\mathbf{u}\| \, \|\mathbf{v}\|}$$

$$= \frac{\langle 4, 3 \rangle \cdot \langle 3, 5 \rangle}{\|\langle 4, 3 \rangle\| \, \|\langle 3, 5 \rangle\|}$$

$$= \frac{27}{5\sqrt{34}}$$

This implies that the angle between the two vectors is

$$\theta = \arccos \frac{27}{5\sqrt{34}} \approx 22.2°,$$

as shown in Figure 3.31.

Rewriting the expression for the angle between two vectors in the form

$$\mathbf{u} \cdot \mathbf{v} = \|\mathbf{u}\| \, \|\mathbf{v}\| \cos \theta \qquad \text{Alternative form of dot product}$$

produces an alternative way to calculate the dot product. From this form, you can see that because $\|\mathbf{u}\|$ and $\|\mathbf{v}\|$ are always positive, $\mathbf{u} \cdot \mathbf{v}$ and $\cos \theta$ will always have the same sign. Figure 3.32 shows the five possible orientations of two vectors.

STUDY T!P

A graphing utility can be used to find the angle between two vectors. The program VECANGL on our website *college.hmco.com*, sketches two vectors $\mathbf{u} = \langle a, b \rangle$ and $\mathbf{v} = \langle c, d \rangle$ in standard position and finds the measure of the angle between them. Use the program to verify Example 4.

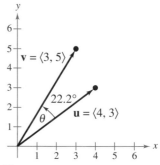

Figure 3.31

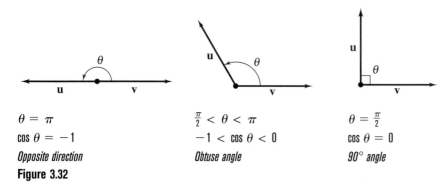

Figure 3.32

Definition of Orthogonal Vectors

The vectors $\mathbf{u}$ and $\mathbf{v}$ are **orthogonal** if $\mathbf{u} \cdot \mathbf{v} = 0$.

The terms *orthogonal* and *perpendicular* mean essentially the same thing—meeting at right angles. Even though the angle between the zero vector and another vector is not defined, it is convenient to extend the definition of orthogonality to include the zero vector. In other words, the zero vector is orthogonal to every vector $\mathbf{u}$ because $\mathbf{0} \cdot \mathbf{u} = 0$.

EXAMPLE 5   Determining Orthogonal Vectors

Are the vectors $\mathbf{u} = \langle 2, -3 \rangle$ and $\mathbf{v} = \langle 6, 4 \rangle$ orthogonal?

Solution

Begin by finding the dot product of the two vectors.

$$\mathbf{u} \cdot \mathbf{v} = \langle 2, -3 \rangle \cdot \langle 6, 4 \rangle$$
$$= 2(6) + (-3)(4)$$
$$= 0$$

The dot product is 0, so the two vectors are orthogonal, as shown in Figure 3.33.

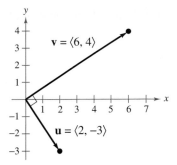

Figure 3.33

A computer simulation of this concept appears in the *Interactive* CD-ROM and *Internet* versions of this text.

Finding Vector Components

You have already seen applications in which two vectors are added to produce a resultant vector. Many applications in physics and engineering pose the reverse problem—decomposing a given vector into the sum of two **vector components.**

Consider a boat on an inclined ramp, as shown in Figure 3.34. The force $\mathbf{F}$ due to gravity pulls the boat *down* the ramp and *against* the ramp. These two orthogonal forces, $\mathbf{w}_1$ and $\mathbf{w}_2$, are vector components of $\mathbf{F}$. That is,

$$\mathbf{F} = \mathbf{w}_1 + \mathbf{w}_2. \qquad \text{Vector components of } \mathbf{F}$$

The negative of component $\mathbf{w}_1$ represents the force needed to keep the boat from rolling down the ramp, whereas $\mathbf{w}_2$ represents the force that the tires must withstand against the ramp. A procedure for finding $\mathbf{w}_1$ and $\mathbf{w}_2$ is shown below.

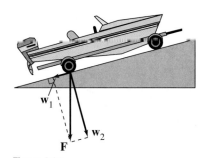

Figure 3.34

Definition of Vector Components

Let $\mathbf{u}$ and $\mathbf{v}$ be nonzero vectors such that

$$\mathbf{u} = \mathbf{w}_1 + \mathbf{w}_2$$

where $\mathbf{w}_1$ and $\mathbf{w}_2$ are orthogonal and $\mathbf{w}_1$ is parallel to (or a scalar multiple of) **v,** as shown in Figure 3.35. The vectors $\mathbf{w}_1$ and $\mathbf{w}_2$ are called **vector components** of $\mathbf{u}$. The vector $\mathbf{w}_1$ is the **projection** of $\mathbf{u}$ onto $\mathbf{v}$ and is denoted by

$$\mathbf{w}_1 = \text{proj}_{\mathbf{v}}\mathbf{u}.$$

The vector $\mathbf{w}_2$ is given by $\mathbf{w}_2 = \mathbf{u} - \mathbf{w}_1$.

From the definition of vector components, you can see that it is easy to find the component $\mathbf{w}_2$ once you have found the projection of $\mathbf{u}$ onto $\mathbf{v}$. To find the projection, you can use the dot product.

Projection of u onto v

Let $\mathbf{u}$ and $\mathbf{v}$ be nonzero vectors. The projection of $\mathbf{u}$ onto $\mathbf{v}$ is

$$\text{proj}_{\mathbf{v}}\mathbf{u} = \left(\frac{\mathbf{u} \cdot \mathbf{v}}{\|\mathbf{v}\|^2} \right)\mathbf{v}.$$

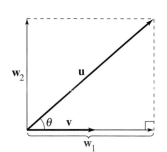

θ is acute.

θ is obtuse.

Figure 3.35

EXAMPLE 6   Decomposing a Vector into Components

Find the projection of $\mathbf{u} = \langle 3, -5 \rangle$ onto $\mathbf{v} = \langle 6, 2 \rangle$. Then write $\mathbf{u}$ as the sum of two orthogonal vectors, one of which is $\text{proj}_{\mathbf{v}}\mathbf{u}$.

Solution

The projection of $\mathbf{u}$ onto $\mathbf{v}$ is

$$\mathbf{w}_1 = \text{proj}_{\mathbf{v}}\mathbf{u} = \left(\frac{\mathbf{u} \cdot \mathbf{v}}{\|\mathbf{v}\|^2}\right)\mathbf{v} = \left(\frac{8}{40}\right)\langle 6, 2 \rangle = \left\langle \frac{6}{5}, \frac{2}{5} \right\rangle$$

as shown in Figure 3.36. The other component, $\mathbf{w}_2$, is

$$\mathbf{w}_2 = \mathbf{u} - \mathbf{w}_1 = \langle 3, -5 \rangle - \left\langle \frac{6}{5}, \frac{2}{5} \right\rangle = \left\langle \frac{9}{5}, -\frac{27}{5} \right\rangle.$$

So, $\mathbf{u} = \mathbf{w}_1 + \mathbf{w}_2 = \left\langle \frac{6}{5}, \frac{2}{5} \right\rangle + \left\langle \frac{9}{5}, -\frac{27}{5} \right\rangle = \langle 3, -5 \rangle.$

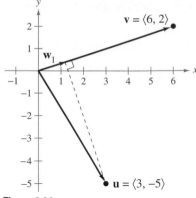

Figure 3.36

EXAMPLE 7   Finding a Force

A 600-pound boat sits on a ramp inclined at 30°, as shown in Figure 3.37. What force is required to keep the boat from rolling down the ramp?

Solution

Because the force due to gravity is vertical and downward, you can represent the gravitational force by the vector

$$\mathbf{F} = -600\mathbf{j}. \qquad \text{Force due to gravity}$$

To find the force required to keep the boat from rolling down the ramp, project $\mathbf{F}$ onto a unit vector $\mathbf{v}$ in the direction of the ramp, as follows.

$$\mathbf{v} = (\cos 30°)\mathbf{i} + (\sin 30°)\mathbf{j} = \frac{\sqrt{3}}{2}\mathbf{i} + \frac{1}{2}\mathbf{j} \qquad \text{Unit vector along ramp}$$

Therefore, the projection of $\mathbf{F}$ onto $\mathbf{v}$ is

$$\mathbf{w}_1 = \text{proj}_{\mathbf{v}}\mathbf{F} = \left(\frac{\mathbf{F} \cdot \mathbf{v}}{\|\mathbf{v}\|^2}\right)\mathbf{v} = (\mathbf{F} \cdot \mathbf{v})\mathbf{v} = (-600)\left(\frac{1}{2}\right)\mathbf{v} = -300\left(\frac{\sqrt{3}}{2}\mathbf{i} + \frac{1}{2}\mathbf{j}\right).$$

The magnitude of this force is 300, and therefore a force of 300 pounds is required to keep the boat from rolling down the ramp.

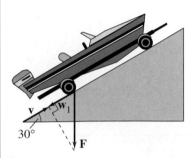

Figure 3.37

Work

The work W done by a constant force $\mathbf{F}$ acting along the line of motion of an object is

$$W = (\text{magnitude of force})(\text{distance})$$

$$= \|\mathbf{F}\| \, \|\overrightarrow{PQ}\|$$

as shown in Figure 3.38(a). If the constant force $\mathbf{F}$ is not directed along the line of motion, you can see from Figure 3.38(b) that the work W done by the force is

$$W = \|\text{proj}_{\overrightarrow{PQ}} \, \mathbf{F}\| \, \|\overrightarrow{PQ}\| = (\cos\theta)\|\mathbf{F}\| \, \|\overrightarrow{PQ}\| = \mathbf{F} \cdot PQ.$$

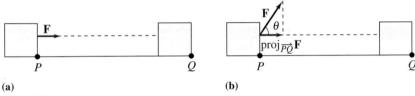

(a)   **(b)**

Figure 3.38

This notion of work is summarized in the following definition.

Definition of Work

The **work** W done by a constant force $\mathbf{F}$ as its point of application moves along the vector $\overrightarrow{PQ}$ is given by either of the following.

1. $W = \|\text{proj}_{\overrightarrow{PQ}} \, \mathbf{F}\| \, \|\overrightarrow{PQ}\|$ Projection form
2. $W = \mathbf{F} \cdot \overrightarrow{PQ}$ Dot product form

EXAMPLE 8   Finding Work

To close a barn's sliding door, a person pulls on a rope with a constant force of 50 pounds at a constant angle of $60°$, as shown in Figure 3.39. Find the work done in moving the door 12 feet to its closed position.

Solution

Using a projection, you can calculate the work as follows.

$$W = \|\text{proj}_{\overrightarrow{PQ}} \, \mathbf{F}\| \, \|\overrightarrow{PQ}\|$$

$$= (\cos 60°)\|\mathbf{F}\| \, \|\overrightarrow{PQ}\|$$

$$= \frac{1}{2}(50)(12)$$

$$= 300 \text{ ft-lb}$$

So, the work done is 300 foot-pounds. You can verify this result by finding the vectors $\mathbf{F}$ and $\overrightarrow{PQ}$ and calculating their dot product.

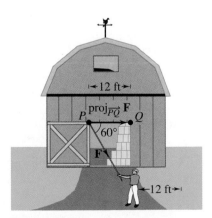

Figure 3.39

3.4 E x e r c i s e s

In Exercises 1–4, find the dot product of u and v.

1. $\mathbf{u} = \langle 3, 6 \rangle$
   $\mathbf{v} = \langle 2, -4 \rangle$

2. $\mathbf{u} = \langle 5, 12 \rangle$
   $\mathbf{v} = \langle -3, 2 \rangle$

3. $\mathbf{u} = 4\mathbf{i} - 7\mathbf{j}$
   $\mathbf{v} = \mathbf{i} - \mathbf{j}$

4. $\mathbf{u} = 3\mathbf{i} + 9\mathbf{j}$
   $\mathbf{v} = 10\mathbf{i} - 3\mathbf{j}$

In Exercises 5–10, use the vectors $\mathbf{u} = \langle 2, 2 \rangle$, $\mathbf{v} = \langle -3, 4 \rangle$, and $\mathbf{w} = \langle 1, -4 \rangle$ to find the indicated quantity. State whether the result is a vector or a scalar.

5. $\mathbf{u} \cdot \mathbf{u}$

6. $\|\mathbf{u}\| - 2$

7. $(\mathbf{u} \cdot \mathbf{v})\mathbf{w}$

8. $(\mathbf{w} \cdot \mathbf{u})\mathbf{v}$

9. $\mathbf{u} \cdot 2\mathbf{v}$

10. $4\mathbf{u} \cdot \mathbf{v}$

In Exercises 11–16, use the dot product to find $\|\mathbf{u}\|$.

11. $\mathbf{u} = \langle -5, 12 \rangle$

12. $\mathbf{u} = \langle 2, -4 \rangle$

13. $\mathbf{u} = 20\mathbf{i} + 25\mathbf{j}$

14. $\mathbf{u} = 16\mathbf{i} - 10\mathbf{j}$

15. $\mathbf{u} = 6\mathbf{j}$

16. $\mathbf{u} = 9\mathbf{i}$

In Exercises 17–24, find the angle θ between the vectors.

17. $\mathbf{u} = \langle -1, 0 \rangle$
   $\mathbf{v} = \langle 0, 2 \rangle$

18. $\mathbf{u} = \langle 4, 4 \rangle$
   $\mathbf{v} = \langle -2, 0 \rangle$

19. $\mathbf{u} = 3\mathbf{i} + 4\mathbf{j}$
   $\mathbf{v} = -2\mathbf{i} + 3\mathbf{j}$

20. $\mathbf{u} = 2\mathbf{i} - 3\mathbf{j}$
   $\mathbf{v} = \mathbf{i} - 2\mathbf{j}$

21. $\mathbf{u} = 2\mathbf{i}$
   $\mathbf{v} = -3\mathbf{j}$

22. $\mathbf{u} = 4\mathbf{j}$
   $\mathbf{v} = -3\mathbf{i}$

23. $\mathbf{u} = \cos\left(\dfrac{\pi}{3}\right)\mathbf{i} + \sin\left(\dfrac{\pi}{3}\right)\mathbf{j}$
   $\mathbf{v} = \cos\left(\dfrac{3\pi}{4}\right)\mathbf{i} + \sin\left(\dfrac{3\pi}{4}\right)\mathbf{j}$

24. $\mathbf{u} = \cos\left(\dfrac{\pi}{4}\right)\mathbf{i} + \sin\left(\dfrac{\pi}{4}\right)\mathbf{j}$
   $\mathbf{v} = \cos\left(\dfrac{2\pi}{3}\right)\mathbf{i} + \sin\left(\dfrac{2\pi}{3}\right)\mathbf{j}$

In Exercises 25–28, use a graphing utility to sketch the vectors and find the degree measure of the angle between the vectors.

25. $\mathbf{u} = 3\mathbf{i} + 4\mathbf{j}$
   $\mathbf{v} = -7\mathbf{i} + 5\mathbf{j}$

26. $\mathbf{u} = -6\mathbf{i} - 3\mathbf{j}$
   $\mathbf{v} = -8\mathbf{i} + 4\mathbf{j}$

27. $\mathbf{u} = 5\mathbf{i} + 5\mathbf{j}$
   $\mathbf{v} = -8\mathbf{i} + 8\mathbf{j}$

28. $\mathbf{u} = 2\mathbf{i} - 3\mathbf{j}$
   $\mathbf{v} = 4\mathbf{i} + 3\mathbf{j}$

In Exercises 29 and 30, use vectors to find the interior angles of the triangle with the given vertices.

29. $(1, 2), (3, 4), (2, 5)$

30. $(-3, 0), (2, 2), (0, 6)$

In Exercises 31 and 32, find $\mathbf{u} \cdot \mathbf{v}$, where θ is the angle between u and v.

31. $\|\mathbf{u}\| = 4, \|\mathbf{v}\| = 10, \ \theta = \dfrac{2\pi}{3}$

32. $\|\mathbf{u}\| = 100, \ \|\mathbf{v}\| = 250, \ \theta = \dfrac{\pi}{6}$

In Exercises 33–38, determine whether u and v are orthogonal, parallel, or neither.

33. $\mathbf{u} = \langle -12, 30 \rangle$
   $\mathbf{v} = \left\langle \frac{1}{2}, -\frac{5}{4} \right\rangle$

34. $\mathbf{u} = \langle 15, 45 \rangle$
   $\mathbf{v} = \langle -5, 12 \rangle$

35. $\mathbf{u} = \frac{1}{4}(3\mathbf{i} - \mathbf{j})$
   $\mathbf{v} = 5\mathbf{i} + 6\mathbf{j}$

36. $\mathbf{u} = \mathbf{j}$
   $\mathbf{v} = \mathbf{i} - 2\mathbf{j}$

37. $\mathbf{u} = 2\mathbf{i} - 2\mathbf{j}$
   $\mathbf{v} = -\mathbf{i} - \mathbf{j}$

38. $\mathbf{u} = 8\mathbf{i} + 4\mathbf{j}$
   $\mathbf{v} = -2\mathbf{i} - \mathbf{j}$

In Exercises 39–42, find the projection of u onto v and the vector component of u orthogonal to v.

39. $\mathbf{u} = \langle 3, 4 \rangle$
   $\mathbf{v} = \langle 8, 2 \rangle$

40. $\mathbf{u} = \langle 4, 2 \rangle$
   $\mathbf{v} = \langle 1, -2 \rangle$

41. $\mathbf{u} = \langle 0, 3 \rangle$
   $\mathbf{v} = \langle 2, 15 \rangle$

42. $\mathbf{u} = \langle -5, -1 \rangle$
   $\mathbf{v} = \langle -1, 1 \rangle$

In Exercises 43–46, use the graph to mentally deter-mine the projection of u onto v. (The coordinates of the terminal points of the vectors in standard position are given.) Use the formula for the projection of u onto v to verify your result.

43.

(6, 4)

(3, 2)   **v**

u

44.

(6, 4)

v

u

(−3, −2)

45.

(−2, 3)   **v**   (6, 4)

u

46.

(6, 4)

v

u   (2, −3)

In Exercises 47–50, find two vectors in opposite direc-tions that are orthogonal to the vector u. (There are many correct answers.)

47. $\mathbf{u} = \langle 4, 7 \rangle$

48. $\mathbf{u} = \langle -8, 1 \rangle$

49. $\mathbf{u} = \frac{1}{2}\mathbf{i} - \frac{3}{4}\mathbf{j}$

50. $\mathbf{u} = -\frac{5}{2}\mathbf{i} - 3\mathbf{j}$

Work   **In Exercises 51 and 52, find the work done in moving a particle from *P* to *Q* if the magnitude and direction of the force are given by v.**

51. $P = (0, 0)$, $Q = (4, 7)$, $\mathbf{v} = \langle 1, 4 \rangle$

52. $P = (1, 3)$, $Q = (-3, 5)$, $\mathbf{v} = -2\mathbf{i} + 3\mathbf{j}$

53. *Revenue*   The vector $\mathbf{u} = \langle 1245, 2600 \rangle$ gives the number of units of two products produced by a com-pany. The vector $\mathbf{v} = \langle 12.20, 8.50 \rangle$ gives the price (in dollars) of each unit, respectively. Find the dot product $\mathbf{u} \cdot \mathbf{v}$, and explain what information it gives.

54. *Revenue*   Repeat Exercise 53 after increasing the prices by 5%. Identify the vector operation used to increase the prices by 5%.

55. *Braking Load*   A truck with a gross weight of 36,000 pounds is parked on a 10° slope. Assume the only force to overcome is that due to gravity.

(a) Find the force required to keep the truck from rolling down the hill.

(b) Find the force perpendicular to the hill.

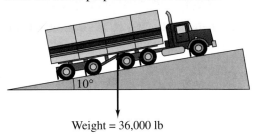

Weight = 36,000 lb

56. *Braking Load*   Rework Exercise 55 for a truck that is parked on a 12° slope.

57. *Work*   A 25-kilogram (245-newton) bag of sugar is lifted 3 meters. Determine the work done.

58. *Work*   Determine the work done by a crane lifting a 2400-pound car 5 feet.

59. *Work*   A force of 45 pounds in the direction of 30° above the horizontal is required to slide an imple-ment across a floor. Find the work done if the imple-ment is dragged 20 feet.

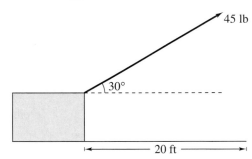

45 lb

30°

20 ft

60. *Work*   A tractor pulls a log 800 meters and the ten-sion in the cable connecting the tractor and log is approximately 1600 kilograms (15,691 newtons). Approximate the work done if the direction of the force is 35° above the horizontal.

61. *Work*   A toy wagon is pulled by exerting a force of 20 pounds on a handle that makes a 25° angle with the horizontal. Find the work done in pulling the wagon 40 feet.

62. *Work*   A mover exerts a horizontal force of 25 pounds on a crate as it is pushed up a ramp that is 12 feet long and inclined at an angle of 20° above the horizontal. Find the work done on the crate.

3 Review Exercises

3.1 **In Exercises 1–10, use the given information to solve the triangle. If two solutions exist, list both.**

1. $A = 12°$, $B = 58°$, $a = 8$
2. $B = 110°$, $C = 30°$, $c = 10.5$
3. $A = 75°$, $a = 2.5$, $b = 16.5$
4. $A = 130°$, $a = 60$, $b = 48$
5. $B = 115°$, $a = 9$, $b = 14.5$
6. $C = 50°$, $a = 25$, $c = 22$
7. $A = 15°$, $a = 5$, $b = 10$
8. $B = 150°$, $a = 64$, $b = 10$
9. $B = 25°$, $a = 6.2$, $b = 4$
10. $A = 74°$, $b = 12.8$, $a = 12.5$

In Exercises 11–14, find the area of the triangle having the indicated angle and sides.

11. $A = 27°$, $b = 5$, $c = 8$
12. $B = 80°$, $a = 4$, $c = 8$
13. $C = 122°$, $b = 18$, $a = 29$
14. $C = 100°$, $a = 120$, $b = 74$

15. **Height** From a certain distance, the angle of elevation to the top of a building is 17°. At a point 50 meters closer to the building, the angle of elevation is 31°. Approximate the height of the building.

16. **Altitude** The angles of elevation to an airplane from two points A and B on level ground are 51° and 68°, respectively. The points A and B are 2.5 miles apart, and the airplane is east of both points in the same vertical plane. Find the altitude of the plane.

17. **Height of a Tree** Find the height of a tree that stands on a hillside of slope 28° (from the horizontal) if, from a point 75 feet down the hill, the angle of elevation to the top of the tree is 45°.

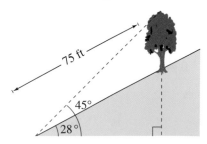

18. **River Width** Determine the width of a river that flows due east, if a surveyor finds that a tree on the opposite bank has a bearing of N 22° 30′ E from a certain point and a bearing of N 15° W from a point 400 feet downstream.

3.2 **In Exercises 19–28, solve the triangle having the indicated sides and angles.**

19. $a = 5$, $b = 8$, $c = 10$
20. $a = 13$, $b = 18$, $c = 26$
21. $a = 6$, $b = 9$, $C = 45°$
22. $B = 90°$, $a = 5$, $c = 12$
23. $B = 110°$, $a = 4$, $c = 4$
24. $B = 12°$, $a = 32$, $c = 36$
25. $a = 42$, $b = 25$, $c = 58$
26. $a = 80$, $b = 60$, $c = 100$
27. $B = 150°$, $a = 10$, $c = 20$
28. $a = 7.5$, $b = 15.0$, $c = 4.5$

29. **Geometry** The lengths of the diagonals of a parallelogram are 10 feet and 16 feet. Find the lengths of the sides of the parallelogram if the diagonals intersect at an angle of 28°.

30. **Geometry** The lengths of the diagonals of a parallelogram are 30 meters and 40 meters. Find the lengths of the sides of the parallelogram if the diagonals intersect at an angle of 34°.

31. **Surveying** To approximate the length of a marsh, a surveyor walks 425 meters from point A to point B. Then the surveyor turns 65° and walks 300 meters to point C. Approximate the length AC of the marsh.

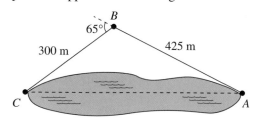

32. *Navigation*   Two planes leave an airport at approximately the same time. One is flying at 425 miles per hour at a bearing of N 5° W, and the other is flying at 530 miles per hour at a bearing of N 67° E. Determine the distance between the planes after flying for 2 hours.

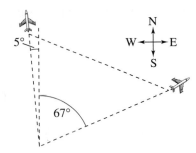

In Exercises 33–36, use Heron's Area Formula to find the area of the triangle with the given side lengths.

33. $a = 4$, $b = 5$, $c = 7$

34. $a = 15$, $b = 8$, $c = 10$

35. $a = 64.8$, $b = 49.2$, $c = 24.1$

36. $a = 8.55$, $b = 5.14$, $c = 12.73$

3.3   **In Exercises 37 and 38, show that u = v.**

37.

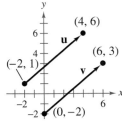

38.

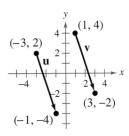

In Exercises 39–44, find the component form of the vector v satisfying the given conditions.

39.

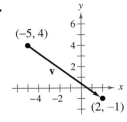

40.

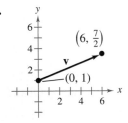

41. Initial point: $(0, 10)$; Terminal point: $(7, 3)$

42. Initial point: $(1, 5)$; Terminal point: $(15, 9)$

43. $\|\mathbf{v}\| = 8$, $\theta = 120°$

44. $\|\mathbf{v}\| = \frac{1}{2}$, $\theta = 225°$

In Exercises 45–50, use the vectors $\mathbf{u} = \langle -1, -3 \rangle$, $\mathbf{v} = \langle -3, 6 \rangle$, and $\mathbf{w} = \langle 4, 5 \rangle$ to perform the given operations.

45. $-\mathbf{w}$

46. $\mathbf{w} - \mathbf{v}$

47. $\mathbf{w} + 2\mathbf{u}$

48. $2\mathbf{w} - 4\mathbf{u}$

49. $3\mathbf{v} + 3\mathbf{w}$

50. $\mathbf{u} - \frac{1}{2}\mathbf{v}$

In Exercises 51–56, find the component form of the specified vector and sketch its graph given that $\mathbf{u} = 6\mathbf{i} - 5\mathbf{j}$ and $\mathbf{v} = 10\mathbf{i} + 3\mathbf{j}$.

51. $\dfrac{1}{\|\mathbf{u}\|}\mathbf{u}$

52. $\dfrac{1}{\|\mathbf{v}\|}\mathbf{v}$

53. $3\mathbf{v}$

54. $\frac{1}{2}\mathbf{v}$

55. $4\mathbf{u} - 5\mathbf{v}$

56. $3\mathbf{v} - 2\mathbf{u}$

In Exercises 57–60, find a unit vector in the direction of the given vector.

57. $\mathbf{u} = \langle 0, -6 \rangle$

58. $\mathbf{v} = \langle -12, -5 \rangle$

59. $\mathbf{v} = 5\mathbf{i} - 2\mathbf{j}$

60. $\mathbf{w} = -7\mathbf{i}$

In Exercises 61–64, find a unit vector in the direction of $\overrightarrow{PQ}$.

61. $P(7, -4)$, $Q(-3, 2)$

62. $P(2, 5)$, $Q(-1, 6)$

63. $P(0, 3)$, $Q(5, -8)$

64. $P(-4, -9)$, $Q(0, -3)$

In Exercises 65 and 66, use the given information to write u as a linear combination of the standard unit vectors i and j.

65. Initial point: $(-8, 3)$

Terminal point: $(1, -5)$

66. Initial point: $(2, -3.2)$

Terminal point: $(-6.4, 10.8)$

In Exercises 67 and 68, write the vector v in the form $\|\mathbf{v}\|[(\cos \theta)\mathbf{i} + (\sin \theta)\mathbf{j}]$.

67. $\mathbf{v} = -10\mathbf{i} + 10\mathbf{j}$

68. $\mathbf{v} = 4\mathbf{i} - \mathbf{j}$

In Exercises 69 and 70, use a graphing utility to graph the vectors and the resultant of the vectors. Find the magnitude and direction of the resultant.

69.   **70.**

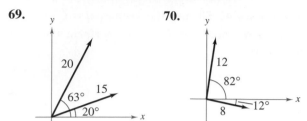

71. Resultant Force Find the direction and magnitude of the resultant of the three forces shown in the graph.

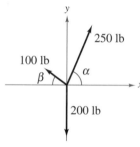

$$\tan \beta = -\frac{3}{4} \qquad \tan \alpha = \frac{12}{5}$$

72. Rope Tension A 180-pound weight is supported by two ropes, as shown in the figure. Find the tension in each rope.

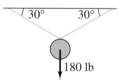

73. Navigation An airplane has an airspeed of 430 miles per hour at a bearing of S 45° E. If the wind velocity is 35 miles per hour in the direction N 30° E, find the groundspeed and the direction of the plane.

74. Navigation An airplane has an airspeed of 724 kilometers per hour at a bearing of N 30° E. If the wind velocity is from the west at 32 kilometers per hour, find the groundspeed and the direction of the plane.

3.4 **In Exercises 75 and 76, find the dot product of u and v.**

75. $\mathbf{u} = 6\mathbf{i} - \mathbf{j}$
   $\mathbf{v} = 2\mathbf{i} + 5\mathbf{j}$

76. $\mathbf{u} = 8\mathbf{i} - 7\mathbf{j}$
   $\mathbf{v} = 3\mathbf{i} - 4\mathbf{j}$

In Exercises 77–80, use the vectors $\mathbf{u} = \langle 6, -3 \rangle$ and $\mathbf{v} = \langle -2, -1 \rangle$ to find the indicated quantity.

77. $(\mathbf{u} \cdot \mathbf{v})\mathbf{u}$ **78.** $\|\mathbf{v}\| - 3$

79. $4\mathbf{u} \cdot \mathbf{v}$ **80.** $\|\mathbf{v}\|^2 + \|\mathbf{u}\|^2$

In Exercises 81–84, find the angle between u and v.

81. $\mathbf{u} = \cos \dfrac{7\pi}{4}\mathbf{i} + \sin \dfrac{7\pi}{4}\mathbf{j}$

   $\mathbf{v} = \cos \dfrac{5\pi}{6}\mathbf{i} + \sin \dfrac{5\pi}{6}\mathbf{j}$

82. $\mathbf{u} = \langle -6, -3 \rangle, \quad \mathbf{v} = \langle 4, 2 \rangle$

83. $\mathbf{u} = \langle 2\sqrt{2}, -4 \rangle, \quad \mathbf{v} = \langle -\sqrt{2}, 1 \rangle$

84. $\mathbf{u} = \langle 3, 1 \rangle, \quad \mathbf{v} = \langle 4, 5 \rangle$

In Exercises 85–88, decide whether the vectors are orthogonal, parallel, or neither.

85. $\mathbf{u} = \langle 39, -12 \rangle$ **86.** $\mathbf{u} = \langle 8, -4 \rangle$
   $\mathbf{v} = \langle -26, 8 \rangle$ $\mathbf{v} = \langle 5, 10 \rangle$

87. $\mathbf{u} = \langle 8, 5 \rangle$ **88.** $\mathbf{u} = \langle -15, 51 \rangle$
   $\mathbf{v} = \langle -2, 4 \rangle$ $\mathbf{v} = \langle 20, -68 \rangle$

In Exercises 89–92, (a) find $\text{proj}_{\mathbf{v}}\mathbf{u}$ and (b) write u as the sum of two vector components.

89. $\mathbf{u} = \langle -4, 3 \rangle, \quad \mathbf{v} = \langle -8, -2 \rangle$

90. $\mathbf{u} = \langle 5, 6 \rangle, \quad \mathbf{v} = \langle 10, 0 \rangle$

91. $\mathbf{u} = \langle 2, 7 \rangle, \quad \mathbf{v} = \langle 1, -1 \rangle$

92. $\mathbf{u} = \langle -3, 5 \rangle, \quad \mathbf{v} = \langle -5, 2 \rangle$

93. Work Determine the work done by a crane lifting an 18,000-pound truck 48 inches.

94. Braking Force A 500-pound motorcycle is headed up a hill inclined at 12°. What force is required to keep the motorcycle from rolling back down the hill when stopped at a red light?

Synthesis

In Exercises 95 and 96, determine whether the statement is true or false. Justify your answer.

95. The Law of Sines is true if one of the angles in the triangle is a right angle.

96. When the Law of Sines is used, the solution is always unique.

97. Think About It What characterizes a vector in the plane?

Chapter Project   *Adding Vectors Graphically*

The psuedo-code at the right can be translated into a program for a graphing utility. (The program ADDVECT for several models of calculators can be found at our website *www.hmco.com*.) The program sketches two vectors

$$\mathbf{u} = a\mathbf{i} + b\mathbf{j} \text{ and } \mathbf{v} = c\mathbf{i} + d\mathbf{j}$$

in standard position. Then, using the parallelogram law for vector addition, the program also sketches the vector sum $\mathbf{u} + \mathbf{v}$. *Before* running the program, you should set values that produce an appropriate viewing window.

(a) Use the program to sketch the sum of the vectors $\mathbf{u} = 5\mathbf{i} + 2\mathbf{j}$ and $\mathbf{v} = -4\mathbf{i} + 3\mathbf{j}$. Set your viewing window as indicated in the graph below. Identify the vectors $\mathbf{u}$, $\mathbf{v}$, and $\mathbf{u} + \mathbf{v}$ in the graph.

Program
- Input a
- Input b
- Input c
- Input d
- Draw a line from $(0, 0)$ to (a, b).
- Draw a line from $(0, 0)$ to (c, d).
- Add $a + c$ and store in e.
- Add $b + d$ and store in f.
- Draw a line from $(0, 0)$ to (e, f).
- Draw a line from (a, b) to (e, f).
- Draw a line from (c, d) to (e, f).
- Pause to view graph.
- End Program

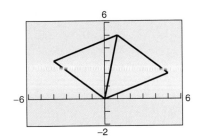

(b) An airplane is headed N 60° W at a speed of 400 miles per hour. The airplane encounters wind of velocity 75 miles per hour in the direction N 40° E. Use the program to find the resultant speed and direction of the airplane.

Questions for Further Exploration

In Questions 1–4, use the program to sketch the sum of the vectors. Use the result to estimate graphically the components of the sum. Then check your result algebraically. (Use $-9 \le x \le 9$ and $-6 \le y \le 6$.)

1. $\mathbf{u} = 3\mathbf{i} + 4\mathbf{j}, \quad \mathbf{v} = -5\mathbf{i} + \mathbf{j}$

2. $\mathbf{u} = 5\mathbf{i} - 4\mathbf{j}, \quad \mathbf{v} = 3\mathbf{i} + 2\mathbf{j}$

3. $\mathbf{u} = -4\mathbf{i} + 4\mathbf{j}, \quad \mathbf{v} = -2\mathbf{i} - 6\mathbf{j}$

4. $\mathbf{u} = 7\mathbf{i} + 3\mathbf{j}, \quad \mathbf{v} = -2\mathbf{i} - 6\mathbf{j}$

5. After encountering the wind, is the airplane in part (b) above traveling at a higher speed or a lower speed? Explain.

6. Consider the airplane described in part (b), headed N 60° W at a speed of 400 miles per hour. What wind velocity, in the direction of N 40° E, will produce a resultant direction of N 50° W? Explain how to use the program listed above to obtain the answer *experimentally*. Then explain how to obtain the answer algebraically.

7. Consider the airplane described in part (b), headed N 60° W at a speed of 400 miles per hour. What wind direction, at a speed of 75 miles per hour, will produce a resultant direction of N 50° W? Explain how to use the program listed above to obtain the answer *experimentally*. Then explain how to obtain the answer algebraically.

3 Chapter Test

Take this test as you would take a test in class. After you are done, check your work against the answers given in the back of the book.

The *Interactive* CD-ROM and *Internet* versions of this text provide answers to the Chapter Tests and Cumulative Tests. They also offer Chapter Pre-Tests (which test key skills and concepts covered in previous chapters) and Chapter Post-Tests, both of which have randomly generated exercises with diagnostic capabilities.

In Exercises 1–6, use the given information to solve the triangle. If two solutions exist, find both.

1. $c = 18.2$, $B = 98.4°$, $A = 37.6°$
2. $a = 14.9$, $b = 23.2$, $c = 10$
3. $A = 58°$, $b = 8.4$, $c = 11.2$
4. $A = 24.6°$, $b = 28$, $a = 15.6$
5. $B = 130°$, $c = 10.1$, $b = 5.2$
6. $A = 150°$, $b = 4.8$, $a = 9.4$

In Exercises 7–9, use Heron's Area Formula to find the area of the triangle.

7. $a = 10$, $b = 18$, $c = 20$
8. $a = 5$, $b = 5$, $c = 4$
9. $a = 1.3$, $b = 4.7$, $c = 5.2$

10. A plane flies 650 kilometers with a bearing of N 34° E from A to B. The plane then flies 1070 kilometers southeast from B to C. Find the bearing of the flight from B to C, as shown in the figure at the right.

11. Find the length of the pond shown at the right.

12. Find the component form and length of the vector **w** that has an initial point $(-8, -12)$ and terminal point $(4, 1)$.

13. Find a unit vector in the direction of the vector $\mathbf{v} = 7\mathbf{i} + 4\mathbf{j}$.

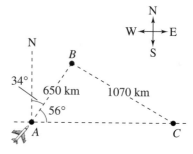

FIGURE FOR 10

In Exercises 14–17, find (a) 2v + u, (b) u − 3v, and (c) 5u − v.

14. $\mathbf{u} = \langle 0, -4 \rangle$, $\mathbf{v} = \langle -2, 4 \rangle$
15. $\mathbf{u} = \langle -2, -3 \rangle$, $\mathbf{v} = \langle 1, 5 \rangle$
16. $\mathbf{u} = \mathbf{i} - \mathbf{j}$, $\mathbf{v} = 6\mathbf{i} + 9\mathbf{j}$
17. $\mathbf{u} = 2\mathbf{i} + 3\mathbf{j}$, $\mathbf{v} = -\mathbf{i} - 2\mathbf{j}$

18. Find the component form of the vector **v** with $\|\mathbf{v}\| = 12$, in the same direction as $\mathbf{u} = \langle 3, -5 \rangle$.

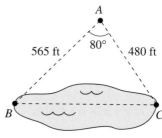

FIGURE FOR 11

19. Forces with magnitudes of 250 pounds and 130 pounds act on an object at angles of 45° and −60°, respectively, with the x-axis. Find the direction and magnitude of the resultant of these forces.

20. Find the angle between the vectors $\mathbf{u} = 7\mathbf{i} + 2\mathbf{j}$ and $\mathbf{v} = -4\mathbf{j}$.

21. Are the vectors $\mathbf{u} = \langle 6, -4 \rangle$ and $\mathbf{v} = \langle 2, -3 \rangle$ orthogonal? Explain.

22. Find the projection of $\mathbf{u} = \langle 6, 7 \rangle$ onto $\mathbf{v} = \langle -5, -1 \rangle$. Then write **u** as the vector sum of two orthogonal vectors.

1–3 Cumulative Test

Take this test to review the material from earlier chapters. After you are done, check your work against the answers given in the back of the book.

1. Consider the angle $\theta = -120°$.

   (a) Sketch the angle in standard position.

   (b) Determine a coterminal angle in the interval $[0°, 360°)$.

   (c) Convert the angle to radian measure.

   (d) Find its reference angle θ'.

   (e) Find the exact values of the six trigonometric functions of θ.

2. Convert the angle of magnitude 2.35 radians to degrees. Round the answer to one decimal place.

3. Find $\cos \theta$ if $\tan \theta = -\frac{4}{3}$ and $\sin \theta < 0$.

In Exercises 4–6, sketch the graph of the function by hand. Use a graphing utility to verify your graph.

4. $f(x) = 3 - 2 \sin \pi x$ 5. $f(x) = \frac{1}{2} \tan\left(x - \frac{\pi}{2}\right)$ 6. $f(x) = \frac{1}{2} \sec(x + \pi)$

7. Find a, b, and c such that the graph of the function $h(x) = a \cos(bx + c)$ matches the graph in the figure at the right.

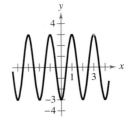

FIGURE FOR 7

In Exercises 8 and 9, find the exact value without using a calculator.

8. $\tan(\arctan 6.7)$ 9. $\tan\left(\arcsin \frac{3}{5}\right)$

10. Write an algebraic expression equivalent to $\sin(\arccos 2x)$.

11. Find the altitude of the triangle shown at the right.

12. Subtract and simplify: $\dfrac{\sin \theta - 1}{\cos \theta} - \dfrac{\cos \theta}{\sin \theta - 1}$.

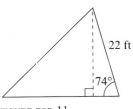

22 ft

74°

FIGURE FOR 11

In Exercises 13 and 14, verify the identity.

13. $\cot^2 \alpha(\sec^2 \alpha - 1) = 1$ 14. $\sin(x + y) \sin(x - y) = \sin^2 x - \sin^2 y$

In Exercises 15 and 16, find all solutions of the equation in the interval $[0, 2\pi)$.

15. $2 \cos^2 \beta - \cos \beta = 0$ 16. $3 \tan \theta - \cot \theta = 0$

17. Approximate the solutions to the equation $\cos^2 x - 5 \cos x - 1 = 0$ in the interval $[0, 2\pi)$.

In Exercises 18 and 19, use a graphing utility to graph the function and approximate its zeros in the interval $[0, 2\pi)$. If possible, find the exact values of the zeros algebraically.

18. $y = \dfrac{1 + \sin x}{\cos x} + \dfrac{\cos x}{1 + \sin x} - 4$ 19. $y = \tan^3 x - \tan^2 x + 3 \tan x - 3$

In Exercises 20 and 21, write the trigonometric expression as an algebraic expression.

20. $\cos(2 \arccos 2x)$

21. $\sin(2 \arctan x)$

22. If $\tan \theta = \dfrac{1}{2}$, find the exact value of $\tan(2\theta)$.

23. If $\tan \theta = \dfrac{4}{3}$, find the exact value of $\sin \dfrac{\theta}{2}$.

24. Use the half-angle formulas to determine the exact values of the sine, cosine, and tangent of $\theta = 67° \, 30'$.

25. Write $\cos 8x + \cos 4x$ as a product.

26. Write the product $5 \sin \dfrac{3\pi}{4} \cdot \cos \dfrac{7\pi}{4}$ as a sum or difference.

In Exercises 27–30, verify the identity.

27. $\tan x(1 - \sin^2 x) = \frac{1}{2} \sin 2x$

28. $\sin 3\theta \sin \theta = \frac{1}{2}(\cos 2\theta - \cos 4\theta)$

29. $\sin 3x \cos 2x = \frac{1}{2}(\sin 5x + \sin x)$

30. $\dfrac{2 \cos 3x}{\sin 4x - \sin 2x} = \csc x$

In Exercises 31–34, use the given information to solve the triangle shown at the right.

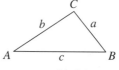

FIGURE FOR 31–34

31. $A = 46°$, $a = 14$, $b = 5$

32. $A = 30°$, $b = 8$, $c = 10$

33. $A = 24°$, $C = 101°$, $a = 10$

34. $a = 24$, $b = 30$, $c = 47$

35. Two sides of a triangle have lengths 14 and 19 inches. Their included angle measures 82°. Find the area of the triangle.

36. Find the area of a triangle with sides 11, 16, and 17 inches.

37. Write the vector $\mathbf{u} = \langle 3, 5 \rangle$ as a linear combination of the standard unit vectors $\mathbf{i}$ and $\mathbf{j}$.

38. Find a unit vector in the direction of $\mathbf{v} = \mathbf{i} + \mathbf{j}$.

39. Find $\mathbf{u} \cdot \mathbf{v}$ for $\mathbf{u} = 3\mathbf{i} + 4\mathbf{j}$ and $\mathbf{v} = \mathbf{i} - 2\mathbf{j}$.

40. Find the projection of $\mathbf{u} = \langle 8, -2 \rangle$ onto $\mathbf{v} = \langle 1, 5 \rangle$, and then find the vector component of $\mathbf{u}$ orthogonal to $\mathbf{v}$.

41. From a point 200 feet from a flagpole, the angles of elevation to the bottom and top of the flag are $16° \, 45'$ and $18°$, respectively. Approximate the height of the flag to the nearest foot.

42. A record single rotates on a turntable at 45 revolutions per minute. Find the angular speed of the record. Then find the speed of the groove that the needle is in when the needle is 3 inches from the center of the record.

43. Write a model for a particle in simple harmonic motion with a displacement of 4 inches and a period of 8 seconds.

44. An airplane's velocity with respect to the air is 500 kilometers per hour with a bearing of N 30° E. The wind at the altitude of the plane has a velocity of 50 kilometers per hour with a bearing of N 60° E. What is the true direction of the plane, and what is its speed relative to the ground?

45. Forces of 60 pounds and 100 pounds have a resultant force of 125 pounds. Find the angle between the two forces.

Complex Numbers

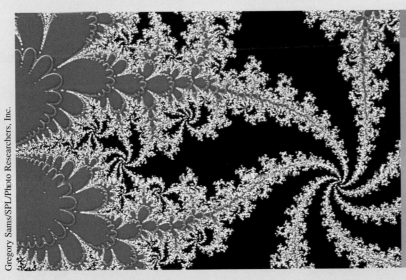

Gregory Sams/SPL/Photo Researchers, Inc.

The Big Picture

In this chapter you will learn how to

☐ perform operations with complex numbers.

☐ determine the number of zeros of polynomial functions.

☐ find the zeros of polynomial functions.

☐ multiply and divide complex numbers written in trigonometric form.

☐ find powers and nth roots of complex numbers.

To add more interest to the picture of a fractal, computer scientists discovered that points not in the set could be assigned a variety of colors, depending on how quickly their sequences diverge.

Important Vocabulary

As you encounter each new vocabulary term in this chapter, add the term and its definition to your notebook glossary.

- imaginary unit i (p. 316)
- complex numbers (p. 316)
- standard form of a complex number (p. 316)
- real part (p. 316)
- imaginary part (p. 316)
- additive identity (p. 317)
- additive inverse (p. 317)
- complex conjugates (p. 319)
- discriminant (p. 324)
- complex plane (p. 330)
- real axis (p. 330)
- imaginary axis (p. 330)
- absolute value of a complex number (p. 330)
- trigonometric form of a complex number (p. 331)
- modulus (p. 331)
- argument (p. 331)
- nth root of a complex number (p. 338)
- nth roots of unity (p. 340)

Additional Resources Text-specific additional resources are available to help you do well in this course. See page xvi for details.

4.1 Complex Numbers

The Imaginary Unit *i*

Some equations have no real solutions. For instance, the quadratic equation $x^2 + 1 = 0$ has no real solution because there is no real number x that can be squared to produce -1. To overcome this deficiency, mathematicians created an expanded system of numbers using the **imaginary unit *i*,** defined as

$$i = \sqrt{-1} \qquad \text{Imaginary unit}$$

where $i^2 = -1$. By adding real numbers to real multiples of this imaginary unit, you obtain the set of **complex numbers.** Each complex number can be written in the **standard form** $a + bi$. For instance, the standard form of the complex number $\sqrt{-9} - 5$ is $-5 + 3i$ because

$$\sqrt{-9} - 5 = \sqrt{3^2(-1)} - 5 = 3\sqrt{-1} - 5 = -5 + 3i.$$

In the standard form $a + bi$, the real number a is called the **real part** of the complex number and the number bi (where b is a real number) is called the **imaginary part** of the complex number.

Definition of a Complex Number

If a and b are real numbers, the number $a + bi$ is a **complex number,** and it is said to be written in **standard form.** If $b = 0$, the number $a + bi = a$ is a real number. If $b \neq 0$, the number $a + bi$ is called an **imaginary number.** A number of the form bi, where $b \neq 0$, is called a **pure imaginary number.**

The set of real numbers is a subset of the set of complex numbers, as shown in Figure 4.1. This is true because every real number a can be written as a complex number using $b = 0$. That is, for every real number a, you can write $a = a + 0i$.

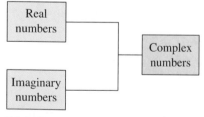

Figure 4.1

Equality of Complex Numbers

Two complex numbers $a + bi$ and $c + di$, written in standard form, are equal to each other

$$a + bi = c + di \qquad \text{Equality of two complex numbers}$$

if and only if $a = c$ and $b = d$.

What You Should Learn:

- How to use the imaginary unit i to write complex numbers
- How to add, subtract, and multiply complex numbers
- How to use complex conjugates to divide complex numbers
- How to use the Quadratic Formula to find complex solutions of quadratic equations

Why You Should Learn It:

Complex numbers are used to model numerous aspects of the natural world, including the impedance of an electrical circuit, as shown in Exercise 80 on page 322.

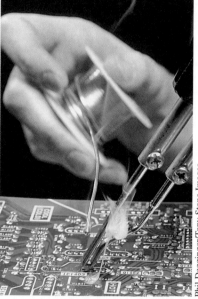

Phil Degginger/Tony Stone Images

Operations with Complex Numbers

To add (or subtract) two complex numbers, you add (or subtract) the real and imaginary parts of the numbers separately.

Addition and Subtraction of Complex Numbers

If $a + bi$ and $c + di$ are two complex numbers written in standard form, their sum and difference are defined as follows.

Sum: $(a + bi) + (c + di) = (a + c) + (b + d)i$

Difference: $(a + bi) - (c + di) = (a - c) + (b - d)i$

The **additive identity** in the complex number system is zero (the same as in the real number system). Furthermore, the **additive inverse** of the complex number $a + bi$ is

$$-(a + bi) = -a - bi. \qquad \text{Additive inverse}$$

So, you have

$$(a + bi) + (-a - bi) = 0 + 0i = 0.$$

Carl Friedrich Gauss (1777–1855) proved that all the roots of any algebraic equation are "numbers" of the form $a + bi$, where a and b are real numbers and i is the square root of -1. These "numbers" are called complex.

Corbis-Bettmann

EXAMPLE 1   Adding and Subtracting Complex Numbers

a. $(3 - i) + (2 + 3i) = 3 - i + 2 + 3i$ Remove parentheses.

$\qquad\qquad\qquad\quad = 3 + 2 - i + 3i$ Group like terms.

$\qquad\qquad\qquad\quad = (3 + 2) + (-1 + 3)i$

$\qquad\qquad\qquad\quad = 5 + 2i$ Write in standard form.

b. $2i + (-4 - 2i) = 2i - 4 - 2i$ Remove parentheses.

$\qquad\qquad\qquad = -4 + 2i - 2i$ Group like terms.

$\qquad\qquad\qquad = -4$ Write in standard form.

c. $3 - (-2 + 3i) + (-5 + i) = 3 + 2 - 3i - 5 + i$

$\qquad\qquad\qquad\qquad\quad = 3 + 2 - 5 - 3i + i$

$\qquad\qquad\qquad\qquad\quad = 0 - 2i$

$\qquad\qquad\qquad\qquad\quad = -2i$

d. $(3 + 2i) + (4 - i) - (7 + i) = 3 + 2i + 4 - i - 7 - i$

$\qquad\qquad\qquad\qquad\qquad\quad = 3 + 4 - 7 + 2i - i - i$

$\qquad\qquad\qquad\qquad\qquad\quad = 0 + 0i$

$\qquad\qquad\qquad\qquad\qquad\quad = 0$

The *Interactive* CD-ROM and *Internet* versions of this text show every example with its solution; clicking on the *Try It!* button brings up similar problems. Guided Examples and Integrated Examples show step-by-step solutions to additional examples. Integrated Examples are related to several concepts in the section.

In Examples 1(b) and 1(d) note that the sum of complex numbers can be a real number.

Many of the properties of real numbers are valid for complex numbers as well. Here are some examples.

Associative Property of Addition and Multiplication
Commutative Property of Addition and Multiplication
Distributive Property of Multiplication over Addition

Notice how these properties are used when two complex numbers are multiplied.

$$(a + bi)(c + di) = a(c + di) + bi(c + di) \qquad \text{Distributive Property}$$

$$= ac + (ad)i + (bc)i + (bd)i^2 \qquad \text{Distributive Property}$$

$$= ac + (ad)i + (bc)i + (bd)(-1) \qquad i^2 = -1$$

$$= ac - bd + (ad)i + (bc)i \qquad \text{Commutative Property}$$

$$= (ac - bd) + (ad + bc)i \qquad \text{Distributive Property}$$

Rather than trying to memorize this multiplication rule, you should simply remember how the distributive property is used to multiply two complex numbers. The procedure is similar to multiplying two polynomials and combining like terms (as in the FOIL Method).

Exploration

Complete the table:

$i^1 = i$	$i^7 = \boxed{}$
$i^2 = -1$	$i^8 = \boxed{}$
$i^3 = -i$	$i^9 = \boxed{}$
$i^4 = 1$	$i^{10} = \boxed{}$
$i^5 = \boxed{}$	$i^{11} = \boxed{}$
$i^6 = \boxed{}$	$i^{12} = \boxed{}$

What pattern do you see? Write a brief description of how you would find i raised to any positive integer power.

EXAMPLE 2   Multiplying Complex Numbers

a. $(i)(-3i) = -3i^2$ Multiply.

$\qquad\qquad = -3(-1)$ $i^2 = -1$

$\qquad\qquad = 3$ Simplify.

b. $2i(3 + 4i) = 6i + 8i^2$ Multiply.

$\qquad\qquad = 6i + 8(-1)$ $i^2 = -1$

$\qquad\qquad = -8 + 6i$ Write in standard form.

c. $(2 - i)(4 + 3i) = 8 + 6i - 4i - 3i^2$ Product of binomials

$\qquad\qquad = 8 + 6i - 4i - 3(-1)$ $i^2 = -1$

$\qquad\qquad = 8 + 3 + 6i - 4i$ Group like terms.

$\qquad\qquad = 11 + 2i$ Write in standard form.

d. $(3 + 2i)(3 - 2i) = 9 - 6i + 6i - 4i^2$ Product of binomials

$\qquad\qquad = 9 - 4(-1)$ $i^2 = -1$

$\qquad\qquad = 9 + 4$ Simplify.

$\qquad\qquad = 13$ Write in standard form.

e. $(3 + 2i)^2 = 9 + 6i + 6i + 4i^2$ Product of binomials

$\qquad\qquad = 9 + 4(-1) + 12i$ $i^2 = -1$

$\qquad\qquad = 9 - 4 + 12i$ Simplify.

$\qquad\qquad = 5 + 12i$ Write in standard form.

Complex Conjugates and Division

Notice in Example 2(d) that the product of two complex numbers can be a real number. This occurs with pairs of complex numbers of the form $a + bi$ and $a - bi$, called **complex conjugates.**

$$(a + bi)(a - bi) = a^2 - abi + abi - b^2i^2$$
$$= a^2 - b^2(-1)$$
$$= a^2 + b^2$$

To find the quotient of $a + bi$ and $c + di$ where c and d are not both zero, multiply the numerator and denominator by the conjugate of the *denominator* to obtain

$$\frac{a + bi}{c + di} = \frac{a + bi}{c + di}\left(\frac{c - di}{c - di}\right)$$
$$= \frac{(ac + bd) + (bc - ad)i}{c^2 + d^2}.$$

EXAMPLE 3   Dividing Complex Numbers

$$\frac{1}{1 + i} = \frac{1}{1 + i}\left(\frac{1 - i}{1 - i}\right)$$ Multiply numerator and denominator by conjugate of denominator.

$$= \frac{1 - i}{1^2 - i^2}$$ Expand.

$$= \frac{1 - i}{1 - (-1)}$$ $i^2 = -1$

$$= \frac{1 - i}{2}$$ Simplify.

$$= \frac{1}{2} - \frac{1}{2}i$$ Write in standard form.

EXAMPLE 4   Dividing Complex Numbers

$$\frac{2 + 3i}{4 - 2i} = \frac{2 + 3i}{4 - 2i}\left(\frac{4 + 2i}{4 + 2i}\right)$$ Multiply numerator and denominator by conjugate of denominator.

$$= \frac{8 + 4i + 12i + 6i^2}{16 - 4i^2}$$ Expand.

$$= \frac{8 - 6 + 16i}{16 + 4}$$ $i^2 = -1$

$$= \frac{2 + 16i}{20}$$ Simplify.

$$= \frac{1}{10} + \frac{4}{5}i$$ Write in standard form.

STUDY T!P

Some graphing utilities can perform operations with complex numbers. For instance, on some graphing utilities, to divide $2 + 3i$ by $4 - 2i$, enter

$$\boxed{(}\,2\,\boxed{+}\,3\,\boxed{i}\,\boxed{)}\,\boxed{\div}$$
$$\boxed{(}\,4\,\boxed{-}\,2\,\boxed{i}\,\boxed{)}\,\boxed{\text{ENTER}}.$$

The display will be as follows.

$$.1 + .8i \quad \text{or} \quad \frac{1}{10} + \frac{4}{5}i$$

Consult your user's manual for specific instructions on performing operations with complex numbers.

Complex Solutions of Quadratic Equations

When using the Quadratic Formula to solve a quadratic equation, you often obtain a result such as $\sqrt{-3}$, which you know is not a real number. By factoring out $i = \sqrt{-1}$, you can write this number in standard form.

$$\sqrt{-3} = \sqrt{3(-1)} = \sqrt{3}\sqrt{-1} = \sqrt{3}\,i$$

The number $\sqrt{3}\,i$ is called the principal square root of -3.

Principal Square Root of a Negative Number

If a is a positive number, the **principal square root** of the negative number $-a$ is defined as

$$\sqrt{-a} = \sqrt{a}\,i.$$

EXAMPLE 5   Writing Complex Numbers in Standard Form

a. $\sqrt{-3}\sqrt{-12} = \sqrt{3}\,i\sqrt{12}\,i = \sqrt{36}\,i^2 = 6(-1) = -6$

b. $\sqrt{-48} - \sqrt{-27} = \sqrt{48}\,i - \sqrt{27}\,i = 4\sqrt{3}\,i - 3\sqrt{3}\,i = \sqrt{3}\,i$

c. $\left(-1 + \sqrt{-3}\right)^2 = \left(-1 + \sqrt{3}\,i\right)^2$

$$= (-1)^2 - 2\sqrt{3}\,i + \left(\sqrt{3}\right)^2(i^2)$$

$$= 1 - 2\sqrt{3}\,i + 3(-1)$$

$$= -2 - 2\sqrt{3}\,i$$

EXAMPLE 6   Complex Solutions of a Quadratic Equation

Solve (a) $x^2 + 4 = 0$ and (b) $3x^2 - 2x + 5 = 0$.

Solution

a. $x^2 + 4 = 0$ Write original equation.

$x^2 = -4$ Subtract 4 from each side.

$x = \pm 2i$ Extract square roots.

b. $3x^2 - 2x + 5 = 0$ Write original equation.

$$x = \frac{-(-2) \pm \sqrt{(-2)^2 - 4(3)(5)}}{2(3)}$$ Quadratic Formula

$$= \frac{2 \pm \sqrt{-56}}{6}$$ Simplify.

$$= \frac{2 \pm 2\sqrt{14}\,i}{6}$$ Write $\sqrt{-56}$ in standard form.

$$= \frac{1}{3} \pm \frac{\sqrt{14}}{3}\,i$$ Write in standard form.

STUDY T!P

The definition of principal square root uses the rule

$$\sqrt{ab} = \sqrt{a}\sqrt{b}$$

for $a > 0$ and $b < 0$. This rule is not valid if *both* a and b are negative. For example,

$$\sqrt{-5}\sqrt{-5} = \sqrt{5(-1)}\sqrt{5(-1)}$$

$$= \sqrt{5}\,i\sqrt{5}\,i$$

$$= \sqrt{25}\,i^2$$

$$= 5i^2$$

$$= -5$$

whereas

$$\sqrt{(-5)(-5)} = \sqrt{25} = 5.$$

To avoid problems with multiplying square roots of negative numbers, be sure to convert to standard form *before* multiplying.

4.1 Exercises

In Exercises 1–4, solve for *a* and *b*.

1. $a + bi = -9 + 4i$ 2. $a + bi = 12 + 5i$
3. $(a - 1) + (b + 3)i = 5 + 8i$
4. $(a + 6) + 2bi = 6 - 5i$

In Exercises 5–14, write in standard form.

5. $4 + \sqrt{-25}$ 6. $3 + \sqrt{-9}$
7. 12 8. 42
9. $-5i + i^2$ 10. $-3i^2 + i$
11. $\left(\sqrt{-75}\right)^2$ 12. $\left(\sqrt{-4}\right)^2$
13. $\sqrt{-0.09}$ 14. $\sqrt{-0.0004}$

In Exercises 15–24, perform the addition or subtraction and write the result in standard form.

15. $(4 + i) + (7 - 2i)$ 16. $(11 - 2i) + (-3 + 6i)$
17. $\left(-1 + \sqrt{-8}\right) + \left(8 - \sqrt{-50}\right)$
18. $\left(7 + \sqrt{-18}\right) - \left(3 + 3\sqrt{2}i\right)$
19. $13i - (14 - 7i)$
20. $22 + (-5 + 8i) + 10i$
21. $-\left(\frac{3}{2} + \frac{5}{2}i\right) + \left(\frac{5}{3} + \frac{11}{3}i\right)$
22. $-\left(\frac{3}{4} + \frac{7}{5}i\right) - \left(\frac{5}{6} - \frac{1}{6}i\right)$
23. $(1.6 + 3.2i) + (-5.8 + 4.3i)$
24. $-(-3.7 - 12.8i) - \left(6.1 - \sqrt{-24.5}\right)$

In Exercises 25–36, perform the operation and write the result in standard form.

25. $\sqrt{-6} \cdot \sqrt{-2}$ 26. $\sqrt{-5} \cdot \sqrt{-10}$
27. $\left(\sqrt{-10}\right)^2$ 28. $\left(\sqrt{-75}\right)^2$
29. $(1 + i)(3 - 2i)$ 30. $(6 - 2i)(2 - 3i)$
31. $6i(5 - 2i)$ 32. $-8i(9 + 4i)$
33. $\left(\sqrt{14} + \sqrt{10}i\right)\left(\sqrt{14} - \sqrt{10}i\right)$
34. $\left(3 + \sqrt{-5}\right)\left(7 - \sqrt{-10}\right)$
35. $(4 + 5i)^2$
36. $(1 - 2i)^2$

The *Interactive* CD-ROM and *Internet* versions of this text contain step-by-step solutions to all odd-numbered Section and Review Exercises. They also provide Tutorial Exercises, which link to Guided Examples for additional help.

37. *Error Analysis*   Describe the error.
$$\sqrt{-6}\sqrt{-6} = \sqrt{(-6)(-6)} = \sqrt{36} = 6 \quad \times$$

38. *Error Analysis*   Describe the error.
$$-i\left(\sqrt{-4} - 1\right) = -i(4i - 1) \quad \times$$
$$= -4i^2 - i$$
$$= 4 - i$$

In Exercises 39–46, find the product of the number and its conjugate.

39. $4 + 3i$ 40. $8 - 12i$
41. $-6 - \sqrt{5}i$ 42. $-3 + \sqrt{2}i$
43. $22i$ 44. $\sqrt{-13}$
45. $3 - \sqrt{-2}$ 46. $1 + \sqrt{8}$

In Exercises 47–58, perform the operation and write the result in standard form.

47. $\dfrac{6}{i}$ 48. $-\dfrac{5}{i}$
49. $\dfrac{4}{4 - 5i}$ 50. $\dfrac{3}{1 - i}$
51. $\dfrac{2 + i}{2 - i}$ 52. $\dfrac{8 - 7i}{1 - 2i}$
53. $\dfrac{6 - 7i}{i}$ 54. $\dfrac{8 + 20i}{2i}$
55. $\dfrac{1}{(4 - 5i)^2}$ 56. $\dfrac{(2 - 3i)(5i)}{2 + 3i}$
57. $\dfrac{2}{1 + i} - \dfrac{3}{1 - i}$ 58. $\dfrac{2i}{2 + i} + \dfrac{5}{2 - i}$

In Exercises 59–68, use the Quadratic Formula to solve the quadratic equation.

59. $x^2 - 2x + 2 = 0$ 60. $x^2 + 6x + 10 = 0$
61. $4x^2 + 16x + 17 = 0$ 62. $9x^2 - 6x + 37 = 0$
63. $4x^2 + 16x + 15 = 0$ 64. $16t^2 - 4t + 3 = 0$
65. $\frac{3}{2}x^2 - 6x + 9 = 0$ 66. $\frac{7}{8}x^2 - \frac{3}{4}x + \frac{5}{16} = 0$
67. $1.4x^2 - 2x - 10 = 0$ 68. $4.5x^2 - 3x + 12 = 0$

69. Express each of the following powers of i as i, $-i$, 1, or -1.

(a) i^{40} (b) i^{25} (c) i^{50} (d) i^{67}

In Exercises 70–77, simplify the complex number and write it in standard form.

70. $-6i^3 + i^2$

71. $4i^2 - 2i^3$

72. $-5i^5$

73. $(-i)^3$

74. $\left(\sqrt{-75}\right)^3$

75. $\left(\sqrt{-2}\right)^6$

76. $\dfrac{1}{i^3}$

77. $\dfrac{1}{(2i)^3}$

78. Cube each complex number. What do you notice?

(a) 2 (b) $-1 + \sqrt{3}i$ (c) $-1 - \sqrt{3}i$

79. Raise each number to the fourth power.

(a) 2 (b) -2 (c) $2i$ (d) $-2i$

80. *Impedance*   The opposition to current in an electrical circuit is called its impedance. The impedance in a parallel circuit with two pathways satisfies the equation

$$\frac{1}{z} = \frac{1}{z_1} + \frac{1}{z_2}$$

when z_1 is the impedance (in ohms) of pathway 1 and z_2 is the impedance (in ohms) of pathway 2. Use the table to determine the impedance of each parallel circuit. (*Hint:* The impedance of each pathway is found by adding the impedance of each component in the pathway.)

	Resistor	Inductor	Capacitor
	—/\/\—	—ooo—	⊣⊢
Symbol	$a\,\Omega$	$b\,\Omega$	$c\,\Omega$
Impedance	a	bi	$-ci$

(a)   (b)

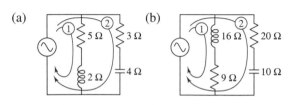

Synthesis

True or False?   **In Exercises 81–83, determine whether the statement is true or false. Justify your answer.**

81. There is no complex number that is equal to its conjugate.

82. $-i\sqrt{6}$ is a solution of $x^4 - x^2 + 14 = 56$.

83. $i^{44} + i^{150} - i^{74} - i^{109} + i^{61} = -1$

84. Prove that the sum of a complex number $a + bi$ and its conjugate is a real number, and that the difference of a complex number $a + bi$ and its conjugate is an imaginary number.

85. Prove that the product of a complex number $a + bi$ and its conjugate is a real number.

Review

In Exercises 86–89 perform the operations and write the result in standard form.

86. $(4 + 3x) + (8 - 6x - x^2)$

87. $(x^3 - 3x^2) - (6 - 2x - 4x^2)$

88. $\left(3x - \frac{1}{2}\right)(x + 4)$

89. $(2x - 5)^2$

In Exercises 90–93, find the x- and y-intercepts of the graph of the equation.

90. $y = -x^2 + 6$

91. $y = x^2 + 2x - 8$

92. $y = |x - 4| + 1$

93. $y = |x| - 1$

94. *Mixture Problem*   A 5-liter container contains a mixture with a concentration of 50%. How much of this mixture must be withdrawn and replaced by 100% concentrate to bring the mixture up to 60% concentration?

95. *Travel*   A business executive drove at an average speed of 100 kilometers per hour on a 200-kilometer trip. Because of heavy traffic, the average speed on the return trip was 80 kilometers per hour. Find the average speed for the round trip.

4.2 Complex Solutions of Equations

The Number of Solutions of a Polynomial Equation

The Fundamental Theorem of Algebra implies that a polynomial equation of degree n has precisely n solutions in the complex number system. These solutions can be real or complex, and they may be repeated.

EXAMPLE 1 Zeros of Polynomial Functions

a. The first-degree equation $x - 2 = 0$ has exactly one solution: $x = 2$.

b. The second-degree equation

$$x^2 - 6x + 9 = 0 \qquad \text{Second-degree equation}$$

$$(x - 3)(x - 3) = 0 \qquad \text{Factor.}$$

has exactly two solutions: $x = 3$ and $x = 3$. (This is called a repeated zero.)

c. The third-degree equation

$$x^3 + 4x = 0 \qquad \text{Third-degree equation}$$

$$x(x - 2i)(x + 2i) = 0 \qquad \text{Factor.}$$

has exactly three solutions: $x = 0$, $x = 2i$, and $x = -2i$.

d. The fourth-degree equation

$$x^4 - 1 = 0 \qquad \text{Fourth-degree equation}$$

$$(x - 1)(x + 1)(x - i)(x + i) = 0 \qquad \text{Factor.}$$

has exactly four solutions: $x = 1$, $x = -1$, $x = i$, and $x = -i$.

You can use a graph to check the number of *real* solutions of an equation. For instance, to check the real solutions of $x^4 - 1 = 0$, sketch the graph of $f(x) = x^4 - 1$. As shown in Figure 4.2, the graph has two x-intercepts, which implies that the equation has two real solutions.

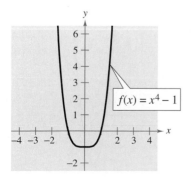

Figure 4.2

Don Smetzer/Tony Stone Images

Every second-degree equation, $ax^2 + bx + c = 0$, has precisely two solutions given by the Quadratic Formula.

$$x = \frac{-b \pm \sqrt{b^2 - 4ac}}{2a}$$

The expression inside the radical, $b^2 - 4ac$, is called the **discriminant.** It can be used to determine whether the solutions are real, repeated, or complex.

1. If $b^2 - 4ac < 0$, the equation has two complex solutions.

2. If $b^2 - 4ac = 0$, the equation has one repeated real solution.

3. If $b^2 - 4ac > 0$, the equation has two real solutions.

EXAMPLE 2   Using the Discriminant

Use the discriminant to find the number of real solutions of each equation.

a. $4x^2 - 20x + 25 = 0$ **b.** $13x^2 + 7x + 2 = 0$ **c.** $5x^2 - 8x = 0$

Solution

a. For this equation, $a = 4$, $b = -20$, and $c = 25$. So, the discriminant is

$$b^2 - 4ac = (-20)^2 - 4(4)(25) = 400 - 400 = 0.$$

Because the discriminant is zero, there is one repeated real solution.

b. For this equation, $a = 13$, $b = 7$, and $c = 2$. So, the discriminant is

$$b^2 - 4ac = 7^2 - 4(13)(2) = 49 - 104 = -55.$$

Because the discriminant is negative, there are two complex solutions.

c. For this equation, $a = 5$, $b = -8$, and $c = 0$. So, the discriminant is

$$b^2 - 4ac = (-8)^2 - 4(5)(0) = 64 - 0 = 64.$$

Because the discriminant is positive, there are two real solutions.

Figure 4.3 shows the graphs of the functions corresponding to the equations in Example 2. Notice that with one repeated solution, the graph touches the x-axis at its x-intercept. With two complex solutions, the graph has no x-intercepts. With two real solutions, the graph crosses the x-axis at its x-intercepts.

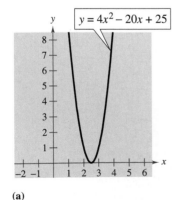

(a)

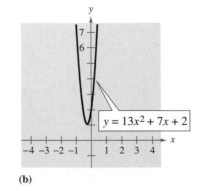

(b)

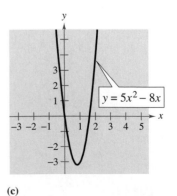

(c)

Figure 4.3

Finding Solutions of Polynomial Equations

EXAMPLE 3   Solving a Quadratic Equation

Solve $x^2 + 2x + 2 = 0$. List complex solutions in standard form.

Solution

Using $a = 1$, $b = 2$, and $c = 2$, you can apply the Quadratic Formula as follows.

$$x = \frac{-b \pm \sqrt{b^2 - 4ac}}{2a} \qquad \text{Quadratic Formula}$$

$$= \frac{-2 \pm \sqrt{2^2 - 4(1)(2)}}{2(1)} \qquad \text{Substitute 1 for } a, \text{ 2 for } b, \text{ and 2 for } c.$$

$$= \frac{-2 \pm \sqrt{-4}}{2} \qquad \text{Simplify.}$$

$$= \frac{-2 \pm 2i}{2} \qquad \text{Simplify.}$$

$$= -1 \pm i \qquad \text{Standard form}$$

So, the two solutions to the quadratic equation are $x = -1 + i$ and $x = -1 - i$.

In Example 3, the two complex solutions are **conjugates.** That is, they are of the form $a \pm bi$. This is not a coincidence, as indicated by the following theorem.

Complex Solutions Occur in Conjugate Pairs

If $a + bi$, $b \neq 0$ is a solution of a polynomial equation with real coefficients, the conjugate $a - bi$ is also a solution of the equation.

Be sure you see that this result is true only if the polynomial has real coefficients. For instance, the result applies to the equation $x^2 + 1 = 0$ but not to the equation $x - i = 0$.

EXAMPLE 4   Solving a Polynomial Equation

Solve $x^4 - x^2 - 20 = 0$.

Solution

$$x^4 - x^2 - 20 = 0$$

$$(x^2 - 5)(x^2 + 4) = 0$$

$$\left(x + \sqrt{5}\right)\left(x - \sqrt{5}\right)(x + 2i)(x - 2i) = 0$$

Setting each factor equal to zero yields the solutions $x = -\sqrt{5}$, $\sqrt{5}$, $-2i$, and $2i$.

Finding Zeros of Polynomial Functions

The problem of finding the zeros of a polynomial function is essentially the same problem as finding the solutions of a polynomial equation. For instance, the zeros of the polynomial function

$$f(x) = 3x^2 - 4x + 5$$

are simply the solutions of the polynomial equation

$$3x^2 - 4x + 5 = 0.$$

EXAMPLE 5   Finding the Zeros of a Polynomial Function

Find all the zeros of

$$f(x) = x^4 - 3x^3 + 6x^2 + 2x - 60$$

given that $1 + 3i$ is a zero of f.

Algebraic Solution

Because complex zeros occur in conjugate pairs, you know that $1 - 3i$ is also a zero of f. This means that both

$$x - (1 + 3i) \qquad \text{and} \qquad x - (1 - 3i)$$

are factors of $f(x)$. Multiplying these two factors produces

$$[x - (1 + 3i)][x - (1 - 3i)] = [(x - 1) - 3i][(x - 1) + 3i]$$
$$= (x - 1)^2 - 9i^2$$
$$= x^2 - 2x + 10.$$

Using long division, you can divide $x^2 - 2x + 10$ into $f(x)$ to obtain the following.

$$
\begin{array}{r}
x^2 - x - 6 \\
x^2 - 2x + 10 \overline{)\, x^4 - 3x^3 + 6x^2 + 2x - 60} \\
\underline{x^4 - 2x^3 + 10x^2} \\
-x^3 - 4x^2 + 2x \\
\underline{-x^3 + 2x^2 - 10x} \\
-6x^2 + 12x - 60 \\
\underline{-6x^2 + 12x - 60} \\
0
\end{array}
$$

Therefore, you have

$$f(x) = (x^2 - 2x + 10)(x^2 - x - 6)$$
$$= (x^2 - 2x + 10)(x - 3)(x + 2)$$

and you can conclude that the zeros of f are $1 + 3i$, $1 - 3i$, 3, and -2.

Graphical Solution

Because complex zeros always occur in conjugate pairs, you know that $1 - 3i$ is also a zero of f. Because the polynomial is a fourth-degree polynomial, you know that there are at most two other zeros of the function. Use a graphing utility to graph

$$y = x^4 - 3x^3 + 6x^2 + 2x - 60$$

as shown in Figure 4.4.

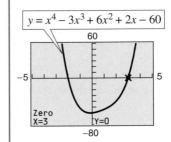

Figure 4.4

You can see that -2 and 3 appear to be x-intercepts of the graph of the function. Use the *zero* or *root* feature or the *zoom* and *trace* features of the graphing utility to confirm that $x = -2$ and $x = 3$ are x-intercepts of the graph. So, you can conclude that the zeros of f are

$$1 + 3i, \quad 1 - 3i, \quad 3, \quad \text{and} \quad -2.$$

EXAMPLE 6 Finding a Polynomial with Given Zeros

Find a fourth-degree polynomial function with real coefficients that has -1, -1, and $3i$ as zeros.

Solution

Because $3i$ is a zero *and* the polynomial is stated to have real coefficients, you know that the conjugate $-3i$ must also be a zero. So, $f(x)$ can be written as

$$f(x) = a(x + 1)(x + 1)(x - 3i)(x + 3i).$$

For simplicity, let $a = 1$ to obtain

$$f(x) = (x^2 + 2x + 1)(x^2 + 9)$$

$$= x^4 + 2x^3 + 10x^2 + 18x + 9.$$

EXAMPLE 7 Finding a Polynomial with Given Zeros

Find a cubic polynomial function f with real coefficients that has 2 and $1 - i$ as zeros, such that $f(1) = 3$.

Solution

Because $1 - i$ is a zero of f, so is $1 + i$. Therefore,

$$f(x) = a(x - 2)[x - (1 - i)][x - (1 + i)]$$

$$= a(x - 2)[(x - 1) + i][(x - 1) - i]$$

$$= a(x - 2)[(x - 1)^2 - i^2]$$

$$= a(x - 2)(x^2 - 2x + 2)$$

$$= a(x^3 - 4x^2 + 6x - 4).$$

To find the value of a, use the fact that $f(1) = 3$ and obtain

$$f(1) = a(1 - 4 + 6 - 4) = 3.$$

So, $a = -3$ and it follows that

$$f(x) = -3(x^3 - 4x^2 + 6x - 4)$$

$$= -3x^3 + 12x^2 - 18x + 12.$$

Writing About Math *Factoring a Polynomial*

Compile a list of all the various techniques for factoring a polynomial that have been covered so far in the text. Give an example illustrating each technique, and write a paragraph discussing when the use of each technique is appropriate.

4.2 E x e r c i s e s

In Exercises 1–4, determine the number of solutions of the equation in the complex number system.

1. $2x^3 + 3x + 1 = 0$

2. $x^6 + 4x^2 + 12 = 0$

3. $50 - 2x^4 = 0$

4. $14 - x + 4x^2 - 7x^5 = 0$

In Exercises 5–12, use the discriminant to determine the number of real solutions of the quadratic equation.

5. $2x^2 - 5x + 5 = 0$ 6. $2x^2 - x - 1 = 0$

7. $\frac{1}{5}x^2 + \frac{6}{5}x - 8 = 0$ 8. $\frac{1}{3}x^2 - 5x + 25 = 0$

9. $2x^2 - x - 15 = 0$ 10. $-2x^2 + 11x - 2 = 0$

11. $x^2 + 2x + 10 = 0$ 12. $x^2 - 4x + 53 = 0$

In Exercises 13–26, solve the equation. List complex solutions in standard form.

13. $x^2 - 5 = 0$ 14. $3x^2 - 1 = 0$

15. $(x + 5)^2 - 6 = 0$ 16. $16 - (x - 1)^2 = 0$

17. $x^2 - 8x + 16 = 0$ 18. $4x^2 + 4x + 1 = 0$

19. $x^2 + 2x + 5 = 0$ 20. $54 + 16x - x^2 = 0$

21. $4x^2 - 4x + 5 = 0$ 22. $4x^2 - 4x + 21 = 0$

23. $230 + 20x - 0.5x^2 = 0$

24. $125 - 30x + 0.4x^2 = 0$

25. $8 + (x + 3)^2 = 0$

26. $6 - (x - 1)^2 = 0$

Graphical and Analytical Analysis **In Exercises 27–30, (a) find all the zeros of the function, (b) use the *zero* or *root* feature of a graphing utility to approximate the real zeros of the function accurate to three decimal places, and (c) describe the relationship between the number of real zeros and the number of *x*-intercepts of the graph.**

27. $f(x) = x^3 - 4x^2 + x - 4$

28. $f(x) = x^3 - 4x^2 - 4x + 16$

29. $f(x) = x^4 + 4x^2 + 4$

30. $f(x) = x^4 - 3x^2 - 4$

In Exercises 31–46, find all the zeros of the function and write the polynomial as a product of linear factors.

31. $f(x) = x^2 + 25$ 32. $f(x) = x^2 - x + 56$

33. $h(x) = x^2 - 4x + 1$ 34. $g(x) = x^2 + 10x + 23$

35. $f(x) = x^4 - 81$ 36. $f(y) = y^4 - 625$

37. $g(x) = x^3 - 6x^2 + 13x - 10$

38. $f(x) = x^3 - 2x^2 - 11x + 52$

39. $h(x) = x^3 - x + 6$

40. $h(x) = x^3 + 9x^2 + 27x + 35$

41. $f(x) = 5x^3 - 9x^2 + 28x + 6$

42. $g(x) = 3x^3 - 4x^2 + 8x + 8$

43. $g(x) = x^4 - 4x^3 + 8x^2 - 16x + 16$

44. $h(x) = x^4 + 6x^3 + 10x^2 + 6x + 9$

45. $f(x) = x^4 + 10x^2 + 9$

46. $f(x) = x^4 + 29x^2 + 100$

In Exercises 47–54, use the given zero to find all the zeros of the function.

	Function	*Zero*
47.	$f(x) = 2x^3 + 3x^2 + 50x + 75$	$5i$
48.	$f(x) = x^3 + x^2 + 9x + 9$	$3i$
49.	$f(x) = 2x^4 - x^3 + 7x^2 - 4x - 4$	$2i$
50.	$g(x) = x^3 - 7x^2 - x + 87$	$5 + 2i$
51.	$g(x) = 4x^3 + 23x^2 + 34x - 10$	$-3 + i$
52.	$h(x) = 3x^3 - 4x^2 + 8x + 8$	$1 - \sqrt{3}i$
53.	$f(x) = x^4 + 3x^3 - 5x^2 - 21x + 22$	$-3 + \sqrt{2}i$
54.	$f(x) = x^3 + 4x^2 + 14x + 20$	$-1 - 3i$

In Exercises 55–60, find a polynomial function with integer coefficients that has the given zeros. (There are many correct answers.)

55. $1, 5i, -5i$ 56. $4, 3i, -3i$

57. $6, -5 + 2i, -5 - 2i$

58. $2, 4 + i, 4 - i$

59. $\frac{2}{3}, -1, 3 + \sqrt{2}i$

60. $-5, -5, 1 + \sqrt{3}i$

61. *Physics* A baseball is thrown upward from ground level with an initial velocity of 48 feet per second, and its height h (in feet) is

$$h = -16t^2 + 48t, \quad 0 \le t \le 3$$

where t is the time (in seconds). Suppose you are told the ball reaches a height of 64 feet. Is this possible?

62. *Economics* The demand equation for a certain product is $p = 140 - 0.0001x$, where p is the unit price (in dollars) of the product and x is the number of units produced and sold. The cost equation for the product is $C = 80x + 150,000$, where C is the total cost (in dollars) and x is the number of units produced. The total profit P obtained by producing and selling x units is $P = xp - C$. You are working in the marketing department of the company that produces this product, and you are asked to determine a price p that will yield a profit of 9 million dollars. Is this possible? Explain.

Synthesis

True or False? **In Exercises 63 and 64, determine whether the statement is true or false. Justify your answer.**

63. It is possible for a third-degree polynomial function with integer coefficients to have no real zeros.

64. If $x = -i$ is a zero of the function

$$f(x) = x^3 + ix^2 + ix - 1,$$

then $x = i$ must also be a zero of f.

Think About It **In Exercises 65–70, determine (if possible) the zeros of the function g if the function f has zeros at $x = r_1$, $x = r_2$, and $x = r_3$.**

65. $g(x) = -f(x)$ **66.** $g(x) = 3f(x)$
67. $g(x) = f(x - 5)$ **68.** $g(x) = f(2x)$
69. $g(x) = 3 + f(x)$ **70.** $g(x) = f(-x)$

71. *Exploration* Use a graphing utility to graph the function $f(x) = x^4 - 4x^2 + k$ for different values of k. Find values of k such that the zeros of f satisfy the specified characteristics. (Some parts do not have unique answers.)

(a) Four real zeros

(b) Two real zeros, each of multiplicity 2

(c) Two real zeros and two complex roots

(d) Four complex zeros

72. *Think About It* Will the answers to Exercise 71 change for the function g?

(a) $g(x) = f(x - 2)$ (b) $g(x) = f(2x)$

73. (a) Find a quadratic function f (with integer coefficients) that has $\pm \sqrt{b}i$ as zeros. Assume that b is a positive integer.

(b) Find a quadratic function f (with integer coefficients) that has $a \pm bi$ as zeros. Assume that b is a positive integer.

74. *Graphical Reasoning* The graph of one of the following functions is shown in the graph. Identify the function shown in the graph. Explain why each of the others is not the correct function. Use a graphing utility to verify your result.

(a) $f(x) = x^2(x + 2)(x - 3.5)$

(b) $g(x) = (x + 2)(x - 3.5)$

(c) $h(x) = (x + 2)(x - 3.5)(x^2 + 1)$

(d) $k(x) = (x + 1)(x + 2)(x - 3.5)$

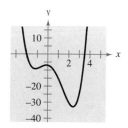

Review

In Exercises 75–80, perform the operation and simplify.

75. $(-3 + 6i) - (8 - 3i)$ **76.** $(12 - 5i) + 16i$

77. $(6 - 2i)(1 + 7i)$ **78.** $(9 - 5i)(9 + 5i)$

79. $\dfrac{1 + i}{1 - i}$ **80.** $(3 + i)^3$

In Exercises 81–86, use the graph of f to graph the function g.

81. $g(x) = f(x - 2)$
82. $g(x) = f(x) - 2$
83. $g(x) = 2f(x)$
84. $g(x) = f(-x)$
85. $g(x) = f(2x)$
86. $g(x) = f\left(\frac{1}{2}x\right)$

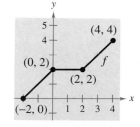

4.3   Trigonometric Form of a Complex Number

The Complex Plane

Just as real numbers can be represented by points on the real number line, you can represent a complex number

$$z = a + bi$$

as the point (a, b) in a coordinate plane (the complex plane). The horizontal axis is called the **real axis** and the vertical axis is called the **imaginary axis,** as shown in Figure 4.5.

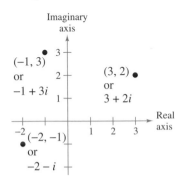

Figure 4.5

The **absolute value of a complex number** $a + bi$ is defined as the distance between the origin $(0, 0)$ and the point (a, b).

Definition of the Absolute Value of a Complex Number

The **absolute value** of the complex number $z = a + bi$ is

$$|a + bi| = \sqrt{a^2 + b^2}.$$

If the complex number $a + bi$ is a real number (that is, if $b = 0$), this definition agrees with that given for the absolute value of a real number

$$|a + 0i| = \sqrt{a^2 + 0^2} = |a|.$$

EXAMPLE 1   Finding the Absolute Value of a Complex Number

Plot $z = -2 + 5i$ and find its absolute value.

Solution

The number is plotted in Figure 4.6. It has an absolute value of

$$|z| = \sqrt{(-2)^2 + 5^2} = \sqrt{29}.$$

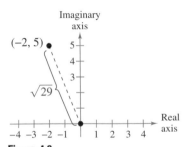

Figure 4.6

What You Should Learn:

- How to plot complex numbers in the complex plane
- How to write trigonometric forms of complex numbers
- How to multiply and divide complex numbers written in trigonometric form

Why You Should Learn It:

You can perform the operations of multiplication and division of complex numbers by learning to write complex numbers in trigonometric form. For instance, in Exercises 59–64 on page 336, you will multiply and divide complex numbers in trigonometric form and in standard form.

Trigonometric Form of a Complex Number

In Section 4.1 you learned how to add, subtract, multiply, and divide complex numbers. To work effectively with *powers* and *roots* of complex numbers, it is helpful to write complex numbers in **trigonometric form.** In Figure 4.7, consider the nonzero complex number $a + bi$. By letting θ be the angle from the positive x-axis (measured counterclockwise) to the line segment connecting the origin and the point (a, b), you can write

$$a = r \cos \theta \qquad \text{and} \qquad b = r \sin \theta$$

where $r = \sqrt{a^2 + b^2}$. Consequently, you have

$$a + bi = (r \cos \theta) + (r \sin \theta)i$$

from which you can obtain the **trigonometric form of a complex number.**

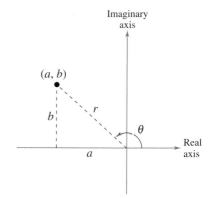

Figure 4.7

> ## Trigonometric Form of a Complex Number
>
> The **trigonometric form** of the complex number $z = a + bi$ is
>
> $$z = r(\cos \theta + i \sin \theta)$$
>
> where $a = r \cos \theta$, $b = r \sin \theta$, $r = \sqrt{a^2 + b^2}$, and $\tan \theta = b/a$. The number r is the **modulus** of z, and θ is called an **argument** of z.

The trigonometric form of a complex number is also called the *polar form*. Because there are infinitely many choices for θ, the trigonometric form of a complex number is not unique. Normally, θ is restricted to the interval $0 \leq \theta < 2\pi$, although on occasion it is convenient to use $\theta < 0$.

EXAMPLE 2   Writing a Complex Number in Trigonometric Form

Write the complex number $z = -2 - 2\sqrt{3}i$ in trigonometric form.

Solution
The absolute value of z is

$$r = \left| -2 - 2\sqrt{3}i \right| = \sqrt{(-2)^2 + \left(-2\sqrt{3}\right)^2} = \sqrt{16} = 4$$

and the angle θ is

$$\tan \theta = \frac{b}{a} = \frac{-2\sqrt{3}}{-2} = \sqrt{3}.$$

Because $\tan(\pi/3) = \sqrt{3}$ and $z = -2 - 2\sqrt{3}i$ lies in Quadrant III, choose θ to be $\theta = \pi + \pi/3 = 4\pi/3$. So, the trigonometric form is

$$z = r(\cos \theta + i \sin \theta) = 4\left(\cos \frac{4\pi}{3} + i \sin \frac{4\pi}{3} \right).$$

See Figure 4.8.

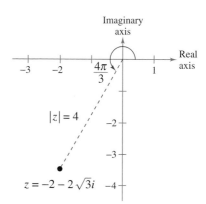

Figure 4.8

EXAMPLE 3 Writing a Complex Number in Trigonometric Form

Write the complex number $z = 6 + 2i$ in trigonometric form.

Solution

The absolute value of z is

$$
\begin{aligned}
r &= |6 + 2i| \\
&= \sqrt{6^2 + 2^2} \\
&= \sqrt{40} \\
&= 2\sqrt{10}
\end{aligned}
$$

and the angle θ is

$$
\tan\theta = \frac{b}{a} = \frac{2}{6} = \frac{1}{3}.
$$

Because θ is in Quadrant I, you can conclude that

$$
\theta = \arctan\frac{1}{3} \approx 0.32175 \text{ radian} \approx 18.4°.
$$

Therefore, the trigonometric form of z is

$$
\begin{aligned}
z &= r(\cos\theta + i\sin\theta) \\
&= 2\sqrt{10}\left[\cos\left(\arctan\frac{1}{3}\right) + i\sin\left(\arctan\frac{1}{3}\right)\right] \\
&\approx 2\sqrt{10}(\cos 18.4° + i\sin 18.4°).
\end{aligned}
$$

This result is illustrated graphically in Figure 4.9.

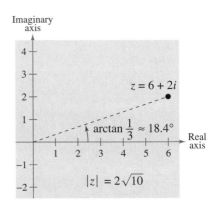

Figure 4.9

EXAMPLE 4 Writing a Complex Number in Standard Form

Write the complex number in standard form $a + bi$.

$$
z = \sqrt{8}\left[\cos\left(-\frac{\pi}{3}\right) + i\sin\left(-\frac{\pi}{3}\right)\right]
$$

Solution

Because $\cos(-\pi/3) = 1/2$ and $\sin(-\pi/3) = -\sqrt{3}/2$, you can write

$$
\begin{aligned}
z &= \sqrt{8}\left[\cos\left(-\frac{\pi}{3}\right) + i\sin\left(-\frac{\pi}{3}\right)\right] \\
&= \sqrt{8}\left(\frac{1}{2} - \frac{\sqrt{3}}{2}i\right) \\
&= 2\sqrt{2}\left(\frac{1}{2} - \frac{\sqrt{3}}{2}i\right) \\
&= \sqrt{2} - \sqrt{6}i.
\end{aligned}
$$

Multiplication and Division of Complex Numbers

The trigonometric form adapts nicely to multiplication and division of complex numbers. Suppose you are given two complex numbers

$$z_1 = r_1(\cos\theta_1 + i\sin\theta_1) \quad \text{and} \quad z_2 = r_2(\cos\theta_2 + i\sin\theta_2).$$

The product of z_1 and z_2 is

$$z_1 z_2 = r_1 r_2(\cos\theta_1 + i\sin\theta_1)(\cos\theta_2 + i\sin\theta_2)$$

$$= r_1 r_2[(\cos\theta_1\cos\theta_2 - \sin\theta_1\sin\theta_2) + i(\sin\theta_1\cos\theta_2 + \cos\theta_1\sin\theta_2)].$$

Using the sum and difference formulas for cosine and sine, you can rewrite this equation as

$$z_1 z_2 = r_1 r_2[\cos(\theta_1 + \theta_2) + i\sin(\theta_1 + \theta_2)].$$

This establishes the first part of the following rule. The second part is left to you (see Exercise 71).

The *Interactive* CD-ROM and *Internet* versions of this text offer a built-in graphing calculator, which can be used with the Examples, Explorations, and Exercises.

Product and Quotient of Two Complex Numbers

Let $z_1 = r_1(\cos\theta_1 + i\sin\theta_1)$ and $z_2 = r_2(\cos\theta_2 + i\sin\theta_2)$ be complex numbers.

$$z_1 z_2 = r_1 r_2[\cos(\theta_1 + \theta_2) + i\sin(\theta_1 + \theta_2)] \qquad \text{Product}$$

$$\frac{z_1}{z_2} = \frac{r_1}{r_2}[\cos(\theta_1 - \theta_2) + i\sin(\theta_1 - \theta_2)], \qquad z_2 \neq 0 \qquad \text{Quotient}$$

Note that this rule says that to *multiply* two complex numbers you multiply moduli and add arguments, whereas to *divide* two complex numbers you divide moduli and subtract arguments.

EXAMPLE 5   Dividing Complex Numbers in Trigonometric Form

Find z_1/z_2 for the following complex numbers.

$$z_1 = 24(\cos 300° + i\sin 300°) \qquad z_2 = 8(\cos 75° + i\sin 75°)$$

Solution

$$\frac{z_1}{z_2} = \frac{24(\cos 300° + i\sin 300°)}{8(\cos 75° + i\sin 75°)}$$

$$= \frac{24}{8}[\cos(300° - 75°) + i\sin(300° - 75°)]$$

$$= 3(\cos 225° + i\sin 225°)$$

$$= 3\left[\left(-\frac{\sqrt{2}}{2}\right) + i\left(-\frac{\sqrt{2}}{2}\right)\right]$$

$$= -\frac{3\sqrt{2}}{2} - \frac{3\sqrt{2}}{2}i$$

EXAMPLE 6   Multiplying Complex Numbers in Trigonometric Form

Find the product of the following complex numbers.

$$z_1 = 2\left(\cos\frac{2\pi}{3} + i\sin\frac{2\pi}{3}\right) \qquad z_2 = 8\left(\cos\frac{11\pi}{6} + i\sin\frac{11\pi}{6}\right)$$

Solution

$$z_1 z_2 = 2\left(\cos\frac{2\pi}{3} + i\sin\frac{2\pi}{3}\right) \cdot 8\left(\cos\frac{11\pi}{6} + i\sin\frac{11\pi}{6}\right)$$

$$= 16\left[\cos\left(\frac{2\pi}{3} + \frac{11\pi}{6}\right) + i\sin\left(\frac{2\pi}{3} + \frac{11\pi}{6}\right)\right]$$

$$= 16\left(\cos\frac{5\pi}{2} + i\sin\frac{5\pi}{2}\right)$$

$$= 16\left(\cos\frac{\pi}{2} + i\sin\frac{\pi}{2}\right)$$

$$= 16[0 + i(1)] = 16i$$

You can check this result by first converting to the standard forms $z_1 = -1 + \sqrt{3}i$ and $z_2 = 4\sqrt{3} - 4i$ and then multiplying algebraically, as in Section 4.1.

$$z_1 z_2 = \left(-1 + \sqrt{3}i\right)\left(4\sqrt{3} - 4i\right)$$

$$= -4\sqrt{3} + 4i + 12i + 4\sqrt{3}$$

$$= 16i$$

> ### STUDY T!P
>
> Some graphing utilities can multiply and divide complex numbers in trigonometric form. If you have access to such a graphing utility, use it to find z_1/z_2 and $z_1 z_2$ in Examples 5 and 6.

Writing About Math   *Multiplying Complex Numbers Graphically*

Discuss how you can graphically approximate the product of the complex numbers. Then, approximate the values of the numbers and check your answers analytically.

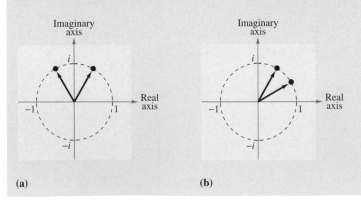

(a)   (b)

4.3 EXERCISES

In Exercises 1–8, plot the complex number and find its absolute value.

1. $-6i$

2. $2i$

3. -4

4. 7

5. $-4 + 4i$

6. $-5 - 12i$

7. $9 - 7i$

8. $10 + 3i$

In Exercises 9–12, write the complex number in trigonometric form.

9.
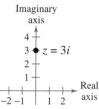

10.
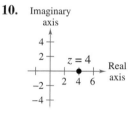

11.
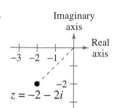

12.
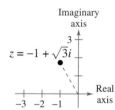

In Exercises 13–28, represent the complex number graphically, and find the trigonometric form of the number.

13. $5 - 5i$

14. $4 + 4i$

15. $\sqrt{3} + i$

16. $-1 + \sqrt{3}i$

17. $-2(1 + \sqrt{3}i)$

18. $\frac{5}{2}(\sqrt{3} - i)$

19. $8i$

20. $-6i$

21. $-7 + 4i$

22. $5 - i$

23. 7

24. 4

25. $1 + 10i$

26. $2\sqrt{2} - i$

27. $-7 - i$

28. $1 + 3i$

In Exercises 29–32, use a graphing utility to represent the complex number in trigonometric form.

29. $5 + 2i$

30. $-3 + i$

31. $3\sqrt{2} - 7i$

32. $-8 - 5\sqrt{3}i$

In Exercises 33–42, represent the complex number graphically and find the standard form of the number.

33. $2(\cos 120° + i \sin 120°)$

34. $5(\cos 135° + i \sin 135°)$

35. $\frac{3}{2}(\cos 330° + i \sin 330°)$

36. $\frac{3}{4}(\cos 315° + i \sin 315°)$

37. $3.75\left(\cos \dfrac{3\pi}{4} + i \sin \dfrac{3\pi}{4}\right)$

38. $8\left(\cos \dfrac{5\pi}{6} + i \sin \dfrac{5\pi}{6}\right)$

39. $4\left(\cos \dfrac{3\pi}{2} + i \sin \dfrac{3\pi}{2}\right)$

40. $9(\cos 0 + i \sin 0)$

41. $3[\cos(18° \, 45') + i \sin(18° \, 45')]$

42. $6[\cos(230° \, 30') + i \sin(230° \, 30')]$

In Exercises 43–46, use a graphing utility to represent the complex number in standard form.

43. $5\left(\cos \dfrac{\pi}{9} + i \sin \dfrac{\pi}{9}\right)$

44. $12\left(\cos \dfrac{3\pi}{5} + i \sin \dfrac{3\pi}{5}\right)$

45. $9(\cos 58° + i \sin 58°)$

46. $4(\cos 216.5° + i \sin 216.5°)$

In Exercises 47–58, perform the operation and leave the result in trigonometric form.

47. $\left[3\left(\cos \dfrac{\pi}{3} + i \sin \dfrac{\pi}{3}\right)\right]\left[4\left(\cos \dfrac{\pi}{6} + i \sin \dfrac{\pi}{6}\right)\right]$

48. $\left[\dfrac{3}{2}\left(\cos \dfrac{\pi}{6} + i \sin \dfrac{\pi}{6}\right)\right]\left[6\left(\cos \dfrac{\pi}{4} + i \sin \dfrac{\pi}{4}\right)\right]$

49. $\left[\frac{5}{3}(\cos 140° + i \sin 140°)\right]\left[\frac{2}{3}(\cos 60° + i \sin 60°)\right]$

50. $\left[\frac{1}{2}(\cos 115° + i \sin 115°)\right]\left[\frac{4}{5}(\cos 300° + i \sin 300°)\right]$

51. $\left[\frac{11}{20}(\cos 290° + i \sin 290°)\right]\left[\frac{2}{5}(\cos 200° + i \sin 200°)\right]$

52. $(\cos 5° + i \sin 5°)(\cos 20° + i \sin 20°)$

53. $\dfrac{\cos 50° + i \sin 50°}{\cos 20° + i \sin 20°}$

54. $\dfrac{5(\cos 4.3 + i \sin 4.3)}{4(\cos 2.1 + i \sin 2.1)}$

55. $\dfrac{2(\cos 120° + i \sin 120°)}{4(\cos 40° + i \sin 40°)}$

56. $\dfrac{\cos\left(\dfrac{7\pi}{4}\right) + i \sin\left(\dfrac{7\pi}{4}\right)}{\cos \pi + i \sin \pi}$

57. $\dfrac{18(\cos 54° + i \sin 54°)}{3(\cos 102° + i \sin 102°)}$

58. $\dfrac{9(\cos 20° + i \sin 20°)}{5(\cos 75° + i \sin 75°)}$

In Exercises 59–64, (a) give the trigonometric form of the complex numbers, (b) perform the indicated operation using the trigonometric form, and (c) perform the indicated operation using the standard form, and then check your result with that of part (b).

59. $(2 + 2i)(1 - i)$

60. $(\sqrt{3} + i)(1 + i)$

61. $-2i(1 + i)$

62. $\dfrac{3 + 4i}{1 - \sqrt{3}i}$

63. $\dfrac{5}{2 + 3i}$

64. $\dfrac{4i}{-4 + 2i}$

In Exercises 65–68, sketch the graph of all complex numbers z satisfying the given condition.

65. $|z| = 2$

66. $|z| = 5$

67. $\theta = \dfrac{\pi}{6}$

68. $\theta = \dfrac{\pi}{4}$

Synthesis

True or False?   In Exercises 69 and 70, determine whether the statement is true or false. Justify your answer.

69. Although the square of the complex number bi is given by $(bi)^2 = -b^2$, the absolute value of the complex number $z = a + bi$ is defined as

$$|a + bi| = \sqrt{a^2 + b^2}.$$

70. The product of two complex numbers, $z_1 = r_1(\cos \theta_1 + i \sin \theta_1)$ and $z_2 = r_2(\cos \theta_2 + i \sin \theta_2)$, is zero only when $r_1 = 0$ or $r_2 = 0$.

71. Given two complex numbers $z_1 = r_1(\cos \theta_1 + i \sin \theta_1)$ and $z_2 = r_2(\cos \theta_2 + i \sin \theta_2)$, $z_2 \neq 0$, show that

$$\dfrac{z_1}{z_2} = \dfrac{r_1}{r_2}[\cos(\theta_1 - \theta_2) + i \sin(\theta_1 - \theta_2)].$$

72. Show that $\bar{z} = r[\cos(-\theta) + i \sin(-\theta)]$ is the complex conjugate of $z = r(\cos \theta + i \sin \theta)$.

73. Use the trigonometric forms of z and $\bar{z}$ in Exercise 72 to find (a) $z\bar{z}$ and (b) $z/\bar{z}$, $\bar{z} \neq 0$.

74. Show that the negative of $z = r(\cos \theta + i \sin \theta)$ is $-z = r[\cos(\theta + \pi) + i \sin(\theta + \pi)]$.

Review

In Exercises 75–80, solve the right triangle shown in the figure. Approximate the result to two decimal places.

75. $A = 22°,\quad a = 8$

76. $B = 66°,\quad a = 33.5$

77. $A = 30°,\quad b = 112.6$

78. $B = 6°,\quad b = 211.2$

79. $A = 42° 15',\quad c = 11.2$

80. $B = 81° 30',\quad c = 6.8$

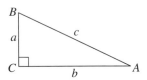

In Exercises 81–84, for the simple harmonic motion described by the trigonometric function, find the maximum displacement and the lowest possible positive value of t for which $d = 0$.

81. $d = 16 \cos \dfrac{\pi}{4}t$

82. $d = \tfrac{1}{8} \cos 12\pi t$

83. $d = \tfrac{1}{16} \sin \tfrac{5}{4}\pi t$

84. $d = \tfrac{1}{12} \sin 60\pi t$

4.4   DeMoivre's Theorem

Powers of Complex Numbers

To raise a complex number to a power, consider repeated use of the multiplication rule.

$$z = r(\cos \theta + i \sin \theta)$$

$$z^2 = r(\cos \theta + i \sin \theta)r(\cos \theta + i \sin \theta) = r^2(\cos 2\theta + i \sin 2\theta)$$

$$z^3 = r^2(\cos 2\theta + i \sin 2\theta)r(\cos \theta + i \sin \theta) = r^3(\cos 3\theta + i \sin 3\theta)$$

$$z^4 = r^4(\cos 4\theta + i \sin 4\theta)$$

$$z^5 = r^5(\cos 5\theta + i \sin 5\theta)$$

$$\vdots$$

This pattern leads to **DeMoivre's Theorem,** which is named after the French mathematician Abraham DeMoivre (1667–1754).

DeMoivre's Theorem

If $z = r(\cos \theta + i \sin \theta)$ is a complex number and n is a positive integer, then

$$z^n = [r(\cos \theta + i \sin \theta)]^n = r^n(\cos n\theta + i \sin n\theta).$$

What You Should Learn:

- How to use DeMoivre's Theorem to find powers of complex numbers
- How to find *n*th roots of complex numbers

Why You Should Learn It:

The trigonometric form of a complex number is useful when finding powers and roots of complex numbers. For instance Exercises 47–54 on page 342 show you how the trigonometric form of complex numbers can help you solve a polynomial equation.

EXAMPLE 1   Finding Powers of a Complex Number

Use DeMoivre's Theorem to find $\left(-1 + \sqrt{3}i\right)^{12}$.

Solution

First convert to trigonometric form using

$$r = \sqrt{(-1)^2 + \left(\sqrt{3}\right)^2} = 2 \qquad\qquad r = \sqrt{a^2 + b^2}$$

and

$$\theta = \arctan \frac{\sqrt{3}}{-1} = \frac{2\pi}{3}. \qquad\qquad \theta = \arctan \frac{b}{a}$$

$$-1 + \sqrt{3}i = 2\left(\cos \frac{2\pi}{3} + i \sin \frac{2\pi}{3}\right) \qquad\qquad a + bi = r(\cos \theta + i \sin \theta)$$

Then, by DeMoivre's Theorem, you have

$$\left(-1 + \sqrt{3}i\right)^{12} = \left[2\left(\cos \frac{2\pi}{3} + i \sin \frac{2\pi}{3}\right)\right]^{12} \qquad z^n = [r(\cos \theta + i \sin \theta)]^n$$

$$= 2^{12}\left[\cos\left(12 \cdot \frac{2\pi}{3}\right) + i \sin\left(12 \cdot \frac{2\pi}{3}\right)\right] \qquad z^n = r^n(\cos n\theta + i \sin n\theta)$$

$$= 4096(\cos 8\pi + i \sin 8\pi)$$

$$= 4096(1 + 0) = 4096.$$

In Figure 4.11, notice that the roots obtained in Example 2 all have a magnitude of 1 and are equally spaced around this unit circle. Also notice that the complex roots occur in conjugate pairs, as discussed in Section 4.2. The n distinct nth roots of 1 are called the **nth roots of unity.**

EXAMPLE 3   Finding the nth Roots of a Complex Number

Find the three cube roots of $z = -2 + 2i$.

Solution

Because z lies in Quadrant II, the trigonometric form for z is

$$z = -2 + 2i = \sqrt{8}\,(\cos 135° + i \sin 135°).$$

By the formula for nth roots, the cube roots have the form

$$\sqrt[6]{8}\left(\cos \frac{135° + 360°k}{3} + i \sin \frac{135° + 360°k}{3}\right).$$

Finally, for $k = 0$, 1, and 2, you obtain the roots

$$\sqrt{2}(\cos 45° + i \sin 45°) = 1 + i$$

$$\sqrt{2}(\cos 165° + i \sin 165°) \approx -1.3660 + 0.3660i \qquad \text{Incremented by } \frac{360°}{3}$$

$$\sqrt{2}(\cos 285° + i \sin 285°) \approx 0.3660 - 1.3660i.$$

Use a graphing utility, set in *parametric* and *radian* modes, to display the graph of

$$X_{1T} = \cos T$$

and

$$Y_{1T} = \sin T.$$

Set the viewing window so that $-1.5 \le X \le 1.5$ and $-1 \le Y \le 1$. Then, using $0 \le T \le 2\pi$, set the "Tstep" to $2\pi/n$ for various values of n. Explain how the graphing utility can be used to obtain the nth roots of unity.

Writing About Math   *A Famous Mathematical Formula*

The famous formula

$$e^{a + bi} = e^a(\cos b + i \sin b)$$

is called Euler's Formula, after the German mathematician Leonhard Euler (1707–1783). Although the interpretation of this formula is beyond the scope of this text, it is included because it gives rise to one of the most wonderful equations in mathematics.

$$e^{\pi i} + 1 = 0$$

This elegant equation relates the five most famous numbers in mathematics—0, 1, π, e, and i—in a single equation. Show how Euler's Formula can be used to derive this equation. Write a short paragraph summarizing your work.

4.4 Exercises

In Exercises 1–24, use DeMoivre's Theorem to find the indicated power of the complex number. Express the result in standard form.

1. $(1 + i)^5$
2. $(2 + 2i)^6$
3. $(-1 + i)^{10}$
4. $(1 - i)^{12}$
5. $(3 - 3i)^5$
6. $(-5 - 5i)^6$
7. $\left(\sqrt{3} - i\right)^3$
8. $\left(\sqrt{3} + i\right)^4$
9. $2\left(\sqrt{3} + i\right)^7$
10. $4\left(1 - \sqrt{3}i\right)^3$
11. $[5(\cos 20° + i \sin 20°)]^3$
12. $[3(\cos 150° + i \sin 150°)]^4$
13. $[5(\cos 3.2 + i \sin 3.2)]^4$
14. $(\cos 0 + i \sin 0)^{20}$
15. $\left(\cos \dfrac{\pi}{4} + i \sin \dfrac{\pi}{4}\right)^{12}$
16. $\left[2\left(\cos \dfrac{\pi}{2} + i \sin \dfrac{\pi}{2}\right)\right]^8$
17. $[3(\cos 15° + i \sin 15°)]^4$
18. $[2(\cos 10° + i \sin 10°)]^8$
19. $[5(\cos 95° + i \sin 95°)]^3$
20. $[4(\cos 110° + i \sin 110°)]^4$
21. $\left[2\left(\cos \dfrac{\pi}{10} + i \sin \dfrac{\pi}{10}\right)\right]^5$
22. $\left[2\left(\cos \dfrac{\pi}{8} + i \sin \dfrac{\pi}{8}\right)\right]^6$
23. $\left[3\left(\cos \dfrac{2\pi}{3} + i \sin \dfrac{2\pi}{3}\right)\right]^3$
24. $\left[3\left(\cos \dfrac{\pi}{12} + i \sin \dfrac{\pi}{12}\right)\right]^5$

Graphical Reasoning   In Exercises 25 and 26, use the graph of the roots of a complex number. (a) Write each of the roots in trigonometric form. (b) Identify the complex number whose roots are given. (c) Use a graphing utility to verify the results of part (b).

25.

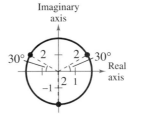

26.

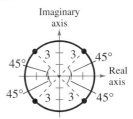

In Exercises 27–46, (a) use the theorem on page 339 to find the indicated roots of the complex number, (b) represent each of the roots graphically, and (c) express each of the roots in standard form.

27. Square roots of $5(\cos 120° + i \sin 120°)$
28. Square roots of $16(\cos 60° + i \sin 60°)$
29. Cube roots of $8\left(\cos \dfrac{2\pi}{3} + i \sin \dfrac{2\pi}{3}\right)$
30. Cube roots of $64\left(\cos \dfrac{\pi}{3} + i \sin \dfrac{\pi}{3}\right)$
31. Fifth roots of $243\left(\cos \dfrac{\pi}{6} + i \sin \dfrac{\pi}{6}\right)$
32. Fifth roots of $32\left(\cos \dfrac{5\pi}{6} + i \sin \dfrac{5\pi}{6}\right)$
33. Square roots of $-25i$
34. Square roots of $-36i$
35. Fourth roots of $81i$
36. Fourth roots of $625i$
37. Cube roots of $-\dfrac{125}{2}\left(1 + \sqrt{3}i\right)$
38. Cube roots of $-4\sqrt{2}(1 - i)$
39. Fourth roots of 16
40. Fourth roots of i
41. Fifth roots of 1
42. Cube roots of 1000
43. Cube roots of -125

Chapter Project   *Creating Fractals with a Graphing Utility*

A **fractal** is a geometric figure that consists of a pattern that is repeated infinitely on a smaller and smaller scale. Fractals are said to be *self-similar* because when you magnify any portion of the fractal, you see a figure that is either identical or similar to the original fractal. In this project you will be introduced to two well-known fractals—the *Koch snowflake* and the *Mandelbrot Set*.

One of the "classic" fractals is the Koch snowflake, named after its creator, the Swedish mathematician Helge von Koch (1870–1924). The construction of this fractal begins with an equilateral triangle whose sides are one unit long. In the first stage, a triangle with sides one-third unit long is added to the center of each side of the original triangle. In the second stage, a triangle with sides one-ninth unit long is added to the center of each side. This process continues without stopping. The first four stages are shown in the figure below.

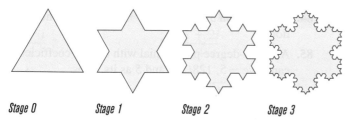

Stage 0   *Stage 1*   *Stage 2*   *Stage 3*

a. Write a formula that describes the side length of each triangle that will be added in the nth stage of construction of the Koch snowflake.

The most famous fractal is called the **Mandelbrot Set,** named for the Polish-born mathematician Benoit Mandelbrot. The construction of the Mandelbrot Set is more intricate than the construction of the Koch snowflake. The construction of this fractal is based on the behavior of the sequence of numbers

$$c, c^2 + c, (c^2 + c)^2 + c, [(c^2 + c)^2 + c]^2 + c, \ldots$$

which is determined by the value of the complex number c. For some values of c, the sequence is *bounded*, and for other values it is *unbounded*. If the sequence is unbounded, the numbers in the sequence get larger and larger without bound (tend to infinity). All values of c for which the sequence is bounded are in the Mandelbrot Set.

The graph at the right shows a picture of the Mandelbrot Set in the complex plane. All numbers in the Mandelbrot Set are plotted as black points, and all numbers that are not in the set are plotted as yellow points.

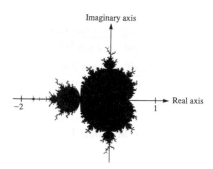

b. Using the sequence on the previous page, determine whether the following complex numbers are in the Mandelbrot Set.

$$c = 2, \quad c = i, \quad c = 1 + i$$

c. You can use a graphing utility to generate the sequence to determine whether a complex number is in the Mandelbrot Set. The pseudocode at the right can be translated into a program for a graphing utility. (The program for several models of graphing calculators can be found at our website *college.hmco.com*.) The program generates the sequence for an entered value c.

Enter the program into your graphing utility and run the program for the complex numbers given in part (b) to verify your answers. Then run the program for $c = -1 + 0.2i$ and determine whether it is in the Mandelbrot Set.

Program
- Enter the real part A
- Enter the imaginary part B
- Store A in C
- Store B in D
- Store 0 in N (number of term)
- Label 1
- Increment N
- Display N
- Display A
- Display B
- Store A in F
- Store B in G
- Store $F^2 - G^2 + C$ in A
- Store $2FG + D$ in B
- Goto Label 1

Questions for Further Exploration

1. Another well-known fractal is called the *Sierpinski Triangle*. The construction of this fractal begins with a triangle. In the first stage, the midpoints of the three sides are used to create the vertices of a new triangle, which is then removed, leaving three triangles. In the second stage, the midpoints of the sides in each new triangle are used to create the vertices of a triangle, which is then removed. Each remaining triangle is similar to the original triangle. The first three stages are shown below. (A program that generates a picture of the Sierpinski Triangle for several models of graphing calculators can be found on our website *college.hmco.com*. Suppose the length of each side of the original triangle is one unit. Write a formula that describes the side length of the triangles that will be generated in the nth stage. Write a formula for the area of the triangles that will be generated at the nth stage.

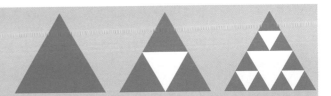

2. Let $c = C + Di$ and let $c^2 + c = F + Gi$. Explain why $(c^2 + c)^2 + c$ is equal to $(F^2 - G^2 + C) + (2FG + D)i$.

In Questions 3–8, use a graphing calculator program or a computer program to determine whether the complex number is in the Mandelbrot Set.

3. $c = 1$

4. $c = -1$

5. $c = -1 + 0.5i$

6. $c = 0.1 + i$

7. $c = 2.1i$

8. $c = 1.9i$

4   Chapter Test

Take this test as you would take a test in class. After you are done, check your work against the answers in the back of the book.

The *Interactive* CD-ROM and *Internet* versions of this text provide answers to the Chapter Tests and Cumulative Tests. They also offer Chapter Pre-Tests (which test key skills and concepts covered in previous chapters) and Chapter Post-Tests, both of which have randomly generated exercises with diagnostic capabilities.

1. Write the complex number $-3 + \sqrt{-81}$ in standard form.

In Exercises 2–5, perform the operations and write the result in standard form.

2. $10i - \left(3 + \sqrt{-25}\right)$

3. $(2 + 6i)^2$

4. $\left(2 + \sqrt{3}i\right)\left(2 - \sqrt{3}i\right)$

5. $\dfrac{5}{2 + i}$

6. Use the Quadratic Formula to solve the equation $2x^2 - 2x + 3 = 0$.

In Exercises 7 and 8, determine the number of solutions of the equation in the complex number system.

7. $x^5 + x^3 - x + 1 = 0$

8. $x^4 - 3x^3 + 2x^2 - 4x - 5 = 0$

In Exercises 9 and 10, find all the zeros of the function.

9. $f(x) = x^3 - 6x^2 + 5x - 30$

10. $f(x) = x^4 - 2x^2 - 24$

In Exercises 11 and 12, use the given zero(s) of the function as an aid in finding all its zeros. Write the polynomial as a product of linear factors.

11. $h(x) = x^4 - 2x^2 - 8$ $-2, 2$

12. $g(v) = 2v^3 - 11v^2 + 22v - 15$ $\frac{3}{2}$

In Exercises 13 and 14, find a polynomial function with integer coefficients that has the given zeros.

13. $0, 3, 3 + i, 3 - i$

14. $1 + \sqrt{6}i, 1 - \sqrt{6}i, 3, 3$

15. Is it possible for a polynomial function with integer coefficients to have exactly one complex zero? Explain.

16. Write the complex number $z = -2 + 2i$ in trigonometric form.

In Exercises 17 and 18, write the complex number in standard form.

17. $100(\cos 240° + i \sin 240°)$

18. $24(\cos 330° + i \sin 330°)$

In Exercises 19 and 20, use DeMoivre's Theorem to find the indicated power of the complex number. Express the result in standard form.

19. $\left[3\left(\cos \dfrac{5\pi}{6} + i \sin \dfrac{5\pi}{6}\right)\right]^8$

20. $(3 - 3i)^6$

21. Find the fourth roots of $128\left(1 + \sqrt{3}i\right)$.

22. Find all solutions of the equation $x^4 - 625i = 0$ and represent the solutions graphically.

Exponential and Logarithmic Functions

5

5.1 Exponential Functions and Their Graphs
5.2 Logarithmic Functions and Their Graphs
5.3 Properties of Logarithms

5.4 Solving Exponential and Logarithmic Equations
5.5 Exponential and Logarithmic Models

Louise Gubb/The Image Works

The Big Picture

In this chapter you will learn how to

❏ recognize, evaluate, and graph exponential and logarithmic functions.
❏ rewrite logarithmic functions with different bases.
❏ use properties of logarithms to evaluate, rewrite, expand, or condense logarithmic expressions.
❏ solve exponential and logarithmic equations.
❏ use exponential growth models, exponential decay models, Gaussian models, logistic models, and logarithmic models to solve real-life problems.
❏ fit exponential and logarithmic models to sets of data.

Personal savings as a percent of disposable income was 3.9% in 1997 and disposable per capita income was $21,969. So, the per capita personal savings was $856.79. (Source: U.S. Bureau of Economic Analysis)

Important Vocabulary

As you encounter each new vocabulary term in this chapter, add the term and its definition to your notebook glossary.

Additional Resources   Text-specific additional resources are available to help you do well in this course. See page xvi for details.

5.1 Exponential Functions and Their Graphs

Exponential Functions

So far, this text has dealt mainly with **algebraic functions,** which include polynomial functions and rational functions. In this chapter you will study two types of nonalgebraic functions—*exponential* functions and *logarithmic* functions. These functions are examples of **transcendental functions.**

Definition of Exponential Function

The **exponential function** f **with base** a is denoted by

$$f(x) = a^x$$

where $a > 0$, $a \neq 1$, and x is any real number.

Note that in the definition of an exponential function, the base $a = 1$ is excluded because it yields $f(x) = 1^x = 1$. This is a constant function, not an exponential function.

You already know how to evaluate a^x for integer and rational values of x. For example, you know that $4^3 = 64$ and $4^{1/2} = 2$. However, to evaluate 4^x for any real number x, you need to interpret forms with *irrational* exponents. For the purposes of this text, it is sufficient to think of

$$a^{\sqrt{2}} \text{ (where } \sqrt{2} \approx 1.41421356)$$

as the number that has the successively closer approximations

$$a^{1.4}, a^{1.41}, a^{1.414}, a^{1.4142}, a^{1.41421}, \ldots$$

Example 1 shows how to use a calculator to evaluate an exponential expression.

EXAMPLE 1 Evaluating Exponential Expressions

Use a calculator to evaluate each expression.

a. $2^{-3.1}$ **b.** $2^{-\pi}$ **c.** $12^{5/7}$ **d.** $(0.6)^{3/2}$

Solution

Number	Graphing Calculator Keystrokes	Display
a. $2^{-3.1}$	2 [∧] [(−)] 3.1 [ENTER]	0.1166291
b. $2^{-\pi}$	2 [∧] [(−)] π [ENTER]	0.1133147
c. $12^{5/7}$	12 [∧] [(] 5 [÷] 7 [)] [ENTER]	5.8998877
d. $(0.6)^{3/2}$	.6 [∧] [(] 3 [÷] 2 [)] [ENTER]	0.4647580

What You Should Learn:

- How to recognize and evaluate exponential functions with base a
- How to graph exponential functions
- How to recognize, evaluate, and graph exponential functions with base e
- How to use exponential functions to model and solve real-life problems

Why You Should Learn It:

Exponential functions are useful in modeling data that increase or decrease quickly. For instance, Exercise 77 on page 361 shows how to use an exponential function to model the amount of defoliation caused by a gypsy moth.

Jenny Hager/The Image Works

Graphs of Exponential Functions

The graphs of all exponential functions have similar characteristics, as shown in Examples 2, 3, and 4.

EXAMPLE 2   Graphs of $y = a^x$

In the same coordinate plane, sketch the graph of each function.

a. $f(x) = 2^x$ **b.** $g(x) = 4^x$

Solution

The table below lists some values for each function, and Figure 5.1 shows the graphs of both functions. Note that both graphs are increasing. Moreover, the graph of $g(x) = 4^x$ is increasing more rapidly than the graph of $f(x) = 2^x$.

x	-2	-1	0	1	2	3
2^x	$\frac{1}{4}$	$\frac{1}{2}$	1	2	4	8
4^x	$\frac{1}{16}$	$\frac{1}{4}$	1	4	16	64

The table in Example 2 was evaluated by hand. You could, of course, use a graphing utility to construct tables with even more values.

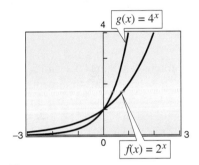

Figure 5.1

EXAMPLE 3   Graphs of $y = a^{-x}$

In the same coordinate plane, sketch the graph of each function.

a. $F(x) = 2^{-x}$ **b.** $G(x) = 4^{-x}$

Solution

The table below lists some values for each function, and Figure 5.2 shows the graphs of both functions. Note that both graphs are decreasing. Moreover, the graph of $G(x) = 4^{-x}$ is decreasing more rapidly than the graph of $F(x) = 2^{-x}$.

x	-3	-2	-1	0	1	2
2^{-x}	8	4	2	1	$\frac{1}{2}$	$\frac{1}{4}$
4^{-x}	64	16	4	1	$\frac{1}{4}$	$\frac{1}{16}$

In Example 3, note that the functions $F(x) = 2^{-x}$ and $G(x) = 4^{-x}$ can be rewritten with positive exponents.

$$F(x) = 2^{-x} = \left(\frac{1}{2}\right)^x \quad \text{and} \quad G(x) = 4^{-x} = \left(\frac{1}{4}\right)^x$$

In general,

$$a^{-x} = \left(\frac{1}{a}\right)^x.$$

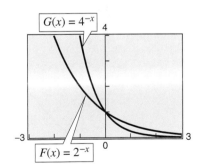

Figure 5.2

Comparing the functions in Examples 2 and 3, observe that

$$F(x) = 2^{-x} = f(-x) \qquad \text{and} \qquad G(x) = 4^{-x} = g(-x).$$

Consequently, the graph of *F* is a reflection (in the *y*-axis) of the graph of *f*, as shown in Figure 5.3(a). The graph of *G* and *g* have the same relationship, as shown in Figure 5.3(b).

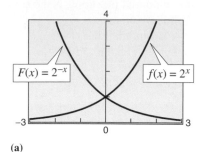

(a)

(b)

Figure 5.3

The graphs in Figures 5.1 and 5.2 are typical of the exponential functions a^x and a^{-x}. They have one *y*-intercept and one horizontal asymptote (the *x*-axis), and they are continuous. The basic characteristics of these exponential functions are summarized in Figure 5.4.

Graph of $y = a^x$, $a > 1$

- Domain: $(-\infty, \infty)$
- Range: $(0, \infty)$
- Intercept: $(0, 1)$
- Increasing
- *x*-axis is a horizontal asymptote ($a^x \to 0$ as $x \to -\infty$)
- Continuous

Graph of $y = a^{-x}$, $a > 1$

- Domain: $(-\infty, \infty)$
- Range: $(0, \infty)$
- Intercept: $(0, 1)$
- Decreasing
- *x*-axis is a horizontal asymptote ($a^{-x} \to 0$ as $x \to \infty$)
- Continuous

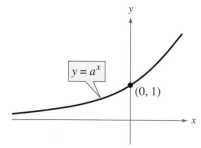

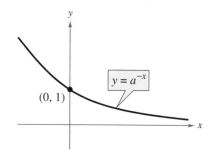

Figure 5.4

Exploration

Use a graphing utility to graph $y = a^x$ with $a = 3, 5,$ and 7 in the same viewing window. (Use a viewing window in which $-2 \le x \le 1$ and $0 \le y \le 2$.) How do the graphs compare with each other? Which graph is on the top in the interval $(-\infty, 0)$? Which is on the bottom? Which graph is on the top in the interval $(0, \infty)$? Which is on the bottom? Repeat this experiment with the graphs of $y = a^x$ for $a = \frac{1}{3}, \frac{1}{5},$ and $\frac{1}{7}$. (Use a viewing window in which $-1 \le x \le 2$ and $0 \le y \le 2$.) What can you conclude about the shape of the graph of $y = a^x$ and the value of *a*?

In the following example, notice how the graph of $y = a^x$ can be used to sketch the graphs of functions of the form

$$f(x) = b \pm a^{x+c}.$$

A computer animation of this example appears in the *Interactive* CD-ROM and *Internet* versions of this text.

EXAMPLE 4   Transformations of Graphs of Exponential Functions

Each of the following graphs is a transformation of the graph of $f(x) = 3^x$, as shown in Figure 5.5.

a. Because $g(x) = 3^{x+1} = f(x + 1)$, the graph of g can be obtained by shifting the graph of f one unit to the left.

b. Because $h(x) = 3^x - 2 = f(x) - 2$, the graph of h can be obtained by shifting the graph of f down two units.

c. Because $k(x) = -3^x = -f(x)$, the graph of k can be obtained by reflecting the graph of f in the x-axis.

d. Because $j(x) = 3^{-x} = f(-x)$, the graph of j can be obtained by reflecting the graph of f in the y-axis.

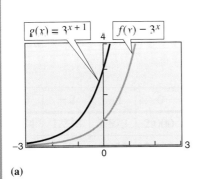

(a)

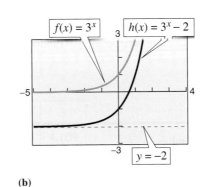

(b)

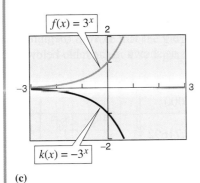

(c)

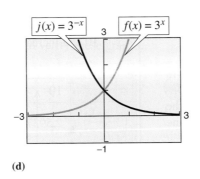

(d)

Figure 5.5

In Figure 5.5, notice that the transformations in parts (a), (c), and (d) keep the x-axis as a horizontal asymptote, but the transformation in part (b) yields a new horizontal asymptote of $y = -2$. Also, be sure to note how the y-intercept is affected by each transformation.

Compound Interest

One of the most familiar examples of exponential growth is that of an investment earning *continuously compounded interest*. Suppose a principal P is invested at an annual interest rate r, compounded once a year. If the interest is added to the principal at the end of the year, the balance is $P_1 = P + Pr = P(1 + r)$. This pattern of multiplying the previous principal by $1 + r$ is then repeated each successive year, as shown in the table.

Time in Years	Balance After Each Compounding
0	$P = P$
1	$P_1 = P(1 + r)$
2	$P_2 = P_1(1 + r) = P(1 + r)(1 + r) = P(1 + r)^2$
$\vdots$	$\vdots$
n	$P_n = P(1 + r)^n$

To accommodate more frequent (quarterly, monthly, or daily) compounding of interest, let n be the number of compoundings per year and let t be the number of years. (The product nt represents the total number of times the interest will be compounded.) Then the interest rate per compounding period is r/n, and the account balance after t years is

$$A = P\left(1 + \frac{r}{n}\right)^{nt}.$$ Amount with n compoundings per year

If you let the number of compoundings n increase without bound, you approach **continuous compounding.** In the formula for n compoundings per year, let $m = n/r$. This produces

$$A = P\left(1 + \frac{r}{n}\right)^{nt} = P\left(1 + \frac{1}{m}\right)^{mrt} = P\left[\left(1 + \frac{1}{m}\right)^m\right]^{rt}.$$

As m increases without bound, you know from Example 5 that $[1 + (1/m)]^m$ approaches e. So, for continuous compounding, it follows that

$$P\left[\left(1 + \frac{1}{m}\right)^m\right]^{rt} \rightarrow P[e]^{rt}$$

and you can write $A = Pe^{rt}$. This result is part of the reason that e is the "natural" choice for a base of an exponential function.

Formulas for Compound Interest

After t years, the balance A in an account with principal P and annual interest rate r (expressed as a decimal) is given by the following formulas.

1. For n compoundings per year: $A = P\left(1 + \frac{r}{n}\right)^{nt}$

2. For continuous compounding: $A = Pe^{rt}$

Exploration

Use the formula

$$A = P\left(1 + \frac{r}{n}\right)^{nt}$$

to calculate the amount in an account when $P = \$3000$, $r = 6\%$, $t = 10$ years, and the number of compoundings is (1) by the day, (2) by the hour, (3) by the minute, and (4) by the second. Use these results to present an argument that increasing the number of compoundings does not mean unlimited growth of the amount in the account.

STUDY TIP

The interest rate r in the formula for compound interest should be written as a decimal. For example, an interest rate of 7% would be written $r = 0.07$.

A computer simulation of this concept appears in the *Interactive* CD-ROM and *Internet* versions of this text.

EXAMPLE 8   Finding the Balance for Compound Interest

A sum of $9000 is invested at an annual interest rate of 8.5%, compounded annually. Find the balance in the account after 3 years.

Algebraic Solution

In this case,

$$P = 9000, r = 8.5\% = 0.085, n = 1, t = 3.$$

Using the formula for compound interest with n compoundings per year, you have

$$A = P\left(1 + \frac{r}{n}\right)^{nt} \qquad \text{Formula for compound interest}$$

$$A = 9000\left(1 + \frac{0.085}{1}\right)^{1(3)} \qquad \text{Substitute values for } P, r, n, \text{ and } t.$$

$$= 9000(1.085)^3 \qquad \text{Simplify.}$$

$$\approx \$11,495.60. \qquad \text{Use a calculator.}$$

So, the balance in the account after 3 years will be about $11,495.60.

Graphical Solution

Substitute values for P, r, and n into the formula for compound interest with n compoundings per year as follows.

$$A = P\left(1 + \frac{r}{n}\right)^{nt} \qquad \text{Formula for compound interest}$$

$$= 9000\left(1 + \frac{0.085}{1}\right)^{(1)t} \qquad \text{Substitute values for } P, r, \text{ and } n.$$

$$= 9000(1 + 0.085)^t \qquad \text{Simplify.}$$

Use a graphing utility to graph $y = 9000(1 + 0.085)^x$, as shown in Figure 5.9. Using the *value* feature or *zoom* and *trace* features, you can approximate the value of y when $x = 3$ to be about 11,495.60. So, the balance in the account after 3 years will be about $11,495.60.

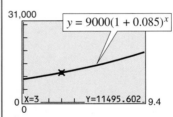

Figure 5.9

EXAMPLE 9   Finding Compound Interest

A total of $12,000 is invested at an annual interest rate of 9%. Find the balance after 5 years if it is compounded

a. quarterly.   **b.** monthly.   **c.** continuously.

Solution

a. For quarterly compoundings, $n = 4$. So, in 5 years at 9%, the balance is

$$A = P\left(1 + \frac{r}{n}\right)^{nt} = 12,000\left(1 + \frac{0.09}{4}\right)^{4(5)} = \$18,726.11.$$

b. For monthly compoundings, $n = 12$. So, in 5 years at 9%, the balance is

$$A = P\left(1 + \frac{r}{n}\right)^{nt} = 12,000\left(1 + \frac{0.09}{12}\right)^{12(5)} = \$18,788.17.$$

c. For continuous compounding, the balance is

$$A = Pe^{rt} = 12,000e^{0.09(5)} = \$18,819.75.$$

Note that continuous compounding yields more than quarterly or monthly compounding.

Other Applications

Exponential functions are used in various other applications.

EXAMPLE 10   Radioactive Decay

Let y represent the mass of a quantity of a radioactive element whose half-life is 25 years. After t years, the mass (in grams) is $y = 10\left(\frac{1}{2}\right)^{t/25}$.

a. What is the initial mass (when $t = 0$)?

b. How much of the initial mass is present after 80 years?

Algebraic Solution

a. $y = 10\left(\frac{1}{2}\right)^{t/25}$ Write original equation.

$\quad = 10\left(\frac{1}{2}\right)^{0/25}$ Substitute 0 for t.

$\quad = 10$ Simplify.

So, the initial mass is 10 grams.

b. $y = 10\left(\frac{1}{2}\right)^{t/25}$ Write original equation.

$\quad = 10\left(\frac{1}{2}\right)^{80/25}$ Substitute 80 for t.

$\quad = 10\left(\frac{1}{2}\right)^{3.2}$ Simplify.

$\quad \approx 1.088$ Use a calculator.

So, about 1.088 grams is present after 80 years.

Graphical Solution

Use a graphing utility to graph $y = 10\left(\frac{1}{2}\right)^{x/25}$.

a. Use the *value* feature or *zoom* and *trace* features of the graphing utility to determine that the value of y when $x = 0$ is 10, as shown in Figure 5.10. So, the initial mass is 10 grams.

b. Use the *value* feature or *zoom* and *trace* features of the graphing utility to determine that the value of y when $x = 80$ is about 1.088, as shown in Figure 5.11. So, about 1.088 grams is present after 80 years.

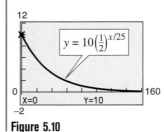

Figure 5.10

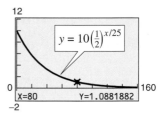

Figure 5.11

EXAMPLE 11   Population Growth

The approximate number of fruit flies in an experimental population after t hours is

$$Q(t) = 20e^{0.03t}, \qquad t \geq 0.$$

a. Find the initial number of fruit flies in the population.

b. How large is the population of fruit flies after 72 hours?

c. Sketch the graph of Q.

Solution

a. To find the initial population, evaluate $Q(t)$ at $t = 0$.

$$Q(0) = 20e^{0.03(0)} = 20e^0 = 20(1) = 20 \text{ flies}$$

b. After 72 hours, the population size is

$$Q(72) = 20e^{0.03(72)} = 20e^{2.16} \approx 173 \text{ flies}.$$

c. The graph of Q is shown in Figure 5.12.

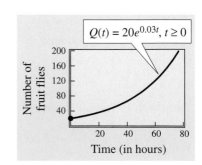

Figure 5.12

5.1 E x e r c i s e s

In Exercises 1–10, use a calculator to evaluate the expression. Round your result to three decimal places.

1. $(3.4)^{6.8}$

2. $5000(2^{-1.5})$

3. $6^{2\pi}$

4. $5^{-\pi}$

5. $\sqrt[3]{7493}$

6. $100^{\sqrt{2}}$

7. $e^{1/2}$

8. $e^{-3/4}$

9. $e^{9.2}$

10. $e^{3.78}$

Think About It **In Exercises 11–14, use properties of exponents to determine which functions (if any) are the same.**

11. $f(x) = 3^{x-2}$

$g(x) = 3^x - 9$

$h(x) = \frac{1}{9}(3^x)$

12. $f(x) = 4^x + 12$

$g(x) = 2^{2x+6}$

$h(x) = 64(4^x)$

13. $f(x) = 16(4^{-x})$

$g(x) = \left(\frac{1}{4}\right)^{x-2}$

$h(x) = 16(2^{-2x})$

14. $f(x) = 5^{-x} + 3$

$g(x) = 5^{3-x}$

$h(x) = -5^{x-3}$

In Exercises 15–22, match the exponential function with its graph. [The graphs are labeled (a), (b), (c), (d), (e), (f), (g), and (h).]

(a)

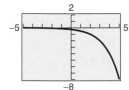

(b)

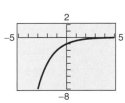

(c)

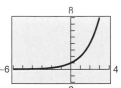

(d)

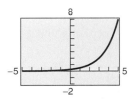

(e)

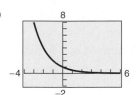

(f)

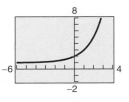

(g)

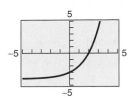

(h)

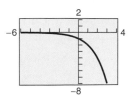

15. $f(x) = 2^x$

16. $f(x) = -2^x$

17. $f(x) = 2^{-x}$

18. $f(x) = -2^{-x}$

19. $f(x) = 2^x - 4$

20. $f(x) = 2^x + 1$

21. $f(x) = -2^{x-2}$

22. $f(x) = 2^{x-2}$

In Exercises 23–26, use the graph of f to describe the transformation that yields the graph of g.

23. $f(x) = 3^x$, $g(x) = 3^{x-5}$

24. $f(x) = -2^x$, $g(x) = 5 - 2^x$

25. $f(x) = \left(\frac{3}{5}\right)^x$, $g(x) = -\left(\frac{3}{5}\right)^{x+4}$

26. $f(x) = 0.3^x$, $g(x) = -0.3^x + 5$

In Exercises 27–34, graph the exponential function by hand. Identify the following features of the graph.

(a) Asymptotes

(b) Intercepts

(c) Increasing or decreasing

27. $g(x) = 5^x$

28. $f(x) = \left(\frac{3}{2}\right)^x$

29. $f(x) = \left(\frac{1}{5}\right)^x = 5^{-x}$

30. $h(x) = \left(\frac{3}{2}\right)^{-x}$

31. $h(x) = 5^{x-2}$

32. $g(x) = \left(\frac{3}{2}\right)^{x+2}$

33. $g(x) = 5^{-x} - 3$

34. $f(x) = \left(\frac{3}{2}\right)^{-x} + 2$

In Exercises 35–44, use a graphing utility to construct a table of values for the function. Then sketch the graph of the function.

35. $f(x) = \left(\frac{5}{2}\right)^x$

36. $f(x) = \left(\frac{5}{2}\right)^{-x}$

37. $f(x) = 6^x$

38. $f(x) = 2^{x-1}$

39. $f(x) = 3^{x+2}$

40. $f(x) = e^{-x}$

41. $f(x) = 3e^{x+4}$

42. $f(x) = 2e^{-0.5x}$

43. $f(x) = 2 + e^{x-5}$

44. $f(x) = 4^{x-3} + 3$

The *Interactive* CD-ROM and *Internet* versions of this text contain step-by-step solutions to all odd-numbered Section and Review Exercises. They also provide Tutorial Exercises, which link to Guided Examples for additional help.

In Exercises 45–52, use a graphing utility to graph the exponential function. Identify any asymptotes of the graph.

45. $y = 2^{-x^2}$

46. $y = 3^{-|x|}$

47. $y = 3^{x-2} + 1$

48. $y = 4^{x+1} - 2$

49. $y = 1.08^{-5x}$

50. $s(t) = 2e^{0.12t}$

51. $s(t) = 3e^{-0.2t}$

52. $g(x) = 1 + e^{-x}$

53. *Exploration* Consider the functions $f(x) = 3^x$ and $g(x) = 4^x$.

(a) Use a graphing utility to complete the table, and use the table to estimate the solution of the inequality $4^x < 3^x$.

x	-1	-0.5	0	0.5	1
$f(x)$					
$g(x)$					

(b) Use a graphing utility to graph $f(x)$ and $g(x)$ in the same viewing window. Use the graphs to solve the inequalities.

(i) $4^x < 3^x$ (ii) $4^x > 3^x$

54. *Exploration* Consider the functions $f(x) = \left(\frac{1}{2}\right)^x$ and $g(x) = \left(\frac{1}{4}\right)^x$.

(a) Use a graphing utility to complete the table, and use the table to estimate the solution of the inequality $\left(\frac{1}{4}\right)^x < \left(\frac{1}{2}\right)^x$.

x	-1	-0.5	0	0.5	1
$f(x)$					
$g(x)$					

(b) Use a graphing utility to graph $f(x)$ and $g(x)$ in the same viewing window. Use the graphs to solve the inequalities.

(i) $\left(\frac{1}{4}\right)^x < \left(\frac{1}{2}\right)^x$ (ii) $\left(\frac{1}{4}\right)^x > \left(\frac{1}{2}\right)^x$

In Exercises 55–58, use a graphing utility to (a) graph the function and (b) find any asymptotes numerically by creating a table of values for the function.

55. $f(x) = \dfrac{8}{1 + e^{-0.5x}}$

56. $g(x) = \dfrac{8}{1 + e^{-0.5/x}}$

57. $f(x) = -\dfrac{6}{2 - e^{0.2x}}$

58. $f(x) = \dfrac{6}{2 - e^{0.2/x}}$

In Exercises 59–62, (a) use a graphing utility to graph the function, (b) use the graph to find the open intervals on which the function is increasing and decreasing, and (c) approximate any relative maximum or minimum values.

59. $f(x) = x^2 e^{-x}$

60. $f(x) = 2x^2 e^{x+1}$

61. $f(x) = x(2^{3-x})$

62. $f(x) = -\left(\frac{1}{2}x\right)3^{x+4}$

Compound Interest **In Exercises 63–66, complete the table to determine the balance A for P dollars invested at rate r for t years and compounded n times per year.**

n	1	2	4	12	365	Continuous
A						

63. $P = \$2500$, $r = 8\%$, $t = 10$ years

64. $P = \$1000$, $r = 6\%$, $t = 10$ years

65. $P = \$2500$, $r = 8\%$, $t = 20$ years

66. $P = \$1000$, $r = 6\%$, $t = 40$ years

Compound Interest **In Exercises 67–70, complete the table to determine the balance A for $12,000 invested at a rate r for t years.**

t	1	10	20	30	40	50
A						

67. $r = 8\%$, compounded continuously

68. $r = 6\%$, compounded continuously

69. $r = 6.5\%$, compounded monthly

70. $r = 7.5\%$, compounded daily

71. *Demand Function* The demand equation for a certain product is

$$p = 5000\left(1 - \frac{4}{4 + e^{-0.002x}}\right)$$

where p is the price and x is the number of units.

(a) Use a graphing utility to graph the demand function for $x > 0$ and $p > 0$.

(b) Find the price p for a demand of $x = 500$ units.

(c) Use the graph in part (a) to approximate the highest price that will still yield a demand of at least 600 units.

(d) Verify your answers to parts (b) and (c) numerically by creating a table of values for the function.

72. *Graphical Reasoning* There are two options for investing $500. The first earns 7% compounded annually, and the second earns 7% simple interest. The figure shows the growth of each investment over a 30-year period.

(a) Identify the two types of investments in the figure. Explain your reasoning.

(b) Verify your answer in part (a) by finding the equations that model the investment growth and using a graphing utility to graph the models.

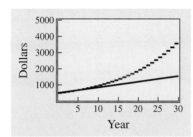

73. *Bacteria Growth* A certain type of bacteria increases according to the model $P(t) = 100e^{0.2197t}$, where t is the time in hours.

(a) Use a graphing utility to graph the model.

(b) Use a graphing utility to approximate $P(0)$, $P(5)$, and $P(10)$.

(c) Verify your answers in part (b) algebraically.

74. *Population Growth* The population of a town increases according to the model $P(t) = 2500e^{0.0293t}$, where t is the time in years, with $t = 0$ corresponding to 1990.

(a) Find the population in 1992, 1995, and 1998.

(b) Use a graphing utility to graph the function for the years 1990 through 2015.

(c) Use a graphing utility to approximate the population in 2005 and 2010.

(d) Verify your answers in part (c) algebraically.

75. *Radioactive Decay* Let Q (in grams) represent the mass of a quantity of radium 226, which has a half-life of 1620 years. The quantity of radium present after t years is

$$Q = 25\left(\tfrac{1}{2}\right)^{t/1620}.$$

(a) Determine the initial quantity (when $t = 0$).

(b) Determine the quantity present after 1000 years.

(c) Use a graphing utility to graph the function over the interval $t = 0$ to $t = 5000$.

(d) When will the quantity of radium be 0 grams? Explain.

76. *Radioactive Decay* Let Q (in grams) represent the mass of a quantity of carbon 14, which has a half-life of 5730 years. The quantity present after t years is $Q = 10\left(\tfrac{1}{2}\right)^{t/5730}$.

(a) Determine the initial quantity (when $t = 0$).

(b) Determine the quantity present after 2000 years.

(c) Sketch the graph of the function over the interval $t = 0$ to $t = 10{,}000$.

77. *Data Analysis* To estimate the amount of defoliation caused by gypsy moths, a forester counts the number x of egg masses on $\frac{1}{40}$ of an acre (circle of radius 18.6 feet) during the fall. The percent of defoliation y the next spring is shown in the table. (Source: USDA Forest Service)

x	0	25	50	75	100
y	12	44	81	96	99

(a) Use a graphing utility to plot the data points.

(b) A model for the data is $y = \dfrac{300}{3 + 17e^{-0.065x}}$.
Use a graphing utility to graph the model in the same viewing window used in part (a). How well does the model fit the data?

(c) Use a graphing utility to create a table comparing the model with the sample data.

(d) Estimate the percent of defoliation if 36 egg masses are counted on $\frac{1}{40}$ acre.

(e) Use the graph to estimate the number of egg masses per $\frac{1}{40}$ acre if approximately $\frac{2}{3}$ of the forest is defoliated the next spring.

78. *Data Analysis* A cup of water at an initial temperature of 78°C is placed in a room at a constant temperature of 21°C. The temperature of the water is measured every 5 minutes for a period of $\frac{1}{2}$ hour. The results are recorded in the table, where t is the time (in minutes) and T is the temperature (in degrees Celsius).

t	0	5	10	15	20	25	30
T	78.0°	66.0°	57.5°	51.2°	46.3°	42.5°	39.6°

(a) Use the regression capabilities of a graphing utility to fit a line to the data. Use the graphing utility to plot the data points and the regression line in the same viewing window. Does the data appear linear? Explain.

(b) Use the regression capabilities of a graphing utility to fit a parabola to the data. Use the graphing utility to plot the data points and the regression parabola in the same viewing window. Does the data appear quadratic? Even though the quadratic model appears to be a "good" fit, explain why it may not be a good model for predicting the temperature of the water when $t = 60$.

(c) The graph of the model should be asymptotic with the temperature of the room. Subtract the room temperature from each of the temperatures in the table. Use a graphing utility to fit an exponential model to the revised data. Add the room temperature to this regression model. Use a graphing utility to plot the original data points and the model in the same viewing window.

(d) Explain why the procedure in part (c) was necessary for finding the exponential model.

79. **Inflation**   If the annual rate of inflation averages 4% over the next 10 years, the approximate cost C of goods or services during any year in that decade will be $C(t) = P(1.04)^t$, where t is the time (in years) and P is the present cost. Assume the price of an oil change for your car is presently \$23.95.

(a) Use a graphing utility to graph the function.

(b) Use the graph in part (a) to approximate the price of an oil change 10 years from now.

(c) Verify your answer in part (b) algebraically.

80. **Depreciation**   After t years, the value of a car that costs \$20,000 is $V(t) = 20,000\left(\frac{3}{4}\right)^t$.

(a) Use a graphing utility to graph the function.

(b) Use a graphing utility to create a table of values that shows the value V for $t = 1$ to 10 years.

Synthesis

True or False?   **In Exercises 81 and 82, determine whether the statement is true or false. Justify your answer.**

81. The x-axis is an asymptote for the graph of $f(x) = 10^x$.

82. $e = \dfrac{271,801}{99,990}$

83. **Exploration**   Use a graphing utility to graph $y_1 = e^x$ and each of the functions $y_2 = x^2$, $y_3 = x^3$, $y_4 = \sqrt{x}$, and $y_5 = |x|$.

(a) Which function increases at the fastest rate for "large" values of x?

(b) Use the result of part (a) to make a conjecture about the rates of growth of $y_1 = e^x$ and $y = x^n$ where n is a natural number and x is "large."

(c) Use the results of parts (a) and (b) to describe what is implied when it is stated that a quantity is growing exponentially.

84. **Conjecture**   Use a graphing utility to graph $f(x) = (1 + 0.5/x)^x$ and $g(x) = e^{0.5}$ in the same viewing window.

(a) What is the relationship between f and g as x increases without bound?

(b) Use the result of part (a) to make a conjecture about the value of $(1 + r/x)^x$ as x increases without bound.

85. **Think About It**   Without a graphing utility, explain why you know $2^{\sqrt{2}}$ is greater than 2, but less than 4.

86. **Think About It**   Which functions are exponential? Why?

(a) $3x$ (b) $3x^2$ (c) 3^x (d) 2^{-x}

87. **Pattern Recognition**   Use a graphing utility to compare the graph of the function $y = e^x$ with the graphs of the following functions.

(a) $y_1 = 1 + \dfrac{x}{1!}$

(b) $y_2 = 1 + \dfrac{x}{1!} + \dfrac{x^2}{2!}$

(c) $y_3 = 1 + \dfrac{x}{1!} + \dfrac{x^2}{2!} + \dfrac{x^3}{3!}$

In your opinion, which function is the best approximation of $y = e^x$?

Review

In Exercises 88–91, determine whether the function has an inverse. If it does, find f^{-1}.

88. $f(x) = 5x - 7$

89. $f(x) = -\frac{2}{3}x + \frac{5}{2}$

90. $f(x) = \sqrt[3]{x + 8}$

91. $f(x) = \sqrt{x^2 + 6}$

In Exercises 92–95, find a polynomial function with integer coefficients that has the given zeros. (There are many correct answers.)

92. $1, 8i, -8i$

93. $i, -i, 4i, -4i$

94. $4, 6 + i, 6 - i$

95. $0, 0, 2, 1 - i$

5.2 Logarithmic Functions and Their Graphs

Logarithmic Functions

In Section P.9, you studied the concept of the inverse of a function. There, you learned if a function has the property such that no horizontal line intersects its graph more than once, the function must have an inverse. By looking back at the graphs of the exponential functions introduced in Section 5.1, you will see that every function of the form

$$f(x) = a^x, \qquad a > 0, a \neq 1$$

passes the Horizontal Line Test and therefore must have an inverse. This inverse function is called the **logarithmic function with base *a*.**

Definition of Logarithmic Function

For $x > 0$ and $0 < a \neq 1$,

$$y = \log_a x \qquad \text{if and only if} \qquad x = a^y.$$

The function

$$f(x) = \log_a x$$

is called the **logarithmic function with base *a*.**

The equations

$$y = \log_a x \qquad \text{Logarithmic form}$$

and

$$x = a^y \qquad \text{Exponential form}$$

are equivalent. The first equation is in logarithmic form and the second is in exponential form.

When evaluating logarithms, remember that *a logarithm is an exponent.* This means that $\log_a x$ is the exponent to which *a* must be raised to obtain *x*. For instance, $\log_2 8 = 3$ because 2 must be raised to the third power to get 8.

What You Should Learn:

- How to recognize and evaluate logarithmic functions with base *a*
- How to graph logarithmic functions
- How to recognize, evaluate, and graph natural logarithmic functions
- How to use logarithmic functions to model and solve real-life problems

Why You Should Learn It:

Logarithmic functions are useful in modeling data that increase or decrease slowly. For instance, Exercise 81 on page 372 shows how to use a logarithmic function to model the minimum required ventilation rates in public school classrooms.

Mark Richards/PhotoEdit

EXAMPLE 1 Evaluating Logarithms

Evaluate each expression.

a. $\log_2 32$ **b.** $\log_3 27$ **c.** $\log_4 2$
d. $\log_{10} \frac{1}{100}$ **e.** $\log_3 1$ **f.** $\log_2 2$

Solution

a. $\log_2 32 = 5$ because $2^5 = 32$. **e.** $\log_3 1 = 0$ because $3^0 = 1$.

b. $\log_3 27 = 3$ because $3^3 = 27$. **f.** $\log_2 2 = 1$ because $2^1 = 2$.

c. $\log_4 2 = \frac{1}{2}$ because $4^{1/2} = \sqrt{4} = 2$.

d. $\log_{10} \frac{1}{100} = -2$ because $10^{-2} = \frac{1}{10^2} = \frac{1}{100}$.

The logarithmic function with base 10 is called the **common logarithmic function.** On most calculators, this function is denoted by $\boxed{\text{LOG}}$. Because $\log_a x$ is the inverse function of a^x, it follows that the domain of $\log_a x$ is the range of a^x, $(0, \infty)$. In other words, $\log_a x$ is defined only if x is positive.

EXAMPLE 2   Evaluating Logarithms on a Calculator

Use a calculator to evaluate each expression.

a. $\log_{10} 10$ **b.** $2 \log_{10} 2.5$ **c.** $\log_{10}(-2)$

Solution

Number	Graphing Calculator Keystrokes	Display
a. $\log_{10} 10$	$\boxed{\text{LOG}}$ 10 $\boxed{\text{ENTER}}$	1
b. $2\log_{10} 2.5$	2 $\boxed{\text{LOG}}$ 2.5 $\boxed{\text{ENTER}}$	0.7958800
c. $\log_{10}(-2)$	$\boxed{\text{LOG}}$ $\boxed{(-)}$ 2 $\boxed{\text{ENTER}}$	ERROR

Note that the calculator displays an error message when you try to evaluate $\log_{10}(-2)$. The reason for this is that the domain of every logarithmic function is the set of *positive* real numbers. In this case, there is no *real* power to which 10 can be raised to get -2.

The following properties follow directly from the definition of the logarithmic function with base a.

Properties of Logarithms

1. $\log_a 1 = 0$ because $a^0 = 1$.
2. $\log_a a = 1$ because $a^1 = a$.
3. $\log_a a^x = x$ and $a^{\log_a x} = x$.   Inverse Properties
4. If $\log_a x = \log_a y$, then $x = y$.   One-to-One Property

---

Library of Functions

The logarithmic function is the inverse of the exponential function. Its domain is the set of positive real numbers and its range is the set of all real numbers. Because of the inverse properties of logarithms and exponents, the exponential equation $a^0 = 1$ implies that $\log_a 1 = 0$.

Consult the Library of Functions Summary inside the front cover for a description of the logarithmic function.

EXAMPLE 3   Using Properties of Logarithms

Solve each equation for x.

a. $\log_2 x = \log_2 3$

b. $\log_4 4 = x$

Solution

a. Using the One-to-One Property (Property 4), you can conclude that $x = 3$.

b. Using Property 2, you can conclude that $x = 1$.

Graphs of Logarithmic Functions

To sketch the graph of $y = \log_a x$, you can use the fact that the graphs of inverse functions are reflections of each other in the line $y = x$.

A computer animation of this example appears in the *Interactive* CD-ROM and *Internet* versions of this text.

EXAMPLE 4   Graphs of Exponential and Logarithmic Functions

In the same coordinate plane, sketch the graph of each function.

a. $f(x) = 2^x$ **b.** $g(x) = \log_2 x$

Solution

a. For $f(x) = 2^x$, construct a table of values.

x	-2	-1	0	1	2	3
2^x	$\frac{1}{4}$	$\frac{1}{2}$	1	2	4	8

By plotting these points and connecting them with a smooth curve, you obtain the graph of $f(x)$ shown in Figure 5.13.

b. Because $g(x) = \log_2 x$ is the inverse of $f(x) = 2^x$, the graph of g is obtained by plotting the points $(2^x, x)$ and connecting them with a smooth curve. The graph of g is a reflection of the graph of f in the line $y = x$, as shown in Figure 5.13.

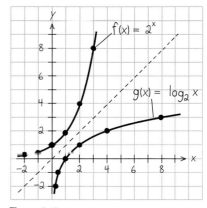

Figure 5.13

Before you can confirm the result of Example 4 using a graphing utility, you need to know how to enter $\log_2 x$. You will learn how to do this using the *change-of-base formula* discussed in Section 5.3.

EXAMPLE 5   Sketching the Graph of a Logarithmic Function

Sketch the graph of the common logarithmic function $f(x) = \log_{10} x$.

Solution

Begin by constructing a table of values. Note that some of the values can be obtained without a calculator using the Inverse Property of logarithms. Others require a calculator. Next, plot the points and connect them with a smooth curve, as shown in Figure 5.14.

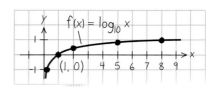

Figure 5.14

	Without Calculator				With Calculator		
x	$\frac{1}{100}$	$\frac{1}{10}$	1	10	2	5	8
$\log_{10} x$	-2	-1	0	1	0.301	0.699	0.903

STUDY T!P

Compare Figure 5.14 with that obtained using a graphing utility. Note that the domain is $(0, \infty)$ and the range is $(-\infty, \infty)$.

In Example 5, you can also sketch the graph of $f(x) = \log_{10} x$ as follows. Evaluate the inverse of f, $g(x) = 10^x$ for several values of x. Plot the points, sketch the graph of g, then reflect the graph in the line $y = x$ to obtain the graph of f.

The nature of the graph in Figure 5.14 is typical of functions of the form $f(x) = \log_a x, a > 1$. They have one x-intercept and one vertical asymptote. Notice how slowly the graph rises for $x > 1$. In Figure 5.14, you would need to move out to $x = 100$ before the graph rose to $y = 2$, and $x = 1000$ before the graph rose to $y = 3$. The basic characteristics of logarithmic graphs are summarized in Figure 5.15.

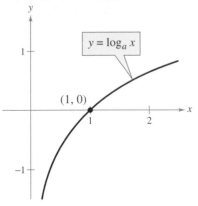

Graph of $y = \log_a x$, $a > 1$

- Domain: $(0, \infty)$
- Range: $(-\infty, \infty)$
- Intercept: $(1, 0)$
- Increasing
- y-axis is a vertical asymptote
  $(\log_a x \to -\infty$ as $x \to 0^+)$
- Continuous
- Reflection of graph of $y = a^x$ in the line $y = x$

Figure 5.15

In Figure 5.15, the vertical asymptote occurs at $x = 0$, where $\log_a x$ is *undefined*.

EXAMPLE 6   Transformations of Graphs of Logarithmic Functions

The graph of each of the following functions is similar to the graph of $f(x) = \log_{10} x$, as shown in Figure 5.16.

a. Because $g(x) = \log_{10}(x - 1) = f(x - 1)$, the graph of g can be obtained by shifting the graph of f one unit to the right.

b. Because $h(x) = 2 + \log_{10} x = 2 + f(x)$, the graph of h can be obtained by shifting the graph of f two units up.

A computer animation of this example appears in the *Interactive* CD-ROM and *Internet* versions of this text.

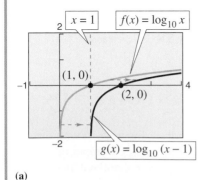

(a)   **(b)**

Figure 5.16

In Figure 5.16, notice that the transformation in part (b) keeps the y-axis as a vertical asymptote, but the transformation in part (a) yields the new vertical asymptote $x = 1$.

The Natural Logarithmic Function

The most widely used base for logarithmic functions is the number e, where

$$e \approx 2.718281828 \ldots .$$

The logarithmic function with base e is the **natural logarithmic function** and is denoted by the special symbol $\ln x$, read as "the natural log of x" or "el en of x."

The Natural Logarithmic Function

The function defined by

$$f(x) = \log_e x = \ln x, \qquad x > 0$$

is called the **natural logarithmic function.**

The *Interactive* CD-ROM and *Internet* versions of this text offer a built-in graphing calculator, which can be used with the Examples, Explorations, and Exercises.

On most calculators, the natural logarithm is denoted by $\boxed{\text{LN}}$, as illustrated in Example 7.

EXAMPLE 7   Evaluating the Natural Logarithmic Function

Use a calculator to evaluate each expression.

a. $\ln 2$ **b.** $\ln 0.3$ **c.** $\ln e^2$ **d.** $\ln(-1)$

Solution

Number	Graphing Calculator Keystrokes	Display
a. $\ln 2$	$\boxed{\text{LN}}$ 2 $\boxed{\text{ENTER}}$	0.6931472
b. $\ln 0.3$	$\boxed{\text{LN}}$ .3 $\boxed{\text{ENTER}}$	-1.2039728
c. $\ln e^2$	$\boxed{\text{LN}}$ $\boxed{e^x}$ 2 $\boxed{\text{ENTER}}$	2
d. $\ln(-1)$	$\boxed{\text{LN}}$ $\boxed{(-)}$ 1 $\boxed{\text{ENTER}}$	ERROR

In Example 7, be sure you see that $\ln(-1)$ gives an error message on most calculators. This occurs because the domain of $\ln x$ is the set of *positive* real numbers. So, $\ln(-1)$ is undefined.

The four properties of logarithms listed on page 364 are also valid for natural logarithms.

Exploration

Because the natural exponential function $y = e^x$ passes the Horizontal Line Test, it has an inverse. Use the *draw inverse* feature of your graphing utility to graph $y = e^x$ and its inverse. Compare this graph with that of the natural logarithmic function. What can you conclude?

Properties of Natural Logarithms

1. $\ln 1 = 0$ because $e^0 = 1$.

2. $\ln e = 1$ because $e^1 = e$.

3. $\ln e^x = x$ and $e^{\ln x} = x$.   Inverse Properties

4. If $\ln x = \ln y$, then $x = y$.   One-to-One Property

EXAMPLE 8   Using Properties of Natural Logarithms

Use the properties of natural logarithms to rewrite each expression.

a. $\ln \dfrac{1}{e}$ **b.** $\ln e^2$ **c.** $\ln e^0$ **d.** $2 \ln e$

Solution

a. $\ln \dfrac{1}{e} = \ln e^{-1} = -1$ Inverse Property

b. $\ln e^2 = 2$ Inverse Property

c. $\ln e^0 = 0$ Property 1

d. $2 \ln e = 2(1) = 2$ Property 2

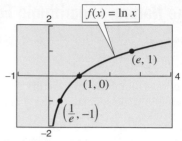

Figure 5.17

The graph of the natural logarithmic function is shown in Figure 5.17. Try using a graphing utility to confirm this graph and to verify that the domain of the natural logarithmic function is $(0, \infty)$.

EXAMPLE 9   Finding the Domains of Logarithmic Functions

Find the domain of each function.

a. $f(x) = \ln(x - 2)$ **b.** $g(x) = \ln(2 - x)$ **c.** $h(x) = \ln x^2$

Algebraic Solution

a. Because $\ln(x - 2)$ is defined only if

$$x - 2 > 0,$$

it follows that the domain of f is $(2, \infty)$.

b. Because $\ln(2 - x)$ is defined only if

$$2 - x > 0,$$

it follows that the domain of g is $(-\infty, 2)$.

c. Because $\ln x^2$ is defined only if

$$x^2 > 0,$$

it follows that the domain of h is all real numbers except $x = 0$.

Graphical Solution

a. Use a graphing utility to graph the equation $y = \ln(x - 2)$ using a decimal setting, as shown in Figure 5.18. Use the *trace* feature to see that the x-coordinates of the points on the graph appear to extend from the right of 2 to $+\infty$. So, you can estimate the domain to be $(2, \infty)$.

b. Use a graphing utility to graph the equation $y = \ln(2 - x)$ using a decimal setting, as shown in Figure 5.19. Use the *trace* feature to see that the x-coordinates of the points on the graph appear to extend from $-\infty$ to the left of 2. So, you can estimate the domain to be $(-\infty, 2)$.

c. Use a graphing utility to graph the equation $y = \ln x^2$ using a decimal setting, as shown in Figure 5.20. Use the *trace* feature to see that the x-coordinates of the points on the graph appear to include all real numbers except $x = 0$. So, you can estimate the domain to be all real numbers except $x = 0$.

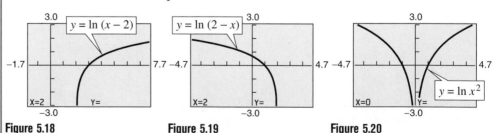

Figure 5.18   **Figure 5.19**   **Figure 5.20**

In Example 9, suppose you had been asked to analyze the function $h(x) = \ln|x - 2|$. How would the domain of this function compare with the domains of the functions given in parts (a) and (b) of the example?

Application

Exponential and logarithmic functions are used to model many situations in real life, as shown in the next example.

EXAMPLE 10 Human Memory Model

Students participating in a psychological experiment attended several lectures on a subject. At the end of the last lecture, and every month for the next year, the students were tested to see how much of the material they remembered. The average scores for the group were given by the *human memory model*

$$f(t) = 75 - 6 \ln(t + 1), \qquad 0 \le t \le 12$$

where t is the time in months.

a. What was the average score on the original $(t = 0)$ exam?

b. What was the average score at the end of $t = 2$ months?

c. What was the average score at the end of $t = 6$ months?

Algebraic Solution

a. The original average score was

$$f(0) = 75 - 6 \ln 1$$

$$= 75 - 6(0)$$

$$= 75.$$

b. After 2 months, the average score was

$$f(2) = 75 - 6 \ln 3$$

$$\approx 75 - 6(1.0986)$$

$$\approx 68.4.$$

c. After 6 months, the average score was

$$f(6) = 75 - 6 \ln 7$$

$$\approx 75 - 6(1.9459)$$

$$\approx 63.3.$$

Graphical Solution

Use a graphing utility to graph the function $y = 75 - 6 \ln(x + 1)$ as shown in Figure 5.21. Then use the *value* or *trace* feature to approximate the following.

a. When $x = 0$, $y = 75$. So, the original average score was 75.

b. When $x = 2$, $y \approx 68.4$. So, the average score after 2 months was about 68.4.

c. When $x = 6$, $y \approx 63.3$. So, the average score after 6 months was about 63.3.

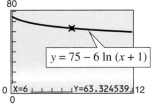

Figure 5.21

Writing About Math *Analyzing a Human Memory Model*

Use a graphing utility to determine how many months it would take for the average score in Example 10 to decrease to 60. Explain your method of solving the problem. Describe another way that you can use a graphing utility to determine the answer. Also, make a statement about the general shape of the model. Would the students forget more quickly soon after the test or as time passes? Explain your reasoning.

5.2 Exercises

In Exercises 1–8, write the logarithmic equation in exponential form. For example, the exponential form of $\log_5 25 = 2$ is $5^2 = 25$.

1. $\log_4 64 = 3$

2. $\log_3 81 = 4$

3. $\log_7 \frac{1}{49} = -2$

4. $\log_{10} \frac{1}{1000} = -3$

5. $\log_{32} 4 = \frac{2}{5}$

6. $\log_{16} 8 = \frac{3}{4}$

7. $\ln 1 = 0$

8. $\ln 4 = 1.386 \ldots$

In Exercises 9–18, write the exponential equation in logarithmic form.

9. $5^3 = 125$

10. $8^2 = 64$

11. $81^{1/4} = 3$

12. $9^{3/2} = 27$

13. $6^{-2} = \frac{1}{36}$

14. $10^{-3} = 0.001$

15. $e^3 = 20.0855 \ldots$

16. $e^x = 4$

17. $e^{2.6} = 13.463 \ldots$

18. $e^\pi = 23.140 \ldots$

In Exercises 19–24, evaluate the expression without using a calculator.

19. $\log_2 16$

20. $\log_{27} 9$

21. $\log_{16}\left(\frac{1}{4}\right)$

22. $\log_2\left(\frac{1}{8}\right)$

23. $\log_{10} 0.01$

24. $\log_{10} 0.1$

In Exercises 25–30, solve the equation for x.

25. $\log_7 x = \log_7 9$

26. $\log_5 5 = x$

27. $\ln e^8 = x$

28. $\ln 1 = \ln x$

29. $\log_6 6^2 = x$

30. $\log_2 2^{-1} = x$

In Exercises 31–42, use a calculator to evaluate the logarithm. Round to three decimal places.

31. $\log_{10} 345$

32. $\log_{10} 145$

33. $\log_{10}\left(\frac{4}{5}\right)$

34. $\log_{10} \frac{25}{2}$

35. $\ln\left(4 + \sqrt{3}\right)$

36. $\ln\left(\sqrt{5} - 2\right)$

37. $\ln \sqrt{42}$

38. $\ln \sqrt{752}$

39. $6 \log_{10} 14.8$

40. $-3 \log_{10} 0.09$

41. $12 \ln 6.4$

42. $-5.5 \ln 34$

In Exercises 43–46, describe the relationship between the graphs of f and g. What is the relationship between the functions f and g?

43. $f(x) = 3^x$
$g(x) = \log_3 x$

44. $f(x) = 5^x$
$g(x) = \log_5 x$

45. $f(x) = e^x$
$g(x) = \ln x$

46. $f(x) = 10^x$
$g(x) = \log_{10} x$

In Exercises 47–52, use the graph of $y = \log_3 x$ to match the given function with its graph. [The graphs are labeled (a), (b), (c), (d), (e), and (f).]

(a)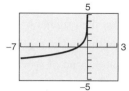

(b)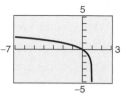

(c)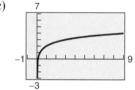

(d)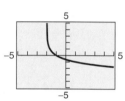

(e)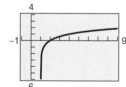

(f)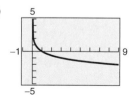

47. $f(x) = \log_3 x + 2$

48. $f(x) = -\log_3 x$

49. $f(x) = -\log_3(x + 2)$

50. $f(x) = \log_3(x - 1)$

51. $f(x) = \log_3(1 - x)$

52. $f(x) = -\log_3(-x)$

In Exercises 53–60, find the domain, vertical asymptote, and x-intercept of the logarithmic function, and sketch its graph by hand. Verify using a graphing utility.

53. $f(x) = \log_4 x$

54. $g(x) = \log_6 x$

55. $h(x) = \log_4(x - 3)$

56. $f(x) = -\log_6(x + 2)$

57. $y = -\log_3 x + 2$

58. $y = \log_5(x - 1) + 4$

59. $f(x) = 6 + \log_6(x - 3)$

60. $f(x) = -\log_3(x + 2) - 4$

In Exercises 61–66, use a graphing utility to graph the logarithmic function. Determine the domain and identify any vertical asymptote and *x*-intercept.

61. $y = \log_{10}\left(\dfrac{x}{5}\right)$

62. $y = \log_{10}(-x)$

63. $f(x) = \ln(x - 2)$

64. $h(x) = \ln(x + 1)$

65. $g(x) = \ln(-x)$

66. $f(x) = \ln(3 - x)$

In Exercises 67–70, (a) use a graphing utility to graph the function, (b) find the domain, (c) use the graph to determine the intervals in which the function is increasing and decreasing, and (d) approximate any relative maximum or minimum values of the function.

67. $f(x) = \dfrac{x}{2} - \ln\dfrac{x}{4}$

68. $g(x) = \dfrac{12 \ln x}{x}$

69. $h(x) = 4x \ln x$

70. $f(x) = \dfrac{x}{\ln x}$

71. *Population Growth* The population of a town will double in

$$t = \frac{10 \ln 2}{\ln 67 - \ln 50} \text{ years.}$$

Find *t*.

72. *Work* The work (in foot-pounds) done in compressing an initial volume of 9 cubic feet of air at a pressure of 15 pounds per square inch to a volume of 3 cubic feet is $W = 19{,}440(\ln 9 - \ln 3)$. Find *W*.

73. *Human Memory Model* Students in a mathematics class were given an exam and then tested monthly with an equivalent exam. The average score for the class was given by the human memory model

$$f(t) = 80 - 17 \log_{10}(t + 1), \quad 0 \le t \le 12$$

where *t* is the time in months.

(a) What was the average score on the original exam ($t = 0$)?

(b) What was the average score after 4 months?

(c) What was the average score after 10 months?

(d) Verify your answers in parts (a), (b), and (c) using a graphing utility.

74. *Investment Time* A principal *P*, invested at $5\frac{1}{2}\%$ and compounded continuously, increases to an amount *K* times the original principal after *t* years, where $t = (\ln K)/0.055$.

(a) Complete the table and interpret your results.

K	1	2	4	6	8	10	12
t							

(b) Use a graphing utility to graph the function.

75. *Data Analysis* The table shows the temperatures *T* (in °F) at which water boils at selected pressures *p* (in pounds per square inch). (Source: Standard Handbook of Mechanical Engineers)

p	5	10	14.696 (1 atm)
T	162.24°	193.21°	212.00°

p	20	30	40
T	227.96°	250.33°	267.25°

p	60	80	100
T	292.71°	312.03°	327.81°

A model that approximates this data is

$$T = 87.97 + 34.96 \ln p + 7.91 \sqrt{p}.$$

(a) Use a graphing utility to plot the data points and graph the model in the same viewing window. How well does the model fit the data?

(b) Use the graph to estimate the pressure required for the boiling point of water to exceed 300°F.

(c) Calculate *T* when the pressure is 74 pounds per square inch. Verify your answer graphically.

76. *Data Analysis* A meteorologist measures the atmospheric pressure *P* (in Pascals) at altitude *h* (in kilometers). The data is shown below.

h	0	5	10	15	20
P	101,293	54,735	23,294	12,157	5069

A model for the data is

$$P = 102{,}303e^{-0.137h}.$$

(a) Use a graphing utility to plot the data points and graph the model in the same viewing window.

(b) Use a graphing utility to plot the points $(h, \ln P)$. Use the regression capabilities of the graphing utility to fit a regression line to the revised data points.

(c) The line in part (b) has the form $\ln P = ah + b$. Write the line in exponential form.

(d) Verify graphically and algebraically that the result of part (c) is equivalent to the given exponential model for the data.

77. ***World Population Growth***   The time t in years for the world population to double if it is increasing at a continuous rate of r is

$$t = \frac{\ln 2}{r}.$$

Complete the table. What do your results imply?

r	0.005	0.010	0.015	0.020	0.025	0.030
t						

78. ***Tractrix***   A person walking along a dock (the y-axis) drags a boat by a 10-foot rope. The boat travels along a path known as a tractrix. The equation of this path is

$$y = 10 \ln\left(\frac{10 + \sqrt{100 - x^2}}{x}\right) - \sqrt{100 - x^2}.$$

(a) Use a graphing utility to obtain a graph of the function. What is the domain of the function?

(b) Identify any asymptotes of the graph.

(c) Determine the position of the person when the x-coordinate of the position of the boat is $x = 2$.

(d) Let $(0, p)$ be the position of the person. Determine p as a function of x, the x-coordinate of the position of the boat.

(e) Use a graphing utility to graph the function p. When does the position of the person change most for a small change in the position of the boat? Explain.

79. ***Sound Intensity***   The relationship between the number of decibels β and the intensity of a sound I in watts per square meter is

$$\beta = 10 \log_{10}\left(\frac{I}{10^{-12}}\right).$$

(a) Determine the number of decibels of a sound with an intensity of 1 watt per square meter.

(b) Determine the number of decibels of a sound with an intensity of 10^{-2} watt per square meter.

(c) The intensity of the sound in part (a) is 100 times as great as that in part (b). Is the number of decibels 100 times as great? Explain.

Ventilation Rates   **In Exercises 80 and 81, use the model**

$$y = 80.4 - 11 \ln x, \quad 100 \le x \le 1500$$

which approximates the minimum required ventilation rate in terms of the air space per child in a public school classroom. In the model, x is the air space per child (in cubic feet) and y is the ventilation rate (in cubic feet per minute).

80. Use a graphing utility to graph the function and approximate the required ventilation rate if there is 300 cubic feet of air space per child.

81. A classroom is designed for 30 students. The air-conditioning system in the room has the capacity to move 450 cubic feet of air per minute.

(a) Determine the ventilation rate per child, assuming that the room is filled to capacity.

(b) Use the graph of Exercise 80 to estimate the air space required per child.

(c) Determine the minimum number of square feet of floor space required for the room if the ceiling height is 30 feet.

Monthly Payment   **In Exercises 82–85, use the model**

$$t = 16.625 \ln\left(\frac{x}{x - 750}\right), \quad x > 750$$

which approximates the length of a home mortgage of \$150,000 at 6% in terms of the monthly payment. In the model, t is the length of the mortgage in years and x is the monthly payment in dollars.

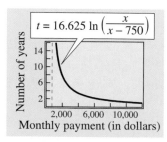

82. Use the model to approximate the length of a $150,000 mortgage (at 6%) if the monthly payment is $897.72.

83. Use the model to approximate the length of a $150,000 mortgage (at 6%) if the monthly payment is $1659.24.

84. Approximate the total amount paid over the term of the mortgage in Exercise 82 with a monthly payment of $897.72. What amount of the total is interest costs?

85. Approximate the total amount paid over the term of the mortgage in Exercise 83 with a monthly payment of $1659.24. What amount of the total is interest costs?

Synthesis

True or False? **In Exercises 86 and 87, determine whether the statement is true or false. Justify your answer.**

86. You can determine the graph of $f(x) = \log_6 x$ by graphing $g(x) = 6^x$ and reflecting it in the x-axis.

87. The graph of $f(x) = \log_3 x$ contains the point $(27, 3)$.

88. *Exploration* The table of values was obtained by evaluating a function. Determine which of the statements may be true and which must be false.

x	1	2	8
y	0	1	3

(a) y is an exponential function of x.

(b) y is a logarithmic function of x.

(c) x is an exponential function of y.

(d) y is a linear function of x.

89. *Exploration* Use a graphing utility to compare the graph of the function $y = \ln x$ with the graphs of the following functions.

(a) $y = x - 1$

(b) $y = (x - 1) - \frac{1}{2}(x - 1)^2$

(c) $y = (x - 1) - \frac{1}{2}(x - 1)^2 + \frac{1}{3}(x - 1)^3$

90. *Finding a Pattern* Identify the pattern of successive polynomials given in Exercise 89. Extend the pattern one more term and compare the graph of the resulting polynomial function with the graph of $y = \ln x$. What do you think the pattern implies?

91. *Numerical and Graphical Analysis* (a) Use a graphing utility to complete the table for the function

$$f(x) = \frac{\ln x}{x}.$$

x	1	5	10	10^2	10^4	10^6
$f(x)$						

(b) Use the table to determine what $f(x)$ approaches as x increases without bound.

(c) Use a graphing utility to check the result of part (b).

Review

In Exercises 92–95, evaluate the expression. Round your answer to three decimal places.

92. e^{12}

93. $e^{7/2}$

94. e^{-5}

95. $6e^{-8}$

In Exercises 96–99, sketch the graph of the function by hand.

96. $f(x) = 2^x + 3$

97. $f(x) = -2^{x-3}$

98. $f(x) = \left(\frac{3}{2}\right)^{x+1}$

99. $f(x) = 5^{-x} - 2$

5.3 Properties of Logarithms

Change of Base

Most calculators have only two types of log keys, one for common logarithms (base 10) and one for natural logarithms (base e). Although common logs and natural logs are the most frequently used, you may occasionally need to evaluate logarithms to other bases. To do this, you can use the following **change-of-base formula.**

Change-of-Base Formula

Let a, b, and x be positive real numbers such that $a \neq 1$ and $b \neq 1$. Then $\log_a x$ can be converted to a different base using any of the following formulas.

Base b	*Base 10*	*Base e*
$\log_a x = \dfrac{\log_b x}{\log_b a}$	$\log_a x = \dfrac{\log_{10} x}{\log_{10} a}$	$\log_a x = \dfrac{\ln x}{\ln a}$

One way to look at the change-of-base formula is that logarithms to base a are simply *constant multiples* of logarithms to base b. The constant multiplier is $1/(\log_b a)$.

EXAMPLE 1 Changing Bases Using Common Logarithms

a. $\log_4 30 = \dfrac{\log_{10} 30}{\log_{10} 4}$ $\qquad \log_a x = \dfrac{\log_{10} x}{\log_{10} a}$

$\qquad\quad \approx \dfrac{1.47712}{0.60206}$ $\qquad$ Use a calculator.

$\qquad\quad \approx 2.4534$ $\qquad$ Use a calculator.

b. $\log_2 x = \dfrac{\log_{10} x}{\log_{10} 2} \approx \dfrac{\log_{10} x}{0.30103} \approx 3.3219 \log_{10} x$

EXAMPLE 2 Changing Bases Using Natural Logarithms

a. $\log_4 30 = \dfrac{\ln 30}{\ln 4}$ $\qquad \log_a x = \dfrac{\ln x}{\ln a}$

$\qquad\quad \approx \dfrac{3.40120}{1.38629}$ $\qquad$ Use a calculator.

$\qquad\quad \approx 2.4535$ $\qquad$ Use a calculator.

b. $\log_2 x = \dfrac{\ln x}{\ln 2} \approx \dfrac{\ln x}{0.693147} \approx 1.4427 \ln x$

In Examples 1 and 2, the result is the same whether common logarithms or natural logarithms are used in the change-of-base formula.

What You Should Learn:

- How to rewrite logarithmic functions with different bases
- How to use properties of logarithms to evaluate or rewrite logarithmic expressions
- How to use properties of logarithms to expand or condense logarithmic expressions
- How to use logarithmic functions to model and solve real-life problems

Why You Should Learn It:

Logarithmic functions can be used to model and solve real-life problems, such as human memory in Exercise 96 on page 379.

Properties of Logarithms

You know from the previous section that the logarithmic function with base a is the *inverse* of the exponential function with base a. So, it makes sense that the properties of exponents should have corresponding properties involving logarithms. For instance, the exponential property $a^0 = 1$ has the corresponding logarithmic property $\log_a 1 = 0$.

Properties of Logarithms

Let a be a positive number such that $a \neq 1$, and let n be a real number. If u and v are positive real numbers, the following properties are true.

1. $\log_a(uv) = \log_a u + \log_a v$ **1.** $\ln(uv) = \ln u + \ln v$

2. $\log_a \dfrac{u}{v} = \log_a u - \log_a v$ **2.** $\ln \dfrac{u}{v} = \ln u - \ln v$

3. $\log_a u^n = n \log_a u$ **3.** $\ln u^n = n \ln u$

For a proof of Property 1, see Appendix A.

EXAMPLE 3   Using Properties of Logarithms

Write the logarithm in terms of $\ln 2$ and $\ln 3$.

a. $\ln 6$ **b.** $\ln \dfrac{2}{27}$

Solution

a. $\ln 6 = \ln(2 \cdot 3)$ Rewrite 6 as $2 \cdot 3$.

$\qquad = \ln 2 + \ln 3$ Property 1

b. $\ln \dfrac{2}{27} = \ln 2 - \ln 27$ Property 2

$\qquad = \ln 2 - \ln 3^3$ Rewrite 27 as 3^3.

$\qquad = \ln 2 - 3 \ln 3$ Property 3

EXAMPLE 4   Using Properties of Logarithms

Use the properties of logarithms to verify that $-\ln \frac{1}{2} = \ln 2$.

Solution

$-\ln \frac{1}{2} = -\ln(2^{-1})$ Rewrite $\frac{1}{2}$ as 2^{-1}.

$\qquad = -(-1) \ln 2$ Property 3

$\qquad = \ln 2$ Simplify.

Try checking this result on your calculator.

John Napier (1550–1617), a Scottish mathematician, developed logarithms as a way to simplify some of the tedious calculations of his day. Beginning in 1594, Napier worked about 20 years on the invention of logarithms. Napier was only partially successful in his quest to simplify tedious calculations. Nonetheless, the development of logarithms was a step forward and received immediate recognition.

Rewriting Logarithmic Expressions

The properties of logarithms are useful for rewriting logarithmic expressions in forms that simplify the operations of algebra. This is true because they convert complicated products, quotients, and exponential forms into simpler sums, differences, and products, respectively.

EXAMPLE 5   Expanding the Logarithm of a Product

Use the properties of logarithms to expand $\log_{10} 5x^3y$.

Solution

$$\log_{10} 5x^3y = \log_{10} 5 + \log_{10} x^3y \qquad \text{Property 1}$$

$$= \log_{10} 5 + \log_{10} x^3 + \log_{10} y \qquad \text{Property 1}$$

$$= \log_{10} 5 + 3\log_{10} x + \log_{10} y \qquad \text{Property 3}$$

EXAMPLE 6   Expanding the Logarithm of a Quotient

Use the properties of logarithms to expand $\ln \dfrac{\sqrt{3x-5}}{7}$.

Solution

$$\ln \frac{\sqrt{3x-5}}{7} = \ln\left[\frac{(3x-5)^{1/2}}{7}\right] \qquad \text{Rewrite rational exponent.}$$

$$= \ln(3x-5)^{1/2} - \ln 7 \qquad \text{Property 2}$$

$$= \frac{1}{2}\ln(3x-5) - \ln 7 \qquad \text{Property 3}$$

In Examples 5 and 6, the properties of logarithms were used to *expand* logarithmic expressions. In Example 7, this procedure is reversed and the properties of logarithms are used to *condense* logarithmic expressions.

EXAMPLE 7   Condensing a Logarithmic Expression

Use the properties of logarithms to condense each logarithmic expression.

a. $\frac{1}{2}\log_{10} x + 3\log_{10}(x+1)$ **b.** $2\ln(x+2) - \ln x$

Solution

a. $\frac{1}{2}\log_{10} x + 3\log_{10}(x+1) = \log_{10} x^{1/2} + \log_{10}(x+1)^3 \qquad \text{Property 3}$

$$= \log_{10}\left[\sqrt{x} \cdot (x+1)^3\right] \qquad \text{Property 1}$$

b. $2\ln(x+2) - \ln x = \ln(x+2)^2 - \ln x \qquad \text{Property 3}$

$$= \ln \frac{(x+2)^2}{x} \qquad \text{Property 2}$$

> ## Exploration
>
> Use a graphing utility to graph the functions
>
> $$y = \ln x - \ln(x-3)$$
>
> and
>
> $$y = \ln \frac{x}{x-3}$$
>
> in the same viewing window. Does the graphing utility show the functions with the same domain? If so, should it? Explain your reasoning.

Application

EXAMPLE 8 Finding a Mathematical Model

The table shows the mean distance from the sun x and the orbital period y of the six planets that are closest to the sun. In the table, the mean distance is given in terms of astronomical units (where the earth's mean distance is defined as 1.0), and the period is given in terms of years. Find an equation that expresses y as a function of x.

Planet	Mercury	Venus	Earth	Mars	Jupiter	Saturn
Mean Distance, x	0.387	0.723	1.0	1.524	5.203	9.539
Period, y	0.241	0.615	1.0	1.881	11.862	29.458

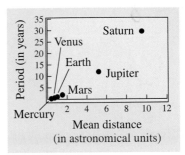

Figure 5.22

Algebraic Solution

The points in the table are plotted in Figure 5.22. From this figure it is not clear how to find an equation that relates y and x. To solve this problem, take the natural log of each of the x and y-values shown in the table. This produces the following results.

Planet	Mercury	Venus	Earth
$\ln x = X$	−0.949	−0.324	0.0
$\ln y = Y$	−1.423	−0.486	0.0

Planet	Mars	Jupiter	Saturn
$\ln x = X$	0.421	1.649	2.255
$\ln y = Y$	0.632	2.473	3.383

Now, by plotting the points in the table, you can see that all six of the points appear to lie in a line.

Choose any two points to determine the slope of the line. Using the two points $(0.421, 0.632)$ and $(0, 0)$, you can determine the slope of the line.

$$m = \frac{0.632 - 0}{0.421 - 0} \approx 1.5 = \frac{3}{2}$$

By the point-slope form, the equation of the line is $Y = \frac{3}{2}X$, where $Y = \ln y$ and $X = \ln x$. You can therefore conclude that $\ln y = \frac{3}{2}\ln x$.

Graphical Solution

The points in the table are plotted in Figure 5.22. From this figure it is not clear how to find an equation that relates y and x. To solve this problem, take the natural log of each of the x- and y-values given in the table. This produces the following results.

Planet	Mercury	Venus	Earth	Mars	Jupiter	Saturn
$\ln x = X$	−0.949	−0.324	0.0	0.421	1.649	2.255
$\ln y = Y$	−1.423	−0.486	0.0	0.632	2.473	3.383

Now, by plotting the points in the table, you can see that all six of the points appear to lie in a line, as shown in Figure 5.23.

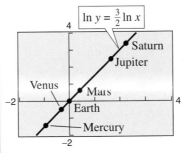

Figure 5.23

Using the regression features of a graphing utility, you can find a linear model for the data $Y = 1.5X = \frac{3}{2}X$, where $Y = \ln y$ and $X = \ln x$. From the model, you can see that the slope of the line is $\frac{3}{2}$. So, you can conclude that $\ln y = \frac{3}{2}\ln x$.

In Example 8, try to convert this to $y = f(x)$ form. You will get a function of the form $y = ax^b$, which is a power model.

5.3 Exercises

In Exercises 1 and 2, use a graphing utility to graph the two functions in the same viewing window. What do the graphs suggest? Explain your reasoning.

1. $f(x) = \log_{10} x$

$g(x) = \dfrac{\ln x}{\ln 10}$

2. $f(x) = \ln x$

$g(x) = \dfrac{\log_{10} x}{\log_{10} e}$

In Exercises 3–10, evaluate the logarithm using the change-of-base formula. Round your result to three decimal places.

3. $\log_3 7$

4. $\log_7 4$

5. $\log_{1/2} 4$

6. $\log_{1/8} 64$

7. $\log_9(0.8)$

8. $\log_{1/3}(0.015)$

9. $\log_{15} 1460$

10. $\log_{20} 135$

In Exercises 11–18, rewrite the logarithm as a multiple of (a) a common logarithm and (b) a natural logarithm.

11. $\log_5 x$

12. $\log_3 x$

13. $\log_{1/5} x$

14. $\log_{1/3} x$

15. $\log_x \dfrac{3}{10}$

16. $\log_x \dfrac{3}{4}$

17. $\log_{2.6} x$

18. $\log_{7.1} x$

In Exercises 19–26, use the change-of-base formula to rewrite the logarithm as a multiple of a logarithm. Then use a graphing utility to sketch the graph.

19. $f(x) = \log_2 x$

20. $f(x) = \log_4 x$

21. $f(x) = \log_{1/2} x$

22. $f(x) = \log_{1/4} x$

23. $f(x) = \log_{11.8} x$

24. $f(x) = \log_{12.4} x$

25. $f(x) = \log_3 x^{1/2}$

26. $f(x) = \log_5 \dfrac{x}{3}$

In Exercises 27–46, use the properties of logarithms to write the expression as a sum, difference, and/or constant multiple of logarithms. (Assume all variables are positive.)

27. $\log_{10} 5x$

28. $\log_{10} 10z$

29. $\log_{10} \dfrac{5}{x}$

30. $\log_{10} \dfrac{y}{2}$

31. $\log_8 x^4$

32. $\log_6 z^{-3}$

33. $\ln \sqrt{z}$

34. $\ln \sqrt[3]{t}$

35. $\ln xyz$

36. $\ln \dfrac{xy}{z}$

37. $\ln \sqrt{a - 1}, \quad a > 1$

38. $\ln\left(\dfrac{x^2 - 1}{x^3}\right), \quad x > 1$

39. $\ln z(z - 1)^2, \quad z > 1$

40. $\ln \sqrt{\dfrac{x^2}{y^3}}$

41. $\ln \sqrt[3]{\dfrac{x}{y}}$

42. $\ln \dfrac{x}{\sqrt{x^2 + 1}}$

43. $\ln \dfrac{x^4 \sqrt{y}}{z^5}$

44. $\ln \sqrt{x^2(x + 2)}$

45. $\log_b \dfrac{x^2}{y^2 z^3}$

46. $\log_b \dfrac{\sqrt{x} y^4}{z^4}$

Graphical Analysis **In Exercises 47 and 48, use a graphing utility to graph the two equations in the same viewing window. What do the graphs suggest? Explain your reasoning.**

47. $y_1 = \ln[x^3(x + 4)], \qquad y_2 = 3 \ln x + \ln(x + 4)$

48. $y_1 = \ln\left(\dfrac{\sqrt{x}}{x - 2}\right), \qquad y_2 = \frac{1}{2} \ln x - \ln(x - 2)$

In Exercises 49–68, write the expression as the logarithm of a single quantity.

49. $\ln x + \ln 4$

50. $\ln y + \ln z$

51. $\log_4 z - \log_4 y$

52. $\log_5 8 - \log_5 t$

53. $2 \log_2(x + 3)$

54. $-6 \log_6 2x$

55. $\frac{1}{3} \log_3 7x$

56. $\frac{5}{2} \log_7(z - 4)$

57. $\ln x - 3 \ln(x + 1)$

58. $2 \ln 8 + 5 \ln z$

59. $\ln(x - 2) - \ln(x + 2)$

60. $3 \ln x + 2 \ln y - 4 \ln z$

61. $\ln x - 2[\ln(x + 2) + \ln(x - 2)]$

62. $4[\ln z + \ln(z + 5)] - 2 \ln(z - 5)$

63. $\frac{1}{3}[2 \ln(x + 3) + \ln x - \ln(x^2 - 1)]$

64. $2[\ln x - \ln(x + 1) - \ln(x - 1)]$

65. $\frac{1}{3}[\ln y + 2 \ln(y + 4)] - \ln(y - 1)$

66. $\frac{1}{2}[\ln(x + 1) + 2 \ln(x - 1)] + 3 \ln x$

67. $2 \ln 3 - \frac{1}{2} \ln(x^2 + 1)$

68. $\frac{3}{2} \ln 5t^6 - \frac{3}{4} \ln t^4$

Graphical Analysis In Exercises 69 and 70, use a graphing utility to graph the two equations in the same viewing window. What do the graphs suggest? Verify your conclusion algebraically.

69. $y_1 = 2[\ln 8 - \ln(x^2 + 1)]$, $y_2 = \ln\left[\dfrac{64}{(x^2 + 1)^2}\right]$

70. $y_1 = \ln x + \frac{1}{3}\ln(x + 1)$, $y_2 = \ln\left(x\sqrt[3]{x + 1}\right)$

Think About It In Exercises 71 and 72, use a graphing utility to graph the two equations in the same viewing window. Are the expressions equivalent? Explain.

71. $y_1 = \ln x^2$, $y_2 = 2\ln x$

72. $y_1 = \frac{1}{4}\ln[x^4(x^2 + 1)]$, $y_2 = \ln x + \frac{1}{4}\ln(x^2 + 1)$

In Exercises 73–86, find the exact value of the logarithm without using a calculator. (If this is not possible, state the reason.)

73. $\log_3 9$

74. $\log_6 \sqrt[3]{6}$

75. $\log_4 16^{3.4}$

76. $\log_5\left(\frac{1}{125}\right)$

77. $\log_2(-4)$

78. $\log_4(-16)$

79. $\log_5 75 - \log_5 3$

80. $\log_4 2 + \log_4 32$

81. $\ln e^3 - \ln e^7$

82. $3\ln e^4$

83. $\log_{10} 0$

84. $\ln 1$

85. $\ln e^{8.5}$

86. $\ln \sqrt[5]{e^3}$

In Exercises 87–94, use the properties of logarithms to simplify the logarithmic expression.

87. $\log_4 8$

88. $\log_5\left(\frac{1}{15}\right)$

89. $\log_7 \sqrt{70}$

90. $\log_2(4^2 \cdot 3^4)$

91. $\log_5\left(\frac{1}{250}\right)$

92. $\log_{10}\left(\frac{9}{300}\right)$

93. $\ln(5e^6)$

94. $\ln \dfrac{6}{e^2}$

95. *Sound Intensity* The relationship between the number of decibels β and the intensity of a sound I in watts per square meter is

$$\beta = 10\log_{10}\left(\frac{I}{10^{-12}}\right).$$

(a) Use the properties of logarithms to write the formula in a simpler form.

(b) Use a graphing utility to complete the table.

I	10^{-4}	10^{-6}	10^{-8}	10^{-10}	10^{-12}	10^{-14}
β						

(c) Verify your answers in part (b) algebraically.

96. *Human Memory Model* Students participating in a psychological experiment attended several lectures. After the last lecture, and every month for the next year, the students were tested to see how much of the material they remembered. The average scores for the group were given by the memory model

$$f(t) = 90 - 15\log_{10}(t + 1), \quad 0 \le t \le 12$$

where t is the time (in months).

(a) Use a graphing utility to graph the function over the specified domain.

(b) What was the average score on the original exam $(t = 0)$?

(c) What was the average score after 6 months?

(d) What was the average score after 12 months?

(e) When would the average score have decreased to 75?

97. *Comparing Models* A cup of water at an initial temperature of 78°C is placed in a room at a constant temperature of 21°C. The temperature of the water is measured every 5 minutes during a half-hour period. The results are recorded as ordered pairs of the form (t, T), where t is the time (in minutes) and T is the temperature (in degrees Celsius).

$(0, 78.0°)$, $(5, 66.0°)$, $(10, 57.5°)$, $(15, 51.2°)$, $(20, 46.3°)$, $(25, 42.5°)$, $(30, 39.6°)$

(a) The graph of the model for the data should be asymptotic with the temperature of the room. Subtract the room temperature from each of the temperatures in the ordered pairs. Use a graphing utility to plot the data points (t, T) and $(t, T - 21)$.

(b) Use the regression capabilities of a graphing utility to fit an exponential model to the revised data. This model will be of the form

$$T - 21 = ab^x.$$

Solve for T and graph the model. Compare the result with the plot of the original data.

(c) Take the natural logarithms of the revised temperatures. Use a graphing utility to plot the points $(t, \ln(T - 21))$ and observe that the points appear linear. Use the regression capabilities of a graphing utility to fit a line to this data. The resulting line has the form

$$\ln(T - 21) = at + b.$$

Use the properties of logarithms to solve for T. Verify that the result is equivalent to the model in part (b).

(d) Fit a rational model to the data. Take the reciprocals of the y-coordinates of the revised data to generate the points

$$\left(t, \frac{1}{T - 21}\right).$$

Use a graphing utility to plot these points and observe that they appear linear. Use the regression capabilities of a graphing utility to fit a line to this data. The resulting line has the form

$$\frac{1}{T - 21} = at + b.$$

Solve for T, and use a graphing utility to graph the rational function and the original data points.

98. *Writing* Write a short paragraph explaining why the transformations of the data in Exercise 97 were necessary to obtain the models. Why did taking the logarithms of the temperatures lead to a linear scatter plot? Why did taking the reciprocals of the temperatures lead to a linear scatter plot?

Synthesis

True or False? **In Exercises 99–104, determine whether the statement is true or false given that $f(x) = \ln x$. Justify your answer.**

99. $f(0) = 0$

100. $f(ax) = f(a) + f(x), \quad a > 0, x > 0$

101. $f(x - 2) = f(x) - f(2), \quad x > 2$

102. $\sqrt{f(x)} = \frac{1}{2}f(x)$

103. If $f(u) = 2f(v)$, then $v = u^2$.

104. If $f(x) < 0$, then $0 < x < 1$.

105. *Think About It* Use a graphing utility to graph

$$f(x) = \ln\frac{x}{2}, \quad g(x) = \frac{\ln x}{\ln 2}, \quad h(x) = \ln x - \ln 2$$

in the same viewing window. Which two functions have identical graphs? Explain why.

106. *Exploration* Approximate the natural logarithms of as many integers as possible between 1 and 20, given that $\ln 2 \approx 0.6931$, $\ln 3 \approx 1.0986$, and $\ln 5 \approx 1.6094$. (Do not use a calculator.)

107. Prove that $\log_b \dfrac{u}{v} = \log_b u - \log_b v$.

108. Prove that $\log_b u^n = n \log_b u$.

Review

In Exercises 109–112, simplify the expression.

109. $\dfrac{24xy^{-2}}{16x^{-3}y}$

110. $\left(\dfrac{2x^2}{3y}\right)^{-3}$

111. $(18x^3y^4)^{-3}(18x^3y^4)^3$

112. $xy(x^{-1} + y^{-1})^{-1}$

In Exercises 113–118, find all solutions of the equation. Be sure to check all your solutions.

113. $x^2 - 6x + 2 = 0$

114. $2x^3 + 20x^2 + 50x = 0$

115. $x^4 - 19x^2 + 48 = 0$

116. $9x^4 - 37x^2 + 4 = 0$

117. $x^3 - 6x^2 - 4x + 24 = 0$

118. $9x^4 - 226x^2 + 25 = 0$

In Exercises 119–126, use a calculator to evaluate the expression. Round the result to three decimal places.

119. $1.6^{-2\pi}$

120. $\sqrt[5]{8251}$

121. $260^{\sqrt{3}}$

122. $170(4^{-1.1})$

123. $\log_{10}(220)$

124. $\log_{10}\left(\frac{7}{5}\right)$

125. $\ln 2.008$

126. $\ln\left(5 - \sqrt{7}\right)$

5.4 Solving Exponential and Logarithmic Equations

Introduction

So far in this chapter, you have studied the definitions, graphs, and properties of exponential and logarithmic functions. In this section, you will study procedures for *solving equations* involving exponential and logarithmic functions.

There are two basic strategies for solving exponential or logarithmic equations. The first is based on the One-to-One Properties and the second is based on the Inverse Properties. For $a > 0$ and $a \neq 1$, the following properties are true for all x and y for which $\log_a x$ and $\log_a y$ are defined.

One-to-One Properties

$a^x = a^y$ if and only if $x = y$.

$\log_a x = \log_a y$ if and only if $x = y$.

Inverse Properties

$a^{\log_a x} = x$

$\log_a a^x = x$

What You Should Learn:

- How to solve simple exponential and logarithmic equations
- How to solve more complicated exponential equations
- How to solve more complicated logarithmic equations
- How to use exponential and logarithmic equations to model and solve real-life problems

Why You Should Learn It:

Exponential and logarithmic equations can be used to model and solve real-life problems. For instance, Exercise 126 on page 391 shows how to use a logarithmic function to model crumple zones for automobile crash tests.

EXAMPLE 1 Solving Simple Exponential and Logarithmic Equations

Original Equation	Rewritten Equation	Solution	Property
a. $2^x = 32$	$2^x = 2^5$	$x = 5$	One-to-One
b. $\ln x - \ln 3 = 0$	$\ln x = \ln 3$	$x = 3$	One-to-One
c. $\left(\frac{1}{3}\right)^x = 9$	$3^{-x} = 3^2$	$x = -2$	One-to-One
d. $e^x = 7$	$\ln e^x = \ln 7$	$x = \ln 7$	Inverse
e. $\ln x = -3$	$e^{\ln x} = e^{-3}$	$x = e^{-3}$	Inverse
f. $\log x = -1$	$10^{\log x} = 10^{-1}$	$x = 10^{-1} = \frac{1}{10}$	Inverse

The strategies used in Example 1 are summarized as follows.

Strategies for Solving Exponential and Logarithmic Equations

1. Rewrite the given equation in a form to use the One-to-One Properties of exponential or logarithmic functions.
2. Rewrite an *exponential* equation in logarithmic form and apply the Inverse Property of logarithmic functions.
3. Rewrite a *logarithmic* equation in exponential form and apply the Inverse Property of exponential functions.

Solving Exponential Equations

EXAMPLE 2   Solving Exponential Equations

Solve each equation.   **a.** $e^x = 72$ **b.** $3(2^x) = 42$

Algebraic Solution

a.

$e^x = 72$	Write original equation.
$\ln e^x = \ln 72$	Take natural log of each side.
$x = \ln 72$	Inverse Property
$x \approx 4.277$	Use a calculator.

The solution is $\ln 72 \approx 4.277$. Check this solution in the original equation.

b.

$3(2^x) = 42$	Write original equation.
$2^x = 14$	Divide each side by 3.
$\log_2 2^x = \log_2 14$	Take log (base 2) of each side.
$x = \log_2 14$	Inverse Property
$x = \dfrac{\ln 14}{\ln 2}$	Change-of-base formula
$x \approx 3.807$	Use a calculator.

The solution is $\log_2 14 \approx 3.807$. Check this solution in the original equation.

Graphical Solution

a. Use a graphing utility to graph the left- and right-hand sides of the equation as $y_1 = e^x$ and $y_2 = 72$ in the same viewing window. Use the *intersect* feature or the *zoom* and *trace* features of the graphing utility to approximate the intersection point, as shown in Figure 5.24. So, the approximate solution is 4.277.

b. Use a graphing utility to graph $y_1 = 3(2^x)$ and $y_2 = 42$ in the same viewing window. Use the *intersect* feature or the *zoom* and *trace* features to approximate the intersection point, as shown in Figure 5.25. So, the approximate solution is 3.807.

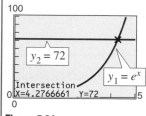

Figure 5.24

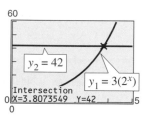

Figure 5.25

EXAMPLE 3   Solving an Exponential Equation

Solve $4e^{2x} = 5$.

Algebraic Solution

$4e^{2x} = 5$	Write original equation.
$e^{2x} = \dfrac{5}{4}$	Divide each side by 4.
$\ln e^{2x} = \ln \dfrac{5}{4}$	Take logarithm of each side.
$2x = \ln \dfrac{5}{4}$	Inverse Property
$x = \dfrac{1}{2} \ln \dfrac{5}{4}$	Solve for x.
$x \approx 0.112$	Use a calculator.

The solution is $\frac{1}{2} \ln \frac{5}{4} \approx 0.112$. Check this solution in the original equation.

Graphical Solution

Rather than graph both sides of the equation you are solving as separate graphs, as you did in Example 2, another way to graphically solve the equation is to first rewrite the equation as $4e^{2x} - 5 = 0$, then use a graphing utility to graph $y = 4e^{2x} - 5$. Use the *zero* or *root* feature or the *zoom* and *trace* features of the graphing utility to approximate the value of x for which $y = 0$. From Figure 5.26, you can see that the zero occurs when $x \approx 0.112$. So, the approximate solution is 0.112.

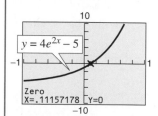

Figure 5.26

EXAMPLE 4   Solving an Exponential Equation

Solve $2(3^{2t-5}) - 4 = 11$.

Solution

$2(3^{2t-5}) - 4 = 11$	Write original equation.
$2(3^{2t-5}) = 15$	Add 4 to each side.
$3^{2t-5} = \frac{15}{2}$	Divide each side by 2.
$\log_3 3^{2t-5} = \log_3 \frac{15}{2}$	Take log (base 3) of each side.
$2t - 5 = \log_3 \frac{15}{2}$	Inverse Property
$2t = 5 + \log_3 7.5$	Add 5 to each side.
$t = \frac{5}{2} + \frac{1}{2}\log_3 7.5$	Divide each side by 2.
$t \approx 3.417$	Use a calculator.

The solution is $\frac{5}{2} + \frac{1}{2}\log_3 7.5 \approx 3.417$. Check this in the original equation.

When an equation involves two or more exponential expressions, you can still use a procedure similar to that demonstrated in the previous three examples. However, the algebra is a bit more complicated.

EXAMPLE 5   Solving an Exponential Equation in Quadratic Form

Solve $e^{2x} - 3e^x + 2 = 0$.

Algebraic Solution

$e^{2x} - 3e^x + 2 = 0$	Write original equation.
$(e^x)^2 - 3e^x + 2 = 0$	Write in quadratic form.
$(e^x - 2)(e^x - 1) = 0$	Factor.
$e^x - 2 = 0$	Set 1st factor equal to 0.
$e^x = 2$	Add 2 to each side.
$x = \ln 2$	Inverse Property
$e^x - 1 = 0$	Set 2nd factor equal to 0.
$e^x = 1$	Add 1 to each side.
$x = \ln 1$	Inverse Property
$x = 0$	Simplify.

The solutions are $\ln 2 \approx 0.693$ and 0. Check these in the original equation.

Graphical Solution

Use a graphing utility to graph $y = e^{2x} - 3e^x + 2$. Use the *zero* or *root* feature or the *zoom* and *trace* features of the graphing utility to approximate the values of x for which $y = 0$. In Figure 5.27, you can see that the zeros occur when $x = 0$ and when $x \approx 0.693$. So, the approximate solutions are 0 and 0.693.

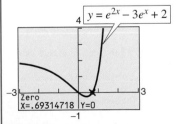

Figure 5.27

Solving Logarithmic Equations

To solve a logarithmic equation, you can write it in exponential form.

$\ln x = 3$	Logarithmic form
$e^{\ln x} = e^3$	Exponentiate each side.
$x = e^3$	Exponential form

This procedure is called *exponentiating* both sides of an equation. It is applied after the logarithmic expression has been isolated.

EXAMPLE 6   Solving Logarithmic Equations

Solve each logarithmic equation.

a. $\ln x = 2$ **b.** $\log_3(5x - 1) = \log_3(x + 7)$

Solution

a.

$\ln x = 2$	Write original equation.
$e^{\ln x} = e^2$	Exponentiate each side.
$x = e^2$	Inverse Property
$x \approx 7.389$	Use a calculator.

The solution is $e^2 \approx 7.389$. Check this in the original equation.

b.

$\log_3(5x - 1) = \log_3(x + 7)$	Write original equation.
$5x - 1 = x + 7$	One-to-One Property
$4x = 8$	Add $-x$ and 1 to each side.
$x = 2$	Divide each side by 4.

The solution is 2. Check this in the original equation.

EXAMPLE 7   Solving a Logarithmic Equation

Solve $5 + 2 \ln x = 4$.

Algebraic Solution

$5 + 2 \ln x = 4$	Write original equation.
$2 \ln x = -1$	Subtract 5 from each side.
$\ln x = -\frac{1}{2}$	Divide each side by 2.
$e^{\ln x} = e^{-1/2}$	Exponentiate each side.
$x = e^{-1/2}$	Inverse property
$x \approx 0.607$	Use a calculator.

The solution is $e^{-1/2} \approx 0.607$. Check this in the original equation.

Graphical Solution

Use a graphing utility to graph $y_1 = 5 + 2 \ln x$ and $y_2 = 4$ in the same viewing window. Use the *intersect* feature or the *zoom* and *trace* features to approximate the intersection point, as shown in Figure 5.28. So, the approximate solution is 0.607.

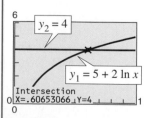

Figure 5.28

EXAMPLE 8 Solving a Logarithmic Equation

Solve $2 \log_5 3x = 4$.

Solution

$2 \log_5 3x = 4$	Write original equation.
$\log_5 3x = 2$	Divide each side by 2.
$5^{\log_5 3x} = 5^2$	Exponentiate each side (base 5).
$3x = 25$	Inverse Property.
$x = \frac{25}{3}$	Divide each side by 3.

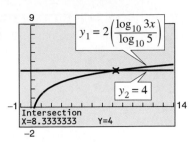

Figure 5.29

The solution is $\frac{25}{3}$. Check this in the original equation. Or, perform a graphical check by graphing

$$y_1 = 2 \log_5 3x = 2\left(\frac{\log_{10} 3x}{\log_{10} 5}\right) \quad \text{and} \quad y_2 = 4$$

in the same viewing window. The two graphs should intersect when $x = \frac{25}{3}$ and $y = 4$, as shown in Figure 5.29.

Because the domain of a logarithmic function generally does not include all real numbers, you should be sure to check for extraneous solutions of logarithmic equations, as shown in the next example.

EXAMPLE 9 Checking for Extraneous Solutions

Solve for x in the equation $\ln(x - 2) + \ln(2x - 3) = 2 \ln x$.

Algebraic Solution

$\ln(x - 2) + \ln(2x - 3) = 2 \ln x$	Write original equation.
$\ln[(x - 2)(2x - 3)] = \ln x^2$	Use properties of logarithms.
$\ln(2x^2 - 7x + 6) = \ln x^2$	
$2x^2 - 7x + 6 = x^2$	One-to-One Property
$x^2 - 7x + 6 = 0$	Write in general form.
$(x - 6)(x - 1) = 0$	Factor.
$x - 6 = 0 \implies x = 6$	Set 1st factor equal to 0.
$x - 1 = 0 \implies x = 1$	Set 2nd factor equal to 0.

Finally, by checking these two "solutions" in the original equation, you can conclude that $x = 1$ is not valid. This is because when $x = 1$, $\ln(x - 2) + \ln(2x - 3) = \ln(-1) + \ln(-1)$, which is invalid because -1 is not in the domain of the natural log function. So, the only solution is 6.

Graphical Solution

First rewrite the original equation as $\ln(x - 2) + \ln(2x - 3) - 2 \ln x = 0$. Then use a graphing utility to graph $y = \ln(x - 2) + \ln(2x - 3) - 2 \ln x$, as shown in Figure 5.30. Use the *zero* or *root* feature or the *zoom* and *trace* features of the graphing utility to determine that 6 is an approximate solution. You can verify that 6 is an exact solution by substituting $x = 6$ in the original equation.

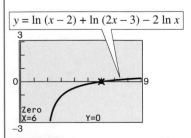

Figure 5.30

EXAMPLE 10   The Change-of-Base Formula

Prove the change-of-base formula: $\log_a x = \dfrac{\log_b x}{\log_b a}$.

Solution

Begin by letting $y = \log_a x$ and writing the equivalent exponential form $a^y = x$. Now, taking the logarithms *with base b* of both sides produces the following.

$$\log_b a^y = \log_b x$$

$$y \log_b a = \log_b x$$

$$y = \frac{\log_b x}{\log_b a}$$

$$\log_a x = \frac{\log_b x}{\log_a a}$$

Approximating Solutions

Equations that involve combinations of algebraic functions, exponential functions, and/or logarithmic functions can be very difficult to solve by algebraic procedures. Here again, you can take advantage of a graphing utility.

EXAMPLE 11   Approximating the Solution of an Equation

Approximate the solution of $\ln x = x^2 - 2$.

Solution

To begin, write the equation so that all terms on one side are equal to 0.

$$\ln x - x^2 + 2 = 0$$

Then use a graphing utility to graph

$$y = -x^2 + 2 + \ln x$$

as shown in Figure 5.31. From this graph, you can see that the equation has two solutions. Next, using the *zero* or *root* feature or the *zoom* and *trace* features, you can approximate the two solutions to be 0.138 and 1.564.

Check

$\ln x = x^2 - 2$	Write original equation.
$\ln(0.138) \stackrel{?}{\approx} (0.138)^2 - 2$	Substitute 0.138 for *x*.
$-1.9805 \approx -1.9810$	Solution checks. ✓
$\ln(1.564) \stackrel{?}{\approx} (1.564)^2 - 2$	Substitute 1.564 for *x*.
$0.4472 \approx 0.4461$	Solution checks. ✓

So, the two solutions 0.138 and 1.564 seem reasonable.

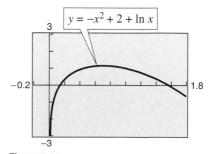

$y = -x^2 + 2 + \ln x$

Figure 5.31

63. 2

64. 1

Applications

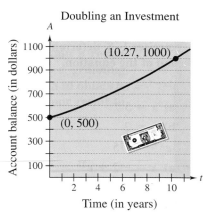

EXAMPLE 12   Doubling an Investment

You have deposited $500 in an account that pays 6.75% interest, compounded continuously. How long will it take your money to double?

Solution

Using the formula for continuous compounding, you can find that the balance in the account is

$$A = Pe^{rt}$$

$$= 500e^{0.0675t}.$$

To find the time required for the balance to double, let $A = 1000$, and solve the resulting equation for t.

$500e^{0.0675t} = 1000$	Substitute 1000 for A.
$e^{0.0675t} = 2$	Divide each side by 500.
$\ln e^{0.0675t} = \ln 2$	Take natural log of each side.
$0.0675t = \ln 2$	Inverse Property
$t = \dfrac{\ln 2}{0.0675}$	Divide each side by 0.0675.
$t \approx 10.27$	Use a calculator.

The balance in the account will double after approximately 10.27 years. This result is demonstrated graphically in Figure 5.32.

Doubling an Investment

A figure with the point (10.27, 1000) and (0, 500) plotted on a curve. Axis labels: Account balance (in dollars), Time (in years).

Figure 5.32

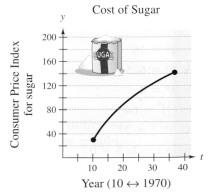

EXAMPLE 13   Consumer Price Index for Sugar

From 1970 to 1997, the Consumer Price Index (CPI) value y for a fixed amount of sugar for the year t can be modeled by the equation

$$y = -171.8 + 87.1 \ln t$$

where $t = 10$ represents 1970 (see Figure 5.33). During which year did the price of sugar reach 4.5 times its 1970 price of 28.8 on the CPI?   (Source: U.S. Bureau of Labor Statistics)

Solution

$-171.8 + 87.1 \ln t = y$	Write original equation.
$-171.8 + 87.1 \ln t = 129.6$	Substitute $(4.5)(28.8) = 129.6$ for y.
$87.1 \ln t = 301.4$	Add 171.8 to each side.
$\ln t = 3.460$	Divide each side by 87.1.
$e^{\ln t} = e^{3.460}$	Exponentiate each side.
$t \approx 31.8$	Inverse Property

The solution is 31.8 years. Because $t = 10$ represents 1970, it follows that the price of sugar reached 4.5 times its 1970 price in late 1991.

Cost of Sugar

A figure showing a curve for the Consumer Price Index for sugar. Axis labels: Consumer Price Index for sugar, Year (10 ↔ 1970).

Figure 5.33

119. $y = 3e^{-2x/3}$

120. $y = 4e^{2x/3}$

121. $y = \ln(x + 3)$

122. $y = 7 - \log_{10}(x + 3)$

123. $y = 2e^{-(x+4)^2/3}$

124. $y = \dfrac{6}{1 + 2e^{-2x}}$

125. *Population*   The population of a town is modeled by

$$P = 12{,}620e^{0.0118t}$$

where $t = 0$ represents 2000. According to this model, when will the population reach 17,000?

126. *Radioactive Decay*   The half-life of radioactive uranium II (^{234}U) is 250,000 years. What percent of the present amount of radioactive uranium II will remain after 5000 years?

127. *Compound Interest*   A deposit of $10,000 is made in a savings account for which the interest is compounded continuously. The balance will double in 12 years.

(a) What is the annual interest rate for this account?

(b) Find the balance after 1 year.

(c) The *effective yield* of a savings plan is the percent increase in the balance after 1 year. Find the effective yield.

128. *Test Scores*   The test scores for a biology test follow a normal distribution modeled by

$$y = 0.0499e^{-(x-71)^2/128}.$$

(a) Use a graphing utility to graph the equation.

(b) From the graph, estimate the average test score.

129. *Typing Speed*   In a typing class, the average number of words per minute N typed after t weeks of lessons was found to be

$$N = \frac{157}{1 + 5.4e^{-0.12t}}.$$

Find the time necessary to type (a) 50 words per minute and (b) 75 words per minute.

130. *Earthquake Magnitude*   On the Richter scale, the magnitude R of an earthquake I is

$$R = \log_{10}\frac{I}{I_0}$$

where $I_0 = 1$ is the minimum intensity used for comparison. Find the intensity per unit of area for the following values of R.

(a) $R = 8.4$ (b) $R = 6.85$ (c) $R = 9.1$

In Exercises 131–134, find the exponential function $y = Ae^{bx}$ that passes through the two points.

131. $(0, 2), (4, 3)$

132. $(0, 2), (5, 1)$

133. $\left(0, \frac{1}{2}\right), (5, 5)$

134. $(0, 4), \left(5, \frac{1}{2}\right)$

135. *Exponential Model*   Use a graphing utility to find an exponential model $y = ab^x$ through the points $(0, 250)$, $(4, 135)$, $(6, 92)$, and $(10, 67)$. Graph the data and the exponential model in the same viewing window.

136. *Net Profit*   The table shows the net profit P (in millions of dollars) of the Bristol-Myers Squibb Company for 1990 through 1998. (Source: Bristol-Myers Squibb Company)

Year	1990	1991	1992	1993	1994
P	1748.0	2056.0	2108.0	2269.0	2330.6

Year	1995	1996	1997	1998
P	2600.0	2850.0	3205.0	3141.0

(a) Make a scatter plot of the data. Let t represent the year, with $t = 0$ corresponding to 1990.

(b) Use a graphing utility to find an exponential model $P = ab^t$ for the data. Let $t = 0$ represent 1990.

Synthesis

True or False?   **In Exercises 137–142, determine whether the equation or statement is true or false. Justify your answer.**

137. $\log_b b^{2x} = 2x$

138. $e^{x-1} = \dfrac{e^x}{e}$

139. $\ln(x + y) = \ln x + \ln y$

140. $\ln(x + y) = \ln(xy)$

141. $\log_{10}\left(\dfrac{10}{x}\right) = 1 - \log_{10} x$

142. The domain of the function $f(x) = \ln x$ is the set of all real numbers.

Chapter Project   *A Graphical Approach to Compound Interest*

In this project, you will use a graphing utility to compare savings plans. For instance, suppose you are depositing $1000 in a savings account and are given the following options.

- 6.2% annual interest rate, compounded annually
- 6.1% annual interest rate, compounded quarterly
- 6.0% annual interest rate, compounded continuously

a. For each option, write a function that gives the balance as a function of the time t (in years).

b. Graph all three functions in the same viewing window. Can you find a viewing window that distinguishes among the graphs of the three functions? If so, describe the viewing window.

c. Find the balances for the three options after 25, 50, 75, and 100 years. Is the option that yields the greatest balance after 25 years the same option that yields the greatest balances after 50, 75, and 100 years? Explain.

d. The *effective yield* of a savings plan is the percent increase in the balance after 1 year. Find the effective yields for the three options listed above. How can the effective yield be used to decide which option is best?

Questions for Further Exploration

1. You deposit $25,000 in an account to accrue interest for 40 years. The account pays 4% compounded annually. Assume that the income tax on the earned interest is 30%. Which of the following plans produces a larger balance after all income tax is paid?

  (a) *Deferred*   The income tax on the interest that is earned is paid in one lump sum at the end of 40 years.

  (b) *Not Deferred*   The income tax on the interest that is earned each year is paid at the end of each year.

2. Which of the following would produce a larger balance? Explain.

  (a) 4.02% annual interest rate, compounded monthly

  (b) 4% annual interest rate, compounded continuously

3. You deposit $1000 in each of two savings accounts. The interest for the accounts is paid according to the two options described in Question 2. How long would it take for the balance in one of the accounts to exceed the balance in the other account by $100? By $100,000?

4. No income tax is due on the interest earned in some types of investments. You deposit $25,000 into an account. Which of the following plans is better? Explain.

  (a) *Tax-Free*   The account pays 5%, compounded annually. There is no income tax on the earned interest.

  (b) *Tax-Deferred*   The account pays 7%, compounded annually. At maturity, the earned interest is taxable at a rate of 40%.

5   Chapter Test

Take this test as you would take a test in class. After you are done, check your work against the answers in the back of the book.

1. Sketch the graph of the function $f(x) = 2^{-x+1} - 3$ by hand.

2. Use a graphing utility to graph f and determine its horizontal asymptotes.

$$f(x) = \frac{1000}{1 + 4e^{-0.2x}}$$

3. Determine the principal that will yield $200,000 when invested at 8% compounded daily for 20 years.

4. Write the logarithmic equation $\log_4 64 = 3$ in exponential form.

5. Sketch the graph of the function $g(x) = \log_3(x - 2)$ by hand.

6. Use the properties of logarithms to expand $\ln\left(\dfrac{6x^2}{\sqrt{x^2 + 1}}\right)$.

The *Interactive* CD-ROM and *Internet* versions of this text provide answers to the Chapter Tests and Cumulative Tests. They also offer Chapter Pre-Tests (that test key skills and concepts covered in previous chapters) and Chapter Post-Tests, both of which have randomly generated exercises with diagnostic capabilities.

In Exercises 7 and 8, evaluate the expression without using a calculator.

7. $\log_5 25$

8. $-2 \ln e^2 + 1$

In Exercises 9–11, solve the equation. Round to three decimal places.

9. $8 + \frac{1}{4}e^{x/2} = 450$

10. $\left(1 + \dfrac{0.06}{4}\right)^{4t} = 3$

11. $3.6 - 5 \ln(x + 4) = 22$

12. A truck that costs $28,000 new has a depreciated value of $20,000 after 1 year. Find the value of the truck when it is 3 years old by using the exponential model $y = ae^{bx}$.

13. The average time between incoming calls at a switchboard is 3 minutes. The probability of waiting less than t minutes for the next incoming call is approximated by the model $F(t) = 1 - e^{-t/3}$. If a call has just come in, find the probability that the next call will come within (a) $\frac{1}{2}$ minute, (b) 2 minutes, and (c) 5 minutes.

14. The population of a certain species t years after it is introduced into a new habitat is $p(t) = \dfrac{1200}{1 + 3e^{-t/5}}$. (a) Determine the initial size of the population. (b) Determine the population after 5 years. (c) After how many years will the population be 800?

15. By observation, identify the equation that corresponds to the graph shown at the right. Explain your reasoning.

   (a) $y = 6e^{-x^2/2}$ (b) $y = \dfrac{6}{1 + e^{-x/2}}$ (c) $y = 6(1 - e^{-x^2/2})$

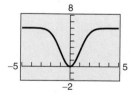

FIGURE FOR 15

16. The numbers of cellular-phone subscribers y (in millions) for the years 1993 through 1997 are (3, 16.0), (4, 24.1), (5, 33.8), (6, 44.0), and (7, 55.3), where x is the time (in years) and $x = 3$ corresponds to 1993. Use a graphing utility to fit an exponential model to the data. Sketch a scatter plot of the data and graph the model in the same viewing window. (Source: Cellular Telecommunications Industry Association)

Topics in Analytic Geometry

Derke O'Hara/Tony Stone Images

The Big Picture

In this chapter you will learn how to

❏ write the standard equations of parabolas, ellipses, and hyperbolas.

❏ analyze and sketch the graphs of parabolas, ellipses, and hyperbolas.

❏ rotate the coordinate axis to eliminate the xy-term in equations of conics and use the discriminant to classify conics.

❏ solve systems of quadratic equations.

❏ rewrite sets of parametric equations as rectangular equations and find sets of parametric equations for graphs.

❏ write equations in polar form.

❏ graph polar equations and recognize special polar graphs.

❏ write equations of conics in polar form.

Ice in the nucleus of a comet is heated and vaporized by the sun. The escaping gas collects dust particles, forming a tail which always points away from the sun.

Important Vocabulary

As you encounter each new vocabulary term in this chapter, add the term and its definition to your notebook glossary.

- conic section or conic (pp. 416, 474)
- parabola (pp. 417, 474)
- directrix (p. 417)
- focus or foci (pp. 417, 424, 433)
- tangent (p. 419)
- ellipse (pp. 424, 474)
- vertices (pp. 424, 433)
- major axis (p. 424)

- center (pp. 424, 433)
- minor axis (p. 424)
- eccentricity (pp. 428, 474)
- hyperbola (pp. 433, 474)
- transverse axis (p. 433)
- asymptotes (p. 435)
- conjugate axis (p. 435)
- discriminant (p. 446)

- parameter (p. 451)
- parametric equations (p. 451)
- plane curve (p. 451)
- orientation (p. 452)
- polar coordinate system (p. 459)
- pole or origin (p. 459)
- polar axis (p. 459)
- polar coordinates (p. 459)

Additional Resources   Text-specific additional resources are available to help you do well in this course. See page xvi for details.

6.1 Introduction to Conics: Parabolas

Conic sections were discovered during the classical Greek period, 600 to 300 B.C. The early Greeks were concerned largely with the geometric properties of conics. It was not until the early 17th century that the broad applicability of conics became apparent, and they then played a prominent role in the early development of calculus.

Each **conic section** (or simply **conic**) is the intersection of a plane and a double-napped cone. Notice in Figure 6.1(a) that in the formation of the four basic conics, the intersecting plane does not pass through the vertex of the cone. When the plane does pass through the vertex, the resulting figure is a *degenerate conic*, as shown in Figure 6.1(b).

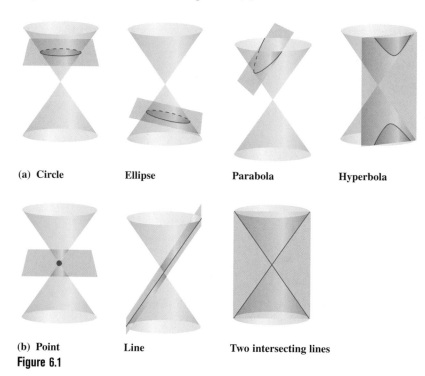

(a) Circle   **Ellipse**   **Parabola**   **Hyperbola**

(b) Point   **Line**   **Two intersecting lines**

Figure 6.1

A computer animation of this concept appears in the *Interactive* CD-ROM and *Internet* versions of this text.

There are several ways to approach the study of conics. You could begin by defining conics in terms of the intersections of planes and cones, as the Greeks did, or you could define them algebraically, in terms of the general second-degree equation

$$Ax^2 + Bxy + Cy^2 + Dx + Ey + F = 0.$$

However, you will study a third approach, in which each of the conics is defined as a **locus** (collection) of points satisfying a geometric property. For example, the definition of a circle as the collection of all points (x, y) that are equidistant from a fixed point (h, k) leads to the standard equation of a circle

$$(x - h)^2 + (y - k)^2 = r^2.$$ Equation of circle

Parabolas

The first type of conic is called a **parabola,** and it is defined as follows.

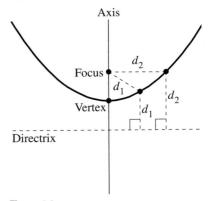

Definition of a Parabola

A **parabola** is the set of all points (x, y) that are equidistant from a fixed line (**directrix**) and a fixed point (**focus**) not on the line.

The midpoint between the focus and the directrix is called the **vertex,** and the line passing through the focus and the vertex is called the **axis** of the parabola. Note in Figure 6.2 that a parabola is symmetric with respect to its axis. Using the definition of a parabola, you can derive the following **standard form of the equation of a parabola** whose directrix is parallel to the *x*-axis or to the *y*-axis. See Appendix A for a proof.

Figure 6.2

Standard Equation of a Parabola

The **standard form of the equation of a parabola** with vertex at (h, k) is as follows.

$(x - h)^2 = 4p(y - k), \ p \neq 0$ Vertical axis; directrix: $y = k - p$

$(y - k)^2 = 4p(x - h), \ p \neq 0$ Horizontal axis; directrix: $x = h - p$

The focus lies on the axis p units (*directed distance*) from the vertex. If the vertex is at the origin $(0, 0)$, the equation takes one of the following forms.

$x^2 = 4py$ Vertical axis

$y^2 = 4px$ Horizontal axis

See Figure 6.3.

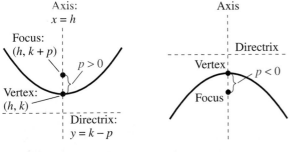

(a) Vertical axis: $p > 0$ **(b) Vertical axis: $p < 0$**
$$(x - h)^2 = 4p(y - k)$$

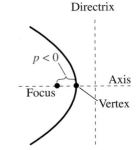

(c) Horizontal axis: $p > 0$ **(d) Horizontal axis: $p < 0$**
$$(y - k)^2 = 4p(x - h)$$

Figure 6.3

EXAMPLE 1   Finding the Standard Equation of a Parabola

Find the standard form of the equation of the parabola with vertex (2, 1) and focus (2, 4).

Solution

Because the axis of the parabola is vertical, consider the equation

$$(x - h)^2 = 4p(y - k)$$

where $h = 2$, $k = 1$, and $p = 4 - 1 = 3$. So, the standard form is

$$(x - 2)^2 = 4(3)(y - 1) = 12(y - 1).$$

You can obtain the more common quadratic form as follows.

$$(x - 2)^2 = 12(y - 1) \qquad \text{Write original equation.}$$
$$x^2 - 4x + 4 = 12y - 12 \qquad \text{Multiply.}$$
$$x^2 - 4x + 16 = 12y \qquad \text{Add 12 to each side.}$$
$$y = \tfrac{1}{12}(x^2 - 4x + 16) \qquad \text{Divide each side by 12.}$$

Try using a graphing utility to confirm the graph of this parabola, as shown in Figure 6.4.

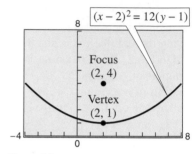

The *Interactive* CD-ROM and *Internet* versions of this text show every example with its solution; clicking on the *Try It!* button brings up similar problems. Guided Examples and Integrated Examples show step-by-step solutions to additional examples. Integrated Examples are related to several concepts in the section.

Figure 6.4

EXAMPLE 2   Finding the Focus of a Parabola

Find the focus of the parabola

$$y = -\frac{1}{2}x^2 - x + \frac{1}{2}.$$

Solution

To find the focus, convert to standard form by completing the square.

$$y = -\frac{1}{2}x^2 - x + \frac{1}{2} \qquad \text{Write original equation.}$$
$$-2y = x^2 + 2x - 1 \qquad \text{Multiply each side by –2.}$$
$$1 - 2y = x^2 + 2x \qquad \text{Group terms.}$$
$$1 + 1 - 2y = x^2 + 2x + 1 \qquad \text{Complete the square.}$$
$$2 - 2y = x^2 + 2x + 1 \qquad \text{Combine like terms.}$$
$$-2(y - 1) = (x + 1)^2 \qquad \text{Write in standard form.}$$

Comparing this equation with

$$(x - h)^2 = 4p(y - k)$$

you can conclude that $h = -1$, $k = 1$, and $p = -\frac{1}{2}$. Because p is negative, the parabola opens downward, as shown in Figure 6.5. Therefore, the focus of the parabola is

$$(h, k + p) = \left(-1, \frac{1}{2}\right). \qquad \text{Focus}$$

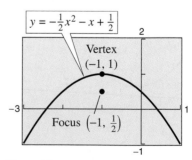

Figure 6.5

EXAMPLE 3   Vertex at the Origin

Find the standard equation of the parabola with vertex at the origin and focus $(2, 0)$.

Solution

The axis of the parabola is horizontal, passing through $(0, 0)$ and $(2, 0)$, as shown in Figure 6.6.

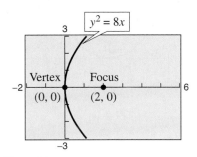

Figure 6.6

So, the standard form is

$$y^2 = 4px$$

where $h = k = 0$ and $p = 2$. Therefore, the equation is

$$y^2 = 8x.$$

STUDY TIP

You can use a graphing utility to confirm the equation found in Example 3. To do this, it helps to split the equation into two parts: $y_1 = \sqrt{8x}$ (upper part) and $y_2 = -\sqrt{8x}$ (lower part).

Applications

A line segment that passes through the focus of a parabola and has endpoints on the parabola is called a **focal chord.** The specific focal chord perpendicular to the axis of the parabola is called the **latus rectum.**

Parabolas occur in a wide variety of applications. For instance, a parabolic reflector can be formed by revolving a parabola around its axis. The resulting surface has the property that all incoming rays parallel to the axis are reflected through the focus of the parabola; this is the principle behind the construction of the parabolic mirrors used in reflecting telescopes. Conversely, the light rays emanating from the focus of a parabolic reflector used in a flashlight are all parallel to one another, as shown in Figure 6.7.

A line is **tangent** to a parabola at a point on the parabola if the line intersects, but does not cross, the parabola at the point. Tangent lines to parabolas have special properties that are related to the use of parabolas in constructing reflective surfaces.

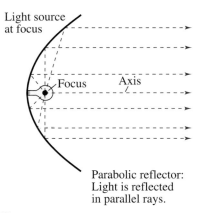

Figure 6.7

Parabolic reflector: Light is reflected in parallel rays.

Reflective Property of a Parabola

The tangent line to a parabola at a point P makes equal angles with the following two lines (see Figure 6.8).

1. The line passing through P and the focus
2. The axis of the parabola

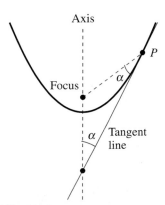

Figure 6.8

EXAMPLE 4   Finding the Tangent Line at a Point on a Parabola

Find the equation of the tangent line to the parabola given by $y = x^2$ at the point $(1, 1)$.

Solution

For this parabola, $p = \frac{1}{4}$ and the focus is $\left(0, \frac{1}{4}\right)$, as shown in Figure 6.9. You can find the y-intercept $(0, b)$ of the tangent line by equating the lengths of the two sides of the isosceles triangle

$$d_1 = \frac{1}{4} - b$$

and

$$d_2 = \sqrt{(1 - 0)^2 + \left[1 - \left(\frac{1}{4}\right)\right]^2} = \frac{5}{4}$$

shown in Figure 6.9. Setting $d_1 = d_2$ produces

$$\frac{1}{4} - b = \frac{5}{4}$$

$$b = -1.$$

So, the slope of the tangent line is

$$m = \frac{1 - (-1)}{1 - 0} = 2$$

and its slope-intercept equation is

$$y = 2x - 1.$$

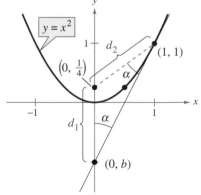

Figure 6.9

Try using a graphing utility to confirm the result of Example 4. By graphing

$$y = x^2 \qquad \text{and} \qquad y = 2x - 1$$

in the same viewing window, you should be able to see that the line touches the parabola at the point $(1, 1)$.

Writing About Math   *Television Antenna Dishes*

Cross sections of television antenna dishes are parabolic in shape. Write a paragraph describing why these dishes are parabolic. Include a graphical representation of your description.

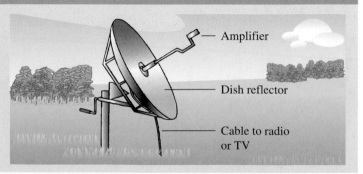

6.1 E x e r c i s e s

In Exercises 1–6, match the equation with its graph. [The graphs are labeled (a), (b), (c), (d), (e), and (f).]

(a)

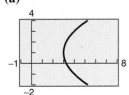

(b)

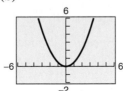

(c)

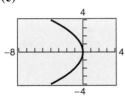

(d)

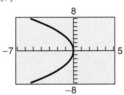

(e)

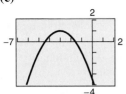

(f)

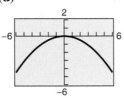

1. $y^2 = -4x$ **2.** $x^2 = 2y$

3. $x^2 = -8y$ **4.** $y^2 = -12x$

5. $(y - 1)^2 = 4(x - 3)$ **6.** $(x + 3)^2 = -2(y - 1)$

In Exercises 7–20, find the vertex, focus, and directrix of the parabola and sketch its graph.

7. $y = \frac{1}{2}x^2$ **8.** $y = -2x^2$

9. $y^2 = -6x$ **10.** $y^2 = 3x$

11. $x^2 + 6y = 0$ **12.** $x + y^2 = 0$

13. $(x - 1)^2 + 8(y + 2) = 0$

14. $(x + 5) + (y - 1)^2 = 0$

15. $\left(x + \frac{3}{2}\right)^2 = 4(y - 2)$ **16.** $\left(x + \frac{1}{2}\right)^2 = 4(y - 1)$

17. $y = \frac{1}{4}(x^2 - 2x + 5)$ **18.** $x = \frac{1}{4}(y^2 + 2y + 33)$

19. $y^2 + 6y + 8x + 25 = 0$

20. $y^2 - 4y - 4x = 0$

The *Interactive* CD-ROM and *Internet* versions of this text contain step-by-step solutions to all odd-numbered Section and Review Exercises. They also provide Tutorial Exercises, which link to Guided Examples for additional help.

In Exercises 21–24, find the vertex, focus, and directrix of the parabola and sketch its graph. Use a graphing utility to verify your graph.

21. $x^2 + 4x + 6y - 2 = 0$

22. $x^2 - 2x + 8y + 9 = 0$

23. $y^2 + x + y = 0$

24. $y^2 - 4x - 4 = 0$

In Exercises 25 and 26, change the equation so that its graph matches the given graph.

25. $y^2 = -6x$ **26.** $y^2 = 9x$

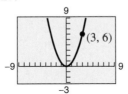 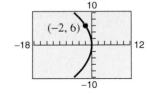

In Exercises 27–38, find the standard form of the equation of the parabola with its vertex at the origin.

27.

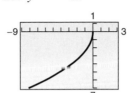

28.

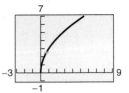

29. Focus: $\left(0, -\frac{3}{2}\right)$ **30.** Focus: $(2, 0)$

31. Focus: $(-2, 0)$ **32.** Focus: $(0, -2)$

33. Directrix: $y = -1$ **34.** Directrix: $y = 3$

35. Directrix: $x = 2$ **36.** Directrix: $x = -3$

37. Horizontal axis and passes through the point $(4, 6)$

38. Vertical axis and passes through the point $(-3, -3)$

In Exercises 39–48, find the standard form of the equation of the parabola.

39.

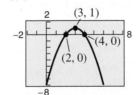

40.

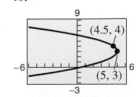

41.

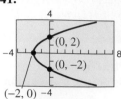

42.

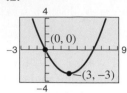

43. Vertex: $(5, 2)$; Focus: $(3, 2)$

44. Vertex: $(-1, 2)$; Focus: $(-1, 0)$

45. Vertex: $(0, 4)$; Directrix: $y = 2$

46. Vertex: $(-2, 1)$; Directrix: $x = 1$

47. Focus: $(2, 2)$; Directrix: $x = -2$

48. Focus: $(0, 0)$; Directrix: $y = 8$

In Exercises 49 and 50, the equations of a parabola and a tangent line to the parabola are given. Use a graphing utility to graph both in the same viewing window. Determine the coordinates of the point of tangency.

Parabola	*Tangent Line*
49. $y^2 - 8x = 0$	$x - y + 2 = 0$
50. $x^2 + 12y = 0$	$x + y - 3 = 0$

In Exercises 51–54, find an equation of the tangent line to the parabola at the given point and find the *x*-intercept of the line.

51. $x^2 = 2y, \quad (4, 8)$

52. $x^2 = 2y, \quad \left(-3, \frac{9}{2}\right)$

53. $y = -2x^2, \quad (-1, -2)$

54. $y = -2x^2, \quad (2, -8)$

55. *Revenue* The revenue R generated by the sale of x units of a product is $R = 265x - \left(\frac{5}{4}\right)x^2$. Use a graphing utility to graph the function and approximate the number of sales that will maximize revenue.

56. *Revenue* The revenue R generated by the sale of x units of a product is $R = 378x - \left(\frac{7}{5}\right)x^2$. Use a graphing utility to graph the function and approximate the number of sales that will maximize revenue.

57. *Satellite Antenna* The receiver in a parabolic television dish antenna is 3.5 feet from the vertex and is located at the focus. Find an equation of a cross section of the reflector. (Assume that the dish is directed upward and that the vertex is at the origin.)

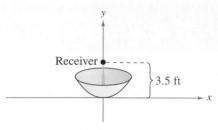

FIGURE FOR 57

58. *Suspension Bridge* Each cable of a suspension bridge is suspended (in the shape of a parabola) between two towers that are 120 meters apart, and the top of each tower is 20 meters above the roadway. The cables touch the roadway midway between the towers.

(a) Draw a diagram for the bridge. Draw a rectangular coordinate system on the bridge with the center of the bridge at the origin. Identify the coordinates of the known points.

(b) Find an equation for the parabolic shape of each cable.

(c) Complete the table by finding the height of the suspension cables above the roadway at a distance of x meters from the center of the bridge.

x	0	20	40	60
y				

59. *Road Design* Roads are often designed with parabolic surfaces to allow rain to drain off. A particular road that is 32 feet wide is 0.4 foot higher in the center than it is on the sides.

(a) Find an equation of the parabola. (Assume that the origin is at the center of the road.)

(b) How far from the center of the road is the road surface 0.1 foot lower than in the middle?

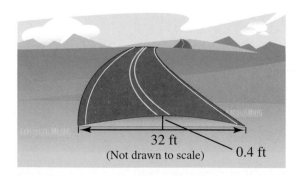

32 ft
(Not drawn to scale)   0.4 ft

60. *Highway Design* Highway engineers design a parabolic curve for an entrance ramp from a straight street to an interstate highway. Find an equation of the parabola.

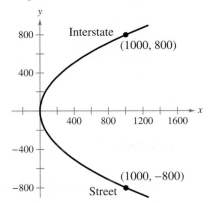

61. *Satellite Orbit* An earth satellite in a 100-mile-high circular orbit around the earth has a velocity of approximately 17,500 miles per hour. If this velocity is multiplied by $\sqrt{2}$, the satellite will have the minimum velocity necessary to escape the earth's gravity and it will follow a parabolic path with the center of the earth as the focus.

(a) Find the escape velocity of the satellite.

(b) Find an equation of its path (assume the radius of the earth is 4000 miles).

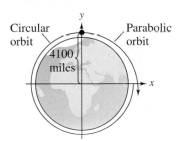

62. *Path of a Projectile* The path of a softball is given by the equation

$$y = -0.08x^2 + x + 4.$$

The coordinates x and y are measured in feet, with $x = 0$ corresponding to the position where the ball was thrown.

(a) Use a graphing utility to graph the trajectory of the softball.

(b) Move the cursor along the path to approximate the highest point and the range of the trajectory.

Projectile Motion **In Exercises 63 and 64, consider the path of a projectile projected horizontally with a velocity of v feet per second at a height of s feet, where the model for the path is**

$$y = -\frac{16}{v^2}t^2 + s.$$

In this model, air resistance is disregarded and y is the height (in feet) of the projectile t seconds after its release.

63. A ball is thrown from the top of a 75-foot tower with a velocity of 32 feet per second.

(a) Find the equation of the parabolic path.

(b) How far does the ball travel horizontally before striking the ground?

64. A bomber flying due east at 550 miles per hour at an altitude of 42,000 feet releases a bomb. Determine the distance the bomb travels horizontally before striking the ground.

Synthesis

True or False? **In Exercises 65 and 66, determine whether the statement is true or false. Justify your answer.**

65. It is possible for a parabola to intersect its directrix.

66. If the vertex and focus of a parabola are on a horizontal line, then the directrix of the parabola is vertical.

Review

In Exercises 67–72, find all x-intercepts and y-intercepts of the quadratic function.

67. $f(x) = (x - 7)^2$

68. $f(x) = (x + 9)^2$

69. $f(x) = (x - 5)^2 - 5$

70. $f(x) = (x + 11)^2 + 12$

71. $f(x) = x^2 - 7x - 1$

72. $f(x) = x^2 + 9x - 22$

In Exercises 73–78, write the quadratic function in standard form by completing the square. Identify the vertex of the function.

73. $f(x) = 3x^2 + 2x - 16$

74. $f(x) = 2x^2 - x - 21$

75. $f(x) = 5x^2 + 34x - 7$

76. $f(x) = -x^2 - 8x - 15$

77. $f(x) = 6x^2 - x - 12$

78. $f(x) = -8x^2 - 34x - 21$

6.2   Ellipses

Introduction

The second type of conic is called an **ellipse.** It is defined as follows.

Definition of an Ellipse

An **ellipse** is the set of all points (x, y) the sum of whose distances from two distinct fixed points (**foci**) is constant. (See Figure 6.10.)

What You Should Learn:

* How to write equations of ellipses in standard form
* How to use properties of ellipses to model and solve real-life problems
* How to find eccentricities of ellipses

Why You Should Learn It:

Ellipses can be used to model and solve many types of real-life problems. For instance, in Exercise 52 on page 431, an ellipse is used to model the orbit of Halley's comet.

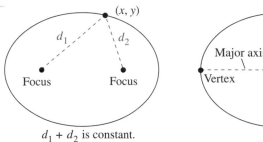
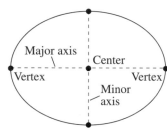

$d_1 + d_2$ is constant.

Figure 6.10

© Royal Observatory, Edinburgh

The line through the foci intersects the ellipse at two points called **vertices.** The chord joining the vertices is the **major axis,** and its midpoint is the **center** of the ellipse. The chord perpendicular to the major axis at the center is the **minor axis** of the ellipse.

To derive the standard form of the equation of an ellipse, consider the ellipse in Figure 6.11 with the following points: center, (h, k); vertices, $(h \pm a, k)$; foci, $(h \pm c, k)$.

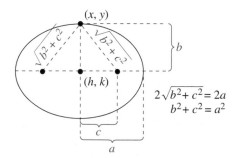

$$2\sqrt{b^2 + c^2} = 2a$$
$$b^2 + c^2 = a^2$$

Figure 6.11

The sum of the distances from any point on the ellipse to the two foci is constant. Using a vertex point, this constant sum is

$$(a + c) + (a - c) = 2a \qquad \text{Length of major axis}$$

or simply the length of the major axis.

Now, if you let (x, y) be *any* point on the ellipse, the sum of the distances between (x, y) and the two foci must also be $2a$. That is,

$$\sqrt{[x - (h - c)]^2 + (y - k)^2} + \sqrt{[x - (h + c)]^2 + (y - k)^2} = 2a.$$

Finally, using Figure 6.11 and $b^2 = a^2 - c^2$, you obtain the following equation of the ellipse.

$$b^2(x - h)^2 + a^2(y - k)^2 = a^2b^2$$

$$\frac{(x - h)^2}{a^2} + \frac{(y - k)^2}{b^2} = 1$$

In the development above, you would obtain a similar equation by starting with a vertical major axis. Both results are summarized as follows.

Standard Equation of an Ellipse

The **standard form of the equation of an ellipse,** with center (h, k) and major and minor axes of lengths $2a$ and $2b$, respectively, where $0 < b < a$, is

$$\frac{(x - h)^2}{a^2} + \frac{(y - k)^2}{b^2} = 1 \qquad \text{Major axis is horizontal.}$$

$$\frac{(x - h)^2}{b^2} + \frac{(y - k)^2}{a^2} = 1. \qquad \text{Major axis is vertical.}$$

The foci lie on the major axis, c units from the center, with $c^2 = a^2 - b^2$. If the center is at the origin $(0, 0)$, the equation takes one of the following forms.

$$\frac{x^2}{a^2} + \frac{y^2}{b^2} = 1 \qquad \text{Major axis is horizontal.}$$

$$\frac{x^2}{b^2} + \frac{y^2}{a^2} = 1 \qquad \text{Major axis is vertical.}$$

STUDY T!P

Don't confuse the equation

$$c^2 = a^2 - b^2$$

with the Pythagorean Theorem—there is a sign difference.

Figure 6.12 shows both the vertical and horizontal orientations for an ellipse.

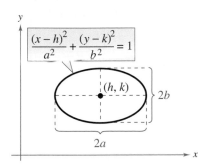

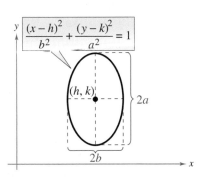

Figure 6.12

Figure 6.13

You can visualize the definition of an ellipse by imagining two thumbtacks placed at the foci, as shown in Figure 6.13. If the ends of a fixed length of string are fastened to the thumbtacks and the string is drawn taut with a pencil, the path traced by the pencil will be an ellipse.

A computer animation of this concept appears in the *Interactive* CD-ROM and *Internet* versions of this text.

EXAMPLE 1   Finding the Standard Equation of an Ellipse

Find the standard form of the equation of the ellipse having foci at $(0, 1)$ and $(4, 1)$, and a major axis of length 6, as shown in Figure 6.14.

Solution

Because the foci occur at $(0, 1)$ and $(4, 1)$, the center of the ellipse is $(2, 1)$ and the distance from the center to one of the foci is $c = 2$. Because $2a = 6$ you know that $a = 3$. Now, from $c^2 = a^2 - b^2$, you have

$$b = \sqrt{a^2 - c^2}$$
$$= \sqrt{9 - 4}$$
$$= \sqrt{5}.$$

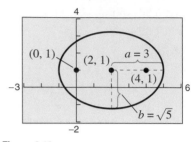

Figure 6.14

Because the major axis is horizontal, the standard equation is

$$\frac{(x - 2)^2}{3^2} + \frac{(y - 1)^2}{\left(\sqrt{5}\right)^2} = 1.$$

EXAMPLE 2   Sketching an Ellipse

Sketch the ellipse given by $4x^2 + y^2 = 36$ and identify the vertices.

Algebraic Solution

$$4x^2 + y^2 = 36 \qquad \text{Write original equation.}$$

$$\frac{4x^2}{36} + \frac{y^2}{36} = \frac{36}{36} \qquad \text{Divide each side by 36.}$$

$$\frac{x^2}{3^2} + \frac{y^2}{6^2} = 1 \qquad \text{Write in standard form.}$$

Because the denominator of the y^2-term is larger than the denominator to the x^2-term, you can conclude that the major axis is vertical. Moreover, because $a = 6$, the vertices are $(0, -6)$ and $(0, 6)$. Finally, because $b = 3$, the endpoints of the minor axis are $(-3, 0)$ and $(3, 0)$, as shown in Figure 6.15.

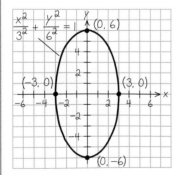

Figure 6.15

Graphical Solution

Solve the equation of the ellipse for y as follows.

$$4x^2 + y^2 = 36$$
$$y^2 = 36 - 4x^2$$
$$y = \pm\sqrt{36 - 4x^2}$$

Then use a graphing utility to graph both $y_1 = \sqrt{36 - 4x^2}$ and $y_2 = -\sqrt{36 - 4x^2}$ in the same viewing window. Be sure to use a square setting. From the graph in Figure 6.16, you can see that the major axis is vertical. You can use the *zoom* and *trace* features to approximate the vertices to be $(0, 6)$ and $(0, -6)$.

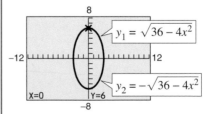

Figure 6.16

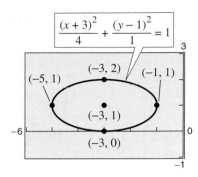

Figure 6.17

EXAMPLE 3 Writing an Equation in Standard Form

Write the equation of the ellipse in standard form and sketch the graph of the ellipse.

$$x^2 + 4y^2 + 6x - 8y + 9 = 0.$$

Solution

To write the given equation in standard form you must complete the square twice. In the fourth step, note that 9 and 4 are added to *both* sides of the equation.

$$x^2 + 4y^2 + 6x - 8y + 9 = 0 \quad \text{Write original equation.}$$

$$\left(x^2 + 6x + \quad\right) + \left(4y^2 - 8y + \quad\right) = -9 \quad \text{Group terms.}$$

$$\left(x^2 + 6x + \quad\right) + 4\left(y^2 - 2y + \quad\right) = -9 \quad \text{Factor 4 out of } y\text{-terms.}$$

$$(x^2 + 6x + 9) + 4(y^2 - 2y + 1) = -9 + 9 + 4(1)$$

$$(x + 3)^2 + 4(y - 1)^2 = 4 \quad \text{Write in completed square form.}$$

$$\frac{(x + 3)^2}{2^2} + \frac{(y - 1)^2}{1^2} = 1 \quad \text{Write in standard form.}$$

Now you see that the center is at $(h, k) = (-3, 1)$. Because the denominator of the x-term is $a^2 = 2^2$, you can locate the endpoints of the major axis two units to the right and left of the center. Similarly, because the denominator of the y-term is $b^2 = 1^2$, you can locate the endpoints of the minor axis one unit up and down from the center. The graph of this ellipse is shown in Figure 6.17.

EXAMPLE 4 Analyzing an Ellipse

Find the center, vertices, and foci of the ellipse $4x^2 + y^2 - 8x + 4y - 8 = 0$.

Solution

By completing the square, you can write the given equation in standard form.

$$4x^2 + y^2 - 8x + 4y - 8 = 0$$

$$\left(4x^2 - 8x + \quad\right) + \left(y^2 + 4y + \quad\right) = 8$$

$$4\left(x^2 - 2x + \quad\right) + \left(y^2 + 4y + \quad\right) = 8$$

$$4(x^2 - 2x + 1) + (y^2 + 4y + 4) = 8 + 4(1) + 4$$

$$4(x - 1)^2 + (y + 2)^2 = 16$$

$$\frac{(x - 1)^2}{2^2} + \frac{(y + 2)^2}{4^2} = 1$$

So, the major axis is vertical, where $h = 1$, $k = -2$, $a = 4$, $b = 2$, and

$$c = \sqrt{16 - 4} = 2\sqrt{3}.$$

Therefore, you have the following.

Center: $(1, -2)$ Vertices: $(1, -6)$ Foci: $\left(1, -2 - 2\sqrt{3}\right)$
                               $(1, 2)$ $\left(1, -2 + 2\sqrt{3}\right)$

The graph of the ellipse is shown in Figure 6.18.

STUDY TIP

You can use a graphing utility to graph an ellipse by graphing the upper and lower portions in the same viewing window. For instance, to graph the ellipse in Example 3, first solve for y to get

$$y_1 = 1 + \sqrt{1 - \frac{(x + 3)^2}{4}}$$

and

$$y_2 = 1 - \sqrt{1 - \frac{(x + 3)^2}{4}}.$$

Use a viewing window in which $-6 \leq x \leq 0$ and $-1 \leq y \leq 3$. You should obtain the graph shown in Figure 6.17.

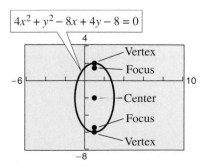

Figure 6.18

Application

Ellipses have many practical and aesthetic uses. For instance, machine gears, supporting arches, and acoustical designs often involve elliptical shapes. The orbits of satellites and planets are also ellipses. Example 5 investigates the elliptical orbit of the moon about the earth.

EXAMPLE 5   An Application Involving an Elliptical Orbit

The moon travels about the earth in an elliptical orbit with the earth at one focus, as shown in Figure 6.19. The major and minor axes of the orbit have lengths of 768,806 kilometers and 767,746 kilometers, respectively. Find the greatest and least distances (the apogee and perigee) from the earth's center to the moon's center.

Solution

Because $2a = 768,806$ and $2b = 767,746$, you have $a = 384,403$ and $b = 383,873$, which implies that

$$c = \sqrt{a^2 - b^2}$$

$$= \sqrt{384,403^2 - 383,873^2}$$

$$\approx 20,179.$$

Therefore, the greatest distance between the center of the earth and the center of the moon is

$$a + c \approx 384,403 + 20,179$$

$$= 404,582 \text{ km}$$

and the least distance is

$$a - c \approx 384,403 - 20,179$$

$$= 364,224 \text{ km}.$$

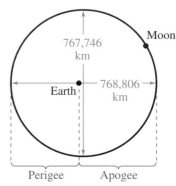

Figure 6.19

Eccentricity

One of the reasons it was difficult for early astronomers to detect that the orbits of the planets are ellipses is that the foci of the planetary orbits are relatively close to their centers, and so the orbits are nearly circular. To measure the ovalness of an ellipse, you can use the concept of **eccentricity.**

Definition of Eccentricity

The **eccentricity** e of an ellipse is given by the ratio

$$e = \frac{c}{a}.$$

Note that $0 < e < 1$ for *every* ellipse.

Exploration

Use a graphing utility to graph the following ellipses in the same viewing window.

$$\frac{x^2}{9} + \frac{y^2}{8} = 1$$

$$\frac{x^2}{25} + \frac{y^2}{9} = 1$$

$$\frac{x^2}{49} + \frac{y^2}{9} = 1$$

Calculate the eccentricity $e = c/a$ for each ellipse. How does the value of e relate to the shape of the ellipse?

To see how this ratio is used to describe the shape of an ellipse, note that because the foci of an ellipse are located along the major axis between the vertices and the center, it follows that

$$0 < c < a.$$

For an ellipse that is nearly circular, the foci are close to the center and the ratio c/a is small [see Figure 6.20(a)]. On the other hand, for an elongated ellipse, the foci are close to the vertices and the ratio c/a is close to 1 [see Figure 6.20(b)].

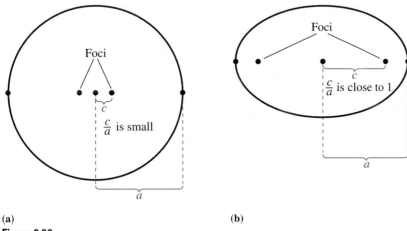

(a)　　　　　　　　　　　　　　　**(b)**

Figure 6.20

The orbit of the moon has an eccentricity of $e = 0.0549$, and the eccentricities of the nine planetary orbits are as follows.

Mercury:	$e = 0.2056$	Saturn:	$e = 0.0543$
Venus:	$e = 0.0068$	Uranus:	$e = 0.0460$
Earth:	$e = 0.0167$	Neptune:	$e = 0.0082$
Mars:	$e = 0.0934$	Pluto:	$e = 0.2481$
Jupiter:	$e = 0.0484$		

Writing About Math *Graphing Ellipses*

Write an equation of an ellipse in standard form and graph it on graph paper. Do not write the equation on your graph. Exchange graphs with another student. Use the graph you receive to reconstruct the equation of the ellipse it represents. Find the eccentricity of the ellipse. Compare your results and write a short paragraph that discusses your findings.

6.2 Exercises

In Exercises 1–6, match the equation with its graph. [The graphs are labeled (a), (b), (c), (d), (e), and (f).]

(a)

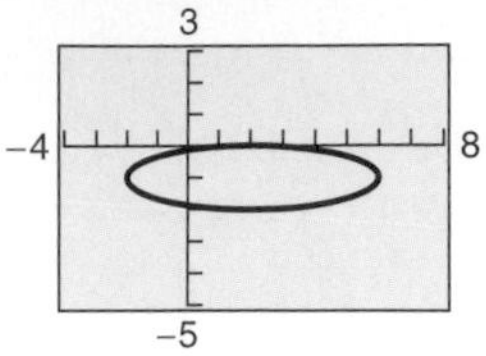

(b)

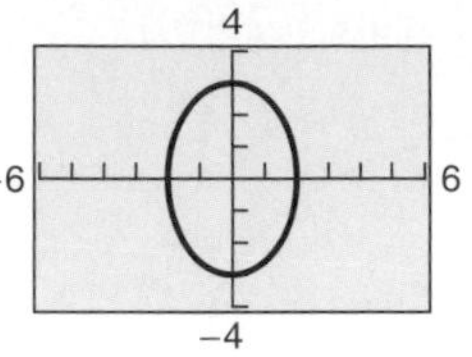

(c)

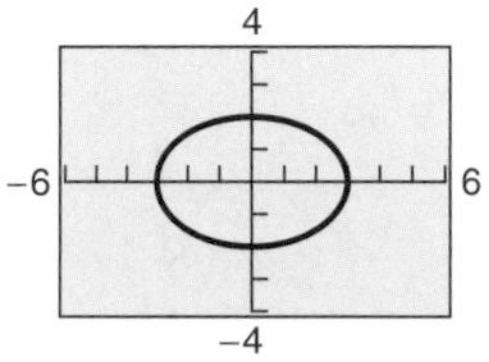

(d)

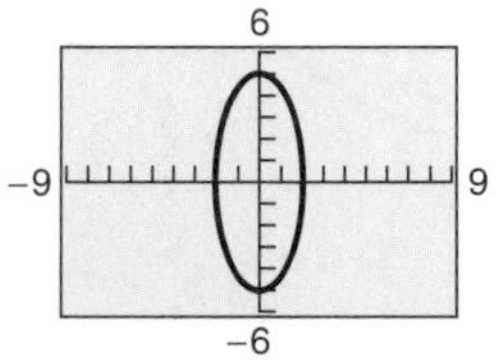

(e)

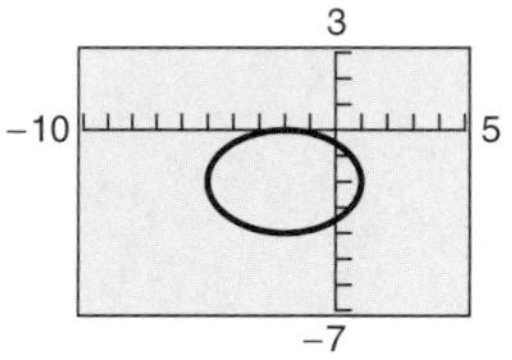

(f)

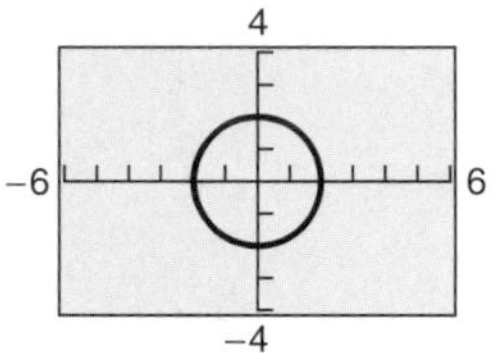

1. $\dfrac{x^2}{4} + \dfrac{y^2}{9} = 1$

2. $\dfrac{x^2}{9} + \dfrac{y^2}{4} = 1$

3. $\dfrac{x^2}{4} + \dfrac{y^2}{25} = 1$

4. $\dfrac{y^2}{4} + \dfrac{x^2}{4} = 1$

5. $\dfrac{(x-2)^2}{16} + (y+1)^2 = 1$

6. $\dfrac{(x+2)^2}{9} + \dfrac{(y+2)^2}{4} = 1$

In Exercises 7–22, find the center, vertices, foci, and eccentricity of the ellipse, and sketch its graph. Use a graphing utility to verify your graph.

7. $\dfrac{x^2}{25} + \dfrac{y^2}{16} = 1$

8. $\dfrac{x^2}{81} + \dfrac{y^2}{144} = 1$

9. $\dfrac{x^2}{5} + \dfrac{y^2}{9} = 1$

10. $\dfrac{x^2}{64} + \dfrac{y^2}{28} = 1$

11. $\dfrac{(x+3)^2}{16} + \dfrac{(y-5)^2}{25} = 1$

12. $\dfrac{(x-4)^2}{12} + \dfrac{(y+3)^2}{16} = 1$

13. $\dfrac{(x+5)^2}{9/4} + (y-1)^2 = 1$

14. $(x+2)^2 + \dfrac{(y+4)^2}{1/4} = 1$

15. $9x^2 + 4y^2 + 36x - 24y + 36 = 0$

16. $9x^2 + 4y^2 - 54x + 40y + 37 = 0$

17. $x^2 + 5y^2 - 8x - 30y - 39 = 0$

18. $3x^2 + y^2 + 18x - 2y - 8 = 0$

19. $6x^2 + 2y^2 + 18x - 10y + 2 = 0$

20. $x^2 + 4y^2 - 6x + 20y - 2 = 0$

21. $16x^2 + 25y^2 - 32x + 50y + 16 = 0$

22. $9x^2 + 25y^2 - 36x - 50y + 61 = 0$

In Exercises 23–26, use a graphing utility to graph the ellipse. Find the center, foci, and vertices. (*Hint:* Use two equations.)

23. $5x^2 + 3y^2 = 15$

24. $3x^2 + 4y^2 = 12$

25. $12x^2 + 20y^2 - 12x + 40y - 37 = 0$

26. $36x^2 + 9y^2 + 48x - 36y + 43 = 0$

In Exercises 27–34, find the standard form of the equation of the ellipse with its center at the origin.

27.

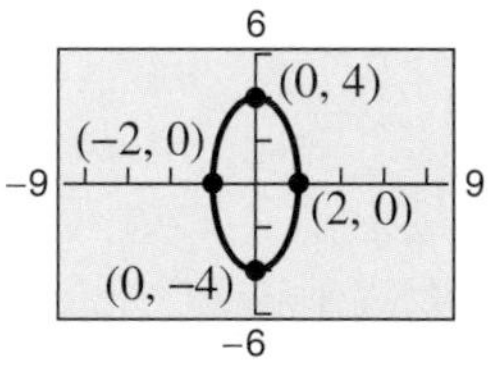

28.

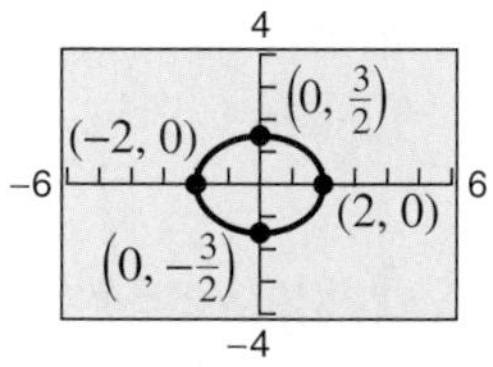

29. Vertices: $(\pm 6, 0)$; Foci: $(\pm 2, 0)$

30. Vertices: $(0, \pm 8)$; Foci: $(0, \pm 4)$

31. Foci: $(\pm 5, 0)$; Major axis of length 12

32. Foci: $(\pm 2, 0)$; Major axis of length 8

33. Vertices: $(0, \pm 5)$; Passes through the point $(4, 2)$

34. Major axis is vertical; Passes through the points $(0, 4)$ and $(2, 0)$

In Exercises 35–46, find the standard form of the equation of the specified ellipse.

35.

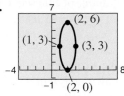

36.

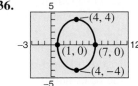

37.

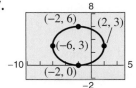

38.

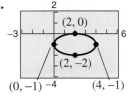

39. Vertices: $(0, 4)$, $(4, 4)$; Minor axis of length 2

40. Foci: $(0, 0)$, $(4, 0)$; Major axis of length 8

41. Foci: $(0, 0)$, $(0, 8)$; Major axis of length 16

42. Center: $(2, -1)$; Vertex: $\left(2, \frac{1}{2}\right)$; Minor axis of length 2

43. Vertices: $(3, 1)$, $(3, 9)$; Minor axis of length 6

44. Center: $(3, 2)$; $a = 3c$; Foci: $(1, 2)$, $(5, 2)$

45. Center: $(0, 4)$; $a = 2c$; Vertices: $(-4, 4)$, $(4, 4)$

46. Vertices: $(5, 0)$, $(5, 12)$; Endpoints of the minor axis: $(0, 6)$, $(10, 6)$

47. Find an equation of the ellipse with vertices $(\pm 5, 0)$ and eccentricity $e = \frac{3}{5}$.

48. Find an equation of the ellipse with vertices $(0, \pm 8)$ and eccentricity $e = \frac{1}{2}$.

49. *Fireplace Arch*   A fireplace arch is to be built in the shape of a semiellipse. The opening is to have a height of 2 feet at the center and a width of 6 feet along the base. The contractor draws the outline of the ellipse using the method described on page 425. Give the required positions of the tacks and the length of the string.

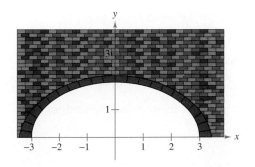

50. *Mountain Travel*   A semielliptical arch over a tunnel for a road through a mountain has a major axis of 80 feet and a height at the center of 30 feet.

(a) Draw a rectangular coordinate system on a sketch of the tunnel with the center of the road entering the tunnel at the origin. Identify the coordinates of the known points.

(b) Find an equation of the elliptical tunnel.

(c) Determine the height of the arch 5 feet from the edge of the tunnel.

51. *Geometry*   The area of the ellipse in the figure is twice the area of the circle. What is the length of the major axis? (*Hint:* $A = \pi ab$ for an ellipse.)

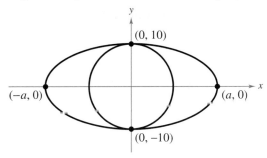

52. *Comet Orbit*   Halley's comet has an elliptical orbit with the sun at one focus. The eccentricity of the orbit is approximately 0.97. The length of the major axis of the orbit is about 36.23 astronomical units. (An astronomical unit is about 93 million miles.) Find an equation for the orbit. Place the center of the orbit at the origin and place the major axis on the x-axis.

53. *Comet Orbit*   The comet Encke has an elliptical orbit with the sun at one focus. Encke ranges from 0.34 to 4.08 astronomical units from the sun. Find an equation of the orbit. Place the center of the orbit at the origin and place the major axis on the x-axis.

54. *Satellite Orbit* The first artificial satellite to orbit earth was *Sputnik I* (launched by Russia in 1957). Its highest point above earth's surface was 938 kilometers, and its lowest point was 212 kilometers. The radius of earth is 6378 kilometers. Find the eccentricity of the orbit.

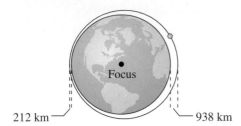

212 km ⌐ Focus ⌐ 938 km

55. *Geometry* A line segment through a focus with endpoints on the ellipse and perpendicular to the major axis is called a **latus rectum** of the ellipse. Therefore, an ellipse has two latera recta. Knowing the length of the latera recta is helpful in sketching an ellipse because it yields other points on the curve. Show that the length of each latus rectum is $2b^2/a$.

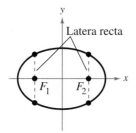

Latera recta

In Exercises 56–59, sketch the graph of the ellipse, making use of the latera recta (see Exercise 55).

56. $\dfrac{x^2}{4} + \dfrac{y^2}{1} = 1$ **57.** $\dfrac{x^2}{9} + \dfrac{y^2}{16} = 1$

58. $9x^2 + 4y^2 = 36$ **59.** $5x^2 + 3y^2 = 15$

Synthesis

True or False? **In Exercises 60–63, determine whether the statement is true or false. Justify your answer.**

60. The graph of $(x^2/4) + y^4 = 1$ is an ellipse.

61. It is easier to distinguish the graph of an ellipse from the graph of a circle if the eccentricity of the ellipse is large (close to 1).

62. The area of a circle with diameter $d = 2r = 8$ is greater than the area of an ellipse with major axis $2a = 8$.

63. It is possible for the foci of an ellipse to occur outside the ellipse.

64. *Think About It* At the beginning of this section it was noted that an ellipse can be drawn using two thumbtacks, a string of fixed length (greater than the distance between the two tacks), and a pencil (see Figure 6.13). If the ends of the string are fastened at the tacks and the string is drawn taut with a pencil, the path traced by the pencil is an ellipse.

(a) What is the length of the string in terms of a?

(b) Explain why the path is an ellipse.

65. *Exploration* The area A of the ellipse

$$\frac{x^2}{a^2} + \frac{y^2}{b^2} = 1 \text{ is } A = \pi ab.$$

For a particular application, $a + b = 20$.

(a) Write the area of the ellipse as a function of a.

(b) Find the equation of an ellipse with an area of 264 square centimeters.

(c) Complete the table and make a conjecture about the shape of the ellipse with a maximum area.

a	8	9	10	11	12	13
A						

(d) Use a graphing utility to graph the area function, and use the graph to make a conjecture about the shape of the ellipse that yields a maximum area.

Review

66. Find a polynomial with integer coefficients that has the zeros 3, $2 + i$, and $2 - i$.

67. Find all the zeros of

$$f(x) = 2x^3 - 3x^2 + 50x - 75$$

if one of the zeros is $x = \frac{3}{2}$.

68. Find all the zeros of the function

$$g(x) = 6x^4 + 7x^3 - 29x^2 - 28x + 20$$

if two of the zeros are $x = \pm 2$.

69. Use a graphing utility to graph the function

$$h(x) = 2x^4 + x^3 - 19x^2 - 9x + 9.$$

Use the graph to approximate the zeros of h.

6.3   Hyperbolas

Introduction

The definition of a hyperbola parallels that of an ellipse. The difference is that for an ellipse, the *sum* of the distances between the foci and a point on the ellipse is fixed; whereas for a hyperbola, the *difference* of these distances is fixed.

Definition of a Hyperbola

A **hyperbola** is the set of all points (x, y) the difference of whose distances from two distinct fixed points (**foci**) is a positive constant. (See Figure 6.21.)

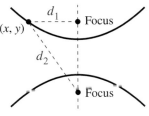
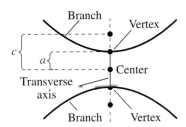

$d_2 - d_1$ is a positive constant.

Figure 6.21

Every hyperbola has two disconnected **branches.** The line through the two foci intersects a hyperbola at its two **vertices.** The line segment connecting the vertices is called the **transverse axis,** and the midpoint of the transverse axis is called the **center** of the hyperbola. The development of the **standard form of the equation of a hyperbola** is similar to that of an ellipse. Note that a, b, and c are related differently for hyperbolas than for ellipses.

Standard Equation of a Hyperbola

The **standard form of the equation of a hyperbola** with center at (h, k) is

$$\frac{(x - h)^2}{a^2} - \frac{(y - k)^2}{b^2} = 1 \qquad \text{Transverse axis is horizontal.}$$

$$\frac{(y - k)^2}{a^2} - \frac{(x - h)^2}{b^2} = 1. \qquad \text{Transverse axis is vertical.}$$

The vertices are a units from the center, and the foci are c units from the center. Moreover, $c^2 = a^2 + b^2$. If the center of the hyperbola is at the origin $(0, 0)$, the equation takes one of the following forms.

$$\frac{x^2}{a^2} - \frac{y^2}{b^2} = 1 \qquad \text{Transverse axis is horizontal.} \qquad \frac{y^2}{a^2} - \frac{x^2}{b^2} = 1 \qquad \text{Transverse axis is vertical.}$$

Figure 6.22 shows both the horizontal and vertical orientations for a hyperbola.

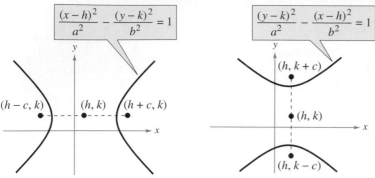

Figure 6.22

EXAMPLE 1 Finding the Standard Equation of a Hyperbola

Find the standard form of the equation of the hyperbola with foci at $(-1, 2)$ and $(5, 2)$, and vertices at $(0, 2)$ and $(4, 2)$.

Solution

By the Midpoint Formula, the center of the hyperbola occurs at the point $(2, 2)$. Furthermore, $c = 3$ and $a = 2$, and it follows that

$$b = \sqrt{c^2 - a^2}$$
$$= \sqrt{3^2 - 2^2}$$
$$= \sqrt{9 - 4}$$
$$= \sqrt{5}.$$

So, the equation of the hyperbola is

$$\frac{(x - 2)^2}{2^2} - \frac{(y - 2)^2}{\left(\sqrt{5}\right)^2} = 1.$$

Figure 6.23 shows the hyperbola.

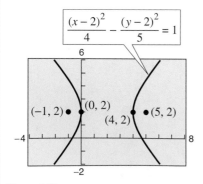

Figure 6.23

Most graphing utilities have a *parametric* mode. Try using *parametric* mode to graph the hyperbola $x = 2 + 2 \sec t$ and $y = 2 + \sqrt{5} \tan t$. How does the result compare with the graph given in Figure 6.23? (Let the parameter vary from $0 \leq t \leq 6.28$, with an increment of t of 0.13.)

Asymptotes of a Hyperbola

Each hyperbola has two **asymptotes** that intersect at the center of the hyperbola, as shown in Figure 6.24. The asymptotes pass through the vertices of a rectangle of dimensions $2a$ by $2b$, with its center at (h, k).

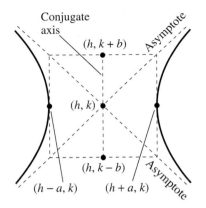

Asymptotes of a Hyperbola

$$y = k \pm \frac{b}{a}(x - h)$$ Asymptotes for horizontal transverse axis

$$y = k \pm \frac{a}{b}(x - h)$$ Asymptotes for vertical transverse axis

The **conjugate axis** of a hyperbola is the line segment of length $2b$ joining $(h, k + b)$ and $(h, k - b)$ if the transverse axis is horizontal, and the line segment of length $2b$ joining $(h + b, k)$ and $(h - b, k)$ if the transverse axis is vertical.

Figure 6.24

EXAMPLE 2 Sketching the Graph of a Hyperbola

Sketch the hyperbola whose equation is $4x^2 - y^2 = 16$.

Algebraic Solution

$$4x^2 - y^2 = 16$$ Write original equation.

$$\frac{4x^2}{16} - \frac{y^2}{16} = \frac{16}{16}$$ Divide each side by 16.

$$\frac{x^2}{2^2} - \frac{y^2}{4^2} = 1$$ Write in standard form.

Because the x^2-term is positive, you can conclude that the transverse axis is horizontal. So, the vertices occur at $(-2, 0)$ and $(2, 0)$, and the endpoints of the conjugate axis occur at $(0, -4)$ and $(0, 4)$. Using these four points, you can sketch the rectangle shown in Figure 6.25(a). Finally, by drawing the asymptotes through the corners of this rectangle, you can complete the sketch, as shown in Figure 6.25(b).

Graphical Solution

Solve the equation of the hyperbola for y as follows.

$$4x^2 - y^2 = 16$$

$$4x^2 - 16 = y^2$$

$$\pm\sqrt{4x^2 - 16} = y$$

Then use a graphing utility to graph $y_1 = \sqrt{4x^2 - 16}$ and $y_2 = -\sqrt{4x^2 - 16}$ in the same viewing window. Be sure to use a square setting. From the graph in Figure 6.26, you can see that the transverse axis is horizontal. You can use the *zoom* and *trace* features to approximate the vertices to be $(-2, 0)$ and $(2, 0)$.

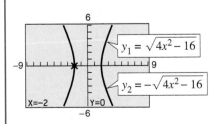

Figure 6.26

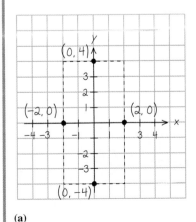

(a)
Figure 6.25

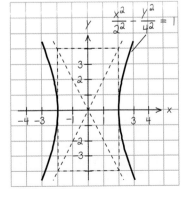

(b)

EXAMPLE 3   Finding the Asymptotes of a Hyperbola

Sketch the hyperbola given by $4x^2 - 3y^2 + 8x + 16 = 0$ and find the equations of its asymptotes.

Solution

$$4x^2 - 3y^2 + 8x + 16 = 0 \qquad \text{Write original equation.}$$

$$4(x^2 + 2x) - 3y^2 = -16 \qquad \text{Subtract 16 from each side; factor.}$$

$$4(x^2 + 2x + 1) - 3y^2 = -16 + 4 \qquad \text{Complete the square.}$$

$$4(x + 1)^2 - 3y^2 = -12 \qquad \text{Write in completed square form.}$$

$$\frac{y^2}{2^2} - \frac{(x + 1)^2}{\left(\sqrt{3}\right)^2} = 1 \qquad \text{Write in standard form.}$$

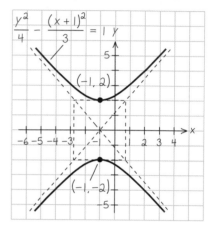

From this equation you can conclude that the hyperbola is centered at $(-1, 0)$ and has vertices at $(-1, 2)$ and $(-1, -2)$, and that the ends of the conjugate axis occur at $\left(-1 - \sqrt{3}, 0\right)$ and $\left(-1 + \sqrt{3}, 0\right)$. To sketch the hyperbola, draw a rectangle through these four points. The asymptotes are the lines passing through the corners of the rectangle, as shown in Figure 6.27. Finally, using $a = 2$ and $b = \sqrt{3}$, you can conclude that the equations of the asymptotes are

$$y = \frac{2}{\sqrt{3}}(x + 1) \qquad \text{and} \qquad y = -\frac{2}{\sqrt{3}}(x + 1).$$

Figure 6.27

If the constant term F in the equation in Example 3 had been $F = 4$ instead of 16, you would have obtained the following degenerate case.

Two Intersecting Lines: $\dfrac{y^2}{4} - \dfrac{(x + 1)^2}{3} = 0$

STUDY T!P

You can use a graphing utility to graph a hyperbola by graphing the upper and lower portions in the same viewing window. For instance, to graph the hyperbola in Example 3, first solve for y to get

$$y_1 = 2\sqrt{1 + \frac{(x + 1)^2}{3}}$$

and

$$y_2 = -2\sqrt{1 + \frac{(x + 1)^2}{3}}.$$

Use a viewing window in which $-8 \le x \le 6$ and $-6 \le y \le 6$. You should obtain the graph shown in Figure 6.28.

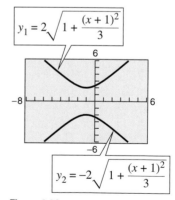

Figure 6.28

EXAMPLE 4   Using Asymptotes to Find the Standard Equation

Find the standard form of the equation of the hyperbola having vertices at $(3, -5)$ and $(3, 1)$ and with asymptotes

$$y = 2x - 8 \quad \text{and} \quad y = -2x + 4$$

as shown in Figure 6.29.

Solution

By the Midpoint Formula, the center of the hyperbola is at $(3, -2)$. Furthermore, the hyperbola has a vertical transverse axis with $a = 3$. From the given equations, you can determine the slopes of the asymptotes to be

$$m_1 = 2 = \frac{a}{b}$$

and

$$m_2 = -2 = -\frac{a}{b}$$

and because $a = 3$, you can conclude that $b = \frac{3}{2}$. So, the standard equation is

$$\frac{(y + 2)^2}{3^2} - \frac{(x - 3)^2}{(3/2)^2} = 1.$$

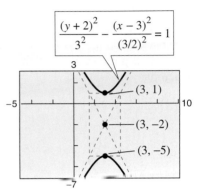

Figure 6.29

As with ellipses, the *eccentricity* of a hyperbola is

$$e = \frac{c}{a} \qquad \text{Eccentricity}$$

and because $c > a$ it follows that $e > 1$. If the eccentricity is large, the branches of the hyperbola are nearly flat. If the eccentricity is close to 1, the branches of the hyperbola are more pointed. See Figure 6.30.

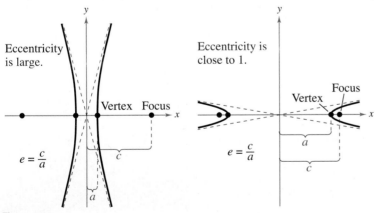

Figure 6.30

Applications

The following application was developed during World War II. It shows how the properties of hyperbolas can be used in radar and other detection systems.

EXAMPLE 5   An Application Involving Hyperbolas

Two microphones, 1 mile apart, record an explosion. Microphone A received the sound 2 seconds before microphone B. Where was the explosion?

Solution

Assuming sound travels at 1100 feet per second, you know that the explosion took place 2200 feet further from B than from A, as shown in Figure 6.31. The locus of all points that are 2200 feet closer to A than to B is one branch of the hyperbola

$$\frac{x^2}{a^2} + \frac{y^2}{b^2} = 1$$

where

$$c = \frac{5280}{2} = 2640$$

and

$$a = \frac{2200}{2} = 1100.$$

So, $b^2 = c^2 - a^2 = 5{,}759{,}600$, and you can conclude that the explosion occurred somewhere on the right branch of the hyperbola

$$\frac{x^2}{1{,}210{,}000} - \frac{y^2}{5{,}759{,}600} = 1.$$

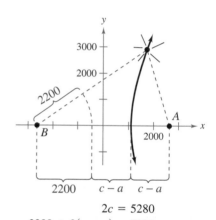

$$2c = 5280$$
$$2200 + 2(c - a) = 5280$$

Figure 6.31

Another interesting application of conic sections involves the orbits of comets in our solar system. Of the 610 comets identified prior to 1970, 245 have elliptical orbits, 295 have parabolic orbits, and 70 have hyperbolic orbits. The center of the sun is a focus of each of these orbits, and each orbit has a vertex at the point where the comet is closest to the sun, as shown in Figure 6.32. Undoubtedly, there have been many comets with parabolic or hyperbolic orbits that have not been identified. You get to see such comets only *once*. Comets with elliptical orbits, such as Halley's comet, are the only ones that remain in our solar system.

If p is the distance between the vertex and the focus in meters, and v is the velocity of the comet at the vertex in meters per second, the type of orbit is determined as follows.

1. Ellipse: $v < \sqrt{2GM/p}$
2. Parabola: $v = \sqrt{2GM/p}$
3. Hyperbola: $v > \sqrt{2GM/p}$

In each of these equations, $M \approx 1.991 \times 10^{30}$ kilograms (the mass of the sun) and $G \approx 6.67 \times 10^{-11}$ cubic meters per kilogram-second squared (the universal gravitational constant).

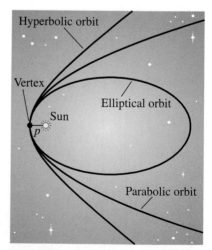

Figure 6.32

Classifying a Conic from Its General Equation

The graph of $Ax^2 + Cy^2 + Dx + Ey + F = 0$ is one of the following.

1. Circle: $A = C$ $A \neq 0$

2. Parabola: $AC = 0$ $A = 0$ or $C = 0$, but not both.

3. Ellipse: $AC > 0$ A and C have like signs.

4. Hyperbola: $AC < 0$ A and C have unlike signs.

The test above is valid *if* the graph is a conic. The test does not apply to equations such as $x^2 + y^2 = -1$, which is not a conic.

EXAMPLE 6   Classifying Conics from General Equations

Classify each graph.

a. $4x^2 - 9x + y - 5 = 0$

b. $4x^2 - y^2 + 8x - 6y + 4 = 0$

c. $2x^2 + 4y^2 - 4x + 12y = 0$

d. $2x^2 + 2y^2 - 8x + 12y + 2 = 0$

Solution

a. For the equation $4x^2 - 9x + y - 5 = 0$, you have

$$AC = 4(0) = 0. \qquad \text{Parabola}$$

So, the graph is a parabola.

b. For the equation $4x^2 - y^2 + 8x - 6y + 4 = 0$, you have

$$AC = 4(-1) < 0. \qquad \text{Hyperbola}$$

So, the graph is a hyperbola.

c. For the equation $2x^2 + 4y^2 - 4x + 12y = 0$, you have

$$AC = 2(4) > 0. \qquad \text{Ellipse}$$

So, the graph is an ellipse.

d. For the equation $2x^2 + 2y^2 - 8x + 12y + 2 = 0$, you have

$$A = C = 2. \qquad \text{Circle}$$

So, the graph is a circle.

The first woman to be credited with detecting a new comet was the English astronomer **Caroline Herschel (1750–1848).** During her long life, Caroline Herschel discovered a total of eight new comets.

The Granger Collection

Writing About Math   *Identifying Equations of Conics*

Use the Internet to research information about the orbits of comets in our solar system. What can you find about the orbits of comets that have been identified since 1970? Write a summary of your results. Identify your source. Does it seem reliable?

6.3 Exercises

In Exercises 1–4, match the equation with its graph. [The graphs are labeled (a), (b), (c), and (d).]

(a)

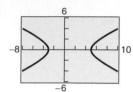

(b)

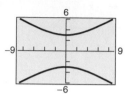

(c)

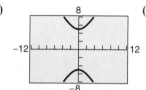

(d)

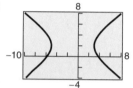

1. $\dfrac{y^2}{9} - \dfrac{x^2}{25} = 1$ **2.** $\dfrac{y^2}{25} - \dfrac{x^2}{9} = 1$

3. $\dfrac{(x-1)^2}{16} - \dfrac{y^2}{4} = 1$ **4.** $\dfrac{(x+1)^2}{16} - \dfrac{(y-2)^2}{9} = 1$

In Exercises 5–18, find the center, vertices, foci, and asymptotes of the hyperbola, and sketch its graph, using the asymptotes as an aid. Use a graphing utility to verify your graph.

5. $x^2 - y^2 = 1$ **6.** $\dfrac{x^2}{9} - \dfrac{y^2}{25} = 1$

7. $\dfrac{y^2}{1} - \dfrac{x^2}{4} = 1$ **8.** $\dfrac{y^2}{9} - \dfrac{x^2}{1} = 1$

9. $\dfrac{y^2}{25} - \dfrac{x^2}{81} = 1$ **10.** $\dfrac{x^2}{36} - \dfrac{y^2}{4} = 1$

11. $\dfrac{(x-1)^2}{4} - \dfrac{(y+2)^2}{1} = 1$

12. $\dfrac{(x+3)^2}{144} - \dfrac{(y-2)^2}{25} = 1$

13. $(y+6)^2 - (x-2)^2 = 1$

14. $\dfrac{(y-1)^2}{1/4} - \dfrac{(x+3)^2}{1/16} = 1$

15. $9x^2 - y^2 - 36x - 6y + 18 = 0$

16. $x^2 - 9y^2 + 36y - 72 = 0$

17. $x^2 - 9y^2 + 2x - 54y - 80 = 0$

18. $16y^2 - x^2 + 2x + 64y + 63 = 0$

In Exercises 19–22, find the center, vertices, foci, and the equations of the asymptotes of the hyperbola. Use a graphing utility to graph the hyperbola and its asymptotes.

19. $2x^2 - 3y^2 = 6$

20. $6y^2 - 3x^2 = 18$

21. $9y^2 - x^2 + 2x + 54y + 62 = 0$

22. $9x^2 - y^2 + 54x + 10y + 55 = 0$

In Exercises 23–40, find the standard form of the equation of the specified hyperbola.

23. Vertices: $(0, \pm 2)$; Foci: $(0, \pm 4)$

24. Vertices: $(\pm 2, 0)$; Foci: $(\pm 5, 0)$

25. Vertices: $(\pm 1, 0)$; Asymptotes: $y = \pm 5x$

26. Vertices: $(0, \pm 3)$; Asymptotes: $y = \pm 3x$

27. Foci: $(0, \pm 8)$; Asymptotes: $y = \pm 4x$

28. Foci: $(\pm 10, 0)$; Asymptotes: $y = \pm \frac{3}{4}x$

29. Vertices: $(2, 0)$, $(6, 0)$; Foci: $(0, 0)$, $(8, 0)$

30. Vertices: $(2, 3)$, $(2, -3)$; Foci: $(2, 5)$, $(2, -5)$

31. Vertices: $(4, 1)$, $(4, 9)$; Foci: $(4, 0)$, $(4, 10)$

32. Vertices: $(-2, 1)$, $(2, 1)$; Foci: $(-3, 1)$, $(3, 1)$

33. Vertices: $(2, 3)$, $(2, -3)$;
Passes through the point $(0, 5)$

34. Vertices: $(-2, 1)$, $(2, 1)$;
Passes through the point $(5, 4)$

35. Vertices: $(0, 4)$, $(0, 0)$;
Passes through the point $\left(\sqrt{5}, 5\right)$

36. Vertices: $(1, 2)$, $(1, -2)$;
Passes through the point $\left(0, \sqrt{5}\right)$

37. Vertices: $(1, 2)$, $(3, 2)$;
Asymptotes: $y = x$, $y = 4 - x$

38. Vertices: $(3, 0)$, $(3, -6)$;
Asymptotes: $y = x - 6$, $y = -x$

39. Vertices: $(0, 2)$, $(6, 2)$;
Asymptotes: $y = \frac{2}{3}x$, $y = 4 - \frac{2}{3}x$

40. Vertices: $(3, 0)$, $(3, 4)$;
Asymptotes: $y = \frac{2}{3}x$, $y = 4 - \frac{2}{3}x$

41. *Sound Location* Three listening stations located at $(3300, 0)$, $(3300, 1100)$, and $(-3300, 0)$ monitor an explosion. If the last two stations detect the explosion 1 second and 4 seconds after the first, respectively, determine the coordinates of the explosion. (Assume that the coordinate system is measured in feet and that sound travels at 1100 feet per second.)

42. *Navigation* Long distance radio navigation for aircraft and ships uses synchronized pulses transmitted by widely separated transmitting stations. These pulses travel at the speed of light (186,000 miles per second). The difference in the times of arrival of these pulses at an aircraft or ship is constant on a hyperbola having the transmitting stations as foci. Assume that two stations, 300 miles apart, are positioned on the rectangular coordinate system at coordinates $(-150, 0)$ and $(150, 0)$, and that a ship is traveling on a path with coordinates $(x, 75)$. Find the x-coordinate of the position of the ship if the time difference between the pulses from the transmitting stations is 1000 microseconds (0.001 second).

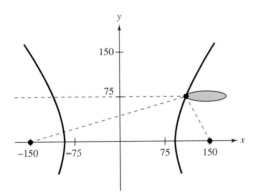

43. *Hyperbolic Mirror* A hyperbolic mirror (used in some telescopes) has the property that a light ray directed at a focus will be reflected to the other focus. The focus of a hyperbolic mirror has coordinates $(24, 0)$. Find the vertex of the mirror if its mount has coordinates $(24, 24)$.

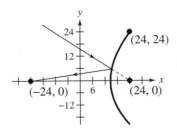

In Exercises 44–51, classify the graph of the equation as a circle, a parabola, an ellipse, or a hyperbola.

44. $x^2 + y^2 - 6x + 4y + 9 = 0$

45. $x^2 + 4y^2 - 6x + 16y + 21 = 0$

46. $4x^2 - y^2 - 4x - 3 = 0$

47. $y^2 - 4y - 4x = 0$

48. $4x^2 + 3y^2 + 8x - 24y + 51 = 0$

49. $4y^2 - 2x^2 - 4y - 8x - 15 = 0$

50. $25x^2 - 10x - 200y - 119 = 0$

51. $4x^2 + 4y^2 - 16y + 15 = 0$

Synthesis

True or False? **In Exercises 52 and 53, determine whether the statement is true or false. Justify your answer.**

52. In the standard form of the equation of a hyperbola, the larger the ratio of b to a, the larger the eccentricity of the hyperbola.

53. In the standard form of the equation of a hyperbola, the trivial solution of two intersecting lines occurs when $b = 0$.

54. *Think About It* Consider a hyperbola centered at the origin with a horizontal transverse axis. Use the definition of a hyperbola to derive its standard form.

55. *Think About It* Explain how the central rectangle of a hyperbola can be used to sketch its asymptotes.

Review

In Exercises 56–59, solve the equation.

56. $x^2 - 10x = 0$

57. $x^3 + x = 0$

58. $100 - (x - 5)^2 = 0$

59. $16x^2 - 40x + 25 = 0$

In Exercises 60–63, evaluate the function at each specified value of the independent variable and simplify.

60. $f(x) = 5x - 8$

    (a) $f(9)$ (b) $f(-4)$ (c) $f(x - 7)$

61. $f(x) = x^2 - 10x$

    (a) $f(4)$ (b) $f(-8)$ (c) $f(x - 4)$

62. $f(x) = \sqrt{x - 12} - 9$

    (a) $f(12)$ (b) $f(40)$ (c) $f\left(-\sqrt{36}\right)$

63. $f(x) = x^4 - x - 5$

    (a) $f(-1)$ (b) $f\left(\frac{1}{2}\right)$ (c) $f\left(2\sqrt{3}\right)$

6.4 Rotation and Systems of Quadratic Equations

Rotation

In the previous section you learned that the equation of a conic with axes parallel to the coordinate axes has a standard form that can be written in the general form

$$Ax^2 + Cy^2 + Dx + Ey + F = 0.$$ Horizontal or vertical axes

In this section you will study the equations of conics whose axes are rotated so that they are not parallel to either the x-axis or the y-axis. The general equation for such conics contains an xy-term.

$$Ax^2 + Bxy + Cy^2 + Dx + Ey + F = 0$$ Equation in xy-plane

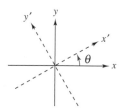

Figure 6.33

To eliminate this xy-term, you can use a procedure called rotation of axes. The objective is to rotate the x- and y-axes until they are parallel to the axes of the conic. The rotated axes are denoted as the x'-axis and the y'-axis, as shown in Figure 6.33. After the rotation, the equation of the conic in the new $x'y'$-plane will have the form

$$A'(x')^2 + C'(y')^2 + D'x' + E'y' + F' = 0.$$ Equation in $x'y'$-plane

Because this equation has no xy-term, you can obtain a standard form by completing the square. The following theorem identifies how much to rotate the axes to eliminate the xy-term and also the equations for determining the new coefficients A', C', D', E', and F'. For a proof of this theorem, see Appendix A.

Rotation of Axes to Eliminate an xy-Term

The general second-degree equation $Ax^2 + Bxy + Cy^2 + Dx + Ey + F = 0$ can be rewritten as

$$A'(x')^2 + C'(y')^2 + D'x' + E'y' + F' = 0$$

by rotating the coordinate axes through an angle θ, where $\cot 2\theta = \dfrac{A - C}{B}$.

The coefficients of the new equation are obtained by making the substitutions

$$x = x' \cos \theta - y' \sin \theta \quad \text{and} \quad y = x' \sin \theta + y' \cos \theta.$$

EXAMPLE 1   Rotation of Axes for a Hyperbola

Rotate the axes to eliminate the xy-term in the equations $xy - 1 = 0$. Then write the equation in standard form.

Solution

Because $A = 0$, $B = 1$, and $C = 0$, you have

$$\cot 2\theta = \frac{A - C}{B} = 0 \quad \Longrightarrow \quad 2\theta = \frac{\pi}{2} \quad \Longrightarrow \quad \theta = \frac{\pi}{4}$$

which implies that

$$x = x' \cos \frac{\pi}{4} - y' \sin \frac{\pi}{4}$$

$$= x'\left(\frac{\sqrt{2}}{2}\right) - y'\left(\frac{\sqrt{2}}{2}\right)$$

$$= \frac{x' - y'}{\sqrt{2}}$$

and

$$y = x' \sin \frac{\pi}{4} + y' \cos \frac{\pi}{4}$$

$$= x'\left(\frac{\sqrt{2}}{2}\right) + y'\left(\frac{\sqrt{2}}{2}\right)$$

$$= \frac{x' + y'}{\sqrt{2}}.$$

The equation in the $x'y'$-system is obtained by substituting these expressions into the equation $xy - 1 = 0$.

$$\left(\frac{x' - y'}{\sqrt{2}}\right)\left(\frac{x' + y'}{\sqrt{2}}\right) - 1 = 0$$

$$\frac{(x')^2 - (y')^2}{2} - 1 = 0$$

$$\frac{(x')^2}{\left(\sqrt{2}\right)^2} - \frac{(y')^2}{\left(\sqrt{2}\right)^2} = 1 \qquad \text{Write in standard form.}$$

In the $x'y'$-system, this is a hyperbola centered at the origin with vertices at $\left(\pm\sqrt{2}, 0\right)$, as shown in Figure 6.34. To find the coordinates of the vertices in the xy-system, substitute the coordinates $\left(\pm\sqrt{2}, 0\right)$ into the equations

$$x = \frac{x' - y'}{\sqrt{2}} \qquad \text{and} \qquad y = \frac{x' + y'}{\sqrt{2}}.$$

This substitution yields the vertices $(1, 1)$ and $(-1, -1)$ in the xy-system. Note also that the asymptotes of the hyperbola have equations $y' = \pm x'$, which correspond to the original x- and y-axes.

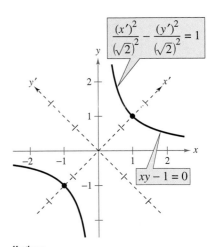

Vertices:
In x'y'-system: $\left(\sqrt{2}, 0\right), \left(-\sqrt{2}, 0\right)$
In xy-system: $(1, 1), (-1, -1)$
Figure 6.34

EXAMPLE 2   Rotation of Axes for an Ellipse

Rotate the axes to eliminate the xy-term in the equation

$$7x^2 - 6\sqrt{3}\,xy + 13y^2 - 16 = 0.$$

Then write the equation in standard form and sketch its graph.

Solution
Because $A = 7$, $B = -6\sqrt{3}$, and $C = 13$, you have

$$\cot 2\theta = \frac{A - C}{B}$$

$$= \frac{7 - 13}{-6\sqrt{3}}$$

$$= \frac{1}{\sqrt{3}}$$

which implies that $\theta = \pi/6$. The equation in the $x'y'$-system is obtained by making the substitutions

$$x = x' \cos \frac{\pi}{6} - y' \sin \frac{\pi}{6}$$

$$= x'\left(\frac{\sqrt{3}}{2}\right) - y'\left(\frac{1}{2}\right)$$

$$= \frac{\sqrt{3}x' - y'}{2}$$

and

$$y = x' \sin \frac{\pi}{6} + y' \cos \frac{\pi}{6}$$

$$= x'\left(\frac{1}{2}\right) + y'\left(\frac{\sqrt{3}}{2}\right)$$

$$= \frac{x' + \sqrt{3}y'}{2}$$

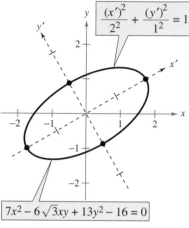

Figure 6.35

Vertices:
In $x'y'$-system: $(\pm2, 0)$, $(0, \pm1)$
In xy-system: $\left(\sqrt{3}, 1\right), \left(-\sqrt{3}, -1\right)$,
$\left(\frac{1}{2}, -\frac{\sqrt{3}}{2}\right), \left(-\frac{1}{2}, \frac{\sqrt{3}}{2}\right)$

into the original equation. So, you have

$$7x^2 - 6\sqrt{3}\,xy + 13y^2 - 16 = 0$$

$$7\left(\frac{\sqrt{3}x' - y'}{2}\right)^2 - 6\sqrt{3}\left(\frac{\sqrt{3}x' - y'}{2}\right)\left(\frac{x' + \sqrt{3}y'}{2}\right) + 13\left(\frac{x' + \sqrt{3}y'}{2}\right)^2 - 16 = 0$$

which simplifies to

$$4(x')^2 + 16(y')^2 - 16 = 0$$

$$4(x')^2 + 16(y')^2 = 16$$

$$\frac{(x')^2}{2^2} + \frac{(y')^2}{1^2} = 1. \qquad \text{Write in standard form.}$$

This is the equation of an ellipse centered at the origin with vertices $(\pm2, 0)$ in the $x'y'$-system, as shown in Figure 6.35.

EXAMPLE 3 Rotation of Axes for a Parabola

Rotate the axes to eliminate the xy-term in the equation

$$x^2 - 4xy + 4y^2 + 5\sqrt{5}y + 1 = 0.$$

Then write the equation in standard form and sketch its graph.

Solution

Because $A = 1$, $B = -4$, and $C = 4$, you have

$$\cot 2\theta = \frac{A - C}{B} = \frac{1 - 4}{-4} = \frac{3}{4}.$$

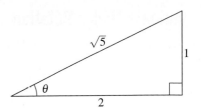

Figure 6.36

Using the identity $\cot 2\theta = (\cot^2 \theta - 1)/(2 \cot \theta)$ produces

$$\cot 2\theta = \frac{3}{4} = \frac{\cot^2 \theta - 1}{2 \cot \theta}$$

from which you obtain the equation

$$4 \cot^2 \theta - 4 = 6 \cot \theta$$

$$4 \cot^2 \theta - 6 \cot \theta - 4 = 0$$

$$(2 \cot \theta - 4)(2 \cot \theta + 1) = 0$$

Considering $0 < \theta < \pi/2$, you have $2 \cot \theta = 4$. So,

$$\cot \theta = 2 \quad \Longrightarrow \quad \theta \approx 26.6°.$$

From the triangle in Figure 6.36, you obtain $\sin \theta = 1/\sqrt{5}$ and $\cos \theta = 2/\sqrt{5}$. So, you use the substitutions

$$x = x'\cos \theta - y'\sin \theta = x'\left(\frac{2}{\sqrt{5}}\right) - y'\left(\frac{1}{\sqrt{5}}\right) = \frac{2x' - y'}{\sqrt{5}}$$

$$y = x'\sin \theta + y'\cos \theta = x'\left(\frac{1}{\sqrt{5}}\right) + y'\left(\frac{2}{\sqrt{5}}\right) = \frac{x' + 2y'}{\sqrt{5}}.$$

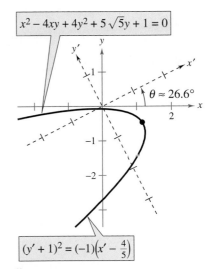

$$(y' + 1)^2 = (-1)\left(x' - \frac{4}{5}\right)$$

Vertex:

In $x'y'$-system: $\left(\frac{4}{5}, -1\right)$

In xy-system: $\left(\frac{13}{5\sqrt{5}}, -\frac{6}{5\sqrt{5}}\right)$

Figure 6.37

Substituting these expressions into the original equation, you have

$$x^2 - 4xy + 4y^2 + 5\sqrt{5}y + 1 = 0$$

$$\left(\frac{2x' - y'}{\sqrt{5}}\right)^2 - 4\left(\frac{2x' - y'}{\sqrt{5}}\right)\left(\frac{x' + 2y'}{\sqrt{5}}\right) + 4\left(\frac{x' + 2y'}{\sqrt{5}}\right)^2 + 5\sqrt{5}\left(\frac{x' + 2y'}{\sqrt{5}}\right) + 1 = 0$$

which simplifies as follows.

$$5(y')^2 + 5x' + 10y' + 1 = 0$$

$$5[(y')^2 + 2y'] = -5x' - 1 \qquad \text{Group terms.}$$

$$5[(y')^2 + 2y' + 1] = -5x' - 1 + 5 \qquad \text{Complete the square.}$$

$$5(y' + 1)^2 = -5x' + 4 \qquad \text{Write in completed square form.}$$

$$(y' + 1)^2 = (-1)\left(x' - \frac{4}{5}\right) \qquad \text{Write in standard form.}$$

The graph of this equation is a parabola with vertex at $\left(\frac{4}{5}, -1\right)$. Its axis is parallel to the x'-axis in the $x'y'$-system, as shown in Figure 6.37.

Invariants Under Rotation

In the rotation of axes theorem listed at the beginning of this section, note that the constant term is the same in both equations—that is, $F' = F$. Such quantities are **invariant under rotation.** The next theorem lists some other rotation invariants.

Rotation Invariants

The rotation of the coordinate axes through an angle θ that transforms the equation $Ax^2 + Bxy + Cy^2 + Dx + Ey + F = 0$ into the form

$$A'(x')^2 + C'(y')^2 + D'x' + E'y' + F' = 0$$

has the following rotation invariants.

1. $F = F'$
2. $A + C = A' + C'$
3. $B^2 - 4AC = (B')^2 - 4A'C'$

You can use the results of this theorem to classify the graph of a second-degree equation *with* an xy-term in much the same way you do for a second-degree equation *without* an xy-term. Note that because $B' = 0$, the invariant $B^2 - 4AC$ reduces to

$$B^2 - 4AC = -4A'C'. \qquad \text{Discriminant}$$

This quantity is called the **discriminant** of the equation

$$Ax^2 + Bxy + Cy^2 + Dx + Ey + F = 0.$$

Now, from the classification procedure given in Section 6.3, you know that the sign of $A'C'$ determines the type of graph for the equation

$$A'(x')^2 + C'(y')^2 + D'x' + E'y' + F' = 0.$$

Consequently, the sign of $B^2 - 4AC$ will determine the type of graph for the original equation, as given in the following classification.

Classification of Conics by the Discriminant

The graph of the equation $Ax^2 + Bxy + Cy^2 + Dx + Ey + F = 0$ is, except in degenerate cases, determined by its discriminant as follows.

1. Ellipse or circle: $B^2 - 4AC < 0$
2. Parabola: $B^2 - 4AC = 0$
3. Hyperbola: $B^2 - 4AC > 0$

For example, in the general equation

$$3x^2 + 7xy + 5y^2 - 6x - 7y + 15 = 0$$

you have $A = 3$, $B = 7$, and $C = 5$. So, the discriminant is

$$B^2 - 4AC = 7^2 - 4(3)(5) = 49 - 60 = -11.$$

Because $-11 < 0$, the graph of the equation is an ellipse or a circle.

EXAMPLE 4   Rotations and Graphing Utilities

For each of the following, classify the graph, use the quadratic formula to solve for y, and then use a graphing utility to graph the equation.

a. $2x^2 - 3xy + 2y^2 - 2x = 0$

b. $x^2 - 6xy + 9y^2 - 2y + 1 = 0$

c. $3x^2 + 8xy + 4y^2 - 7 = 0$

Solution

a. Because $B^2 - 4AC = 9 - 16 < 0$, the graph is a circle or an ellipse. Solve for y as follows.

$$2x^2 - 3xy + 2y^2 - 2x = 0 \qquad \text{Write original equation.}$$

$$2y^2 - 3xy + (2x^2 - 2x) = 0 \qquad \text{Quadratic form } ay^2 + by + c = 0$$

$$y = \frac{-(-3x) \pm \sqrt{(-3x)^2 - 4(2)(2x^2 - 2x)}}{2(2)}$$

$$y = \frac{3x \pm \sqrt{x(16 - 7x)}}{4}$$

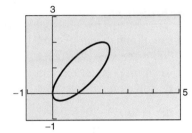

Figure 6.38

Graph both of the equations to obtain the ellipse in Figure 6.38.

$$y = \frac{3x + \sqrt{x(16 - 7x)}}{4} \qquad \text{Top half of ellipse}$$

$$y = \frac{3x - \sqrt{x(16 - 7x)}}{4} \qquad \text{Bottom half of ellipse}$$

b. Because $B^2 - 4AC = 36 - 36 = 0$, the graph is a parabola.

$$x^2 - 6xy + 9y^2 - 2y + 1 = 0 \qquad \text{Write original equation.}$$

$$9y^2 - (6x + 2)y + (x^2 + 1) = 0 \qquad \text{Quadratic form } ay^2 + by + c = 0$$

$$y = \frac{(6x + 2) \pm \sqrt{(6x + 2)^2 - 4(9)(x^2 + 1)}}{18}$$

$$y = \frac{3x + 1 \pm \sqrt{2(3x - 4)}}{9}$$

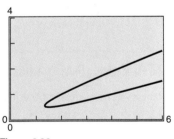

Figure 6.39

Graphing the resulting two equations gives the parabola in Figure 6.39.

c. Because $B^2 - 4AC = 64 - 48 > 0$, the graph is a hyperbola.

$$3x^2 + 8xy + 4y^2 - 7 = 0 \qquad \text{Write original equation.}$$

$$4y^2 + 8xy + (3x^2 - 7) = 0 \qquad \text{Quadratic form } ay^2 + by + c = 0$$

$$y = \frac{-8x \pm \sqrt{(8x)^2 - 4(4)(3x^2 - 7)}}{8}$$

$$y = \frac{-2x \pm \sqrt{x^2 + 7}}{2}$$

The graph of the resulting two equations yields the hyperbola in Figure 6.40.

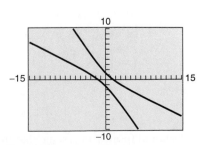

Figure 6.40

Systems of Quadratic Equations

To find the points of intersection of two conics, you can use elimination or substitution, as demonstrated in Examples 5 and 6.

EXAMPLE 5 Solving a Quadratic System

Solve the system of quadratic equations.

$$\begin{cases} x^2 + y^2 - 16x + 39 = 0 & \text{Equation 1} \\ x^2 - y^2 - 9 = 0 & \text{Equation 2} \end{cases}$$

Algebraic Solution

You can eliminate the y^2-term
by adding the two equations.
The resulting equation can
then be solved for x.

$$2x^2 - 16x + 30 = 0$$

$$2(x - 3)(x - 5) = 0$$

There are two real solutions:
$x = 3$ and $x = 5$. The corresponding y-values are $y = 0$
and $y = \pm 4$. So, the graphs
have three points of intersection:

$$(3, 0), (5, 4), \text{ and } (5, -4).$$

Graphical Solution

Begin by solving each equation for y as follows.

$$y = \pm\sqrt{-x^2 + 16x - 39} \qquad y = \pm\sqrt{x^2 - 9}$$

Use a graphing utility to graph all four equations $y_1 = \sqrt{-x^2 + 16x - 39}$, $y_2 = -\sqrt{-x^2 + 16x - 39}$, $y_3 = \sqrt{x^2 - 9}$, and $y_4 = -\sqrt{x^2 - 9}$ in the same viewing window. In Figure 6.41, you can see that the graphs appear to intersect at the points $(3, 0)$, $(5, 4)$, and $(5, -4)$. Use the *intersect* feature to confirm this.

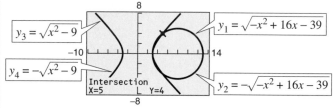

Figure 6.41

EXAMPLE 6 Solving a Quadratic System by Substitution

Solve the system of quadratic equations.

$$\begin{cases} x^2 + 4y^2 - 4x - 8y + 4 = 0 & \text{Equation 1} \\ x^2 + 4y - 4 = 0 & \text{Equation 2} \end{cases}$$

Solution

Because Equation 2 has no y^2-term, solve the equation for y to obtain
$y = 1 - (1/4)x^2$. Next, substitute this into Equation 1 and solve for x.

$$x^2 + 4\left(1 - \frac{1}{4}x^2\right)^2 - 4x - 8\left(1 - \frac{1}{4}x^2\right) + 4 = 0$$

$$x^2 + 4 - 2x^2 + \frac{1}{4}x^4 - 4x - 8 + 2x^2 + 4 = 0$$

$$x^4 + 4x^2 - 16x = 0$$

$$x(x - 2)(x^2 + 2x + 8) = 0$$

In factored form, you can see that the equation has two real solutions: $x = 0$ and
$x = 2$. The corresponding values of y are $y = 1$ and $y = 0$. This implies that the
solutions of the system of equations are $(0, 1)$ and $(2, 0)$, as shown in Figure 6.42.

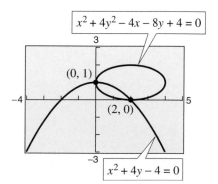

Figure 6.42

6.4   E x e r c i s e s

In Exercises 1–4, the $x'y'$-coordinate system has been rotated θ degrees from the xy-coordinate system. The coordinates of a point on the xy-coordinate system are given. Find the coordinates of the point on the rotated coordinate system.

1. $\theta = 90°, (0, 4)$

2. $\theta = 45°, (3, 3)$

3. $\theta = 30°, (1, 6)$

4. $\theta = 60°, (5, 1)$

In Exercises 5–16, rotate the axes to eliminate the xy-term in the equation. Then write the equation in standard form. Sketch the graph of the equation, showing both sets of axes.

5. $xy + 1 = 0$

6. $xy - 2 = 0$

7. $x^2 - 8xy + y^2 + 1 = 0$

8. $xy + x - 2y + 3 = 0$

9. $xy - 2y - 4x = 0$

10. $13x^2 + 6\sqrt{3}xy + 7y^2 - 16 = 0$

11. $5x^2 - 6xy + 5y^2 - 12 = 0$

12. $2x^2 - 3xy - 2y^2 + 10 = 0$

13. $3x^2 - 2\sqrt{3}xy + y^2 + 2x + 2\sqrt{3}y = 0$

14. $16x^2 - 24xy + 9y^2 - 60x - 80y + 100 = 0$

15. $9x^2 + 24xy + 16y^2 + 90x - 130y = 0$

16. $9x^2 + 24xy + 16y^2 + 80x - 60y = 0$

In Exercises 17–22, use a graphing utility to graph the conic. Determine the angle θ through which the axes are rotated. Explain how you used the graphing utility to obtain the graph.

17. $x^2 + xy + y^2 = 12$

18. $x^2 - 4xy + 2y^2 = 10$

19. $17x^2 + 32xy - 7y^2 = 75$

20. $40x^2 + 36xy + 25y^2 = 52$

21. $32x^2 + 50xy + 7y^2 = 52$

22. $4x^2 - 12xy + 9y^2 + \left(4\sqrt{13} - 12\right)x$
$- \left(6\sqrt{13} + 8\right)y = 91$

In Exercises 23–28, match the graph with its equation. [The graphs are labeled (a), (b), (c), (d), (e), and (f).]

(a)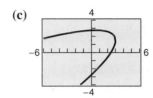

(b)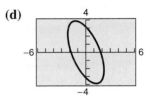

(c)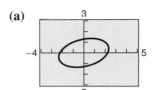

(d)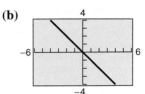

(e)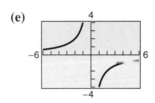

(f)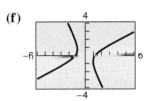

23. $xy + 4 = 0$

24. $x^2 + 2xy + y^2 = 0$

25. $-2x^2 + 3xy + 2y^2 + 3 = 0$

26. $x^2 - xy + 3y^2 - 5 = 0$

27. $3x^2 + 2xy + y^2 - 10 = 0$

28. $x^2 - 4xy + 4y^2 + 10x - 30 = 0$

In Exercises 29–36, (a) use the discriminant to classify the graph, (b) use the quadratic formula to solve for y, and (c) use a graphing utility to graph the equation.

29. $16x^2 - 24xy + 9y^2 - 30x - 40y = 0$

30. $x^2 - 8xy - 2y^2 - 6 = 0$

31. $15x^2 - 8xy + 7y^2 - 45 = 0$

32. $2x^2 + 4xy + 5y^2 + 3x - 4y - 20 = 0$

33. $x^2 - 6xy - 5y^2 + 4x - 22 = 0$

34. $36x^2 - 60xy + 25y^2 + 9y = 0$

35. $x^2 + 4xy + 4y^2 - 5x - y - 3 = 0$

36. $x^2 + xy + 4y^2 + x + y - 4 = 0$

In Exercises 37–40, sketch (if possible) the graph of the degenerate conic.

37. $y^2 - 16x^2 = 0$

38. $x^2 + y^2 - 2x + 6y + 10 = 0$

39. $x^2 + 2xy + y^2 - 4 = 0$

40. $x^2 - 10xy + y^2 = 0$

In Exercises 41–48, solve the system of quadratic equations algebraically by the method of elimination. Then verify your results by using a graphing utility to graph the equations and find any points of intersection of the graphs.

41. $\begin{cases} -x^2 + y^2 + 4x - 6y + 4 = 0 \\ x^2 + y^2 - 4x - 6y + 12 = 0 \end{cases}$

42. $\begin{cases} -x^2 - y^2 - 8x + 20y - 7 = 0 \\ x^2 + 9y^2 + 8x + 4y + 7 = 0 \end{cases}$

43. $\begin{cases} -4x^2 - y^2 - 16x + 24y - 16 = 0 \\ 4x^2 + y^2 + 40x - 24y + 208 = 0 \end{cases}$

44. $\begin{cases} x^2 - 4y^2 - 20x - 64y - 172 = 0 \\ 16x^2 + 4y^2 - 320x + 64y + 1600 = 0 \end{cases}$

45. $\begin{cases} x^2 - y^2 - 12x + 16y - 64 = 0 \\ x^2 + y^2 - 12x - 16y + 64 = 0 \end{cases}$

46. $\begin{cases} x^2 + 4y^2 - 2x - 8y + 1 = 0 \\ -x^2 + 2x - 4y - 1 = 0 \end{cases}$

47. $\begin{cases} -16x^2 - y^2 + 24y - 80 = 0 \\ 16x^2 + 25y^2 - 400 = 0 \end{cases}$

48. $\begin{cases} 16x^2 - y^2 + 16y - 128 = 0 \\ y^2 - 48x - 16y - 32 = 0 \end{cases}$

In Exercises 49–54, solve the system of quadratic equations algebraically by the method of substitution. Then verify your results by using a graphing utility to graph the equations and find any points of intersection of the graphs.

49. $\begin{cases} x^2 + y^2 - 4 = 0 \\ 3x - y^2 = 0 \end{cases}$

50. $\begin{cases} 4x^2 + 9y^2 - 36y = 0 \\ x^2 + 9y - 27 = 0 \end{cases}$

51. $\begin{cases} x^2 + 2y^2 - 4x + 6y - 5 = 0 \\ -x + y - 4 = 0 \end{cases}$

52. $\begin{cases} x^2 + 2y^2 - 4x + 6y - 5 = 0 \\ x^2 - 4x - y + 4 = 0 \end{cases}$

53. $\begin{cases} xy + x - 2y + 3 = 0 \\ x^2 + 4y^2 - 9 = 0 \end{cases}$

54. $\begin{cases} 5x^2 - 2xy + 5y^2 - 12 = 0 \\ x + y - 1 = 0 \end{cases}$

Synthesis

True or False?   In Exercises 55 and 56, determine whether the statement is true or false. Justify your answer.

55. The graph of $x^2 + xy + ky^2 + 6x + 10 = 0$, where k is any constant less than $\frac{1}{4}$, is a hyperbola.

56. After using a rotation of axes to eliminate the xy-term from an equation of the form

$$Ax^2 + Bxy + Cy^2 + Dx + Ey + F = 0$$

the coefficients of the x^2- and y^2-terms remain A and B, respectively.

57. Show that the equation $x^2 + y^2 = r^2$ is invariant under rotation of axes.

58. Find the lengths of the major and minor axes of the ellipse in Exercise 10.

Review

In Exercises 59–62, sketch the graph of the function. Identify any intercepts and asymptotes.

59. $y = 2^{x-1}$

60. $y = 3^{x+2}$

61. $y = \ln(x - 1)$

62. $y = \ln(x + 1)$

In Exercises 63–66, solve the equation. Round your answer to three decimal places.

63. $4^{3-x} = 726$

64. $\dfrac{4500}{4 + e^{2x}} = 50$

65. $\ln x = -6$

66. $\ln\sqrt{x + 10} = 1$

6.5 Parametric Equations

Plane Curves

Up to this point, you have been representing a graph by a single equation involving *two* variables such as x and y. In this section, you will study situations in which it is useful to introduce a *third* variable to represent a curve in the plane.

To see the usefulness of this procedure, consider the path followed by an object that is propelled into the air at an angle of 45°. If the initial velocity of the object is 48 feet per second, it can be shown that the object follows the parabolic path

$$y = -\frac{x^2}{72} + x \qquad \text{Rectangular equation}$$

as shown in Figure 6.43. However, this equation does not tell the whole story. Although it does tell us *where* the object has been, it doesn't tell us *when* the object was at a given point (x, y) on the path. To determine this time, you can introduce a third variable t, which is called a **parameter.** It is possible to write both x and y as functions of t to obtain the **parametric equations**

$$x = 24\sqrt{2}\,t \qquad \text{Parametric equation for } x$$

$$y = -16t^2 + 24\sqrt{2}\,t. \qquad \text{Parametric equation for } y$$

From this set of equations you can determine that at time $t = 0$, the object is at the point $(0, 0)$. Similarly, at time $t = 1$, the object is at the point $(24\sqrt{2}, 24\sqrt{2} - 16)$, and so on.

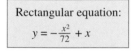

Rectangular equation:
$$y = -\frac{x^2}{72} + x$$

Parametric equations:
$$x = 24\sqrt{2}\,t$$
$$y = -16t^2 + 24\sqrt{2}\,t$$

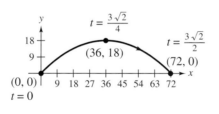

Curvilinear motion: two variables for position, one variable for time
Figure 6.43

For this particular motion problem, x and y are continuous functions of t, and the resulting path is a **plane curve.** (Recall that a *continuous function* is one whose graph can be traced without lifting the pencil from the paper.)

Definition of a Plane Curve

If f and g are continuous functions of t on an interval I, the set of ordered pairs $(f(t), g(t))$ is a **plane curve** C. The equations

$$x = f(t) \qquad \text{and} \qquad y = g(t)$$

are **parametric equations** for C, and t is the **parameter.**

What You Should Learn:

- How to evaluate sets of parametric equations for given values of the parameter
- How to graph curves that are represented by sets of parametric equations
- How to rewrite sets of parametric equations as single rectangular equations by eliminating the parameter
- How to find sets of parametric equations for graphs

Why You Should Learn It:

Parametric equations are useful for modeling the path of an object. For instance, in Exercise 59 on page 458, a set of parametric equations is used to model the path of a baseball.

Jonathan Daniel/Allsport

A computer animation of this concept appears in the *Interactive* CD-ROM and *Internet* versions of this text.

Sketching a Plane Curve

One way to sketch a curve represented by a pair of parametric equations is to plot points in the *xy*-plane. Each set of coordinates (x, y) is determined from a value chosen for the parameter *t*. By plotting the resulting points in the order of *increasing* values of *t*, you trace the curve in a specific direction. This is called the **orientation** of the curve.

EXAMPLE 1   Sketching a Plane Curve

Sketch the curve given by the parametric equations

$$x = t^2 - 4 \quad \text{and} \quad y = \frac{t}{2}, \quad -2 \le t \le 3.$$

Describe the orientation of the curve.

Solution

Using values of *t* in the given interval, the parametric equations yield the points (x, y) shown in the table.

t	-2	-1	0	1	2	3
x	0	-3	-4	-3	0	5
y	-1	$-\frac{1}{2}$	0	$\frac{1}{2}$	1	$\frac{3}{2}$

By plotting these points in the order of increasing *t*, you obtain the curve shown in Figure 6.44(a). The arrows on the curve indicate its orientation as *t* increases from -2 to 3. So, if a particle were moving on this curve, it would start at $(0, -1)$ and then move along the curve to the point $\left(5, \frac{3}{2}\right)$.

The graph shown in Figure 6.44(a) does not define *y* as a function of *x*. This points out one benefit of parametric equations—they can be used to represent graphs that are more general than graphs of functions.

Two different sets of parametric equations can have the same graph. For example, the set of parametric equations

$$x = 4t^2 - 4 \quad \text{and} \quad y = t, \quad -1 \le t \le \frac{3}{2}$$

has the same graph as the set given in Example 1. [See Figure 6.44(b).] However, by comparing the values of *t* in Figures 6.44(a) and (b), you can see that this second graph is traced out more *rapidly* (considering *t* as time) than the first graph. So, in applications, different parametric representations can be used to represent various *speeds* at which objects travel along a given path.

Another way to display a curve represented by a pair of parametric equations is to use a graphing utility, as shown in Example 2.

Library of Functions

Parametric equations consist of a pair of functions $x = f(t)$ and $y = g(t)$, each of which is a function of the parameter *t*. These equations define a plane curve, which might not be the graph of a function, as in Example 1. Most graphing utilities have a *parametric* mode.

A computer animation of this example appears in the *Interactive* CD-ROM and *Internet* versions of this text.

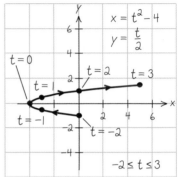

(a)

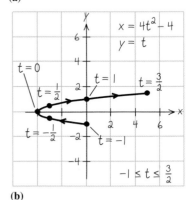

(b)

Figure 6.44

EXAMPLE 2   Using a Graphing Utility in Parametric Mode

Use a graphing utility to graph the curves represented by the parametric equations. For which curve is y a function of x? (Use $-4 \leq t \leq 4$.)

a. $x = t^2$ **b.** $x = t$ **c.** $x = t^2$
 $y = t^3$ $y = t^3$ $y = t$

Solution

Begin by setting the graphing utility to *parametric* mode. When choosing a viewing window, you must set not only minimum and maximum values of x and y but also minimum and maximum values of t.

a. Enter the parametric equations for x and y.

  $X_{1T} = T^2$, $Y_{1T} = T^3$

The curve is shown in Figure 6.45(a). From the graph, you can see that y *is not* a function of x.

b. Enter the parametric equations for x and y.

  $X_{1T} = T$, $Y_{1T} = T^3$

The curve is shown in Figure 6.45(b). From the graph, you can see that y *is a* function of x.

c. Enter the parametric equations for x and y.

  $X_{1T} = T^2$, $Y_{1T} = T$

The curve is shown in Figure 6.45(c). From the graph, you can see that y *is not* a function of x.

The *Interactive* CD-ROM and *Internet* versions of this text offer a built-in graphing calculator, which can be used with the Examples, Explorations, and Exercises.

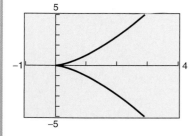

(a)

(b)

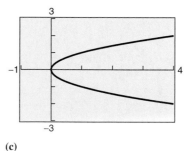

(c)

Figure 6.45

Exploration

Use a graphing utility set in *parametric* mode to graph the curve

  $X_{1T} = T$ and $Y_{1T} = 1 - T^2$.

Set the viewing window so that $-4 \leq x \leq 4$ and $-12 \leq y \leq 2$. Now, graph the curve with various settings for t. Use the following.

a. $0 \leq t \leq 3$ **b.** $-3 \leq t \leq 0$ **c.** $-3 \leq t \leq 3$

Compare the curves given by the different t settings. Repeat this experiment using $X_{1T} = -T$. How does this change the results?

Eliminating the Parameter

Many curves that are represented by sets of parametric equations have graphs that can also be represented by rectangular equations (in x and y). The process of finding the rectangular equation is called **eliminating the parameter.**

Parametric equations	$\Rightarrow$	Solve for t in one equation.	$\Rightarrow$	Substitute in second equation.	$\Rightarrow$	Rectangular equation

$$x = t^2 - 4 \qquad\qquad t = 2y \qquad\qquad x = (2y)^2 - 4 \qquad x = 4y^2 - 4$$

$$y = \tfrac{1}{2}t$$

After eliminating the parameter, you can recognize that the curve is a parabola with a horizontal axis and vertex at $(-4, 0)$.

Converting equations from parametric to rectangular form can change the ranges of x and y. In such cases, you should restrict x and y in the rectangular equation so that its graph matches the graph of the parametric equations.

EXAMPLE 3 Eliminating the Parameter

Identify the curve represented by the equations

$$x = \frac{1}{\sqrt{t + 1}} \quad \text{and} \quad y = \frac{t}{t + 1}.$$

Solution

Solving for t in the equation for x produces

$$x^2 = \frac{1}{t + 1} \quad \text{or} \quad \frac{1}{x^2} = t + 1$$

which implies that $t = (1/x^2) - 1$. Substituting in the equation for y, you obtain

$$y = \frac{t}{t + 1}$$

$$= \frac{\left(\dfrac{1}{x^2}\right) - 1}{\left(\dfrac{1}{x^2}\right) - 1 + 1}$$

$$= \frac{\dfrac{1 - x^2}{x^2}}{\left(\dfrac{1}{x^2}\right)} \cdot \frac{x^2}{x^2}$$

$$= 1 - x^2.$$

From the rectangular equation, you can recognize the curve to be a parabola that opens downward and has its vertex at $(0, 1)$, as shown in Figure 6.46(a). The rectangular equation is defined for all values of x. The parametric equation for x, however, is defined only when $t > -1$. From the graph of the parametric equation, you can see that x is always positive, as shown in Figure 6.46(b). So, you should restrict the domain of x to positive values, as shown in Figure 6.46(c).

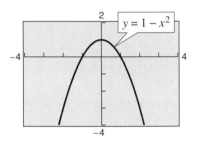

(a)

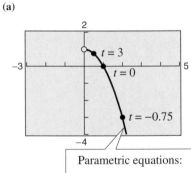

Parametric equations:

$$x = \frac{1}{\sqrt{t + 1}}, \, y = \frac{t}{t + 1}$$

(b)

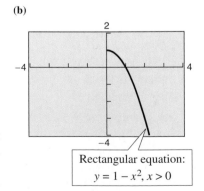

Rectangular equation:

$$y = 1 - x^2, \, x > 0$$

(c)

Figure 6.46

It is not necessary for the parameter in a set of parametric equations to represent time. The next example uses an *angle* as the parameter.

EXAMPLE 4   Eliminating the Parameter

Sketch the curve represented by $x = 3 \cos \theta$ and $y = 4 \sin \theta$, $0 \le \theta \le 2\pi$, by eliminating the parameter.

Solution
Begin by solving for $\cos \theta$ and $\sin \theta$ in the given equations.

$$\cos \theta = \frac{x}{3} \quad \text{and} \quad \sin \theta = \frac{y}{4} \qquad \text{Solve for } \cos \theta \text{ and } \sin \theta.$$

Make use of the identity $\sin^2 \theta + \cos^2 \theta = 1$ to form an equation involving only x and y.

$$\cos^2 \theta + \sin^2 \theta = 1 \qquad \text{Pythagorean identity}$$

$$\left(\frac{x}{3}\right)^2 + \left(\frac{y}{4}\right)^2 = 1 \qquad \text{Substitute.}$$

$$\frac{x^2}{9} + \frac{y^2}{16} = 1 \qquad \text{Rectangular equation}$$

From this rectangular equation, you can see that the graph is an ellipse centered at $(0, 0)$, with vertices at $(0, 4)$ and $(0, -4)$, and minor axis of length $2b = 6$, as shown in Figure 6.47. Note that the elliptic curve is traced out *counterclockwise* as θ varies from 0 to 2π.

Finding Parametric Equations for a Graph

How can you determine a set of parametric equations for a given graph or a given physical description? From the discussion following Example 1, you know that such a representation is not unique. This is further demonstrated in Example 5.

EXAMPLE 5   Finding Parametric Equations for a Given Graph

Find a set of parametric equations to represent the graph of $y = 1 - x^2$ using the following parameters. **a.** $t = x$ **b.** $t = 1 - x$

Solution
a. Letting $t = x$, you obtain the parametric equations $x = t$ and $y = 1 - t^2$.

The graph of these equations is shown in Figure 6.48(a).

b. Letting $t = 1 - x$, you obtain the following parametric equations.

$$x = 1 - t \qquad \text{Parametric equation for } x$$

$$y = 1 - (1 - t)^2 \qquad \text{Substitute } 1 - t \text{ for } x.$$

$$= 2t - t^2 \qquad \text{Parametric equation for } y$$

The graph of these equations is shown in Figure 6.48(b). In this figure, note how the resulting curve is oriented by the increasing values of t. In Figure 6.48(a), the curve has the opposite orientation.

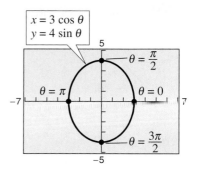

Exploration

In Example 4 you made use of the trigonometric identity $\sin^2 \theta + \cos^2 \theta = 1$ to sketch an ellipse. Which trigonometric identity would you use to obtain the graph of a hyperbola? Sketch the curve represented by $x = 3 \sec \theta$ and $y = 4 \tan \theta$, $0 \le \theta \le 2\pi$, by eliminating the parameter.

Figure 6.47

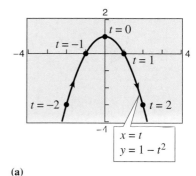

(a)

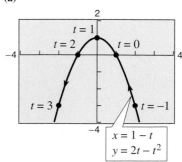

(b)

Figure 6.48

In Exercises 1–8, match the equation with its graph.
[The graphs are labeled (a), (b), (c), (d), (e), (f), (g),
and (h).]

(a)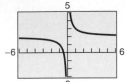

(b)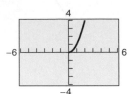

(c)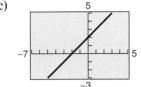

(d)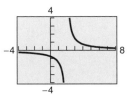

(e)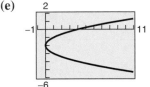

(f)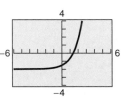

(g)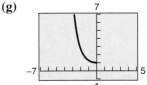

(h)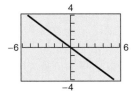

1. $x = t$
$y = t + 2$

2. $x = t$
$y = -\frac{3}{4}t$

3. $x = \sqrt{t}$
$y = t$

4. $x = t^2$
$y = t - 2$

5. $x = \dfrac{1}{t}$

$y = t + 2$

6. $x = \dfrac{1}{2}t$

$y = \dfrac{3}{t - 4}$

7. $x = \ln t$
$y = \frac{1}{2}t - 2$

8. $x = -2\sqrt{t}$
$y = e^t$

9. Consider the parametric equations

$x = \sqrt{t}$ and $y = 2 - t$.

(a) Complete the table.

t	0	1	2	3	4
x					
y					

(b) Plot the points (x, y) generated in part (a) and sketch a graph of the parametric equations.

(c) Use a graphing utility to graph the curve represented by the parametric equations.

(d) Find the rectangular equation by eliminating the parameter. Sketch its graph. How do the graphs differ from those in parts (b) and (c)?

10. Consider the parametric equations

$x = 4 \cos^2 \theta$ and $y = 2 \sin \theta$.

(a) Complete the table.

θ	$-\dfrac{\pi}{2}$	$-\dfrac{\pi}{4}$	0	$\dfrac{\pi}{4}$	$\dfrac{\pi}{2}$
x					
y					

(b) Plot the points (x, y) generated in part (a) and sketch a graph of the parametric equations.

(c) Use a graphing utility to graph the curve represented by the parametric equations.

(d) Find the rectangular equation by eliminating the parameter. Sketch its graph. How do the graphs differ from those in parts (b) and (c)?

In Exercises 11–26, sketch the curve represented by
the parametric equations (indicate the direction of the
curve). Use a graphing utility to confirm your result.
Then eliminate the parameter and write the corre-
sponding rectangular equation whose graph repre-
sents the curve.

11. $x = t$
$y = -4t$

12. $x = t$
$y = \frac{1}{2}t$

13. $x = 3t + 1$
$y = 2t - 1$

14. $x = 3 - 2t$
$y = 2 + 3t$

15. $x = \frac{1}{4}t$
$y = t^2$

16. $x = t$
$y = t^3$

17. $x = t + 5$
$y = t^2$

18. $x = \sqrt{t}$
$y = 1 - t$

19. $x = 2t$
$y = |t - 2|$

20. $x = |t - 1|$
$y = t + 2$

21. $x = 3 \cos \theta$
$y = 3 \sin \theta$

22. $x = \cos \theta$
$y = 3 \sin \theta$

23. $x = e^{-t}$
$y = e^{3t}$

24. $x = e^{2t}$
$y = e^t$

25. $x = t^3$
$y = 3 \ln t$

26. $x = \ln 2t$
$y = 2t^2$

In Exercises 27–34, use a graphing utility to graph the curve represented by the parametric equations.

27. $x = 4 \sin 2\theta$
$y = 2 \cos 2\theta$

28. $x = \cos \theta$
$y = 2 \sin 2\theta$

29. $x = 4 + 2 \cos \theta$
$y = -1 + \sin \theta$

30. $x = 4 + 2 \cos \theta$
$y = -1 + 2 \sin \theta$

31. $x = 4 \sec \theta$
$y = 3 \tan \theta$

32. $x = \sec \theta$
$y = \tan \theta$

33. $x = t/2$
$y = \ln(t^2 + 1)$

34. $x = 10 - 0.01e^t$
$y = 0.4t^2$

In Exercises 35 and 36, determine how the plane curves differ from each other.

35. (a) $x = t$
$y = 2t + 1$

(b) $x = \cos \theta$
$y = 2 \cos \theta + 1$

(c) $x = e^{-t}$
$y = 2e^{-t} + 1$

(d) $x = e^t$
$y = 2e^t + 1$

36. (a) $x = t$
$y = t$

(b) $x = t^2$
$y = t^2$

(c) $x = -t$
$y = -t$

(d) $x = t^3$
$y = t^3$

In Exercises 37–40, eliminate the parameter and obtain the standard form of the rectangular equation.

37. Line through (x_1, y_1) and (x_2, y_2):
$x = x_1 + t(x_2 - x_1)$
$y = y_1 + t(y_2 - y_1)$

38. Circle: $x = h + r \cos \theta$
$y = k + r \sin \theta$

39. Ellipse: $x = h + a \cos \theta$
$y = k + b \sin \theta$

40. Hyperbola: $x = h + a \sec \theta$
$y = k + b \tan \theta$

In Exercises 41–46, use the results of Exercises 37–40 to find a set of parametric equations for the line or conic.

41. Line: Passes through $(0, 0)$ and $(5, -2)$

42. Line: Passes through $(1, 4)$ and $(5, -2)$

43. Circle: Center: $(2, 1)$; Radius: 4

44. Circle: Center: $(-3, 1)$; Radius: 3

45. Ellipse: Vertices: $(\pm 5, 0)$; Foci: $(\pm 4, 0)$

46. Hyperbola: Vertices: $(0, \pm 1)$; Foci: $(0, \pm 2)$

In Exercises 47 and 48, find two different sets of parametric equations for the given rectangular equation.

47. $y = 3x - 2$

48. $y = x^2$

In Exercises 49–54, use a graphing utility to obtain a graph of the curve represented by the parametric equations.

49. Cycloid: $x = 2(\theta - \sin \theta)$
$y = 2(1 - \cos \theta)$

50. Prolate cycloid: $x = 2\theta - 4 \sin \theta$
$y = 2 - 4 \cos \theta$

51. Hypocycloid: $x = 3 \cos^3 \theta$
$y = 3 \sin^3 \theta$

52. Curtate cycloid: $x = 2\theta - \sin \theta$
$y = 2 - \cos \theta$

53. Witch of Agnesi: $x = 2 \cot \theta$
$y = 2 \sin^2 \theta$

54. Folium of Descartes: $x = \dfrac{3t}{1 + t^3}$

$y = \dfrac{3t^2}{1 + t^3}$

In Exercises 55–58, match the parametric equations with the correct graph. [The graphs are labeled (a), (b), (c), and (d).]

(a)

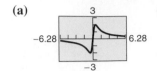

(b)

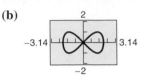

(c)

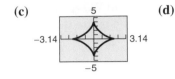

(d)

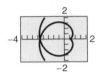

55. Lissajous curve: $x = 2 \cos \theta, y = \sin 2\theta$

56. Evolute of ellipse: $x = 2 \cos^3 \theta, y = 4 \sin^3 \theta$

57. Involute of circle: $x = \frac{1}{2}(\cos \theta + \theta \sin \theta)$

$$y = \frac{1}{2}(\sin \theta - \theta \cos \theta)$$

58. Serpentine curve: $x = \frac{1}{2} \cot \theta, y = 4 \sin \theta \cos \theta$

Projectile Motion **In Exercises 59 and 60, consider a projectile launched at a height h feet above the ground at an angle θ with the horizontal. If the initial velocity is v_0 feet per second, the path of the projectile is modeled by the parametric equations**

$$x = (v_0 \cos \theta)t \quad \text{and} \quad y = h + (v_0 \sin \theta)t - 16t^2.$$

59. *Baseball* The center-field fence in a ballpark is 10 feet high and 400 feet from home plate. The baseball is hit 3 feet above the ground. It leaves the bat at an angle of θ degrees with the horizontal at a speed of 100 miles per hour.

(a) Write a set of parametric equations for the path of the baseball.

(b) Use a graphing utility to sketch the path of the baseball for $\theta = 15°$. Is the hit a home run?

(c) Use a graphing utility to sketch the path of the baseball for $\theta = 23°$. Is the hit a home run?

(d) Find the minimum angle required for the hit to be a home run.

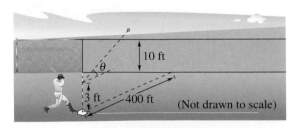

(Not drawn to scale)

60. *Football* The quarterback of a football team releases a pass at a height of 7 feet above the playing field, and the football is caught by a receiver at a height of 4 feet, 30 yards directly downfield. The pass is released at an angle of 35° with the horizontal.

(a) Write a set of parametric equations for the path of the football.

(b) Find the speed of the football when it is released.

(c) Use a graphing utility to graph the path of the football and approximate its maximum height.

(d) Find the time the receiver has to position himself after the quarterback releases the football.

Synthesis

True or False? **In Exercises 61 and 62, determine whether the statement is true or false. Justify your answer.**

61. The two sets of parametric equations $x = t$, $y = t^2 + 1$ and $x = 3t$, $y = 9t^2 + 1$ correspond to the same rectangular equation.

62. The graph of the parametric equations $x = t^2$ and $y = t^2$ is the line $y = x$.

63. *Think About It* The graph of the parametric equations $x = t^3$ and $y = t - 1$ is shown below. Would the graph change for the equations $x = (-t^3)$ and $y = -t - 1$? If so, how would it change?

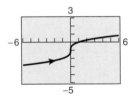

Review

In Exercises 64–67, find all solutions of the equation.

64. $5x^2 + 8 = 0$

65. $x^2 - 6x + 4 = 0$

66. $4x^2 + 4x - 11 = 0$

67. $x^4 - 18x^2 + 18 = 0$

In Exercises 68–71, solve graphically the system of equations.

68. $\begin{cases} 5x - 7y = 11 \\ -3x + y = -13 \end{cases}$

69. $\begin{cases} 3x + 5y = 9 \\ 4x - 2y = -14 \end{cases}$

70. $\begin{cases} 3x - 2y = 8 \\ 2x + y = -3 \\ x - 3y = 16 \end{cases}$

71. $\begin{cases} 5x + 7y = 4 \\ x - 2y = 7 \\ 8x - 2y = 20 \end{cases}$

6.6 Polar Coordinates

Introduction

So far, you have been representing graphs of equations as collections of points (x, y) on the rectangular coordinate system, where x and y represent the directed distances from the coordinate axes to the point (x, y). In this section, you will study a second system called the **polar coordinate system.**

To form the polar coordinate system in the plane, fix a point O, called the **pole** (or **origin**), and construct from O an initial ray called the **polar axis,** as shown in Figure 6.49. Then each point P in the plane can be assigned **polar coordinates** (r, θ) as follows.

1. $r = directed\ distance$ from O to P
2. $\theta = directed\ angle$, counterclockwise from polar axis to segment $\overline{OP}$

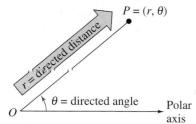

Figure 6.49

What You Should Learn:

- How to plot points and find multiple representations of points in the polar coordinate system
- How to convert points from rectangular to polar form and vice versa
- How to convert equations from rectangular to polar form and vice versa

Why You Should Learn It:

Polar coordinates offer a different mathematical perspective on graphing. For instance, in Exercises 5–12 on page 463, you see that a polar coordinate can be written in more than one way.

A computer animation of this concept appears in the *Interactive* CD-ROM and *Internet* versions of this text.

EXAMPLE 1   Plotting Points in the Polar Coordinate System

a. The point $(r, \theta) = (2, \pi/3)$ lies two units from the pole on the terminal side of the angle $\theta = \pi/3$, as shown in Figure 6.50(a).

b. The point $(r, \theta) = (3, -\pi/6)$ lies three units from the pole on the terminal side of the angle $\theta = -\pi/6$, as shown in Figure 6.50(b).

c. The point $(r, \theta) = (3, 11\pi/6)$ coincides with the point $(3, -\pi/6)$, as shown in Figure 6.50(c).

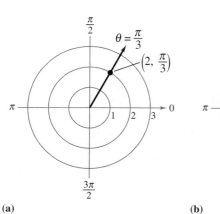

(a)

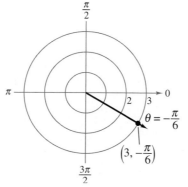

(b)

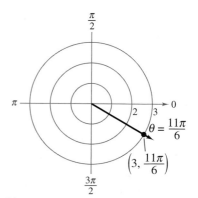

(c)

Figure 6.50

In rectangular coordinates, each point (x, y) has a unique representation. This is not true for polar coordinates. For instance, the coordinates (r, θ) and $(r, 2\pi + \theta)$ represent the same point, as illustrated in Example 1. Another way to obtain multiple representations of a point is to use negative values for r. Because r is a *directed distance*, the coordinates (r, θ) and $(-r, \theta + \pi)$ represent the same point. In general, the point (r, θ) can be represented as

$$(r, \theta) = (r, \theta \pm 2n\pi) \qquad \text{or} \qquad (r, \theta) = (-r, \theta \pm (2n + 1)\pi)$$

where n is any integer. Moreover, the pole is represented by $(0, \theta)$, where θ is any angle.

EXAMPLE 2 Multiple Representation of Points

Plot the point $(3, -3\pi/4)$ and find three additional polar representations of this point, using $-2\pi < \theta < 2\pi$.

Solution
The point is shown in Figure 6.51. Three other representations are as follows.

$$\left(3, -\frac{3\pi}{4} + 2\pi\right) = \left(3, \frac{5\pi}{4}\right) \qquad \text{Add } 2\pi \text{ to } \theta.$$

$$\left(-3, -\frac{3\pi}{4} - \pi\right) = \left(-3, -\frac{7\pi}{4}\right) \qquad \text{Replace } r \text{ by } -r; \text{ subtract } \pi \text{ from } \theta.$$

$$\left(-3, -\frac{3\pi}{4} + \pi\right) = \left(-3, \frac{\pi}{4}\right) \qquad \text{Replace } r \text{ by } -r; \text{ add } \pi \text{ to } \theta.$$

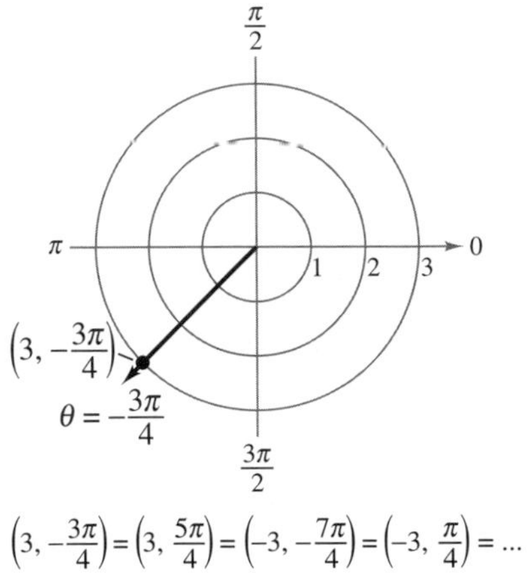

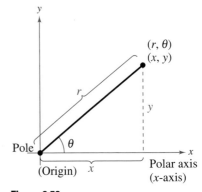

$$\left(3, -\tfrac{3\pi}{4}\right) = \left(3, \tfrac{5\pi}{4}\right) = \left(-3, -\tfrac{7\pi}{4}\right) = \left(-3, \tfrac{\pi}{4}\right) = \cdots$$

Figure 6.51

Coordinate Conversion

To establish the relationship between polar and rectangular coordinates, let the polar axis coincide with the positive x-axis and the pole with the origin, as shown in Figure 6.52. Because (x, y) lies on a circle of radius r, it follows that $r^2 = x^2 + y^2$. Moreover, for $r > 0$, the definitions of the trigonometric functions imply that

$$\tan \theta = \frac{y}{x}, \qquad \cos \theta = \frac{x}{r}, \qquad \text{and} \qquad \sin \theta = \frac{y}{r}.$$

You can show that the same relationships hold for $r < 0$.

Figure 6.52

Coordinate Conversion

The polar coordinates (r, θ) are related to the rectangular coordinates (x, y) as follows.

$$x = r \cos \theta \qquad \text{and} \qquad \tan \theta = \frac{y}{x}$$

$$y = r \sin \theta \qquad \qquad r^2 = x^2 + y^2$$

EXAMPLE 3   Polar-to-Rectangular Conversion

Convert each point to rectangular coordinates. (See Figure 6.53.)

a. $(2, \pi)$ **b.** $\left(\sqrt{3}, \dfrac{\pi}{6}\right)$

Solution

a. For the point $(r, \theta) = (2, \pi)$, you have

$$x = r \cos \theta = 2 \cos \pi = -2$$

and

$$y = r \sin \theta = 2 \sin \pi = 0.$$

The rectangular coordinates are $(x, y) = (-2, 0)$.

b. For the point $(r, \theta) = \left(\sqrt{3}, \pi/6\right)$, you have

$$x = \sqrt{3} \cos \frac{\pi}{6} = \sqrt{3}\left(\frac{\sqrt{3}}{2}\right) = \frac{3}{2}$$

and

$$y = \sqrt{3} \sin \frac{\pi}{6} = \sqrt{3}\left(\frac{1}{2}\right) = \frac{\sqrt{3}}{2}.$$

The rectangular coordinates are $(x, y) = \left(3/2, \sqrt{3}/2\right)$.

EXAMPLE 4   Rectangular-to-Polar Conversion

Convert each point to polar coordinates.

a. $(-1, 1)$ **b.** $(0, 2)$

Solution

a. For the second-quadrant point $(x, y) = (-1, 1)$, you have

$$\tan \theta = \frac{y}{x} = \frac{1}{-1} = -1$$

$$\theta = \frac{3\pi}{4}.$$

Because θ lies in the same quadrant as (x, y), use positive r.

$$r = \sqrt{x^2 + y^2} = \sqrt{(-1)^2 + (1)^2} = \sqrt{2}$$

So, *one* set of polar coordinates is $(r, \theta) = \left(\sqrt{2}, 3\pi/4\right)$, as shown in Figure 6.54(a).

b. Because the point $(x, y) = (0, 2)$ lies on the positive y-axis, choose

$$\theta = \pi/2 \quad \text{and} \quad r = 2.$$

This implies that one set of polar coordinates is $(r, \theta) = (2, \pi/2)$, as shown in Figure 6.54(b).

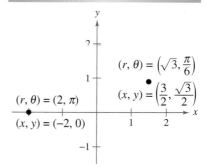

Exploration

Set your graphing utility to *polar* mode. Then graph the equation $r = 3$. (Use a viewing window of $0 \leq \theta \leq 2\pi$, $-6 \leq x \leq 6$, and $-4 \leq y \leq 4$.) You should obtain a circle of radius 3.

a. Use the *trace* feature to cursor around the circle. Can you locate the point $(3, 5\pi/4)$?

b. Can you locate other names for the point $(3, 5\pi/4)$? If so, explain how you did it.

Figure 6.53

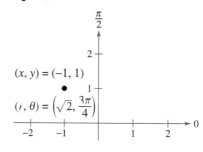

(a)

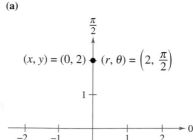

(b)
Figure 6.54

Equation Conversion

By comparing Examples 3 and 4, you see that point conversion from the polar to the rectangular system is straightforward, whereas point conversion from the rectangular to the polar system is more involved. For equations, the opposite is true. To convert a rectangular equation to polar form, you simply replace x by $r \cos \theta$ and y by $r \sin \theta$. For instance, the rectangular equation $y = x^2$ can be written in polar form as follows.

$$y = x^2 \qquad \text{Rectangular equation}$$

$$r \sin \theta = (r \cos \theta)^2 \qquad \text{Polar equation}$$

$$r = \sec \theta \tan \theta \qquad \text{Simplest form}$$

On the other hand, converting a polar equation to rectangular form requires considerable ingenuity.

Example 5 demonstrates several polar-to-rectangular conversions that enable you to sketch the graphs of some polar equations.

EXAMPLE 5   Converting Polar Equations to Rectangular Form

Describe the graph of each polar equation and find the corresponding rectangular equation.

a. $r = 2$ **b.** $\theta = \dfrac{\pi}{3}$ **c.** $r = \sec \theta$

Solution

a. The graph of the polar equation $r = 2$ consists of all points that are two units from the pole. In other words, this graph is a circle centered at the origin with a radius of 2, as shown in Figure 6.55(a). You can confirm this by converting to rectangular form, using the relationship $r^2 = x^2 + y^2$.

$$r = 2 \quad \blacktriangleright \quad r^2 = 2^2 \quad \blacktriangleright \quad x^2 + y^2 = 2^2$$

Polar equation   Rectangular equation

b. The graph of the polar equation $\theta = \pi/3$ consists of all points on the line that make an angle of $\pi/3$ with the positive x-axis, as shown in Figure 6.55(b). To convert to rectangular form, you make use of the relationship $\tan \theta = y/x$.

$$\theta = \frac{\pi}{3} \quad \blacktriangleright \quad \tan \theta = \sqrt{3} \quad \blacktriangleright \quad y = \sqrt{3}x$$

Polar equation   Rectangular equation

c. The graph of the polar equation $r = \sec \theta$ is not evident by simple inspection, so you convert to rectangular form by using the relationship $r \cos \theta = x$.

$$r = \sec \theta \quad \blacktriangleright \quad r \cos \theta = 1 \quad \blacktriangleright \quad x = 1$$

Polar equation   Rectangular equation

Now you see that the graph is a vertical line, as shown in Figure 6.55(c).

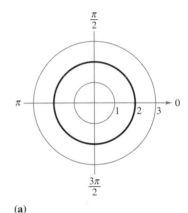

(a)

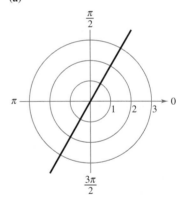

(b)

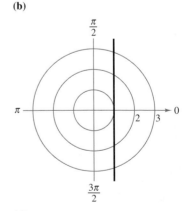

(c)

Figure 6.55

6.6 EXERCISES

In Exercises 1–4, a point in polar coordinates is given. Find the corresponding rectangular coordinates for the point.

1. $\left(4, \dfrac{\pi}{2}\right)$

2. $\left(4, \dfrac{3\pi}{2}\right)$

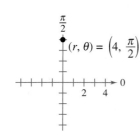

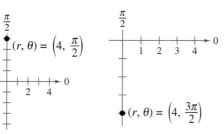

3. $\left(-1, \dfrac{5\pi}{4}\right)$

4. $\left(2, -\dfrac{\pi}{4}\right)$

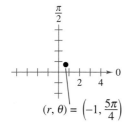

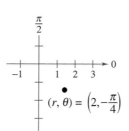

In Exercises 5–12, plot the point given in polar coordinates and find three additional polar representations of the point, using $-2\pi < \theta < 2\pi$.

5. $\left(4, \dfrac{2\pi}{3}\right)$

6. $\left(1, \dfrac{7\pi}{4}\right)$

7. $\left(5, -\dfrac{5\pi}{3}\right)$

8. $\left(-3, -\dfrac{7\pi}{6}\right)$

9. $\left(\sqrt{3}, \dfrac{5\pi}{6}\right)$

10. $\left(5\sqrt{2}, -\dfrac{11\pi}{6}\right)$

11. $\left(\dfrac{3}{2}, -\dfrac{3\pi}{2}\right)$

12. $\left(-\dfrac{7}{8}, -\dfrac{\pi}{6}\right)$

In Exercises 13–22, plot the point given in polar coordinates and find the corresponding rectangular coordinates for the point.

13. $\left(4, -\dfrac{\pi}{3}\right)$

14. $\left(2, \dfrac{7\pi}{6}\right)$

15. $\left(-1, -\dfrac{3\pi}{4}\right)$

16. $\left(-3, -\dfrac{2\pi}{3}\right)$

17. $\left(0, -\dfrac{7\pi}{6}\right)$

18. $\left(0, \dfrac{5\pi}{4}\right)$

19. $\left(32, \dfrac{5\pi}{2}\right)$

20. $\left(18, -\dfrac{3\pi}{2}\right)$

21. $\left(\sqrt{2}, 2.36\right)$

22. $(-3, -1.57)$

In Exercises 23–26, use a graphing utility to find the rectangular coordinates for the point given in polar coordinates.

23. $\left(2, \dfrac{3\pi}{4}\right)$

24. $\left(-2, \dfrac{7\pi}{6}\right)$

25. $(-4.5, 1.3)$

26. $(8.25, 3.5)$

In Exercises 27–36, the rectangular coordinates of a point are given. Plot the point and find *two* sets of polar coordinates for the point for $0 \le \theta < 2\pi$.

27. $(-7, 0)$

28. $(0, -5)$

29. $(1, 1)$

30. $(-3, -3)$

31. $(-3, 4)$

32. $(3, -1)$

33. $\left(-\sqrt{3}, -\sqrt{3}\right)$

34. $(2, -2)$

35. $(4, 6)$

36. $(5, 12)$

In Exercises 37–44, use a graphing utility to find one set of polar coordinates for the point given in rectangular coordinates.

37. $(3, -2)$

38. $(-4, 1)$

39. $\left(\sqrt{3}, 2\right)$

40. $\left(3\sqrt{2}, 3\sqrt{2}\right)$

41. $\left(\dfrac{5}{2}, \dfrac{4}{3}\right)$

42. $\left(\dfrac{11}{4}, -\dfrac{5}{8}\right)$

43. $(0, -5)$

44. $(-8, 0)$

In Exercises 45–54, convert the rectangular equation to polar form. Assume $a > 0$.

45. (a) $x^2 + y^2 = 49$ (b) $x^2 + y^2 = a^2$

46. (a) $x^2 + y^2 - 6x = 0$ (b) $x^2 + y^2 - 8y = 0$

47. (a) $x^2 + y^2 - 2ax = 0$ (b) $x^2 + y^2 - 2ay = 0$

48. (a) $y = 4$ (b) $y = b$

49. (a) $x = 12$ (b) $x = a$

50. (a) $3x - 6y + 2 = 0$ (b) $4x + 7y - 2 = 0$

51. (a) $xy = 4$ (b) $2xy = 1$

52. (a) $y = x$ (b) $y^2 = 2x$

53. (a) $y^2 = x^3$ (b) $x^2 = y^3$

54. (a) $(x^2 + y^2)^2 - 9(x^2 - y^2) = 0$

(b) $y^2 - 8x - 16 = 0$

In Exercises 55–70, convert the polar equation to rectangular form.

55. $r = 4 \sin \theta$ **56.** $r = 4 \cos \theta$

57. $\theta = \dfrac{\pi}{6}$ **58.** $\theta = \dfrac{5\pi}{3}$

59. $r = 4$ **60.** $r = 10$

61. $r = -3 \csc \theta$ **62.** $r = 2 \sec \theta$

63. $r^2 = \cos \theta$ **64.** $r^2 = \sin 2\theta$

65. $r = 2 \sin 3\theta$ **66.** $r = 3 \cos 2\theta$

67. $r = \dfrac{1}{1 - \cos \theta}$ **68.** $r = \dfrac{2}{1 + \sin \theta}$

69. $r = \dfrac{6}{2 - 3 \sin \theta}$ **70.** $r = \dfrac{6}{2 \cos \theta - 3 \sin \theta}$

In Exercises 71–76, describe the graph of the polar equation and find the corresponding rectangular equation. Sketch its graph.

71. $r = 3$ **72.** $r = 8$

73. $\theta = \dfrac{\pi}{4}$ **74.** $\theta = \dfrac{5\pi}{6}$

75. $r = 3 \sec \cdot \theta$ **76.** $r = 2 \csc \theta$

Synthesis

True or False? **In Exercises 77 and 78, determine whether the statement is true or false. Justify your answer.**

77. If (r_1, θ_1) and (r_2, θ_2) represent the same point in the polar coordinate system, then $|r_1| = |r_2|$.

78. If (r, θ_1) and (r, θ_2) represent the same point in the polar coordinate system, then $\theta_1 = \theta_2 + 2\pi n$ for some integer n.

79. *Think About It*

(a) Show that the distance between the points (r_1, θ_1) and (r_2, θ_2) is

$$\sqrt{r_1^2 + r_2^2 - 2r_1 r_2 \cos(\theta_1 - \theta_2)}.$$

(b) Describe the position of the points relative to

each other if $\theta_1 = \theta_2$. Simplify the distance formula for this case. Is the simplification what you expected? Explain.

(c) Simplify the distance formula if $\theta_1 - \theta_2 = 90°$. Is the simplification what you expected? Explain.

(d) Choose two points on the polar coordinate system and find the distance between them. Then choose different polar representations of the same two points and apply the distance formula again. Discuss the result.

80. *Exploration*

(a) Set the viewing window of your graphing utility to rectangular coordinates and locate the cursor at any position off the coordinate axes. Move the cursor horizontally and observe any changes in the displayed coordinates of the points. Explain the changes. Now repeat the process moving the cursor vertically.

(b) Set the viewing window of your graphing utility to polar coordinates and locate the cursor at any position off the coordinate axes. Move the cursor horizontally and observe any changes in the displayed coordinates of the points. Explain the changes. Now repeat the process moving the cursor vertically.

(c) Explain why the results of parts (a) and (b) are not the same.

Review

In Exercises 81–86, solve the equation algebraically. Round the result to three decimal places.

81. $e^x = 19$ **82.** $6e^x = 47$

83. $10^x = 84$ **84.** $5^{2x} = 60$

85. $\ln x = 4$ **86.** $4 \ln 4x = 18$

In Exercises 87–92, solve the trigonometric equation.

87. $4\sqrt{3} \tan \theta - 3 = 1$

88. $6 \cos x - 2 = 1$

89. $12 \sin^2 \theta = 9$

90. $9 \csc^2 x - 10 = 2$

91. $2 \cot x = 5 \cos \dfrac{\pi}{2}$

92. $\sqrt{2} \sec \theta = 2 \csc \dfrac{\pi}{4}$

6.7 Graphs of Polar Equations

Introduction

In previous chapters you spent a lot of time learning how to sketch graphs in rectangular coordinates. You began with the basic point-plotting method. Then you used sketching aids such as a graphing utility, symmetry, intercepts, asymptotes, periods, and shifts to further investigate the nature of the graph. This section approaches curve sketching in the polar coordinate system similarly.

EXAMPLE 1   Graphing a Polar Equation by Point Plotting

Sketch the graph of the polar equation $r = 4 \sin \theta$.

Solution

The sine function is periodic, so you can get a full range of r-values by considering values of θ in the interval $0 \le \theta \le 2\pi$, as shown in the table.

θ	0	$\dfrac{\pi}{6}$	$\dfrac{\pi}{3}$	$\dfrac{\pi}{2}$	$\dfrac{2\pi}{3}$	$\dfrac{5\pi}{6}$	π	$\dfrac{7\pi}{6}$	$\dfrac{3\pi}{2}$	$\dfrac{11\pi}{6}$	2π
r	0	2	$2\sqrt{3}$	4	$2\sqrt{3}$	2	0	-2	-4	-2	0

If you plot these points as shown in Figure 6.56, it appears that the graph is a circle of radius 2 whose center is at the point $(x, y) = (0, 2)$.

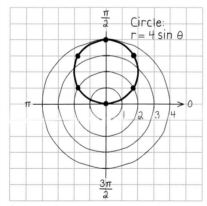

Figure 6.56

You can confirm the graph found in Example 1 in three ways.

1. *Convert to Rectangular Form*   Multiply both sides of the polar equation by r and convert the result to rectangular form.

2. *Use a Polar Coordinate Mode*   Set your graphing utility to *polar* mode and graph the polar equation. (Use $0 \le \theta \le 2\pi$, $-6 \le x \le 6$, and $-4 \le y \le 4$.)

3. *Use a Parametric Mode*   Set your graphing utility to *parametric* mode and graph $x = (4 \sin t) \cos t$ and $y = (4 \sin t) \sin t$.

What You Should Learn:

- How to graph polar equations by point plotting
- How to use symmetry, zeros, and maximum r-values as graphing aids
- How to recognize special polar graphs

Why You Should Learn It:

Several common figures, such as the circle in Exercise 3 on page 472, are easier to graph in the polar coordinate system than in the rectangular coordinate system.

Most graphing utilities have a *polar-coordinate* graphing mode. If yours doesn't, you can use the following parametric conversion to graph a polar equation.

Polar Equations in Parametric Form

The graph of the polar equation $r = f(\theta)$ can be written in parametric form, using t as a parameter, as follows.

$$x = f(t) \cos t \qquad \text{and} \qquad y = f(t) \sin t.$$

Symmetry

In Figure 6.56, note that as θ increases from 0 to 2π the graph is traced out twice. Moreover, note that the graph is *symmetric with respect to the line* $\theta = \pi/2$. Had you known about this symmetry and retracing ahead of time, you could have used fewer points.

Symmetry with respect to the line $\theta = \pi/2$ is one of three important types of symmetry to consider in polar curve sketching. (See Figure 6.57.)

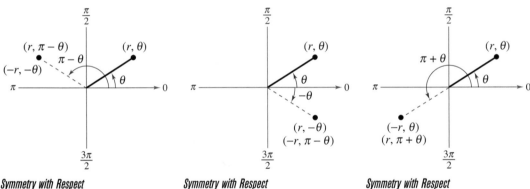

Symmetry with Respect to the Line $\theta = \pi/2$

Symmetry with Respect to the Polar Axis

Symmetry with Respect to the Pole

Figure 6.57

Tests for Symmetry on Polar Coordinates

The graph of a polar equation is symmetric with respect to the following if the given substitution yields an equivalent equation.

1. The line $\theta = \pi/2$:   Replace (r, θ) by $(r, \pi - \theta)$ or $(-r, -\theta)$.

2. The polar axis: Replace (r, θ) by $(r, -\theta)$ or $(-r, \pi - \theta)$.

3. The pole: Replace (r, θ) by $(r, \pi + \theta)$ or $(-r, \theta)$.

EXAMPLE 2 Using Symmetry to Sketch a Polar Graph

Use symmetry to sketch the graph of $r = 3 + 2 \cos \theta$.

Solution

Replacing (r, θ) by $(r, -\theta)$ produces

$$r = 3 + 2 \cos(-\theta)$$

$$= 3 + 2 \cos \theta.$$

So, you can conclude that the curve is symmetric with respect to the polar axis. Plotting the points in the table and using polar axis symmetry, you obtain the graph shown in Figure 6.58. This graph is called a **limaçon.**

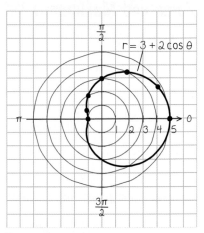

Figure 6.58

θ	0	$\dfrac{\pi}{6}$	$\dfrac{\pi}{3}$	$\dfrac{\pi}{2}$	$\dfrac{2\pi}{3}$	$\dfrac{5\pi}{6}$	π
r	5	$3 + \sqrt{3}$	4	3	2	$3 - \sqrt{3}$	1

Use a graphing utility to confirm this graph.

The three tests for symmetry in polar coordinates on page 466 are sufficient to guarantee symmetry, but they are not necessary. For instance, Figure 6.59 shows the graph of

$$r = \theta + 2\pi. \qquad \textit{Spiral of Archimedes}$$

From the figure, you can see that the graph is symmetric with respect to the line $\theta = \pi/2$. Yet the tests on page 466 fail to indicate symmetry because neither of the following replacements yields an equivalent equation.

Original Equation	*Replacement*	*New Equation*
$r = \theta + 2\pi$	(r, θ) by $(-r, -\theta)$	$-r = -\theta + 2\pi$
$r = \theta + 2\pi$	(r, θ) by $(r, \pi - \theta)$	$r = -\theta + 3\pi$

The equations discussed in Examples 1 and 2 are of the form

$$r = 4 \sin \theta = f(\sin \theta)$$

and

$$r = 3 + 2 \cos \theta = g(\cos \theta).$$

The graph of the first equation is symmetric with respect to the line $\theta = \pi/2$, and the graph of the second equation is symmetric with respect to the polar axis. This observation can be generalized to yield the following *quick test for symmetry.*

1. The graph of $r = f(\sin \theta)$ is symmetric with respect to the line $\theta = \pi/2$.
2. The graph of $r = g(\cos \theta)$ is symmetric with respect to the polar axis.

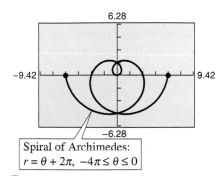

Spiral of Archimedes:
$r = \theta + 2\pi, \; -4\pi \le \theta \le 0$

Figure 6.59

Zeros and Maximum *r*-Values

Two additional aids to sketching graphs of polar equations involve knowing the θ-values for which $|r|$ is maximum and knowing the θ-values for which $r = 0$. In Example 1, the maximum value of $|r|$ for $r = 4 \sin \theta$ is $|r| = 4$, and this occurs when $\theta = \pi/2$ (see Figure 6.56). Moreover, $r = 0$ when $\theta = 0$.

EXAMPLE 3   Finding Maximum *r*-Values of a Polar Graph

Find the maximum value of r for the graph of $r = 1 - 2 \cos \theta$.

Graphical Solution

Because the polar equation is of the form

$$r = 1 - 2 \cos \theta = g(\cos \theta)$$

you know the graph is symmetric with respect to the polar axis. You can confirm this by graphing the polar equation, as shown in Figure 6.60. (In the graph, θ varies from 0 to 2π.) To find the maximum r-value for the graph, use your graphing utility's *trace* feature. When you do this, you should find that the graph has a maximum r-value of 3. This value of r occurs when $\theta = \pi$. In the graph, note that the point $(3, \pi)$ is farthest from the pole.

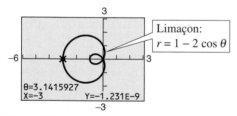

Figure 6.60

Numerical Solution

To approximate the maximum value of r for the graph of $r = 1 - 2 \cos \theta$, use the *table* feature of a graphing utility to create a table that begins at $\theta = 0$ and increments by $\pi/12$, as shown in Figure 6.61. From the table, the maximum value of r appears to be 3 when $\theta = 3.1416 \approx \pi$.

θ	r1
2.0944	2
2.3562	2.4142
2.618	2.7321
2.8798	2.9319
3.1416	**3**
3.4034	2.9319
3.6652	2.7321
θ=3.14159265359	

Figure 6.61

By creating a second table that begins at $\theta = \pi/2$ and increments by $\pi/24$, as shown in Figure 6.62, the maximum value of r still appears to be 3 when $\theta = 3.1416 \approx \pi$.

θ	r1
2.7489	2.8478
2.8798	2.9319
3.0107	2.9829
3.1416	**3**
3.2725	2.9829
3.4034	2.9319
3.5343	2.8478
θ=3.14159265359	

Figure 6.62

Note how the negative r-values determine the *inner loop* of the graph in Figure 6.60. This type of graph is a limaçon.

Exploration

The graph of the polar equation $r = e^{\cos \theta} - 2 \cos 4\theta + \sin^5(\theta/12)$ is called *the butterfly curve*, as shown in Figure 6.63.

a. The graph at the right was produced using $0 \le \theta \le 2\pi$. Does this show the entire graph? Explain your reasoning.

b. Use the *trace* feature of your graphing calculator to approximate the maximum r-value of the graph. Does this value change if you use $0 \le \theta \le 4\pi$ instead of $0 \le \theta \le 2\pi$? Explain.

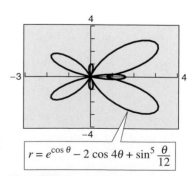

$$r = e^{\cos \theta} - 2 \cos 4\theta + \sin^5 \frac{\theta}{12}$$

Figure 6.63

Some curves reach their zeros and maximum *r*-values at more than one point. Example 4 shows how to handle this situation.

EXAMPLE 4   Analyzing a Polar Graph

Analyze the graph of $r = 2 \cos 3\theta$.

Solution

Symmetry	With respect to the polar axis
Maximum value of $\lvert r \rvert$	$\lvert r \rvert = 2$ when $3\theta = 0,\ \pi,\ 2\pi,\ 3\pi$
	or $\theta = 0,\ \pi/3,\ 2\pi/3,\ \pi$
Zeros of r	$r = 0$ when $3\theta = \pi/2,\ 3\pi/2,\ 5\pi/2$
	or $\theta = \pi/6,\ \pi/2,\ 5\pi/6$

θ	0	$\dfrac{\pi}{12}$	$\dfrac{\pi}{6}$	$\dfrac{\pi}{4}$	$\dfrac{\pi}{3}$	$\dfrac{5\pi}{12}$	$\dfrac{\pi}{2}$
r	2	$\sqrt{2}$	0	$-\sqrt{2}$	-2	$-\sqrt{2}$	0

By plotting these points and using the specified symmetry, zeros, and maximum values, you can obtain the graph shown in Figure 6.64. This graph is called a **rose curve,** and each loop on the graph is called a *petal*. Note how the entire curve is generated as θ increases from 0 to π.

A computer animation of this example appears in the *Interactive* CD-ROM and *Internet* versions of this text.

$0 \le \theta \le \dfrac{\pi}{6}$

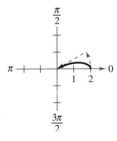

$0 \le \theta \le \dfrac{\pi}{3}$

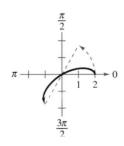

$0 \le \theta \le \dfrac{\pi}{2}$

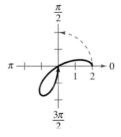

$0 \le \theta \le \dfrac{2\pi}{3}$

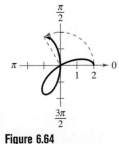

$0 \le \theta \le \dfrac{5\pi}{6}$

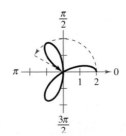

$0 \le \theta \le \pi$

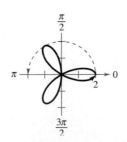

Figure 6.64

Exploration

Notice that the rose curve in Example 4 has three petals. How many petals does the rose curve $r = 2 \cos 4\theta$ have? Experiment with other rose curves and determine the number of petals for the curves $r = 2 \cos n\theta$ and $r = 2 \sin n\theta$, where n is a positive integer.

Special Polar Graphs

Several important types of graphs have equations that are simpler in polar form than in rectangular form. For example, the circle

$$r = 4 \sin \theta$$

in Example 1 has the more complicated rectangular equation

$$x^2 + (y - 2)^2 = 4.$$

The following list gives several other types of graphs that have simple polar equations.

$\dfrac{a}{b} < 1$	$\dfrac{a}{b} = 1$	$1 < \dfrac{a}{b} < 2$	$\dfrac{a}{b} \geq 2$	
Limaçon with inner loop	Cardioid (heart-shaped)	Dimpled limaçon	Convex limaçon	*Limaçons* $r = a \pm b \cos \theta$ $r = a \pm b \sin \theta$ $(a > 0, b > 0)$
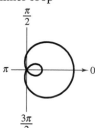	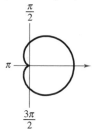	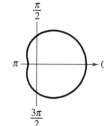	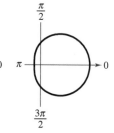	
$r = a \cos n\theta$ Rose curve	$r = a \cos n\theta$ Rose curve	$r = a \sin n\theta$ Rose curve	$r = a \sin n\theta$ Rose curve	*Rose Curves* n petals if n is odd $2n$ petals if n is even $(n \geq 2)$
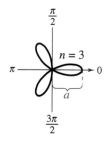	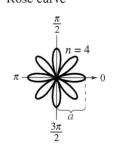	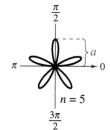	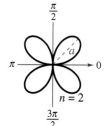	
$r = a \cos \theta$ Circle	$r = a \sin \theta$ Circle	$r^2 = a^2 \sin 2\theta$ Lemniscate	$r^2 = a^2 \cos 2\theta$ Lemniscate	*Circles and Lemniscates*
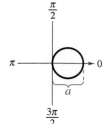	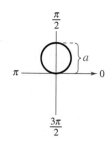	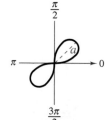	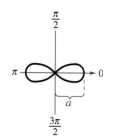	

EXAMPLE 5   Analyzing a Rose Curve

Analyze the graph of

$$r = 3 \cos 2\theta.$$

Solution

Begin with an analysis of the basic features of the graph.

Type of curve	Rose curve with $2n = 4$ petals
Symmetry	With respect to polar axis, the line $\theta = \pi/2$, and the pole
Maximum value of $|r|$	$|r| = 3$ when $\theta = 0, \pi/2, \pi, 3\pi/2$
Zeros of r	$r = 0$ when $\theta = \pi/4, 3\pi/4$

Using a graphing utility (with $0 \le \theta \le 2\pi$), you can obtain the graph shown in Figure 6.65.

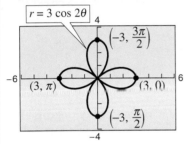

Figure 6.65

EXAMPLE 6   Analyzing a Lemniscate

Analyze the graph

$$r^2 = 9 \sin 2\theta.$$

Solution

Begin with an analysis of the basic features of the graph.

Type of curve	Lemniscate
Symmetry	With respect to pole
Maximum value of $|r|$	$|r| = 3$ when $\theta = \pi/4$
Zeros of r	$r = 0$ when $\theta = 0, \pi/2$

Using a graphing utility (with $r = \sqrt{9 \sin 2\theta}$ and $0 \le \theta \le 2\pi$), you can obtain the graph shown in Figure 6.66.

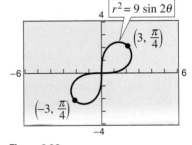

Figure 6.66

Writing About Math   *Heart to Bell*

Use a graphing utility to graph the polar equation

$$r = \cos 5\theta + n \cos \theta$$

for $0 \le \theta < \pi$ for the integers $n = -5$ to $n = 5$. As you graph these equations, you should see the graph change shape from a heart to a bell. Write a short paragraph explaining what values of n produce the heart portion of the curve and what values of n produce the bell.

6.7 Exercises

In Exercises 1–6, identify the type of polar graph.

1.

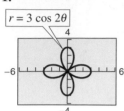

$r = 3 \cos 2\theta$

2.

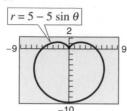

$r = 5 - 5 \sin \theta$

3.

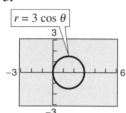

$r = 3 \cos \theta$

4.

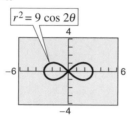

$r^2 = 9 \cos 2\theta$

5.

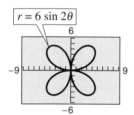

$r = 6 \sin 2\theta$

6.

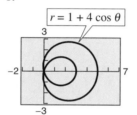

$r = 1 + 4 \cos \theta$

In Exercises 7–16, test for symmetry with respect to $\theta = \pi/2$, the polar axis, and the pole.

7. $r = 10 + 4 \cos \theta$ **8.** $r = 16 \cos 3\theta$

9. $r = \dfrac{6}{1 + \sin \theta}$ **10.** $r = \dfrac{4}{1 - \cos \theta}$

11. $r = 6 \sin \theta$ **12.** $r = 4 - \sin \theta$

13. $r = 4 \sec \theta \csc \theta$ **14.** $r = 2 \csc \theta \cos \theta$

15. $r^2 = 25 \sin 2\theta$ **16.** $r^2 = 25 \cos 4\theta$

In Exercises 17–20, find the maximum value of $|r|$ and any zeros of r. Verify your answers numerically.

17. $r = 10 - 10 \sin \theta$ **18.** $r = 6 + 12 \cos \theta$

19. $r = 4 \cos 3\theta$ **20.** $r = 5 \sin 2\theta$

In Exercises 21–38, sketch the graph of the polar equation. Use a graphing utility to confirm your graph.

21. $r = 5$ **22.** $r = 2$

23. $\theta = \dfrac{\pi}{6}$ **24.** $\theta = -\dfrac{5\pi}{3}$

25. $r = 3 \sin \theta$ **26.** $r = 3 \cos \theta$

27. $r = 3(1 - \cos \theta)$ **28.** $r = 2(1 - \sin \theta)$

29. $r = 3 - 4 \cos \theta$ **30.** $r = 5 - 4 \sin \theta$

31. $r = 6 + \sin \theta$ **32.** $r = 4 + 5 \cos \theta$

33. $r = 5 \cos 3\theta$ **34.** $r = -\sin 5\theta$

35. $r = 7 \sin 2\theta$ **36.** $r = 3 \cos 5\theta$

37. $r = \dfrac{\theta}{2}$ **38.** $r = \theta$

In Exercises 39–54, use a graphing utility to graph the polar equation. Describe your viewing window.

39. $r = \dfrac{\theta}{4}$ **40.** $r = -\dfrac{\theta}{3}$

41. $r = 6 \cos \theta$ **42.** $r = \cos 2\theta$

43. $r = 2(3 - \sin \theta)$ **44.** $r = 6 - 4 \sin \theta$

45. $r = 3 - 6 \cos \theta$ **46.** $r = 2(3 - 2 \sin \theta)$

47. $r = \dfrac{3}{\sin \theta - 2 \cos \theta}$ **48.** $r = \dfrac{6}{2 \sin \theta - 3 \cos \theta}$

49. $r^2 = 4 \cos 2\theta$ **50.** $r^2 = 4 \sin \theta$

51. $r = 4 \sin \theta \cos^2 \theta$ **52.** $r = 2 \cos(3\theta - 2)$

53. $r = 4 \csc \theta + 5$ **54.** $r = 4 - \sec \theta$

In Exercises 55–62, use a graphing utility to graph the polar equation. Find an interval for θ for which the graph is traced *only once*.

55. $r = 3 - 2 \cos \theta$ **56.** $r = 2(1 - 2 \sin \theta)$

57. $r = 2 + \sin \theta$ **58.** $r = 4 + 3 \cos \theta$

59. $r = 2 \cos\left(\dfrac{3\theta}{2}\right)$ **60.** $r = 3 \sin\left(\dfrac{5\theta}{2}\right)$

61. $r^2 = 4 \sin 2\theta$ **62.** $r^2 = \dfrac{1}{\theta}$

In Exercises 63–66, use a graphing utility to graph the polar equation and show that the indicated line is an asymptote of the graph.

Name of Graph	Polar Equation	Asymptote
63. Conchoid	$r = 2 - \sec\theta$	$x = -1$
64. Conchoid	$r = 2 + \csc\theta$	$y = 1$
65. Hyperbolic spiral	$r = \dfrac{2}{\theta}$	$y = 2$
66. Strophoid	$r = 2\cos 2\theta \sec\theta$	$x = -2$

Synthesis

True or False? **In Exercises 67–70, determine whether the statement is true or false. Justify your answer.**

67. The point with polar coordinate $\left(6, \dfrac{11\pi}{6}\right)$ lies on the graph of $r = 2\sin\theta + 5$.

68. The graph of $r = 4\cos 8\theta$ is a rose curve with 8 petals.

69. The graph of $r = 10\sin 5\theta$ is a rose curve with 10 petals.

70. A rose curve will always have symmetry with respect to the line $\theta = \pi/2$.

71. *Graphical Reasoning* Use a graphing utility to graph the polar equation

$$r = 6[1 + \cos(\theta - \phi)]$$

for (a) $\phi = 0$, (b) $\phi = \pi/4$, and (c) $\phi = \pi/2$. Use the graphs to describe the effect of the angle ϕ. Write the equation as a function of $\sin\theta$ for part (c).

72. The graph of $r = f(\theta)$ is rotated about the pole through an angle ϕ. Show that the equation of the rotated graph is $r = f(\theta - \phi)$.

73. Consider the graph of $r = f(\sin\theta)$.

   (a) Show that if the graph is rotated counterclockwise $\pi/2$ radians about the pole, the equation of the rotated graph is $r = f(-\cos\theta)$.

   (b) Show that if the graph is rotated counterclockwise π radians about the pole, the equation of the rotated graph is $r = f(-\sin\theta)$.

   (c) Show that if the graph is rotated counterclockwise $3\pi/2$ radians about the pole, the equation of the rotated graph is $r = f(\cos\theta)$.

In Exercises 74–76, use the results of Exercise 72 and 73.

74. Write an equation for the limaçon $r = 2 - \sin\theta$ after it has been rotated through the given angle.

   (a) $\dfrac{\pi}{4}$ (b) $\dfrac{\pi}{2}$ (c) π (d) $\dfrac{3\pi}{2}$

75. Write an equation for the rose curve $r = 2\sin 2\theta$ after it has been rotated through the given angle.

   (a) $\dfrac{\pi}{6}$ (b) $\dfrac{\pi}{2}$ (c) $\dfrac{2\pi}{3}$ (d) π

76. Sketch the graph of each equation.

   (a) $r = 1 - \sin\theta$ (b) $r = 1 - \sin\left(\theta - \dfrac{\pi}{4}\right)$

77. *Exploration* Use a graphing utility to graph the polar equation $r = 2 + k\cos\theta$ for $k = 0$, $k = 1$, $k = 2$, and $k = 3$. Identify each graph.

78. *Exploration* Consider the polar equation $r = 3\sin k\theta$.

   (a) Use a graphing utility to graph the equation for $k = 1.5$. Find the interval for θ for which the graph is traced only once.

   (b) Use a graphing utility to graph the equation for $k = 2.5$. Find the interval for θ for which the graph is traced only once.

   (c) Is it possible to find an interval for θ for which the graph is traced only once for any rational number k? Explain.

Review

In Exercises 79–82, find the value of the trigonometric function given that u and v are in Quadrant IV and $\sin u = -\frac{3}{5}$ and $\cos v = 1/\sqrt{2}$.

79. $\cos(u + v)$

80. $\sin(u + v)$

81. $\cos(u - v)$

82. $\sin(u - v)$

In Exercises 83 and 84, find the exact values of sin 2u, cos 2u, and tan 2u using the double-angle formulas.

83. $\sin u = \dfrac{4}{5}$, $\dfrac{\pi}{2} < u < \pi$

84. $\tan u = -\sqrt{3}$, $\dfrac{3\pi}{2} < u < 2\pi$

6.8   Polar Equations of Conics

Alternative Definition of Conics

In this chapter, you have learned that the rectangular equations of ellipses and hyperbolas take simple forms when the origin lies at the *center*. As it happens, there are many important applications of conics in which it is more convenient to use one of the *foci* as the origin for the coordinate system. For example, the sun lies at one focus of the earth's orbit. Similarly the light source of a parabolic reflector lies at its focus. In this section you will learn that polar equations of conics take simple forms if one of the foci lies at the pole.

To begin, consider the following alternative definition of a conic that uses the concept of eccentricity.

Alternative Definition of a Conic

The locus of a point in the plane which moves so that its distance from a fixed point (focus) is in constant ratio to its distance from a fixed line (directrix) is a **conic.** The constant ratio is the **eccentricity** of the conic and is denoted by e. Moreover, the conic is an **ellipse** if $e < 1$, a **parabola** if $e = 1$, and a **hyperbola** if $e > 1$.

In Figure 6.67, note that for each type of conic, the pole corresponds to the fixed point (focus) given in the definition.

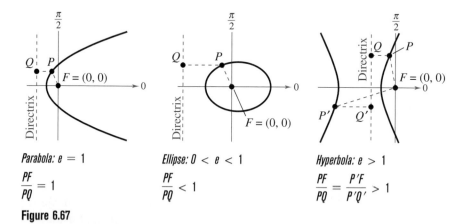

Parabola: $e = 1$
$$\frac{PF}{PQ} = 1$$

Ellipse: $0 < e < 1$
$$\frac{PF}{PQ} < 1$$

Hyperbola: $e > 1$
$$\frac{PF}{PQ} = \frac{P'F}{P'Q'} > 1$$

Figure 6.67

Polar Equations of Conics

The benefit of locating a focus of a conic at the pole is that the equation of the conic takes on a simpler form. A proof of the polar form is given in Appendix A.

What **You Should Learn:**

- How to define conics in terms of eccentricities
- How to write equations of conics in polar form
- How to use equations of conics in polar form to model real-life problems

Why **You Should Learn It:**

The elliptical orbits of planets and satellites can be modeled with polar equations. Exercise 39 on page 479 shows a polar equation of a planetary orbit.

NASA

Polar Equations of Conics

The graph of a polar equation of the form

1. $r = \dfrac{ep}{1 \pm e \cos \theta}$ **2.** $r = \dfrac{ep}{1 \pm e \sin \theta}$

is a conic, where $e > 0$ is the eccentricity and $|p|$ is the distance between the focus (pole) and the directrix.

Equations of the form $r = \dfrac{ep}{1 \pm e \cos \theta}$ Vertical directrix

correspond to conics with vertical directrices and equations of the form

$r = \dfrac{ep}{1 \pm e \sin \theta}$ Horizontal directrix

correspond to conics with horizontal directrices. Moreover, the converse is also true—that is, any conic with a focus at the pole and having a horizontal or vertical directrix can be represented by one of the given equations.

A computer simulation of this concept appears in the *Interactive* CD-ROM and *Internet* versions of this text.

EXAMPLE 1   Determining a Conic from Its Equation

Determine the type of conic represented by the equation $r = \dfrac{15}{3 - 2 \cos \theta}$.

Algebraic Solution

To determine the type of conic, rewrite the equation in the form $r = ep/(1 \pm e \cos \theta)$.

$r = \dfrac{15}{3 - 2 \cos \theta}$

$= \dfrac{5}{1 - (2/3) \cos \theta}$ Divide numerator and denominator by 3.

From this form you can conclude that the graph is an ellipse with $e = \frac{2}{3}$.

Graphical Solution

Use a graphing utility in *polar* mode to graph $r = \dfrac{15}{3 - 2 \cos \theta}$.

Be sure to use a square setting. From the graph in Figure 6.68, you can see that the conic appears to be an ellipse.

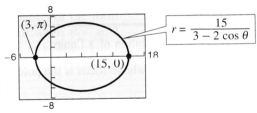

Figure 6.68

For the ellipse in Figure 6.68, the major axis is horizontal and the vertices lie at $(r, \theta) = (15, 0)$ and $(r, \theta) = (3, \pi)$. So, the length of the *major* axis is $2a = 18$. To find the length of the *minor* axis, you can use the equations $e = c/a$ and $b^2 = a^2 - c^2$ to conclude that

$b^2 = a^2 - c^2 = a^2 - (ea)^2 = a^2(1 - e^2)$.   Ellipse

Because $e = \frac{2}{3}$, you have $b^2 = 9^2[1 - (2/3)^2] = 45$, which implies that $b = \sqrt{45} = 3\sqrt{5}$. So, the length of the minor axis is $2b = 6\sqrt{5}$. A similar analysis for hyperbolas yields

$b^2 = c^2 - a^2 = (ea)^2 - a^2 = a^2(e^2 - 1)$.   Hyperbola

Looking for solutions to your math problems? These study aids offer more than just the answers!

Two technology options offer complete solutions to every odd-numbered problem, help you practice and assess your skills, and provide problem-solving tools.

Interactive Trigonometry: A Graphing Approach 2.0 CD-ROM

- Enhance your understanding with step-by-step solutions to all odd-numbered exercises in the text.
- Assess your skill levels with diagnostic pre- and post-tests for each chapter.
- Strengthen your skills with tutorial exercises accompanied by examples and diagnostics.
- Visualize and graph with a built-in Meridian Graphing Calculator Emulator.
- Enjoy learning with additional interactive features.

Internet Trigonometry: A Graphing Approach 1.0

A subscription to this web site offers all of the CD-ROM features listed above, plus:

- Chat rooms for peer support
- Bulletin boards for sharing ideas and insights on each chapter

To purchase Interactive Trigonometry: A Graphing Approach 2.0 CD-ROM:

- Visit Houghton Mifflin's College Store at **college.hmco.com** or contact your campus bookstore.

To subscribe online to Internet Trigonometry: A Graphing Approach 1.0:

- Visit Houghton Mifflin's College Division web site at **college.hmco.com** and select mathematics.

Use these print supplements for added practice and convenient support.

Study and Solutions Guide to accompany
Trigonometry: A Graphing Approach, 3rd Edition

- Work through step-by-step solutions for all odd-numbered exercises in the text.
- Test your skills by taking practice tests with accompanying solutions.
- Find useful study strategies designed to help you succeed.

Student Success Organizer
Ask your instructor about this new study aid.

- Use its practical format to guide you step-by-step through difficult concepts.
- Enhance your organizational skills for approaching problems and assignments.

To purchase these print supplements:

- Visit Houghton Mifflin's College Store at **college.hmco.com** or contact your campus bookstore.

Larson • Hostetler

Answers to Odd-Numbered Exercises and Tests

Chapter P

Section P.1 *(page 9)*

1. (a) $5, 1$ (b) $-9, 5, 0, 1, -4, -1$
(c) $-9, -\frac{7}{2}, 5, \frac{2}{3}, 0, 1, -4, -1$ (d) $\sqrt{2}$

3. (a) $1, 20$ (b) $-13, 1, -10, 20$
(c) $2.01, 0.666\ldots, -13, 1, -10, 20$
(d) $0.010110111\ldots$

5. (a) $\frac{6}{3}, 3$ (b) $\frac{6}{3}, -2, 3, -3$
(c) $-\frac{1}{3}, \frac{6}{3}, -7.5, -2, 3, -3$ (d) $-\pi, \frac{1}{2}\sqrt{2}$

7. 0.625 **9.** $0.\overline{123}$ **11.** $\frac{23}{5}$ **13.** $\frac{13}{2}$

15. $-\frac{183}{100}$ **17.** $-1 < 2.5$

19. $-4 > -8$ **21.** $\frac{3}{2} < 7$

23. $\frac{5}{6} > \frac{2}{3}$

25. (a) $x \le 5$ is the set of all real numbers less than or equal to 5.
(b) (c) Unbounded

27. (a) $x < 0$ is the set of all negative real numbers.
(b) (c) Unbounded

29. (a) $x \ge 4$ is the set of all real numbers greater than or equal to 4.
(b) (c) Unbounded

31. (a) $-2 < x < 2$ is the set of all real numbers greater than -2 and less than 2.
(b) (c) Bounded

33. (a) $-1 \le x < 0$ is the set of all negative real numbers greater than or equal to -1.
(b) (c) Bounded

35. $\frac{127}{90}, \frac{584}{413}, \frac{7071}{5000}, \sqrt{2}, \frac{47}{33}$ **37.** $x < 0; (-\infty, 0)$

39. $y \ge 0; [0, \infty)$ **41.** $12 \le c \le 32; [12, 32]$

43. $W > 45; (45, \infty)$

45. The set of all real numbers greater than -6

47. 10 **49.** $\pi - 3 \approx 0.1416$ **51.** -1 **53.** -9

55. 1 for $x > -2$; undefined for $x = -2$; -1 for $x < -2$

57. $|-3| > -|-3|$ **59.** $-5 = -|5|$

61. $-|-2| = -|2|$ **63.** 4 **65.** 51

67. $\frac{5}{2}$ **69.** $\frac{128}{75}$

71. (a) $-A$ is negative. (b) $B - A$ is negative.

73. $|x - 5| \le 3$ **75.** $|y - 0| \ge 6$ **77.** 11 miles

79. The temperature dropped $23°$.

81. $|\$113{,}356 - \$112{,}700| = \$656 > \$500$
$0.05(\$112{,}700) = \$5635$
Because the actual expenses differ from the budget by more than \$500, there is failure to meet the "budget variance test."

83. $|\$37{,}335 - \$37{,}640| = \$305 < \$500$
$0.05(\$37{,}640) = \$1882$
Because the difference between the actual expenses and the budget is less than \$500 and less than 5% of the budgeted amount, there is compliance with the "budget variance test."

85. $y = \$92.5$ billion, $|y - x| = \$0.3$ billion
There was a surplus of \$0.3 billion.

87. $y = \$517.1$ billion, $|y - x| = \$73.8$ billion
There was a deficit of \$73.8 billion.

89. $y = \$1351.8$ billion, $|y - x| = \$163.9$ billion
There was a deficit of \$163.9 billion.

91. Terms: $7x, 4$; Coefficient: 7

93. Terms: $3x^2, -8x, -11$; Coefficients: $3, -8$

95. Terms: $4x^3, \frac{x}{2}, -5$; Coefficients: $4, \frac{1}{2}$

97. (a) -10 (b) -6 **99.** (a) 14 (b) 2

101. (a) Division by 0 is undefined. (b) 0

103. Commutative Property of Addition

105. Multiplicative Inverse Property

107. Distributive Property

109. Multiplicative Identity Property

111. Associative and Commutative Properties of Multiplication

113. 0 **115.** Division by 0 is undefined. **117.** $\frac{1}{2}$

119. $\dfrac{3}{8}$ **121.** $\dfrac{11x}{12}$ **123.** 48   **125.** -2.57

127. 1.56

129. (a)

n	1	0.5	0.01	0.0001	0.000001
$5/n$	5	10	500	50,000	5,000,000

(b) $5/n$ approaches ∞ as n approaches 0.

131. (a) No. If u is negative while v is positive, or vice versa, the expressions will not be equal.

(b) $|u + v| \le |u| + |v|$

133. Answers will vary. Natural numbers are the integers from 1 to infinity. A rational number can be expressed as the ratio of two integers; an irrational number cannot.

135. False. $\dfrac{3 + 5}{4} = 2 = \dfrac{3}{4} + \dfrac{5}{4}$, but $\dfrac{4}{3} + \dfrac{4}{5} = \dfrac{32}{15} \ne \dfrac{4}{3 + 5}$.

Section P.2 *(page 23)*

1.

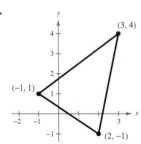

3.

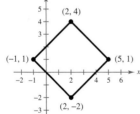

5. A: $(2, 6)$; B: $(-6, -2)$; C: $(4, -4)$; D: $(-3, 2)$

7. $(-3, 4)$ **9.** $(-5, -5)$ **11.** Quadrant IV

13. Quadrant II   **15.** Quadrant III or IV

17. Quadrant III   **19.** Quadrants I and III

21.

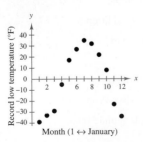

23.

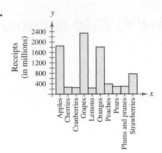

25. 8   **27.** 5   **29.** 13

31. (a)

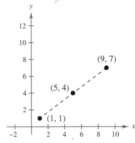

(b) 10

(c) $(5, 4)$

33. (a)

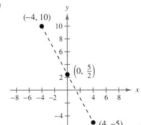

(b) 17

(c) $\left(0, \tfrac{5}{2}\right)$

35. (a)

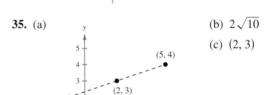

(b) $2\sqrt{10}$

(c) $(2, 3)$

37. (a)

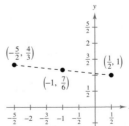

(b) $\dfrac{\sqrt{82}}{3}$

(c) $\left(-1, \dfrac{7}{6}\right)$

39. (a)

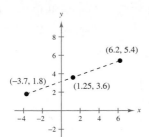

(b) $\sqrt{110.97}$

(c) $(1.25, 3.6)$

41. (a)

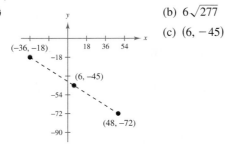

(b) $6\sqrt{277}$

(c) $(6, -45)$

43. $10^2 + 3^2 = \left(\sqrt{109}\right)^2$ **45.** \$630,000

47. (a) Yes   (b) Yes   **49.** (a) No   (b) Yes

51. (a) No   (b) Yes   **53.** (a) Yes   (b) Yes

55.

x	-1	0	1	$\frac{3}{2}$	2
y	5	3	1	0	-1

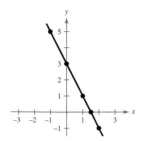

57.

x	-1	0	1	2	3
y	3	0	-1	0	3

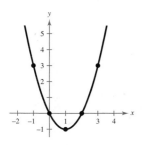

59.

x	0	1	2	3	4
y	1	2	3	2	1

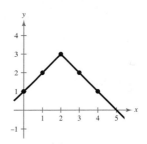

61. (d)   **63.** (f)   **65.** (a)

67.

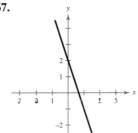

69.

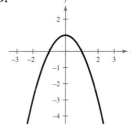

71.

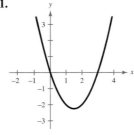

73.

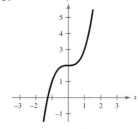

75.

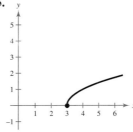

77.

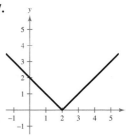

79.

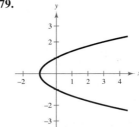

81.

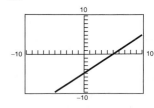

Intercepts: $(5, 0)$, $(0, -5)$

83.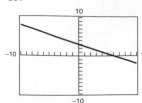

Intercepts: $(6, 0), (0, 3)$

85.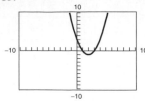

Intercepts: $(3, 0), (1, 0), (0, 3)$

87.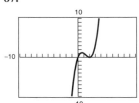

Intercepts: $(0, 0), (2, 0)$

89.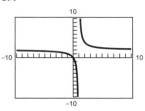

Intercept: $(0, 0)$

91.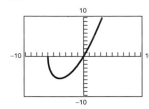

Intercepts: $(-6, 0), (0, 0)$

93.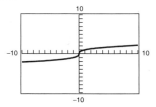

Intercept: $(0, 0)$

95.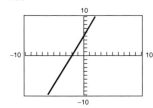

The standard setting gives a more complete graph.

97.

Xmin = -5
Xmax = 5
Xscl = 1
Ymin = -30
Ymax = 10
Yscl = 5

99. $x^2 + y^2 = 9$ **101.** $(x - 2)^2 + (y + 1)^2 = 16$

103. $(x + 1)^2 + (y - 2)^2 = 5$

105. $(x - 3)^2 + (y - 4)^2 = 25$

107.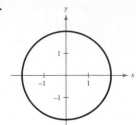

Center: $(0, 0)$

Radius = 2

109.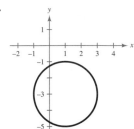

Center: $(1, -3)$

Radius = 2

111.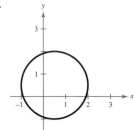

Center: $\left(\frac{1}{2}, \frac{1}{2}\right)$

Radius = $\frac{3}{2}$

113. $y_1 = \sqrt{64 - x^2}$

$y_2 = -\sqrt{64 - x^2}$

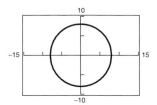

115. $y_1 = 2 + \sqrt{16 - (x - 1)^2}$

$y_2 = 2 - \sqrt{16 - (x - 1)^2}$

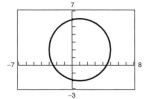

117. \$1002.6 million

119. (a) about 55.9 years, about 77.2 years

(b) 1968

(c) about 78.2 years, about 78.5 years

121. True. The lengths of two sides are $\sqrt{149}$.

123. -2 and 5; a is one and b is the other.

Section P.3 *(page 39)*

1. (a) Yes (b) No (c) No (d) No

3. (a) Yes (b) No (c) No (d) No

5. (a) No (b) No (c) No (d) Yes

7. Identity **9.** Identity **11.** Conditional **13.** $-\frac{96}{23}$

15. -4 **17.** $-\frac{6}{5}$ **19.** 10 **21.** 4 **23.** 5

25. $\frac{11}{6}$ **27.** $\frac{5}{3}$ **29.** No solution

31. $(5, 0), (0, -5)$ **33.** $(-2, 0), (1, 0), (0, -2)$

35. $(-2, 0), (0, 0)$ **37.** $(-2, 0), (6, 0), (0, -2)$

39. $(1, 0), \left(0, \frac{1}{2}\right)$

41.

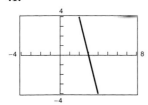

43.

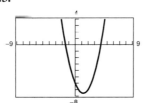

45.

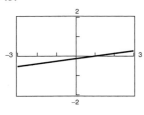

47.

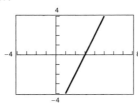

$(3, 0)$

49.

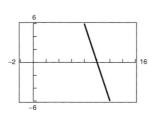

51. $\frac{15}{4}$ **53.** $\frac{89}{13}$

$(10, 0)$

55. 6 **57.** 3, 12 **59.** 1, -1.2 **61.** $-\frac{5}{8}$

63. 0.5, -3, 3 **65.** -0.717, 2.107 **67.** -1.333

69. $-1, 7$ **71.** $(1, 1)$ **73.** $(-1, 3), (2, 6)$

75. $(4, 1)$ **77.** $(1.449, 1.899), (-3.449, -7.899)$

79. $(0, 0), (-2, 8), (2, 8)$ **81.** $0, -\frac{1}{2}$ **83.** $4, -2$

85. $3, -\frac{1}{2}$ **87.** $2, -6$ **89.** $\pm\sqrt{7}$; ± 2.65

91. $12 \pm 3\sqrt{2}$; 16.24, 7.76 **93.** $\frac{1}{2} \pm \frac{3}{2}\sqrt{2}$; -1.62, 2.62

95. 2 **97.** $-8, 4$ **99.** $-3 \pm \sqrt{7}$ **101.** $1 \pm \dfrac{\sqrt{6}}{3}$

103. $1 \pm \sqrt{3}$ **105.** $-4 \pm 2\sqrt{5}$ **107.** $-\frac{3}{2}, -\frac{5}{2}$

109. $1 \pm \sqrt{2}$ **111.** $6, -12$ **113.** $\frac{1}{2} \pm \sqrt{3}$

115. $0, \pm\dfrac{3\sqrt{2}}{2}$ **117.** ± 3 **119.** $-3, 0$

121. $3, 1, -1$ **123.** $\pm\sqrt{3}, \pm 1$ **125.** $\pm\frac{1}{2}, \pm 4$

127. $-\frac{1}{5}, -\frac{1}{3}$ **129.** $\frac{1}{4}$ **131.** 26 **133.** 0 **135.** 9

137. $-59, 69$ **139.** 1 **141.** $4, -5$

143. $-\dfrac{3 \pm \sqrt{21}}{6}$ **145.** $2, -\dfrac{3}{2}$ **147.** $3, -2$

149. $\sqrt{3}, -3$

151. (a)

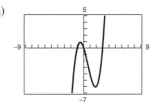

(b) and (c) $x = 0, 3, -1$

153. (a)

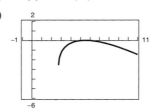

(b) and (c) $x = 5, 6$

155. (a)

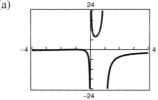

(b) and (c) $x = -1$

157. (a)

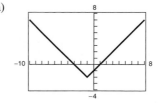

(b) and (c) $x = 1, -3$

159. (a) The time and quantity when utilization of cucumbers equaled that of nectarines and peaches

 (b) $(3.84, 5.46)$

 (c) Answers will vary.

161. (a)

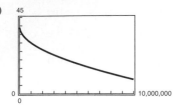

 (b) $7,307,025$

163. False. Two linear equations could also have an infinite number of points of intersection.

Section P.4   *(page 51)*

1. (a) L_2 (b) L_3 (c) L_1

3.

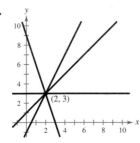

5. $\frac{3}{2}$ **7.** 0 **9.** -4

11. $m = -\frac{5}{2}$ **13.** m is undefined.

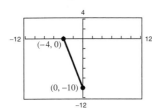

15. $(0, 1), (3, 1), (-1, 1)$ **17.** $(-6, 2), (-4, 6), (-3, 8)$

19. $(3, -4), (5, -3), (9, -1)$

21. Perpendicular   **23.** Parallel

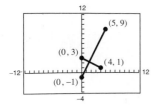
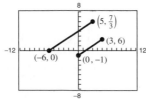

25. (a) $m = 5$;

 Intercept: $(0, 3)$

 (b)

27. (a) m is undefined.

 There is no y-intercept.

 (b)

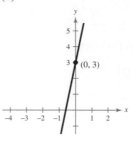

29. (a) $m = 0$;

 Intercept: $\left(0, -\frac{5}{3}\right)$

 (b)

31. (a) $m = -\frac{7}{6}$;

 Intercept: $(0, 5)$

 (b)

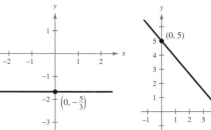

33. $3x - y - 2 = 0$ **35.** $2x + y = 0$

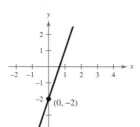

37. $x + 3y - 4 = 0$ **39.** $x - 6 = 0$

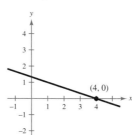

41. $3x + y = 0$

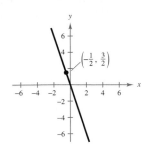

43. $3x + 5y - 10 = 0$

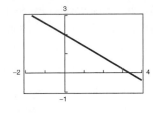

45. $x + 8 = 0$

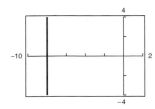

47. $x + 2y - 3 = 0$

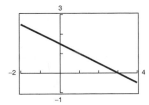

49. $30x + 25y + 18 = 0$

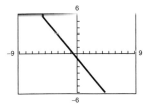

51. $2x - 5y + 1 = 0$

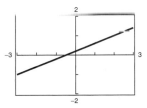

53.

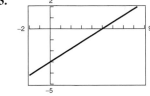

x- and *y*-intercepts

55. $3x + 2y - 6 = 0$ **57.** $12x + 3y + 2 = 0$

59.

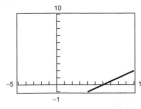

The second setting gives a more complete graph.

61.

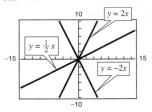

$y = \frac{1}{2}x$ and $y = -2x$ are perpendicular.

63.

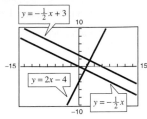

$y = -\frac{1}{2}x$ and $y = -\frac{1}{2}x + 3$ are parallel. Both are perpendicular to $y = 2x - 4$.

65. (a) $2x - y - 3 = 0$ (b) $x + 2y - 4 = 0$

67. (a) $-6x - 8y + 3 = 0$ (b) $96x - 72y + 127 = 0$

69. (a) $10x - 10y + 43 = 0$ (b) $10x + 10y - 93 = 0$

71. $3x - 2y - 1 = 0$

73. (a) Sales increase of $135

(b) No sales increase

(c) Sales decrease of $40

75. (a) Greatest increases per share: 1990 and 1996

Greatest decrease per share: 1997

(b) $37x - 1000y + 943 = 0$

(c) Slope is the average increase per share per year.

(d) $1.46. Answers will vary.

77. 16,667 feet   **79.** $125t - V + 2415 = 0$

81. $2000t + V - 22,400 = 0$

83. (b); slope $= -10$; the amount owed decreases by $10 per week.

85. (a); slope $= 0.25$; expenses increase by $0.25 per mile.

87. $F = \frac{9}{5}C + 32$ **89.** $39,500

91. (a) $V = -175t + 875$

(b)

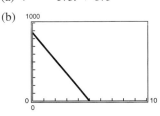

t	0	1	2	3	4	5
V	875	700	525	350	175	0

93. (a) $C = 16.75t + 36,500$ (b) $R = 27t$

(c) $P = 10.25t - 36,500$ (d) $t \approx 3561$ hours

95. (a) Answers will vary. Example: $y = 92.84t + 487.82$

    (b) Answers will vary. Example: $1,601,000

    (c) Average increase per year

97. False. The line through $(10, -3)$ and $(2, -9)$ is $y = \frac{3}{4}x + \frac{9}{4}$, and $\left(-12, -\frac{37}{2}\right)$ is not on this line.

99. -4

101. No. The slopes of two perpendicular lines have opposite signs (assuming that neither line is vertical or horizontal).

Section P.5 *(page 64)*

1. Yes   **3.** No

5. Yes. Each input value is matched with one output value.

7. No. The same input value is matched with two different output values.

9. (a) Function

    (b) Not a function because the element 1 in A corresponds to two elements, -2 and 1, in B.

    (c) Function

    (d) Not a function because the element 2 in A corresponds to no element in B.

11. Each is a function. To each year there corresponds one and only one circulation.

13. Not a function   **15.** Function   **17.** Function

19. Not a function   **21.** Function   **23.** Not a function

25. (a) $\dfrac{1}{5}$ (b) 1   (c) $\dfrac{1}{4t + 1}$ (d) $\dfrac{1}{x + c + 1}$

27. (a) -1 (b) -9 (c) $2x - 5$

29. (a) 0   (b) -0.75 (c) $x^2 + 2x$

31. (a) 1   (b) 2.5   (c) $3 - 2|x|$

33. (a) $-\dfrac{1}{9}$ (b) Undefined   (c) $\dfrac{1}{y^2 + 6y}$

35. (a) 1   (b) -1 (c) 1   **37.** (a) -1 (b) 2   (c) 6

39.

x	-2	-1	0	1	2
$f(x)$	1	-2	-3	-2	1

41.

t	-5	-4	-3	-2	-1
$h(t)$	1	$\frac{1}{2}$	0	$\frac{1}{2}$	1

43.

x	-2	-1	0	1	2
$f(x)$	5	$\frac{9}{2}$	4	1	0

45. 5   **47.** $\frac{4}{3}$ **49.** ± 3 **51.** ± 4 **53.** 2, -1

55. 0, 3   **57.** All real numbers x

59. All real numbers t except $t = 0$

61. All real numbers y such that $y \geq 10$

63. All real numbers x such that $-1 \leq x \leq 1$

65. All real numbers x except $x = 0, -2$

67. All real numbers s such that $s \neq 4$ and $s \geq 1$

69. All real numbers x except $x = 0$

71. $\{(-2, 4), (-1, 1), (0, 0), (1, 1), (2, 4)\}$

73. $\{(-2, 0), (-1, 1), (0, \sqrt{2}), (1, \sqrt{3}), (2, 2)\}$

75. $g(x) = -2x^2; c = -2$ **77.** $r(x) = \dfrac{32}{x}; c = 32$

79. $2, c \neq 0$ **81.** $3 + h, h \neq 0$

83. $3x^2 + 3xc + c^2, c \neq 0$ **85.** $-\dfrac{1}{t}, t \neq 1$

87. $A = \dfrac{C^2}{4\pi}$ **89.** $A = \dfrac{s^2}{2}$

91. (a)

Height, x	Width	Volume, V
1	$24 - 2(1)$	$1[24 - 2(1)]^2 = 484$
2	$24 - 2(2)$	$2[24 - 2(2)]^2 = 800$
3	$24 - 2(3)$	$3[24 - 2(3)]^2 = 972$
4	$24 - 2(4)$	$4[24 - 2(4)]^2 = 1024$
5	$24 - 2(5)$	$5[24 - 2(5)]^2 = 980$
6	$24 - 2(6)$	$6[24 - 2(6)]^2 = 864$

Maximum when $x = 4$

    (b) $V = x(24 - 2x)^2, 0 < x < 12$

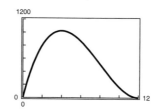

    (c) $x = 9, V = 324; x = 10, V = 160$

93. $A = \dfrac{x^2}{2(x - 2)}, x > 2$

95. (a) $V = x^2y$ (b) $0 < x < 27$

        $= x^2(108 - 4x)$

        $= 108x^2 - 4x^3$

    (c)   (d) $x = 18$ in., $y = 36$ in.

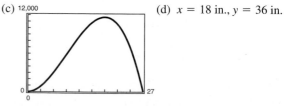

97. (a) $C = 12.30x + 98,000$ (b) $R = 17.98x$

   (c) $P = 5.68x - 98,000$

99. (a) $R = \dfrac{240n - n^2}{20}$

   (b)

n	90	100	110	120	130	140	150
$R(n)$	\$675	\$700	\$715	\$720	\$715	\$700	\$675

   The revenue is maximum when $n = 120$.

   (c)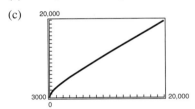

   The revenue is maximum when $n = 120$.

101. (a)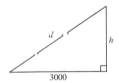

   (b) $h = \sqrt{d^2 - 3000^2}$; $d \geq 3000$

   (c)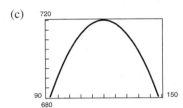

   Xmin = 3000
   Xmax = 20,000
   Xscl = 1000
   Ymin = 0
   Ymax = 20,000
   Yscl = 1000

   (d) ≈ 9539 feet

103. False. Range is $[-1, \infty)$.

105. No. 3 corresponds to both u and v.

107. Function notation is a convenient way of referencing the value of a function for a specific domain value.

Section P.6   *(page 78)*

1. Domain: $(-\infty, \infty)$; Range: $(-\infty, 1]$

3. Domain: $(-\infty, -1], [1, \infty)$; Range: $[0, \infty)$

5. Domain: $(-\infty, \infty)$; Range: $[0, \infty)$

7.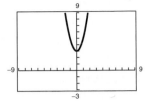

   Domain: $(-\infty, \infty)$; Range: $[3, \infty)$

9.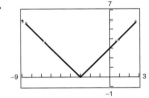

   Domain: $[1, \infty)$; Range: $[0, \infty)$

11.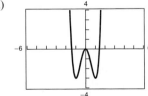

   Domain: $(-\infty, \infty)$; Range: $[0, \infty)$

13. Function. Graph the given function over the window shown in the figure.

15. Not a function. Solve for y and graph the resulting two functions.

17. Function. Solve for y and graph the resulting function.

19. (a) Increasing on $(-\infty, \infty)$ (b) Odd function

21. (a) Increasing on $(-\infty, 0), (2, \infty)$

   Decreasing on $(0, 2)$

   (b) Neither even nor odd

23. (a)

   (b) Increasing on $(-1, 0), (1, \infty)$

   Decreasing on $(-\infty, -1), (0, 1)$

   (c) Even function

25. (a)

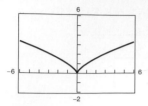

(b) Decreasing on $(-\infty, 0)$; Increasing on $(0, \infty)$

(c) Even function

27. (a)

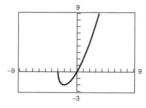

(b) Increasing on $(-2, \infty)$; Decreasing on $(-3, -2)$

(c) Neither even nor odd

29. (a)

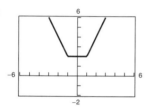

(b) Decreasing on $(-\infty, -1)$; Constant on $(-1, 1)$;
Increasing on $(1, \infty)$

(c) Even function

31.

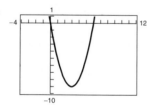

Relative minimum: $(3, -9)$

33.

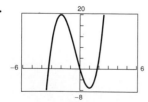

Relative minimum: $(1, -7)$
Relative maximum: $(-2, 20)$

35.

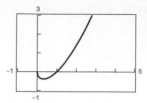

Minimum: $(0.33, -0.38)$

37. (a) Answers will vary.

(b) Relative minimum at $(2, -9)$

(c) Answers will vary.

39. (a) Answers will vary.

(b) Relative minimum at $(1.63, -8.71)$
Relative maximum at $(-1.63, 8.71)$

(c) Answers will vary.

41. (a) Answers will vary.   (b) Relative minimum at $(4, 0)$

(c) Answers will vary.

43.

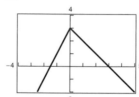

45.

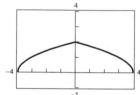

47. Neither even nor odd   **49.** Odd function

51. Odd function   **53.** Even function

55. (a) $\left(\frac{3}{2}, 4\right)$ (b) $\left(\frac{3}{2}, -4\right)$

57. (a) $(-4, 9)$ (b) $(-4, -9)$

59. (a) $(-x, -y)$ (b) $(-x, y)$

61. Even function   **63.** Neither even nor odd

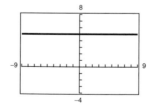

65. Even function   **67.** Neither even nor odd

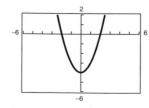

69. Neither even nor odd   **71.** Neither even nor odd

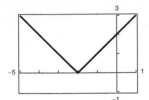
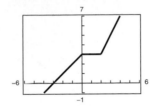

73. $(-\infty, 4]$ **75.** $(-\infty, -3], [3, \infty)$

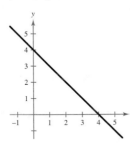
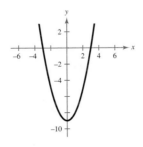

77. $[-1, 1]$ **79.** $[-2, \infty)$

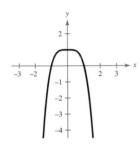
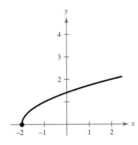

81. $f(x) < 0$ for all x

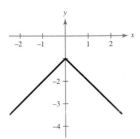

83.

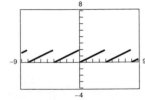

Domain: $(-\infty, \infty)$

Range: $[0, 2)$

Sawtooth pattern

85. (a) Answers will vary.

(b)
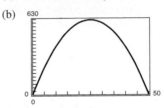

(c) 625 square meters; 25×25 meters

87. (a) C_2 is the appropriate model. The cost of the first minute is \$1.05 and the cost increases \$0.38 when the next minute begins, etc.

(b)
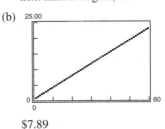

     \$7.89

89. $h = -x^2 + 4x - 3, \ 1 \le x \le 3$

91. $h = 2x - x^2, \ 0 \le x \le 2$ **93.** $L = \frac{1}{2}y^2, 0 \le y \le 4$

95. (a) $y = 1.473x^3 - 16.411x^2 + 31.24x - 95.2$

(b) Domain: $[0, 7]$

(c)
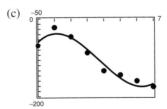

(d) Most accurate: 1992; Least accurate: 1991

(e) Yes. The cubic curve is starting an upswing for future years.

97. False. Counterexample: $f(x) = \sqrt{1 + x^2}$

99. If $y = a_{2n+1}x^{2n+1} + a_{2n-1}x^{2n-1} + \cdots + a_3x^3 + a_1x,$ each exponent is odd. Then

$$f(-x) = -a_{2n+1}x^{2n+1} - a_{2n-1}x^{2n-1} - $$
$$\cdots - a_3x^3 - a_1x,$$

which is equal to $-f(x)$. Therefore, by definition, the original function is odd.

101. (a) Even. g is a reflection in the x-axis.

(b) Even. g is a reflection in the y-axis.

(c) Even. g is a vertical shift downward.

(d) Neither even nor odd. g is shifted to the right and reflected in the x-axis.

103. No. x is not a function of y because horizontal lines can be drawn to intersect the graph twice, so each y-value corresponds to two distinct x-values if $-5 < y < 5$.

Section P.7 *(page 88)*

1.

3.
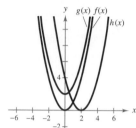

5.
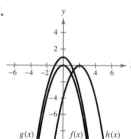

7.
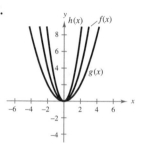

9.
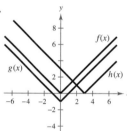

11.
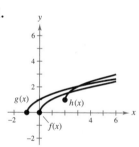

13. (a)
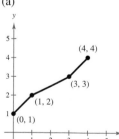

(b)
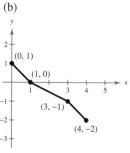

(c)
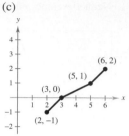

(d)
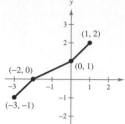

(e)
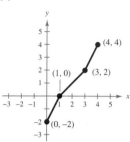

(f)
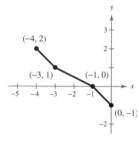

15. Vertical shrink of $y = x$
$y = \frac{1}{2}x$

17. Constant function
$y = 7$

19. Reflection in the x-axis and a vertical shift of $y = \sqrt{x}$
$y = 1 - \sqrt{x}$

21. Horizontal shift of $y = |x|$
$y = |x + 2|$

23. Vertical shift of $y = x^2$
$y = x^2 - 1$

25. Reflection in the x-axis of $y = x^3$ followed by a vertical shift
$y = 1 - x^3$

27. Vertical shift 2 units upward

29. Horizontal shift 2 units to the right

31. Vertical stretch   **33.** Horizontal shift 2 units to the left

35. Reflection in the x-axis   **37.** Vertical shrink

39. Reflection in the x-axis and vertical shift 4 units upward

41. Horizontal shift 2 units to the left and vertical shrink

43. Vertical shrink and vertical shift 2 units upward

45.
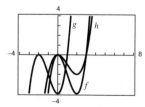

g is a horizontal shift and h is a vertical shrink.

47.

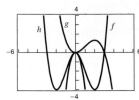

g is a vertical shrink and a reflection in the *x*-axis and *h* is a reflection in the *y*-axis.

49. $g(x) = -(x^3 - 3x^2) + 1$

51. (a) $f(x) = x^2$

(b) Reflection in the *x*-axis and vertical shift 12 units upward

(c)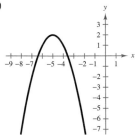

(d) $g(x) = 12 - f(x)$

53. (a) $f(x) = x^2$

(b) Horizontal shift 5 units to the left, reflection in the *x*-axis, and vertical shift 2 units upward

(c)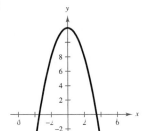

(d) $g(x) = 2 - f(x + 5)$

55. (a) $f(x) = x^2$

(b) Horizontal shift 4 units to the right, vertical stretch, and vertical shift 3 units upward

(c)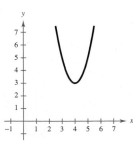

(d) $g(x) = 3 + 2f(x - 4)$

57. (a) $f(x) = x^3$

(b) Vertical shift 7 units upward

(c)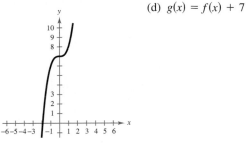

(d) $g(x) = f(x) + 7$

59. (a) $f(x) = x^3$

(b) Horizontal shift 1 unit to the right and vertical shift 2 units upward

(c)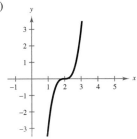

(d) $g(x) = f(x - 1) + 2$

61. (a) $f(x) = x^3$

(b) Horizontal shift 2 units to the right and vertical stretch

(c)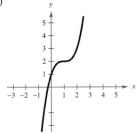

(d) $g(x) = 3f(x - 2)$

63. (a) $f(x) = |x|$

(b) Reflection in the *x*-axis and vertical shift 2 units downward

(c)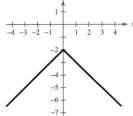

(d) $g(x) = -f(x) - 2$

65. (a) $f(x) = |x|$

(b) Horizontal shift 4 units to the left, reflection in the x-axis, and vertical shift 8 units upward

(c)

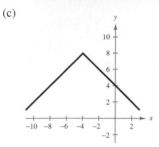

(d) $g(x) = -f(x + 4) + 8$

67. (a) $f(x) = |x|$

(b) Horizontal shift 1 unit to the right, reflection in the x-axis, and vertical stretch

(c)

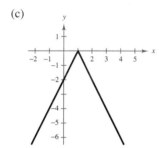

(d) $g(x) = -2f(x - 1)$

69. (a) $f(x) = \sqrt{x}$

(b) Horizontal shift 9 units to the right

(c)

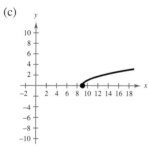

(d) $g(x) = f(x - 9)$

71. (a) $f(x) = \sqrt{x}$

(b) Reflection in the y-axis, horizontal shift 7 units to the right, and vertical shift 2 units downward

(c)

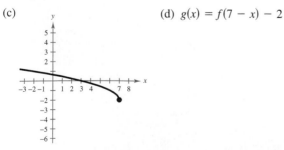

(d) $g(x) = f(7 - x) - 2$

73. (a) $f(x) = \sqrt{x}$

(b) Horizontal shift 1 unit to the right and vertical stretch 4 units

(c)

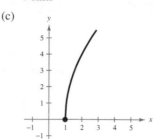

(d) $g(x) = 4f(x - 1)$

75. (a)

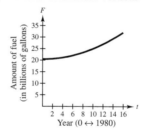

(b) $P(x) = 55 + 20x - 0.5x^2$; vertical shift

(c) $P(x) = 80 + \dfrac{1}{5}x - \dfrac{x^2}{20{,}000}$; horizontal stretch

77. (a) Vertical shrink and vertical shift

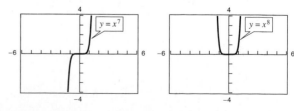

(b) $G(t) = 0.04t^2 + 0.8t + 24.46$; $G(t) = F(t + 10)$

79. (a) To each time t there corresponds one and only one temperature T.

(b) $60°, 72°$

(c) All the temperature changes would be 1 hour later.

(d) The temperature would be decreased by 1 degree.

81. False. The point $(-1, 28)$ does not lie on the graph of $g(x) = -(x - 6)^2 + 3$.

83. $y = x^7$ resembles the cubic graph and $y = x^8$ resembles the quadratic graph. Both are steeper on $(-\infty, -1)$ and $(1, \infty)$ and both are closer to 0 on $-1 < x < 1$.

85.

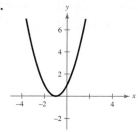

87.

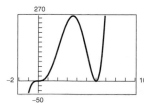

89.

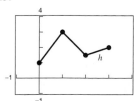

Section P.8 *(page 98)*

1. (a) $2x$ (b) 2 (c) $x^2 - 1$ (d) $\dfrac{x+1}{x-1}$, $x \neq 1$

(e) All $x \neq 1$

3. (a) $x^2 - x + 1$ (b) $x^2 + x - 1$ (c) $x^2 - x^3$

(d) $\dfrac{x^2}{1-x}$, $x \neq 1$ (e) All $x \neq 1$

5. (a) $x^2 + 5 + \sqrt{1-x}$ (b) $x^2 + 5 - \sqrt{1-x}$

(c) $(x^2 + 5)\sqrt{1-x}$ (d) $\dfrac{x^2+5}{\sqrt{1-x}}$, $x < 1$

(e) $x < 1$

7. (a) $\dfrac{x+1}{x^2}$ (b) $\dfrac{x-1}{x^2}$ (c) $\dfrac{1}{x^3}$ (d) x, $x \neq 0$

(e) $x \neq 0$

9. 9 **11.** 5 **13.** 0 **15.** 26 **17.** $4t^2 - 2t + 5$

19. $-125t^3 - 100t^2 - 5t - 4$ **21.** $\dfrac{t^2+1}{-t-4}$

23.

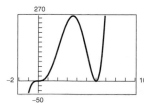

25.

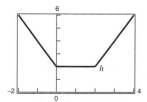

27.

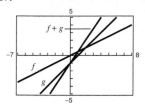

29.

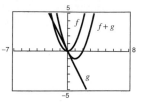

31.

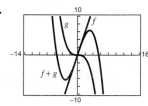

$f(x), 0 \leq x \leq 2$;

$g(x), x > 6$

33.

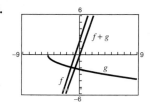

$f(x), 0 \leq x \leq 2$;

$f(x), x > 6$

35. (a) $(x-1)^2$ (b) $x^2 - 1$

37. (a) $20 - 3x$ (b) $-3x$

39. (a) $(f \circ g)(x) = \sqrt{x^2 + 4}$

$(g \circ f)(x) = x + 4$, $x \geq -4$

(b)

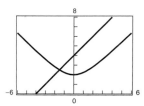

Not equal

41. (a) $(f \circ g)(x) = x - \frac{8}{3}$; $(g \circ f)(x) = x - 8$

(b)

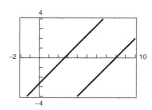

Not equal

43. (a) $(f \circ g)(x) = x^4$; $(g \circ f)(x) = x^4$

(b)

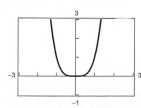

Equal

45. (a) $(f \circ g)(x) = 24 - 5x$; $(g \circ f)(x) = -5x$

(b) $24 - 5x \neq -5x$

(c)

x	0	1	2	3
$g(x)$	4	3	2	1
$(f \circ g)(x)$	24	19	14	9

x	0	1	2	3
$f(x)$	4	9	14	19
$(g \circ f)(x)$	0	-5	-10	-15

47. (a) $(f \circ g)(x) = \sqrt{x^2 + 1}$; $(g \circ f)(x) = x + 1,\ x \geq -6$

(b) $x + 1 \neq \sqrt{x^2 + 1}$

(c)

x	0	1	2	3
$g(x)$	-5	-4	-1	4
$(f \circ g)(x)$	1	$\sqrt{2}$	$\sqrt{5}$	$\sqrt{10}$

x	0	1	2	3
$f(x)$	$\sqrt{6}$	$\sqrt{7}$	$\sqrt{8}$	3
$(g \circ f)(x)$	1	2	3	4

49. (a) $(f \circ g)(x) = |2x + 2|$; $(g \circ f)(x) = 2|x + 3| - 1$

(b) $(f \circ g)(x) = \begin{cases} 2x + 2, & x \geq -1 \\ -2x - 2, & x < -1 \end{cases}$

$(g \circ f)(x) = \begin{cases} 2x + 5, & x \geq -3 \\ -2x - 7, & x < -3 \end{cases}$

$(f \circ g)(x) \neq (g \circ f)(x)$

(c)

x	0	-1	-3	-5
$g(x)$	-1	-3	-7	-11
$(f \circ g)(x)$	2	0	4	8

x	0	-1	-3	-5
$f(x)$	3	2	0	2
$(g \circ f)(x)$	5	3	-1	3

51. (a) 3   (b) 0   **53.** (a) 0   (b) 4

55. (a) 0   (b) 4   **57.** $f(x) = x^2, g(x) = 2x + 1$

59. $f(x) = \sqrt[3]{x}, g(x) = x^2 - 4$

61. $f(x) = \dfrac{1}{x}, g(x) = x + 2$

63. $f(x) = x^2 + 2x, g(x) = x + 4$

65. (a) $x \geq 0$ (b) All real numbers   (c) All real numbers

67. (a) All real numbers except $x = 0$ (b) All real numbers

(c) All real numbers except $x = -3$

69. (a) All real numbers except $x = 0$ (b) All real numbers

(c) All real numbers except $x = 1$

71. $3, h \neq 0$ **73.** $-2x - h,\ h \neq 0$

75. $\dfrac{-4}{x(x + h)},\ h \neq 0$

77. $\dfrac{2}{\sqrt{2(x + h) + 1} + \sqrt{2x + 1}},\ h \neq 0$

79. (a) $T = \frac{3}{4}x + \frac{1}{15}x^2$

(b)

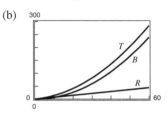

(c) B. For example, $B(60) = 240$ while $R(60)$ is only 45.

81. $y_1 = -0.59x^2 + 7.66x + 144.9$

$y_2 = 16.58x + 245.06$

$y_3 = 1.84x + 21.92$

83. $(A \circ r)(t) = 0.36\pi t^2$

$A \circ r$ represents the area of the circle at time t.

85. (a) $(C \circ x)(t) = 3000t + 750$

$C \circ x$ represents the cost after t production hours.

(b)

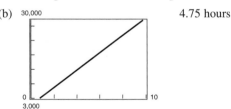

4.75 hours

87. $g(f(x))$ represents 3 percent of an amount over \$500,000.

89.

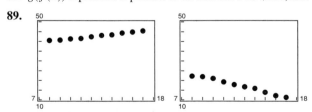

Both data sets appear to be linear.

$y_1 = 0.57x + 35.6$; $y_2 = -1.29x + 33.3$

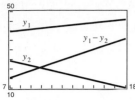

The difference between the morning and evening newspaper circulations is increasing.

91. True   **93.** Odd (Proofs will vary.)

95. $\frac{1}{2}[f(x) + f(-x)] + \frac{1}{2}[f(x) - f(-x)]$

$\qquad = \frac{1}{2}[f(x) + f(-x) + f(x) - f(-x)]$

$\qquad = \frac{1}{2}[2f(x)]$

$\qquad = f(x)$

Section P.9   *(page 109)*

1. (c)   **3.** (a)   **5.** $f^{-1}(x) = \frac{1}{8}x$

7. $f^{-1}(x) = x - 10$ **9.** $f^{-1}(x) = \frac{1}{2}(x - 1)$

11. $f^{-1}(x) = x^3$

13. (a) $f(g(x)) = f\left(\dfrac{x}{2}\right) = 2\left(\dfrac{x}{2}\right) = x$

$\qquad g(f(x)) = g(2x) = \dfrac{(2x)}{2} = x$

(b)

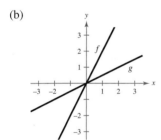

15. (a) $f(g(x)) = f\left(\dfrac{x-1}{5}\right) = 5\left(\dfrac{x-1}{5}\right) + 1 = x$

$\qquad g(f(x)) = g(5x + 1) = \dfrac{(5x + 1) - 1}{5} = x$

(b)

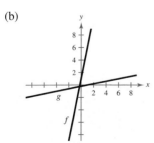

17. $f(g(x)) = f(\sqrt[3]{x}) = (\sqrt[3]{x})^3 = x$

$\qquad g(f(x)) = g(x^3) = \sqrt[3]{x^3} = x$

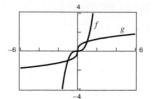

Reflections in the line $y = x$

19. $f(g(x)) = f(x^2 + 4), \quad x \geq 0$

$\qquad = \sqrt{(x^2 + 4) - 4} = x$

$\qquad g(f(x)) = g(\sqrt{x - 4})$

$\qquad\qquad = (\sqrt{x - 4})^2 + 4 = x$

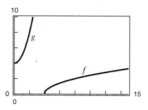

Reflections in the line $y = x$

21. $f(g(x)) = f(\sqrt[3]{1 - x}) = 1 - (\sqrt[3]{1 - x})^3 = x$

$\qquad g(f(x)) = g(1 - x^3) = \sqrt[3]{1 - (1 - x^3)} = x$

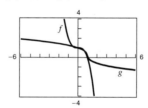

Reflections in the line $y = x$

23. (a) $f(g(x)) = f\left(-\dfrac{2x + 6}{7}\right)$

$\qquad\qquad = -\dfrac{7}{2}\left(-\dfrac{2x + 6}{7}\right) - 3 = x$

$\qquad g(f(x)) = g\left(-\dfrac{7}{2}x - 3\right)$

$\qquad\qquad = -\dfrac{2\left(-\frac{7}{2}x - 3\right) + 6}{7} = x$

(b)

x	0	2	-2	6
$f(x)$	-3	-10	4	-24

x	-3	-10	4	-24
$g(x)$	0	2	-2	6

25. (a) $f(g(x)) = f(\sqrt[3]{x - 5}) = (\sqrt[3]{x - 5})^3 + 5 = x$

       $g(f(x)) = g(x^3 + 5) = \sqrt[3]{(x^3 + 5) - 5} = x$

(b)

x	0	1	-1	-2	4
$f(x)$	5	6	4	-3	69

x	5	6	4	-3	69
$g(x)$	0	1	-1	-2	4

27. (a) $f(g(x)) = f(8 + x^2)$

              $= -\sqrt{(8 + x^2) - 8}$

              $= -\sqrt{x^2} = -(-x) = x,\ x \le 0$

       $g(f(x)) = g(-\sqrt{x - 8})$

              $= 8 + (-\sqrt{x - 8})^2$

              $= 8 + (x - 8) = x,\ x \ge 8$

(b)

x	8	9	12	15
$f(x)$	0	-1	-2	$-\sqrt{7}$

x	0	-1	-2	$-\sqrt{7}$
$g(x)$	8	9	12	15

29.

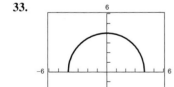

One-to-one

31.

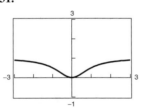

Not one-to-one

33.

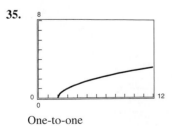

Not one-to-one

35.

One-to-one

37.

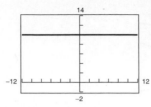

Not one-to-one

39.

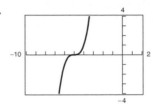

One-to-one

41.

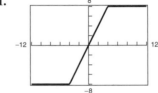

Not one-to-one

43. $f^{-1}(x) = \dfrac{x + 3}{2}$

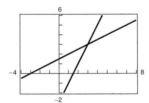

Reflections in the line $y = x$

45. $f^{-1}(x) = \sqrt[5]{x}$

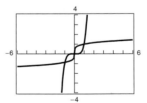

Reflections in the line $y = x$

47. $f^{-1}(x) = x^2, x \geq 0$

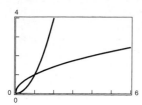

Reflections in the line $y = x$

49. $f^{-1}(x) = \sqrt{4 - x^2}, 0 \leq x \leq 2$

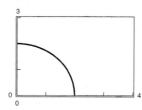

Reflections in the line $y = x$

51. $f^{-1}(x) = x^3 + 1$

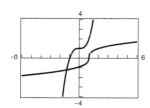

Reflections in the line $y = x$

53. $f^{-1}(x) = \dfrac{4}{x}$

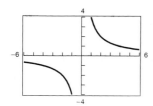

Reflections in the line $y = x$

55. Not one-to-one

57. $f^{-1}(x) = \dfrac{5x - 4}{3}$ **59.** $f^{-1}(x) = \sqrt{x} - 3, x \geq 0$

61. Not one-to-one **63.** $f^{-1}(x) = \dfrac{x^2 - 3}{2}, x \geq 0$

65. Not one-to-one **67.** $f^{-1}(x) = \dfrac{x - b}{a}$

69. $y = \sqrt{x} + 2, x \geq 0$ **71.** $y = x - 2, x \geq 0$

73.

x	-4	-2	2	3
$f^{-1}(x)$	-2	-1	1	3

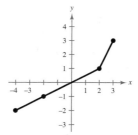

75. (a) and (b)

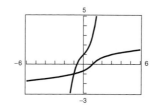

(c) Inverse function because it satisfies the Vertical Line Test

77. (a) and (b)

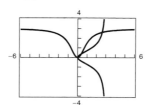

(c) Not an inverse function because it does not satisfy the Vertical Line Test

79. 32 **81.** 600 **83.** $2\sqrt[3]{x + 3}$

85. $\dfrac{x + 1}{2}$ **87.** $\dfrac{x + 1}{2}$

89. (a) $y = \dfrac{x - 8}{0.75}$

$y = $ number of units produced

$x = $ hourly wage

(b)

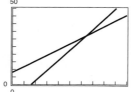

(c) $15.50 (d) 19 units

91. (a) Yes

(b) $f^{-1}(t)$ represents the year new car sales totaled $\$t$ billion.

(c) 5, or 1995

(d) No. The inverse is not a function because f is not one-to-one.

93. True   **95.** Answers will vary.

Review Exercises *(page 114)*

1. (a) 11   (b) 11, -14 (c) 11, -14, $-\frac{8}{9}$, $\frac{5}{2}$, 0.4

(d) $\sqrt{6}$

3. $0.8\overline{3} < 0.875$

5. The set of all real numbers less than or equal to 7.

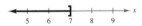

7. 122   **9.** $|x - 7| \geq 4$ **11.** $|y + 30| < 5$

13. (a) -13 (b) 27   **15.** (a) -18 (b) -12

17. Associative Property of Addition

19. Commutative Property of Multiplication

21. $\dfrac{14}{9}$ **23.** $\dfrac{1}{24}$ **25.** $\dfrac{47x}{60}$

27.

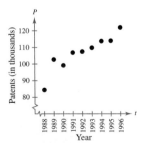

Quadrant IV

29.

Quadrant II

31. IV

33. (a)

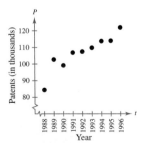

(b) The number of patents issued has had a fairly steady increase since 1988.

35.

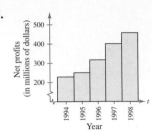

37. $d \approx 9.93$

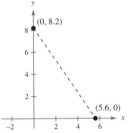

39.

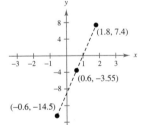

Opposite sides have equal lengths of $\sqrt{10}$ and $5\sqrt{2}$.

41.

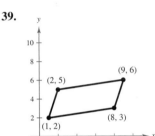

43.

x	-2	0	2	3	4
y	3	2	1	$\frac{1}{2}$	0

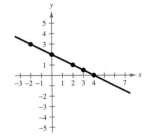

45.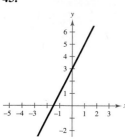

47.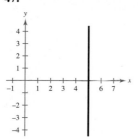

49.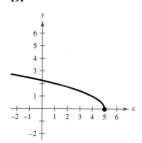

51.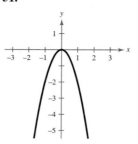

53.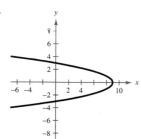

55.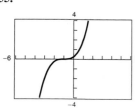

57.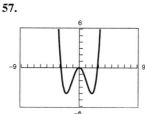

Intercepts: $(-1, 0)$, $\left(0, \frac{1}{4}\right)$ Intercepts: $(0, 0)$, $\left(\pm 2\sqrt{2}, 0\right)$

59.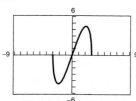

61.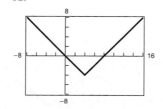

Intercepts: $(0, 0)$, $(\pm 3, 0)$ Intercepts: $(0, 0)$, $(8, 0)$

63. $(x - 3)^2 + (y + 1)^2 = 68$

65. (a)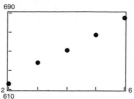

(b) $y = 16.7x + 585$

(c)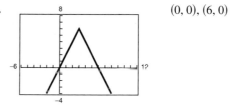

(d) 2000: \$752; 2002: \$785

67. $x = 6$ **69.** $x = -3$ **71.** $(20, 0)$, $(0, -4)$

73. $(0, 25)$, $(5, 0)$, $(-5, 0)$

75.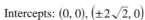

$(0, 0)$, $(6, 0)$

77. $x = 9.4$ **79.** $x \approx -2.722$ **81.** No solution

83. $(5, 2)$ **85.** $(-1, 8)$ **87.** $-\frac{5}{2}, 3$ **89.** $\pm\frac{5}{4}$

91. $-3 \pm 2\sqrt{3}$ **93.** $\dfrac{-1 \pm \sqrt{61}}{2}$ **95.** $0, -\dfrac{13}{2}$

97. $0, \dfrac{1}{6}$ **99.** $0, \dfrac{3}{2}$ **101.** 66   **103.** $\dfrac{38 + 5\sqrt{3}}{24}$

105. 79   **107.** $\frac{7}{3}$ **109.** $-2, 0$ **111.** $-5, 2$

113. $2, 3$

115.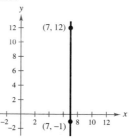

m is undefined.

117.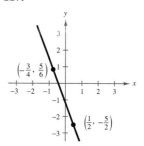

$m = -\frac{8}{3}$

119.

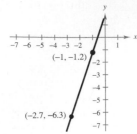

$m = 3$

121. $t = \frac{11}{4}$ **123.** $t = -\frac{53}{3}$

125. (a) $3x + 2y - 1 = 0$ (b) $(-1, 2), (1, -1), (-5, 8)$

127. (a) $2x + 3y - 6 = 0$ (b) $(0, 2), (6, -2), (-3, 4)$

129. (a) $4x - y + 30 = 0$ (b) $(0, 30), (5, 50), (10, 70)$

131. (a) $32x + 40y - 35 = 0$

   (b) $\left(3, -\frac{61}{40}\right), \left(\frac{35}{32}, 0\right), \left(-1, \frac{67}{40}\right)$

133. (a) $y = 8$ (b) $(0, 8), (-3, 8), (5, 8)$

135. (a) $x = 5$ (b) $(5, 1), (5, 2), (5, -5)$

137. (a) $x = 0$

   (b)

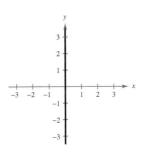

139. (a) $y = -\frac{12}{5}x - \frac{14}{5}$

   (b)

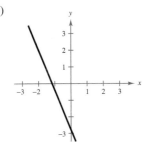

141. (a) $y = -\frac{4}{3}x + \frac{22}{3}$

   (b)

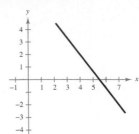

143. $V = 5.15t + 72.95$

145. (a) $2x + 3y + 7 = 0$

   (b) $3x - 2y + 30 = 0$

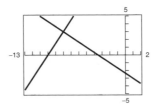

147. (a) $y = -4$

   (b) $x = 3$

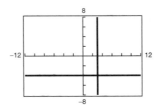

149. (a) Not a function because element u in A corresponds to two elements, 2 and -2, in B.

   (b) Function

   (c) Function

   (d) Not a function because element w in A corresponds to two elements, -2 and 2, in B.

151. Function   **153.** Not a function

155. (a) 16   (b) $(t + 1)^{4/3}$ (c) $\frac{15}{7}$ (d) $x^{4/3}$

157. $(-\infty, \infty)$ **159.** $(-\infty, -8]$ and $[0, \infty)$

161. All real numbers x except $x = -\frac{4}{3}$

163.

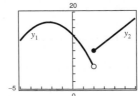

1985: \$12.61 billion

1990: \$14.16 billion

1995: \$16.208 billion

165. Domain: $\left(-\infty, -\dfrac{1}{\sqrt{2}}\right], \left[\dfrac{1}{\sqrt{2}}, \infty\right)$; Range: $[0, \infty)$

167. Domain: All real numbers; Range: $[0, \infty)$

169. (a)

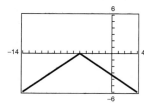

(b) Function

171. (a)

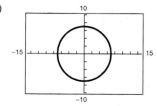

(b) Not a function

173. Increasing on $(3, \infty)$;

Decreasing on $(-\infty, -3)$

175. Increasing on $(-8, \infty)$;

Decreasing on $(-\infty, -8)$

177. Relative minimum: $\left(\dfrac{1}{2}, -\dfrac{5}{4}\right)$

179. Relative maximum: $(0, -1)$;

Relative minimum: $(2.67, -10.48)$

181.

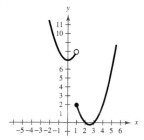

183. Neither even nor odd

185. Absolute value function $f(x) = |x|$

Vertical shift 3 units upward

$g(x) = |x| + 3$

187. Square root function $f(x) = \sqrt{x}$

Horizontal shift 3 units to the right and reflection in x-axis

$g(x) = -\sqrt{x - 3}$

189.

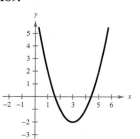

191.

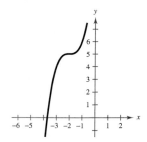

193.

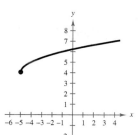

195.

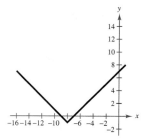

197.

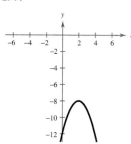

199.

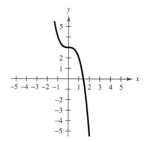

201.

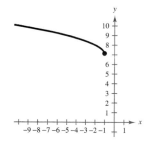

203.

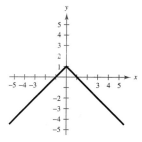

205.

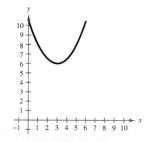

207.

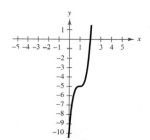

209.   **211.**

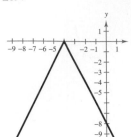   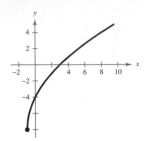

213. 70   **215.** -4 **217.** $\frac{1}{5}$

219. $\sqrt{7}$ **221.** $\sqrt{110}$

223. $\$164.4$ billion

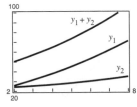

225. $f^{-1}(x) = 12x$

$f(f^{-1}(x)) = f(12x) = \frac{1}{12}(12x) = x$

$f^{-1}(f(x)) = f^{-1}\left(\frac{1}{12}x\right) = 12\left(\frac{1}{12}x\right) = x$

227. $f^{-1}(x) = x - 5$

$f(f^{-1}(x)) = f(x - 5) = (x - 5) + 5 = x$

$f^{-1}(f(x)) = f^{-1}(x + 5) = (x + 5) - 5 = x$

229.

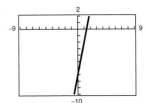

$f^{-1}(x) = \dfrac{x + 7}{5}$

231.

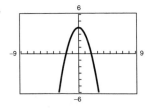

$f^{-1}(x)$ does not exist.

233.

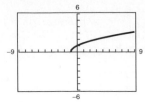

$f^{-1}(x) = x^2 - 1, x \geq 0$

235.

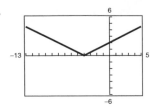

$f^{-1}(x)$ does not exist.

237. $f^{-1}(x) = \dfrac{8x - 3}{7}$ **239.** $f^{-1}(x) = \sqrt[3]{x + 2}$

241. $f^{-1}(x) = -\dfrac{x^2}{16} + 6, x \geq 0$

243. False. Definition of a function

245. True. For odd n, $f(x) = x^n$ is a one-to-one function.

247. $x - 1$ is not equal to $1 - x$.

$\dfrac{x - 1}{-(x - 1)} = \dfrac{1}{-1} = -1$

249. The 2 was not raised to the fourth power.

$(2x)^4 = 16x^4$

251. An equation that is true for every real number in the domain of the variable is called an identity. An equation that is true for just some of the real numbers in the domain of the variable is called a conditional equation.

253. Vertical lines of the form $x = c$ are not functions because an infinite number of y values correspond with the value of $x = c$.

Chapter Test   *(page 121)*

1. $-\frac{10}{3} > -|-4|$ **2.** 56   **3.** (a) 0   (b) $\frac{39}{4}$

4.

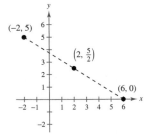

Midpoint: $\left(2, \frac{5}{2}\right)$; Distance $\sqrt{89}$

5.

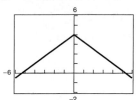

6.

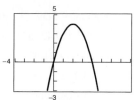

(c)

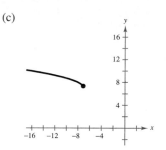

$(0, 4), \left(-\frac{16}{3}, 0\right), \left(\frac{16}{3}, 0\right)$ $(0, 0), (4, 0)$

7.

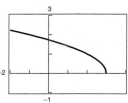

8.

$(x - 4)^2 + (y + 1)^2 = 34$

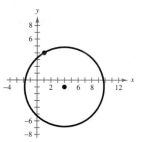

$\left(0, \sqrt{3}\right), (3, 0)$

21. (a) $x^2 - \sqrt{2 - x}, \ (-\infty, 2]$

(b) $\dfrac{x^2}{\sqrt{2 - x}}, \ (-\infty, 2)$

(c) $2 - x, \ (-\infty, 2]$

(d) $2 - x^2, \ [0, \infty)$

22. $f^{-1}(x) = \sqrt{x + 1} - 2$

Chapter 1

Section 1.1   *(page 131)*

1. 2   **3.** −3

5. (a) Quadrant I   (b) Quadrant III

7. (a) Quadrant IV   (b) Quadrant II

9. (a) Quadrant III   (b) Quadrant II

11. (a)   (b)

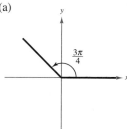 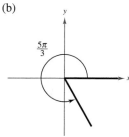

13. (a)   (b)

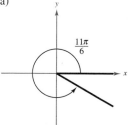 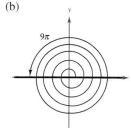

9. $x = 1, \ x = 9$ **10.** $x = 2$ **11.** $x = -\frac{5}{2}, \ x = \frac{11}{4}$

12. $2x - 5y + 20 = 0$

13. No. To some x there corresponds more than one value of y.

14. 8   **15.** $10 - \sqrt{6 - t}$ **16.** $\dfrac{\sqrt{3 - x} - 1}{2 - x}$

17. Increasing: $(-2, 0), (2, \infty)$

Decreasing: $(-\infty, -2), (0, 2)$

18. Increasing: $(-2, 2)$

Constant: $(-\infty, -2), (2, \infty)$

19. (a) $f(x) = x^3$

(b) Horizontal shift right 5 units, reflection in x-axis, vertical stretch, and then a vertical shift 3 units upward

(c)

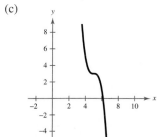

20. (a) $f(x) = \sqrt{x}$

(b) Reflection in y-axis, horizontal shift 7 units left, and a vertical shift 7 units up

15. (a) $\dfrac{25\pi}{12}, -\dfrac{23\pi}{12}$ (b) $\dfrac{8\pi}{3}, -\dfrac{4\pi}{3}$

17. (a) $\dfrac{5\pi}{4}, -\dfrac{3\pi}{4}$ (b) $\dfrac{28\pi}{15}, -\dfrac{32\pi}{15}$

19. (a) Complement: $\dfrac{\pi}{6}$; Supplement: $\dfrac{2\pi}{3}$

(b) Complement: none; Supplement: $\dfrac{\pi}{4}$

21. (a) Complement: none; Supplement: none

(b) Complement: none; Supplement: $\dfrac{\pi}{2}$

23. $210°$ **25.** $-45°$

27. (a) Quadrant II   (b) Quadrant IV

29. (a) Quadrant III   (b) Quadrant I

31. (a)   (b)

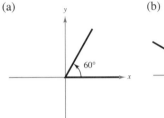   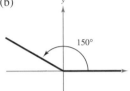

33. (a)   (b)

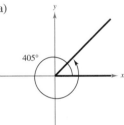   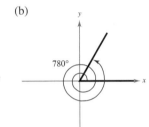

35. (a) $412°, -308°$ (b) $324°, -396°$

37. (a) $660°, -60°$ (b) $590°, -130°$

39. (a) Complement: $66°$ (b) Complement: none
        Supplement: $156°$ Supplement: $54°$

41. (a) Complement: $11°$ (b) Complement: none
        Supplement: $101°$ Supplement: $30°$

43. (a) $\dfrac{\pi}{6}$ (b) $\dfrac{5\pi}{6}$ **45.** (a) $-\dfrac{\pi}{9}$ (b) $-\dfrac{4\pi}{3}$

47. 2.007 **49.** -3.776 **51.** 11.205 **53.** -0.014

55. (a) $270°$ (b) $-210°$ **57.** (a) $420°$ (b) $-39°$

59. $25.714°$ **61.** $562.5°$ **63.** $-756°$

65. $-114.592°$ **67.** (a) $64.75°$ (b) $-124.5°$

69. (a) $85.308°$ (b) $-408.274°$

71. (a) $280°\,36'$ (b) $-115°\,48'$

73. (a) $257°\,49'\,52''$ (b) $-205°\,7'\,8''$

75. $\frac{6}{5}$ rad   **77.** $4\frac{4}{7}$ rad   **79.** $\frac{8}{15}$ rad   **81.** $2\frac{12}{29}$ rad

83. 14π inches ≈ 43.98 inches

85. 4π meters ≈ 12.57 meters   **87.** 1141.02 miles

89. 0.094 rad $\approx 5.39°$ **91.** $\frac{5}{12}$ rad $\approx 23.87°$

93. (a) $540° \approx 9.42$ rad   (b) $900° \approx 15.71$ rad

(c) $1260° \approx 21.99$ rad

95. (a) 448.2 rev/min

(b) 2816 rad/min

97. 20.16π in./sec

99. False. A radian is larger: 1 rad $\approx 57.3°$.

101. True. The sum of the angles is π, or $180°$.

103. Two angles in standard position are coterminal angles if they have the same initial and terminal sides.

105. $\dfrac{50\pi}{3}$ square meters

107. (a) $A = 0.4r^2,\ r > 0;\ s = 0.8r,\ r > 0$

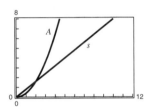

The area function changes more rapidly for $r > 1$ because it is quadratic and the arc length function is linear.

(b) $A = 50\theta,\ 0 < \theta < 2\pi;\ s = 10\theta,\ 0 < \theta < 2\pi$

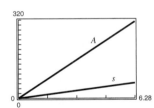

109. Decreasing on $\left(-\infty, -\dfrac{\sqrt{3}}{2}\right)$ and $\left(0, \dfrac{\sqrt{3}}{2}\right)$, increasing on $\left(-\dfrac{\sqrt{3}}{2}, 0\right)$ and $\left(\dfrac{\sqrt{3}}{2}, \infty\right)$

111.   **113.**

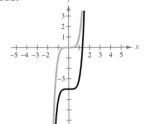   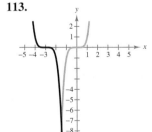

Section 1.2   *(page 140)*

1. $\sin\theta = \frac{15}{17}$

$\cos\theta = -\frac{8}{17}$

$\tan\theta = -\frac{15}{8}$

$\csc\theta = \frac{17}{15}$

$\sec\theta = -\frac{17}{8}$

$\cot\theta = -\frac{8}{15}$

3. $\sin\theta = -\frac{5}{13}$

$\cos\theta = \frac{12}{13}$

$\tan\theta = -\frac{5}{12}$

$\csc\theta = -\frac{13}{5}$

$\sec\theta = \frac{13}{12}$

$\cot\theta = -\frac{12}{5}$

5. $\left(\dfrac{\sqrt{2}}{2}, \dfrac{\sqrt{2}}{2}\right)$ **7.** $\left(-\dfrac{\sqrt{3}}{2}, -\dfrac{1}{2}\right)$ **9.** $\left(-\dfrac{1}{2}, -\dfrac{\sqrt{3}}{2}\right)$

11. $(0, -1)$

13. $\sin\dfrac{\pi}{4} = \dfrac{\sqrt{2}}{2}$

$\cos\dfrac{\pi}{4} = \dfrac{\sqrt{2}}{2}$

$\tan\dfrac{\pi}{4} = 1$

15. $\sin\left(-\dfrac{\pi}{6}\right) = -\dfrac{1}{2}$

$\cos\left(-\dfrac{\pi}{6}\right) = \dfrac{\sqrt{3}}{2}$

$\tan\left(-\dfrac{\pi}{6}\right) = -\dfrac{\sqrt{3}}{3}$

17. $\sin\left(-\dfrac{7\pi}{4}\right) = \dfrac{\sqrt{2}}{2}$

$\cos\left(-\dfrac{7\pi}{4}\right) = \dfrac{\sqrt{2}}{2}$

$\tan\left(-\dfrac{7\pi}{4}\right) = 1$

19. $\sin\dfrac{11\pi}{6} = -\dfrac{1}{2}$

$\cos\dfrac{11\pi}{6} = \dfrac{\sqrt{3}}{2}$

$\tan\dfrac{11\pi}{6} = -\dfrac{\sqrt{3}}{3}$

21. $\sin\left(-\dfrac{3\pi}{2}\right) = 1$

$\cos\left(-\dfrac{3\pi}{2}\right) = 0$

$\tan\left(-\dfrac{3\pi}{2}\right)$ is undefined.

23. $\sin\dfrac{3\pi}{4} = \dfrac{\sqrt{2}}{2}$ $\csc\dfrac{3\pi}{4} = \sqrt{2}$

$\cos\dfrac{3\pi}{4} = -\dfrac{\sqrt{2}}{2}$ $\sec\dfrac{3\pi}{4} = -\sqrt{2}$

$\tan\dfrac{3\pi}{4} = -1$ $\cot\dfrac{3\pi}{4} = -1$

25. $\sin\dfrac{\pi}{2} = 1$ $\csc\dfrac{\pi}{2} = 1$

$\cos\dfrac{\pi}{2} = 0$ $\sec\dfrac{\pi}{2}$ is undefined.

$\tan\dfrac{\pi}{2}$ is undefined. $\cot\dfrac{\pi}{2} = 0$

27. $\sin\left(-\dfrac{\pi}{3}\right) = -\dfrac{\sqrt{3}}{2}$ $\csc\left(-\dfrac{\pi}{3}\right) = -\dfrac{2\sqrt{3}}{3}$

$\cos\left(-\dfrac{\pi}{3}\right) = \dfrac{1}{2}$ $\sec\left(-\dfrac{\pi}{3}\right) = 2$

$\tan\left(-\dfrac{\pi}{3}\right) = -\sqrt{3}$ $\cot\left(-\dfrac{\pi}{3}\right) = -\dfrac{\sqrt{3}}{3}$

29. $\sin 5\pi = \sin\pi = 0$ **31.** $\cos\dfrac{8\pi}{3} = \cos\dfrac{2\pi}{3} = -\dfrac{1}{2}$

33. $\cos(-3\pi) = \cos(-\pi) = -1$

35. $\sin\left(-\dfrac{9\pi}{4}\right) = \sin\dfrac{7\pi}{4} = -\dfrac{\sqrt{2}}{2}$ **37.** (a) $-\dfrac{1}{3}$ (b) -3

39. (a) $-\dfrac{1}{5}$ (b) -5 **41.** (a) $\dfrac{4}{5}$ (b) $-\dfrac{4}{5}$

43. 0.7071   **45.** 1.0378   **47.** -0.1288

49. 1.3940   **51.** -1.4486 **53.** (a) -1 (b) -0.4

55. (a) 0.25, 2.89   (b) 1.82, 4.46

57. (a) 156.25 feet   (b) 239.39 feet   (c) 270.63 feet

59. (a) 0.2500 foot   (b) 0.0177 foot   (c) -0.2475 foot

61. $0.0707 = \cos 1.5 \neq 2\cos 0.75 - 1.4634$

63. False. $\sin(-t) = -\sin t$ means that the function is odd, not that the sine of a negative angle is a negative number.

65. (a) y-axis   (b) $\sin t_1 = \sin(\pi - t_1)$

(c) $\cos(\pi - t_1) = -\cos t_1$

67. Answers will vary.   **69.** It is an even function.

71. $f^{-1}(x) = \sqrt[3]{4(x-1)}$ **73.** $f^{-1}(x) = \dfrac{x}{2-x}, \; x < 2$

75. (a) 10   (b) $(-1, 3)$

77. (a) $\dfrac{\sqrt{733}}{4}$ (b) $\left(-\dfrac{21}{8}, \dfrac{5}{12}\right)$

79.   **81.**

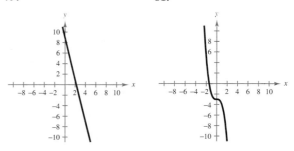

83.

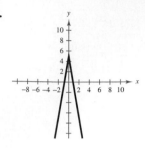

85. (a) -17 (b) 1 (c) $-x^2 + 3x + 1$

87. (a) 6 (b) -55 (c) $\frac{1}{9}$

89. $x \geq 5, \ x \neq 7$ **91.** All real numbers

93. $-h - 6, h \neq 0$

Section 1.3   *(page 150)*

1. $\sin \theta = \frac{3}{5}$ $\csc \theta = \frac{5}{3}$

$\cos \theta = \frac{4}{5}$ $\sec \theta = \frac{5}{4}$

$\tan \theta = \frac{3}{4}$ $\cot \theta = \frac{4}{3}$

3. $\sin \theta = \frac{8}{17}$

$\cos \theta = \frac{15}{17}$

$\tan \theta = \frac{8}{15}$

$\csc \theta = \frac{17}{8}$

$\sec \theta = \frac{17}{15}$

$\cot \theta = \frac{15}{8}$

5. $\sin \theta = \frac{3\sqrt{13}}{13}$

$\cos \theta = \frac{2\sqrt{13}}{13}$

$\tan \theta = \frac{3}{2}$

$\csc \theta = \frac{\sqrt{13}}{3}$

$\sec \theta = \frac{\sqrt{13}}{2}$

$\cot \theta = \frac{2}{3}$

7. $\sin \theta = \frac{1}{3}$

$\cos \theta = \frac{2\sqrt{2}}{3}$

$\tan \theta = \frac{\sqrt{2}}{4}$

$\csc \theta = 3$

$\sec \theta = \frac{3\sqrt{2}}{4}$

$\cot \theta = 2\sqrt{2}$

9. $\sin \theta = \frac{3}{5}$

$\cos \theta = \frac{4}{5}$

$\tan \theta = \frac{3}{4}$

$\csc \theta = \frac{5}{3}$

$\sec \theta = \frac{5}{4}$

$\cot \theta = \frac{4}{3}$

The triangles are similar and corresponding sides are proportional.

The triangles are similar and corresponding sides are proportional.

11. $\cos \theta = \frac{\sqrt{11}}{6}$

$\tan \theta = \frac{5\sqrt{11}}{11}$

$\csc \theta = \frac{6}{5}$

$\sec \theta = \frac{6\sqrt{11}}{11}$

$\cot \theta = \frac{\sqrt{11}}{5}$

13. $\sin \theta = \frac{\sqrt{15}}{4}$

$\cos \theta = \frac{1}{4}$

$\tan \theta = \sqrt{15}$

$\csc \theta = \frac{4\sqrt{15}}{15}$

$\cot \theta = \frac{\sqrt{15}}{15}$

15. $\sin \theta = \frac{3\sqrt{10}}{10}$

$\cos \theta = \frac{\sqrt{10}}{10}$

$\csc \theta = \frac{\sqrt{10}}{3}$

$\sec \theta = \sqrt{10}$

$\cot \theta = \frac{1}{3}$

17. $\sin \theta = \frac{4\sqrt{97}}{97}$

$\cos \theta = \frac{9\sqrt{97}}{97}$

$\tan \theta = \frac{4}{9}$

$\csc \theta = \frac{\sqrt{97}}{4}$

$\sec \theta = \frac{\sqrt{97}}{9}$

19. (a) $\sqrt{3}$ (b) $\frac{1}{2}$ (c) $\frac{\sqrt{3}}{2}$ (d) $\frac{\sqrt{3}}{3}$

21. (a) $\frac{1}{3}$ (b) $\frac{2\sqrt{2}}{3}$ (c) $\frac{\sqrt{2}}{4}$ (d) 3

23. (a) 4 (b) $\pm\frac{\sqrt{15}}{4}$ (c) $\pm\frac{\sqrt{15}}{15}$ (d) $\frac{1}{4}$

25.–31. Answers will vary.

33. (a) $\dfrac{1}{2}$ (b) $\dfrac{\sqrt{3}}{3}$ **35.** (a) 1   (b) $\dfrac{\sqrt{2}}{2}$

37. (a) $\dfrac{\sqrt{3}}{2}$ (b) 2   **39.** (a) 0.4226   (b) 0.4226

41. (a) 1.3499   (b) 1.3432

43. (a) 5.0273   (b) 0.1989

45. (a) 1.1884   (b) 0.5463

47. (a) $30° = \dfrac{\pi}{6}$ (b) $30° = \dfrac{\pi}{6}$

49. (a) $60° = \dfrac{\pi}{3}$ (b) $45° = \dfrac{\pi}{4}$

51. (a) $60° = \dfrac{\pi}{3}$ (b) $45° = \dfrac{\pi}{4}$

53. (a) $55° \approx 0.960$ (b) $89° \approx 1.553$

55. (a) $50° \approx 0.873$ (b) $25° \approx 0.436$

57. $35\sqrt{3}$ **59.** $\dfrac{38\sqrt{3}}{3}$ **61.** 11.5

63. (a)   (b) $\dfrac{6}{3} = \dfrac{h}{135}$

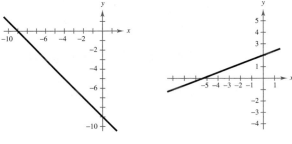

h

132   $\dfrac{6}{3}$

Not drawn to scale

(c) 270 feet

65. 160.03 feet

67. (a)

θ 20 $3\frac{1}{3}$

(b) $\sin \theta = \dfrac{3\frac{1}{3}}{20} = \dfrac{1}{6}$

(c) 9.59°   (d) 9.59°

69. 425.3 meters; 431.9 meters

71. 6.57 centimeters

73. $\sin 75° \approx 0.97$

$\cos 75° \approx 0.26$

$\tan 75° \approx 3.73$

$\csc 75° \approx 1.04$

$\sec 75° \approx 3.86$

$\cot 75° \approx 0.27$

75. True, $\sec x = \csc(90° - x)$ **77.** True for all θ

79.

θ	0°	20°	40°	60°	80°
$\cos \theta$	1	0.9397	0.7660	0.5000	0.1736
$\sin(90° - \theta)$	1	0.9397	0.7660	0.5000	0.1736

$\cos \theta = \sin(90° - \theta)$

θ and $90° - \theta$ are complementary angles.

81.   **83.**

x-intercept: -9 x-intercept: $-5\frac{1}{3}$

y-intercept: -9 y-intercept: 2

85. Quadrant II   **87.** Quadrant I

Section 1.4   *(page 160)*

1. (a) $\sin \theta = \frac{3}{5}$ (b) $\sin \theta = -\frac{15}{17}$

$\cos \theta = \frac{4}{5}$ $\cos \theta = -\frac{8}{17}$

$\tan \theta = \frac{3}{4}$ $\tan \theta = \frac{15}{8}$

$\csc \theta = \frac{5}{3}$ $\csc \theta = -\frac{17}{15}$

$\sec \theta = \frac{5}{4}$ $\sec \theta = -\frac{17}{8}$

$\cot \theta = \frac{4}{3}$ $\cot \theta = \frac{8}{15}$

3. (a) $\sin \theta = -\dfrac{1}{2}$ (b) $\sin \theta = \dfrac{\sqrt{2}}{2}$

$\cos \theta = -\dfrac{\sqrt{3}}{2}$ $\cos \theta = -\dfrac{\sqrt{2}}{2}$

$\tan \theta = \dfrac{\sqrt{3}}{3}$ $\tan \theta = -1$

$\csc \theta = -2$ $\csc \theta = \sqrt{2}$

$\sec \theta = -\dfrac{2\sqrt{3}}{3}$ $\sec \theta = -\sqrt{2}$

$\cot \theta = \sqrt{3}$ $\cot \theta = -1$

5. $\sin \theta = \frac{24}{25}$ **7.** $\sin \theta = -\frac{12}{13}$

$\cos \theta = \frac{7}{25}$ $\cos \theta = \frac{5}{13}$

$\tan \theta = \frac{24}{7}$ $\tan \theta = -\frac{12}{5}$

$\csc \theta = \frac{25}{24}$ $\csc \theta = -\frac{13}{12}$

$\sec \theta = \frac{25}{7}$ $\sec \theta = \frac{13}{5}$

$\cot \theta = \frac{7}{24}$ $\cot \theta = -\frac{5}{12}$

9. $\sin\theta = \dfrac{5\sqrt{29}}{29}$

$\cos\theta = -\dfrac{2\sqrt{29}}{29}$

$\tan\theta = -\dfrac{5}{2}$

$\csc\theta = \dfrac{\sqrt{29}}{5}$

$\sec\theta = -\dfrac{\sqrt{29}}{2}$

$\cot\theta = -\dfrac{2}{5}$

11. $\sin\theta = \dfrac{9\sqrt{85}}{85}$

$\cos\theta = -\dfrac{2\sqrt{85}}{85}$

$\tan\theta = -\dfrac{9}{2}$

$\csc\theta = \dfrac{\sqrt{85}}{9}$

$\sec\theta = -\dfrac{\sqrt{85}}{2}$

$\cot\theta = -\dfrac{2}{9}$

13. Quadrant III **15.** Quadrant II **17.** Quadrant I

19. $\sin\theta = \frac{3}{5}$

$\cos\theta = -\frac{4}{5}$

$\tan\theta = -\frac{3}{4}$

$\csc\theta = \frac{5}{3}$

$\sec\theta = -\frac{5}{4}$

$\cot\theta = -\frac{4}{3}$

21. $\sin\theta = -\frac{15}{17}$

$\cos\theta = \frac{8}{17}$

$\tan\theta = -\frac{15}{8}$

$\csc\theta = -\frac{17}{15}$

$\sec\theta = \frac{17}{8}$

$\cot\theta = -\frac{8}{15}$

23. $\sin\theta = \dfrac{\sqrt{3}}{2}$

$\cos\theta = -\dfrac{1}{2}$

$\tan\theta = -\sqrt{3}$

$\csc\theta = \dfrac{2\sqrt{3}}{3}$

$\sec\theta = -2$

$\cot\theta = -\dfrac{\sqrt{3}}{3}$

25. $\sin\theta = 0$

$\cos\theta = -1$

$\tan\theta = 0$

$\csc\theta$ is undefined.

$\sec\theta = -1$

$\cot\theta$ is undefined.

27. $\sin\theta = \dfrac{\sqrt{2}}{2}$

$\cos\theta = -\dfrac{\sqrt{2}}{2}$

$\tan\theta = -1$

$\csc\theta = \sqrt{2}$

$\sec\theta = -\sqrt{2}$

$\cot\theta = -1$

29. $\sin\theta = -\dfrac{2\sqrt{5}}{5}$

$\cos\theta = -\dfrac{\sqrt{5}}{5}$

$\tan\theta = 2$

$\csc\theta = -\dfrac{\sqrt{5}}{2}$

$\sec\theta = -\sqrt{5}$

$\cot\theta = \dfrac{1}{2}$

31. -1 **33.** 0 **35.** 1 **37.** Undefined

39. $\theta' = 28°$

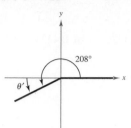

41. $\theta' = 65°$

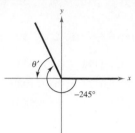

43. $\theta' = 68°$

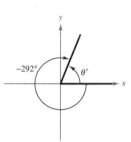

45. $\theta' = \dfrac{\pi}{3}$

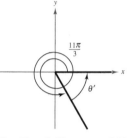

47. $\theta' = 3.5 - \pi$

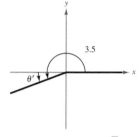

49. $\theta' = 3.68 - \pi \approx 0.5384$

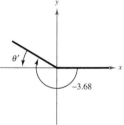

51. $\sin(225°) = -\dfrac{\sqrt{2}}{2}$

$\cos(225°) = -\dfrac{\sqrt{2}}{2}$

$\tan(225°) = 1$

53. $\sin(-750°) = -\dfrac{1}{2}$

$\cos(-750°) = \dfrac{\sqrt{3}}{2}$

$\tan(-750°) = -\dfrac{\sqrt{3}}{3}$

55. $\sin(-240°) = \dfrac{\sqrt{3}}{2}$

$\cos(-240°) = -\dfrac{1}{2}$

$\tan(-240°) = -\sqrt{3}$

57. $\sin\dfrac{5\pi}{3} = -\dfrac{\sqrt{3}}{2}$

$\cos\dfrac{5\pi}{3} = \dfrac{1}{2}$

$\tan\dfrac{5\pi}{3} = -\sqrt{3}$

59. $\sin\left(-\dfrac{\pi}{6}\right) = -\dfrac{1}{2}$

$\cos\left(-\dfrac{\pi}{6}\right) = \dfrac{\sqrt{3}}{2}$

$\tan\left(-\dfrac{\pi}{6}\right) = -\dfrac{\sqrt{3}}{3}$

61. $\sin\dfrac{11\pi}{4} = \dfrac{\sqrt{2}}{2}$

$\cos\dfrac{11\pi}{4} = -\dfrac{\sqrt{2}}{2}$

$\tan\dfrac{11\pi}{4} = -1$

63. $\sin\left(-\frac{7\pi}{6}\right) = \frac{1}{2}$

$\cos\left(-\frac{7\pi}{6}\right) = -\frac{\sqrt{3}}{2}$

$\tan\left(-\frac{7\pi}{6}\right) = -\frac{\sqrt{3}}{3}$

65. 0.1736   **67.** 1.7321

69. -0.3420 **71.** 5.7588   **73.** 0.6052

75. 0.3640   **77.** -2.9238

79. (a) $30° = \frac{\pi}{6}, 150° = \frac{5\pi}{6}$ (b) $210° = \frac{7\pi}{6}, 330° = \frac{11\pi}{6}$

81. (a) $60° = \frac{\pi}{3}, 120° = \frac{2\pi}{3}$ (b) $135° = \frac{3\pi}{4}, 315° = \frac{7\pi}{4}$

83. (a) $150° = \frac{5\pi}{6}, 210° = \frac{7\pi}{6}$ (b) $120° = \frac{2\pi}{3}, 240° = \frac{4\pi}{3}$

85. $54.99°, 125.01°$ **87.** $33.12°, 213.12°$

89. $144.78°, 215.22°$ **91.** $208.64°, 331.36°$

93. $10°, 350°$ **95.** $50°, 230°$ **97.** $\frac{4}{5}$

99. $-\frac{\sqrt{13}}{2}$ **101.** $\frac{8}{5}$

103. (a) $25.2°F$ (b) $65.1°F$ (c) $50.8°F$

105. (a) 12 miles   (b) 6 miles   (c) 6.9 miles

107. False. $\theta' \neq 180° - \theta$

109. (a)

θ	0°	20°	40°	60°	80°
$\sin\theta$	0	0.3420	0.6428	0.8660	0.9848
$\sin(180° - \theta)$	0	0.3420	0.6428	0.8660	0.9848

(b) $\sin\theta = \sin(180° - \theta)$

111. Answers will vary.

113. $10x - y + 38 = 0$ **115.** $30x + 11y - 34 = 0$

117.   **119.**

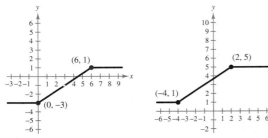

121.

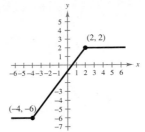

123. $\sin\theta = \frac{3\sqrt{73}}{73}$

$\cos\theta = \frac{8\sqrt{73}}{73}$

$\tan\theta = \frac{3}{8}$

$\csc\theta = \frac{\sqrt{73}}{3}$

$\sec\theta = \frac{\sqrt{73}}{8}$

125. $\sin\theta = \frac{7}{25}$

$\cos\theta = \frac{24}{25}$

$\csc\theta = \frac{25}{7}$

$\sec\theta = \frac{25}{24}$

$\cot\theta = \frac{24}{7}$

Section 1.5   *(page 170)*

1. Period: π; Amplitude: 3

3. Period: 4π; Amplitude: $\frac{5}{2}$

5. Period: 2; Amplitude: $\frac{2}{3}$

7. Period: 2π; Amplitude: 2

9. Period: $\frac{\pi}{3}$; Amplitude: 3

11. Period: 3π; Amplitude: $\frac{1}{4}$

13. Period: $\frac{1}{2}$; Amplitude: 3

15. g is a shift of f π units to the right.

17. g is a reflection of f about the x-axis.

19. g is a reflection of f in the x-axis and five times the amplitude of f.

21. g is the graph of f shifted 4 units up.

23. g has twice the amplitude of f.

25. g is a horizontal shift of f π units to the right.

27.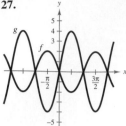

29.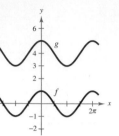

47.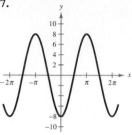

49.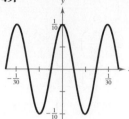

31.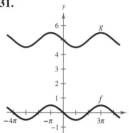

33.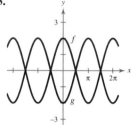

51.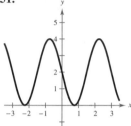

53.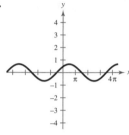

35.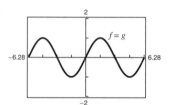

55.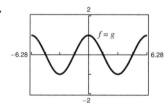

Amplitude: 2

Period: $\dfrac{\pi}{2}$

57.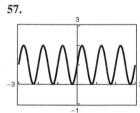

Amplitude: 1

Period: 1

37.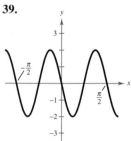

59.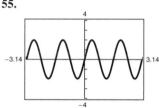

Amplitude: 5

Period: π

39.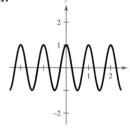

41.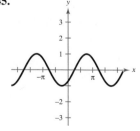

61.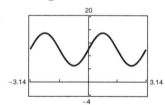

Amplitude: $\dfrac{1}{100}$

Period: $\dfrac{1}{60}$

63. $f(x) = 4 - 4\cos x$

65. $f(x) = 1 - 6\cos x$

43.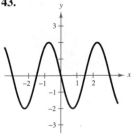

45.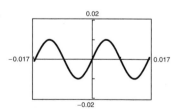

67. $f(x) = -3\sin(2x)$

69. $f(x) = \sin\!\left(x - \dfrac{\pi}{4}\right)$

71.

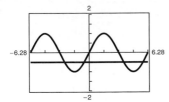

$$x = -\frac{\pi}{6}, -\frac{5\pi}{6}, \frac{7\pi}{6}, \frac{11\pi}{6}$$

73.

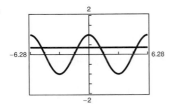

$$x = \pm\frac{\pi}{4}, \pm\frac{7\pi}{4}$$

75. (a)

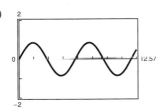

(b) 6 seconds

(c) 10 cycles per minute

77.

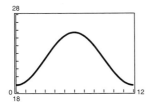

Maximum sales: June

Minimum sales: December

79. (a)

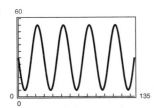

(b) Minimum height: 5 feet

   Maximum height: 55 feet

81. (a) 365 days. The cycle is one year.

(b) 30.35 gallons per day; Both terms of the model; Average of the minimum and maximum of the model

(c)

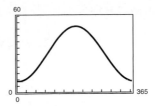

Consumption exceeds 40 gallons per day from the beginning of May through part of September.

83. (a) $p = 2.55 \cos\left(\dfrac{\pi t}{6} - \dfrac{\pi}{12}\right) + 3.45$

(b)

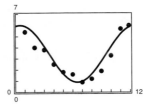

The model fits the data fairly well.

85. True

87.

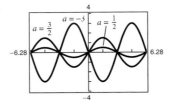

Amplitude changes

89.

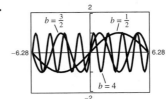

Period changes

91.

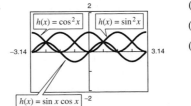

(a) Even

(b) Even

(c) Odd

93. g is the periodic function and f is the parabola.

95. 26   **97.** $\frac{1}{3}$

99. $-20°$ **101.** $-27.502°$

Section 1.6 *(page 181)*

1. (g), 4π **3.** (f), $\dfrac{\pi}{2}$ **5.** (b), 2   **7.** (e), 2π

9.
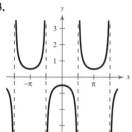

11.
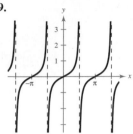

13.
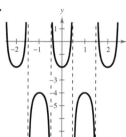

15.
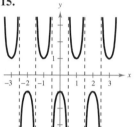

17.
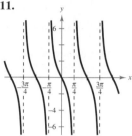

19.
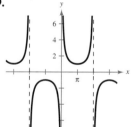

21.
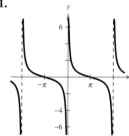

23.
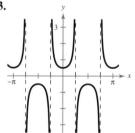

25.
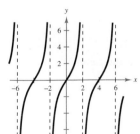

27.
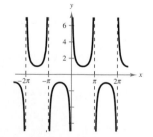

29.
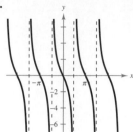

31.
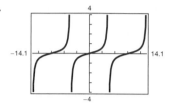

33.
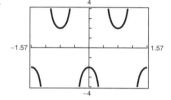

35.
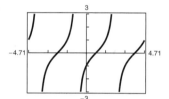

37.
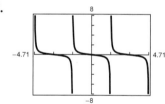

39.
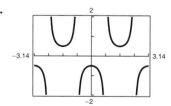

41. $-\dfrac{7\pi}{4},\ -\dfrac{3\pi}{4},\ \dfrac{\pi}{4},\ \dfrac{5\pi}{4}$ **43.** $-\dfrac{4\pi}{3},\ -\dfrac{2\pi}{3},\ \dfrac{2\pi}{3},\ \dfrac{4\pi}{3}$

45. Even

47.

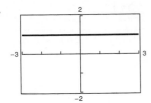

Not equivalent; y_1 is undefined at $x = 0$.

49.

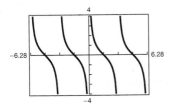

Equivalent

51. (d); as x approaches 0, $f(x)$ approaches 0.

53. (b); as x approaches 0, $g(x)$ approaches 0.

55.

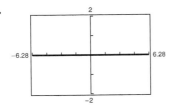

Equal

57.

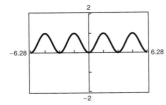

Equal

59.

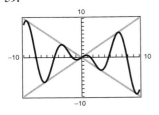

As $x \to \infty$, $f(x)$ oscillates and approaches $-\infty$ and ∞.

61.

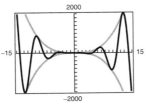

As $x \to \infty$, $f(x)$ oscillates and approaches $-\infty$ and ∞.

63.

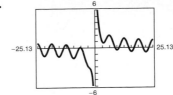

Approaches $\pm\infty$

65.

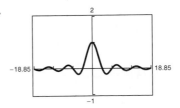

Approaches 1

67.

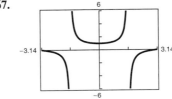

Approaches 1

69. $d = 5 \cot x$

71. As the predator population increases, the number of prey decreases. When the number of prey is small, the number of predators decreases.

73. (a)

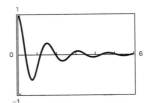

(b) Damped cosine wave; oscillates and goes to 0 as t increases.

75. (a)

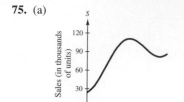

(b) According to the graph, the greatest number of sales occurs when $t = 6$, or in June. The least number of sales is in January.

77. True   **79.** True

81. As x approaches π from the left, f approaches ∞. As x approaches π from the right, f approaches $-\infty$.

83. (a)

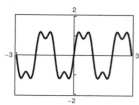   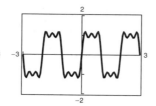

(b) $y_3 = \dfrac{4}{\pi}\left[\sin(\pi x) + \dfrac{1}{3}\sin(3\pi x)\right.$

$\left. + \dfrac{1}{5}\sin(5\pi x) + \dfrac{1}{7}\sin(7\pi x)\right]$

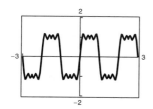

(c) $y_4 = \dfrac{4}{\pi}\left[\sin(\pi x) + \dfrac{1}{3}\sin(3\pi x) + \dfrac{1}{5}\sin(5\pi x)\right.$

$\left. + \dfrac{1}{7}\sin(7\pi x) + \dfrac{1}{9}\sin(9\pi x)\right]$

85. Not one-to-one   **87.** Yes. $f^{-1}(x) = x^3 + 5$

89. $\sin\theta = \dfrac{9\sqrt{277}}{277}$ $\csc\theta = \dfrac{\sqrt{277}}{9}$

$\cos\theta = \dfrac{14\sqrt{277}}{277}$ $\sec\theta = \dfrac{\sqrt{277}}{14}$

$\tan\theta = \dfrac{9}{14}$ $\cot\theta = \dfrac{14}{9}$

Section 1.7   *(page 191)*

1. (a)

x	-1	-0.8	-0.6	-0.4	-0.2
y	-1.5708	-0.9273	-0.6435	-0.4115	-0.2014

x	0	0.2	0.4	0.6	0.8	1
y	0	0.2014	0.4115	0.6435	0.9273	1.5708

(b)

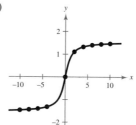

(c)

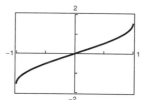

(d) Intercept: $(0, 0)$; Symmetric about the origin

3. (a)

x	-10	-8	-6	-4	-2
y	-1.4711	-1.4464	-1.4056	-1.3258	-1.1071

x	0	2	4	6	8	10
y	0	1.1071	1.3258	1.4056	1.4464	1.4711

(b)

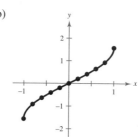

(c)

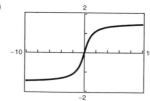

(d) $y = \pm\dfrac{\pi}{2}$

5. $\arctan(-1) = -\dfrac{\pi}{4}$ **7.** $\sin\left(-\dfrac{\pi}{2}\right) = -1$

9. (a) 1.047   (b) 1.571   **11.** (a) 2.356   (b) -0.785

13. (a) 2.094   (b) 0.785   **15.** (a) -1.571 (b) 0

17. $\pi, \dfrac{2\pi}{3}, \dfrac{\sqrt{3}}{2}$ **19.** (a) -0.85 (b) 2.35

21. (a) 0.42   (b) 1.20   **23.** (a) 0.78   (b) 1.36

25.

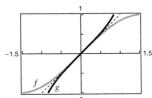

27. $\theta = \arccos \dfrac{4}{x}$

29. $\theta = \arctan \dfrac{x+1}{10}$ **31.** 35   **33.** -0.1 **35.** $\dfrac{\pi}{2}$

37. $-\dfrac{\pi}{4}$ **39.** $\dfrac{5}{4}$ **41.** $-\dfrac{13}{12}$ **43.** $-\dfrac{3\sqrt{55}}{55}$ **45.** $\dfrac{11}{6}$

47. $\dfrac{x}{\sqrt{x^2 + 1}}$ **49** $\dfrac{1}{\sqrt{2x - x^2}}$ **51.** x

53. $\dfrac{\sqrt{r^2 - (x - h)^2}}{r}$

55.

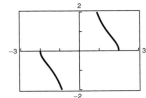

$x = 0$

57. $\dfrac{x}{6}$ **59.** $\dfrac{\sqrt{4x - x^2}}{x - 2}$

61.

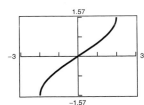

63.

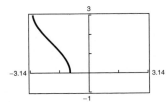

65.

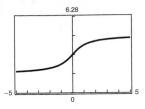

67.

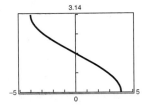

69. $5\sin\left(\pi t + \arctan \dfrac{4}{3}\right)$

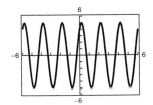

The two forms are equivalent.

71. (a) $\theta = \arctan \dfrac{s}{750}$ (b) 0.49 rad, 1.13 rad

73. (a) 0.45 rad   (b) 24.39 feet

75. (a) $\dfrac{\pi}{4}$ (b) $\dfrac{\pi}{2}$ (c) 1.25   (d) 2.03

77. (a) $\theta = \arctan \dfrac{x}{20}$ (b) 0.24 rad, 0.54 rad

79. False. Inverse trigonometric functions do not have the same relationships that trigonometric functions have.

81.

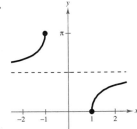

83. (a) $\dfrac{\pi}{4}$

     (b) 0

     (c) $\dfrac{5\pi}{6}$

     (d) $\dfrac{\pi}{6}$

85.–89. Answers will vary.

91. $\sin(-585°) = \dfrac{\sqrt{2}}{2}$ **93.** $\sin\left(-\dfrac{19\pi}{4}\right) = -\dfrac{\sqrt{2}}{2}$

     $\cos(-585°) = -\dfrac{\sqrt{2}}{2}$ $\cos\left(-\dfrac{19\pi}{4}\right) = -\dfrac{\sqrt{2}}{2}$

     $\tan(-585°) = -1$ $\tan\left(-\dfrac{19\pi}{4}\right) = 1$

95.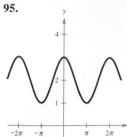

97.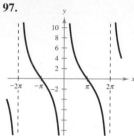

Section 1.8 (page 201)

1. $a \approx 4.66$
$c \approx 11.03$
$B = 65°$

3. $a \approx 8.26$
$c \approx 25.38$
$A = 19°$

5. $c \approx 17.09$
$A \approx 20.56°$
$B \approx 69.44°$

7. $a \approx 37.95$
$A \approx 64.62°$
$B \approx 25.38°$

9. $a \approx 91.34$
$b \approx 420.70$
$B = 77°45'$

11. 2.56 inches   **13.** 6.30 feet

15. (a) $L = 60 \cot \theta$

(b)

θ	10°	20°	30°	40°	50°
L	340	165	104	72	50

(c) No. Cotangent is not a linear function.

17. (a) $h = 20 \sin \theta$

(b)

θ	60°	65°	70°	75°	80°
h	17.3	18.1	18.8	19.3	19.7

19. (a)

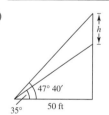

(b) $h = 50(\tan 47° 40' - \tan 35°)$ (c) 19.9 feet

21. ≈ 2090 feet   **23.** $\approx 38.3°$ **25.** $\approx 15.5°$

27. ≈ 5099 feet   **29.** ≈ 0.66 miles

31. ≈ 437 miles north; ≈ 700 miles east

33. (a) N 58° E   (b) ≈ 68.82 meters   **35.** N 55° W

37. ≈ 1933.3 feet   **39.** 17,054 feet ≈ 3.23 miles

41. 78.69°   **43.** $\approx 35.3°$ **45.** $y = \sqrt{3}\,r$

47. ≈ 28.2 inches   **49.** $a \approx 12.2, b \approx 7$

51. (a) 4   (b) 4   (c) $\frac{1}{16}$

53. (a) $\frac{1}{16}$ (b) 70   (c) $\frac{1}{140}$ **55.** $y = 8 \sin \pi t$

57. $y = 3 \cos\left(\dfrac{10\pi t}{9}\right)$ **59.** $\omega = 528\pi$

61. (a) 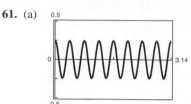   (b) $\dfrac{\pi}{8}$ seconds

(c) $\dfrac{\pi}{32}$ seconds

63. (a)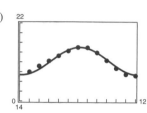

(b) 12 months. Yes, 1 period is 1 year.

(c) 1.41 hours. 1.41 represents the maximum change in time from the average time ($d = 18.09$) of sunset.

65. (a)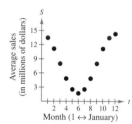

(b)

$S = 8 + 6.3 \cos\left(\dfrac{\pi t}{6}\right)$; The model is a good fit.

(c) 12 months. Sales of outerwear is seasonal.

(d) Maximum displacement of 6.3 from the average sales of 8

67. False. $a = \dfrac{22.56}{\tan(48.1°)}$

69. 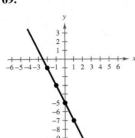 **71.**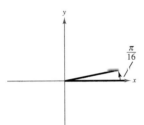

73. $x^2 + y^2 = 81$ **75.** $(x - 10)^2 + (y - 1)^2 = 169$

77. 1.44 radians or 82.53°

79. -0.11 radian or $-6.32°$

Review Exercises   *(page 208)*

1. 0.5 rad   **3.** 4.5 rad

5. (a) 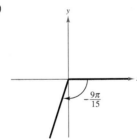 (b) Quadrant I
(c) $\dfrac{33\pi}{16}, -\dfrac{31\pi}{16}$

7. (a) 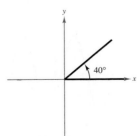 (b) Quadrant III
(c) $\dfrac{21\pi}{15}, -\dfrac{39\pi}{15}$

9. Complement: $\dfrac{3\pi}{8}$, Supplement: $\dfrac{7\pi}{8}$

11. Complement: $\dfrac{\pi}{5}$; Supplement: $\dfrac{7\pi}{10}$ **13.** 128.57°

15. $-200.54°$

17. (a) 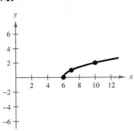 (b) Quadrant I
(c) 400°, $-320°$

19. (a) 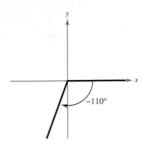 (b) Quadrant III
(c) 250°, $-470°$

21. Complement: 82°; Supplement: 172°

23. Complement: none; Supplement: 9°   **25.** 135.28°

27. 5.38°   **29.** 135° 17′ 24″   **31.** $-85°21'36''$

33. 8.3776   **35.** -0.5890 **37.** 2.083 rad

39. 48.171 meters   **41.** 0.094 rad $\approx 5.39°$

43. $\left(-\dfrac{1}{2}, \dfrac{\sqrt{3}}{2}\right)$ **45.** $\left(-\dfrac{\sqrt{3}}{2}, \dfrac{1}{2}\right)$

47. $\sin\dfrac{7\pi}{6} = -\dfrac{1}{2}$ **49.** $\sin\left(-\dfrac{2\pi}{3}\right) = -\dfrac{\sqrt{3}}{2}$

$\cos\dfrac{7\pi}{6} = -\dfrac{\sqrt{3}}{2}$ $\cos\left(-\dfrac{2\pi}{3}\right) = -\dfrac{1}{2}$

$\tan\dfrac{7\pi}{6} = \dfrac{\sqrt{3}}{3}$ $\tan\left(-\dfrac{2\pi}{3}\right) = \sqrt{3}$

$\csc\dfrac{7\pi}{6} = -2$ $\csc\left(-\dfrac{2\pi}{3}\right) = -\dfrac{2\sqrt{3}}{2}$

$\sec\dfrac{7\pi}{6} = -\dfrac{2\sqrt{3}}{3}$ $\sec\left(-\dfrac{2\pi}{3}\right) = -2$

$\cot\dfrac{7\pi}{6} = \sqrt{3}$ $\cot\left(-\dfrac{2\pi}{3}\right) = \dfrac{\sqrt{3}}{3}$

51. $\sin\dfrac{11\pi}{4} = \sin\dfrac{3\pi}{4} = \dfrac{\sqrt{2}}{2}$

53. $\sin\left(-\dfrac{17\pi}{6}\right) = \sin\left(-\dfrac{5\pi}{6}\right) = -\dfrac{1}{2}$ **55.** -0.89 **57.** $\dfrac{1}{2}$

59. $\sin\theta = \dfrac{5\sqrt{61}}{61}$ **61.** $\sin\theta = \dfrac{\sqrt{65}}{9}$

$\cos\theta = \dfrac{6\sqrt{61}}{61}$ $\cos\theta = \dfrac{4}{9}$

$\tan\theta = \dfrac{5}{6}$ $\tan\theta = \dfrac{\sqrt{65}}{4}$

$\csc\theta = \dfrac{\sqrt{61}}{5}$ $\csc\theta = \dfrac{9\sqrt{65}}{65}$

$\sec\theta = \dfrac{\sqrt{61}}{6}$ $\sec\theta = \dfrac{9}{4}$

$\cot\theta = \dfrac{6}{5}$ $\cot\theta = \dfrac{4\sqrt{65}}{65}$

63. Answers will vary.   **65.** (a) 0.1045   (b) 0.1045

67. (a) 0.7071   (b) 1.4142   **69.** 9.2 meters

71. $\sin \theta = \frac{4}{5}$ $\csc \theta = \frac{5}{4}$

    $\cos \theta = \frac{3}{5}$ $\sec \theta = \frac{5}{3}$

    $\tan \theta = \frac{4}{3}$ $\cot \theta = \frac{3}{4}$

73. $\sin \theta = \dfrac{2\sqrt{53}}{53}$ $\csc \theta = \dfrac{\sqrt{53}}{2}$

    $\cos \theta = -\dfrac{7\sqrt{53}}{53}$ $\sec \theta = -\dfrac{\sqrt{53}}{7}$

    $\tan \theta = -\dfrac{2}{7}$ $\cot \theta = -\dfrac{7}{2}$

75. $\sin \theta = \dfrac{15\sqrt{241}}{241}$

    $\cos \theta = \dfrac{4\sqrt{241}}{241}$

    $\tan \theta = \dfrac{15}{4}$

    $\csc \theta = \dfrac{\sqrt{241}}{15}$

    $\sec \theta = \dfrac{\sqrt{241}}{4}$

    $\cot \theta = \dfrac{4}{15}$

77. $\sin \theta = -\dfrac{\sqrt{11}}{6}$

    $\cos \theta = \dfrac{5}{6}$

    $\tan \theta = -\dfrac{\sqrt{11}}{5}$

    $\csc \theta = -\dfrac{6\sqrt{11}}{11}$

    $\cot \theta = -\dfrac{5\sqrt{11}}{11}$

79. $\cos \theta = -\dfrac{\sqrt{55}}{8}$

    $\tan \theta = -\dfrac{3\sqrt{55}}{55}$

    $\csc \theta = \dfrac{8}{3}$

    $\sec \theta = -\dfrac{8\sqrt{55}}{55}$

    $\cot \theta = -\dfrac{\sqrt{55}}{3}$

81. $\sqrt{3}$ **83.** $\dfrac{\sqrt{3}}{2}$ **85.** $-\dfrac{\sqrt{2}}{2}$ **87.** 0.65

89. 3.24   **91.** 84°   **93.** $\dfrac{\pi}{5}$

95. $\sin 240° = -\dfrac{\sqrt{3}}{2}$ **97.** $\sin(-210°) = \dfrac{1}{2}$

    $\cos 240° = -\dfrac{1}{2}$ $\cos(-210°) = -\dfrac{\sqrt{3}}{2}$

    $\tan 240° = \sqrt{3}$ $\tan(-210°) = -\dfrac{\sqrt{3}}{3}$

99. $\sin\left(-\dfrac{9\pi}{4}\right) = -\dfrac{\sqrt{2}}{2}$ **101.** $\sin\left(-\dfrac{\pi}{2}\right) = -1$

    $\cos\left(-\dfrac{9\pi}{4}\right) = \dfrac{\sqrt{2}}{2}$ $\cos\left(-\dfrac{\pi}{2}\right) = 0$

    $\tan\left(-\dfrac{9\pi}{4}\right) = -1$ $\tan\left(-\dfrac{\pi}{2}\right)$ is undefined.

103. $\left(-\dfrac{1}{2}, \dfrac{\sqrt{3}}{2}\right)$

    $\sin \dfrac{2\pi}{3} = \dfrac{\sqrt{3}}{2}$

    $\cos \dfrac{2\pi}{3} = -\dfrac{1}{2}$

    $\tan \dfrac{2\pi}{3} = -\sqrt{3}$

105. $\left(-\dfrac{\sqrt{3}}{2}, -\dfrac{1}{2}\right)$

    $\sin \dfrac{7\pi}{6} = -\dfrac{1}{2}$

    $\cos \dfrac{7\pi}{6} = -\dfrac{\sqrt{3}}{2}$

    $\tan \dfrac{7\pi}{6} = \dfrac{\sqrt{3}}{3}$

107.

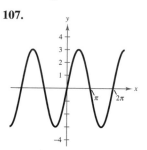

109.

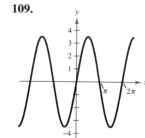

111. Period: 2; Amplitude: 5

113. Period: π; Amplitude: 3.4

115.

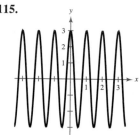

117.

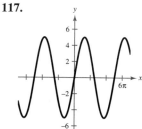

119.

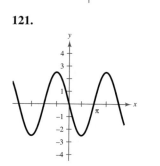

121.

123.

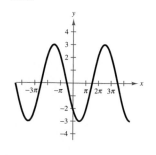

125.

127. $f(x) = -2 \cos\left(x - \frac{\pi}{4}\right)$ **129.** $f(x) = -4 \cos\left(2x - \frac{\pi}{2}\right)$

131.

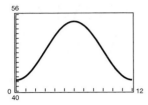

Maximum sales: June; Minimum sales: December

133. $f(x) = \frac{1}{2} \tan\left(\frac{1}{2}x\right)$

135.

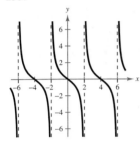

137.

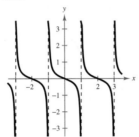

139.

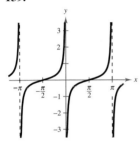

141.

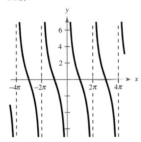

143.

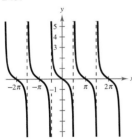

145.

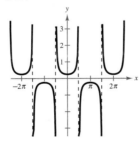

147.

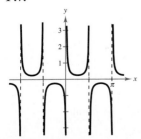

149.

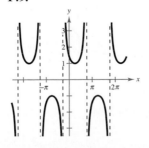

151.

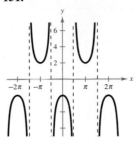

153.

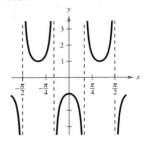

155.

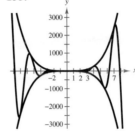

157.

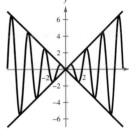

159. (a) $\dfrac{\pi}{2}$ (b) Does not exist   **161.** (a) $\dfrac{\pi}{4}$ (b) $\dfrac{5\pi}{6}$

163. (a) 1.137   (b) 0.682

165. (a) -1.488 (b) 1.523   **167.** $\theta = \arcsin\left(\dfrac{x+3}{16}\right)$

169. $\dfrac{\sqrt{-x^2 + 2x}}{-x^2 + 2x}$ **171.** $\dfrac{2\sqrt{4 - 2x^2}}{4 - x^2}$

173. About 0.071 kilometer or 71 meters

175. ≈ 612 miles north and ≈ 634 miles west

177. $y = 3 \cos\left(\dfrac{2\pi t}{15}\right)$

179. False. The expressions are not equal.

181. False. $3\pi/4$ is not in the range of the arctangent function.

183. Increases. The linear velocity is proportional to the radius.

Chapter Test   *(page 214)*

1. (a)

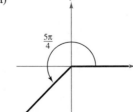

(b) $\dfrac{13\pi}{4}, -\dfrac{3\pi}{4}$

(c) $225°$

2. 3000 radians per minute

3. $\sin \theta = \dfrac{4\sqrt{17}}{17}$

$\cos \theta = -\dfrac{\sqrt{17}}{17}$

$\tan \theta = -4$

$\csc \theta = \dfrac{\sqrt{17}}{4}$

$\sec \theta = -\sqrt{17}$

$\cot \theta = -\dfrac{1}{4}$

4. $\sin \theta = \pm\dfrac{11\sqrt{157}}{157}$

$\cos \theta = \pm\dfrac{6\sqrt{157}}{157}$

$\csc \theta = \pm\dfrac{\sqrt{157}}{11}$

$\sec \theta = \pm\dfrac{\sqrt{157}}{6}$

$\cot \theta = \dfrac{6}{11}$

5. $\theta' = 70°$

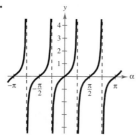

6. Quadrant III   **7.** $150°, 210°$ **8.** $1.33, 1.81$

9. $\sin \theta = -\dfrac{\sqrt{11}}{6}$

$\cos \theta = \dfrac{5}{6}$

$\tan \theta = -\dfrac{\sqrt{11}}{5}$

$\csc \theta = -\dfrac{6\sqrt{11}}{11}$

$\cot \theta = -\dfrac{5\sqrt{11}}{11}$

10.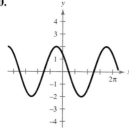

11.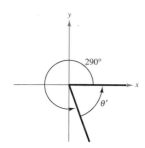

12.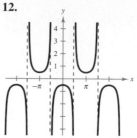

13.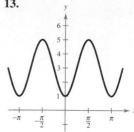

14.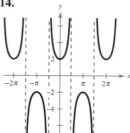

15.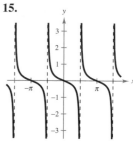

16.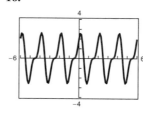

17.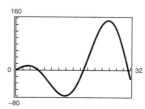

Period: 2

Not periodic

18. $y = -2 \sin\left(\dfrac{x}{2} - \dfrac{\pi}{4}\right)$ **19.** $\dfrac{\sqrt{5}}{2}$

20.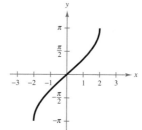

21.

22.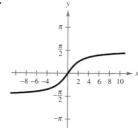

23. S 34.5° W   **24.** ≈ 2290 feet ≈ 0.434 mile

Chapter 2

Section 2.1   *(page 221)*

1. $\tan x = \sqrt{3}$

$\csc x = \dfrac{2\sqrt{3}}{3}$

$\sec x = 2$

$\cot x = \dfrac{\sqrt{3}}{3}$

3. $\cos \theta = \dfrac{\sqrt{2}}{2}$

$\tan \theta = -1$

$\csc \theta = -\sqrt{2}$

$\cot \theta = -1$

5. $\sin x = -\dfrac{7}{25}$

$\cos x = -\dfrac{24}{25}$

$\csc x = -\dfrac{25}{7}$

$\cot x = \dfrac{24}{7}$

7. $\cos \phi = -1$

$\tan \phi = 0$

$\csc \phi$ is undefined.

$\cot \phi$ is undefined.

9. $\sin x = \dfrac{2}{3}$

$\cos x = -\dfrac{\sqrt{5}}{3}$

$\csc x = \dfrac{3}{2}$

$\sec x = -\dfrac{3\sqrt{5}}{5}$

$\cot x = -\dfrac{\sqrt{5}}{2}$

11. $\sin \theta = -\dfrac{4\sqrt{17}}{17}$

$\cos \theta = -\dfrac{\sqrt{17}}{17}$

$\csc \theta = -\dfrac{\sqrt{17}}{4}$

$\sec \theta = -\sqrt{17}$

$\cot \theta = \dfrac{1}{4}$

13. $\cos \theta = 0$

$\tan \theta$ is undefined.

$\csc \theta = -1$

$\sec \theta$ is undefined.

15. $1, 1$ **17.** $\infty, 0$ **19.** (d)   **21.** (a)

23. (e)   **25.** (b)   **27.** (f)   **29.** (e)

31. $\cos x$ **33.** $\cos^2 \phi$ **35.** $\cos x$

37. 1 **39.** $-\tan x$ **41.** $\cot x$ **43.** $1 + \sin y$

45.–59. Answers will vary.

61. $\cos^2 x$ **63.** $\sin^2 x \tan^2 x$ **65.** $\sec^4 x$

67. $\sin^2 x - \cos^2 x$ **69.** $1 + 2 \sin x \cos x$

71. $\tan^2 x$ **73.** $2 \csc^2 x$ **75.** $2 \sec x$

77. $1 + \cos y$ **79.** $3(\sec x + \tan x)$

81.

x	0.2	0.4	0.6	0.8
y_1	0.1987	0.3894	0.5646	0.7174
y_2	0.1987	0.3894	0.5646	0.7174

x	1.0	1.2	1.4
y_1	0.8415	0.9320	0.9854
y_2	0.8415	0.9320	0.9854

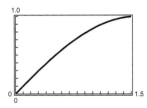

$y_1 = y_2$

83.

x	0.2	0.4	0.6	0.8
y_1	1.2230	1.5085	1.8958	2.4650
y_2	1.2230	1.5085	1.8958	2.4650

x	1.0	1.2	1.4
y_1	3.4082	5.3319	11.6814
y_2	3.4082	5.3319	11.6814

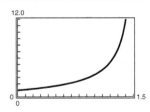

$y_1 = y_2$

85. $\csc x$ **87.** $\tan x$ **89.** $5 \cos \theta$ **91.** $3 \tan \theta$

93. $5 \sec \theta$ **95.** $0 \le \theta \le \pi$

97. $0 \le \theta < \dfrac{\pi}{2}, \dfrac{3\pi}{2} < \theta < 2\pi$

99. (a) $\csc^2(132°) - \cot^2(132°) \approx 1.8107 - 0.8107 = 1$

  (b) $\csc^2\left(\dfrac{2\pi}{7}\right) - \cot^2\left(\dfrac{2\pi}{7}\right) \approx 1.63596 - 0.63596 = 1$

101. (a) $\cos(90° - 80°) = \cos(10°) \approx 0.9848$;

  $\sin(80°) \approx 0.9848$

  (b) $\cos\left(\dfrac{\pi}{2} - 0.8\right) \approx 0.7174$; $\sin(0.8) \approx 0.7174$

103. Answers will vary.

105. False. $5 \sec \theta = \dfrac{5}{\cos \theta} \ne \dfrac{1}{5 \cos \theta}$

107. False. $\sin \theta \csc \phi = \dfrac{\sin \theta}{\sin \phi} \ne 1$ when $\theta \ne \phi$.

109. $\sin \theta = \pm \sqrt{1 - \cos^2 \theta}$

$\tan \theta = \pm \dfrac{\sqrt{1 - \cos^2 \theta}}{\cos \theta}$

$\csc \theta = \pm \dfrac{1}{\sqrt{1 - \cos^2 \theta}}$

$\sec \theta = \dfrac{1}{\cos \theta}$

$\cot \theta = \pm \dfrac{\cos \theta}{\sqrt{1 - \cos^2 \theta}}$

The sign depends on the choice of θ.

111. $\theta = 341°, \theta' = 19°$　　**113.** $\theta = -212°, \theta' = 32°$

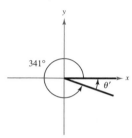　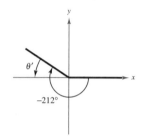

115. $\theta = \dfrac{35\pi}{6}, \theta' = \dfrac{\pi}{6}$

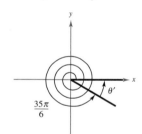

117.　　　　　　　　**119.**

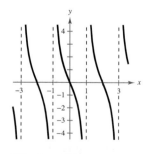　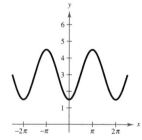

121. $B = 62°$
$C = 90°$
$a = 9.39$
$b = 17.66$

123. $A = 60.37°$
$B = 29.63°$
$C = 90°$
$a = 10.9$

Section 2.2　*(page 229)*

1.–9. Answers will vary.

11.

x	0.2	0.4	0.6	0.8
y_1	4.835	2.1785	1.2064	0.6767
y_2	4.835	2.1785	1.2064	0.6767

x	1.0	1.2	1.4
y_1	0.3469	0.1409	0.0293
y_2	0.3469	0.1409	0.0293

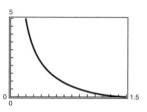

13.

x	0.2	0.4	0.6	0.8
y_1	4.835	2.1785	1.2064	0.6767
y_2	4.835	2.1785	1.2064	0.6767

x	1.0	1.2	1.4
y_1	0.3469	0.1409	0.0293
y_2	0.3469	0.1409	0.0293

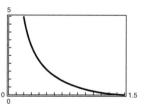

15.

x	0.2	0.4	0.6	0.8
y_1	5.0335	2.5679	1.771	1.394
y_2	5.0335	2.5679	1.771	1.394

x	1.0	1.2	1.4
y_1	1.1884	1.0729	1.0148
y_2	1.1884	1.0729	1.0148

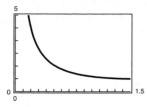

17.

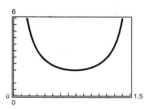

x	0.2	0.4	0.6	0.8
y_1	5.1359	2.7880	2.1458	2.0009
y_2	5.1359	2.7880	2.1458	2.0009

x	1.0	1.2	1.4
y_1	2.1995	2.9609	5.9704
y_2	2.1995	2.9609	5.9704

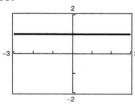

19. $\cot(-x) = -\cot(x)$ so,

$(1 + \tan x)[1 + \cos(-x)] = \tan x - \cot x.$

21. $(\tan^2 x)^2$

23.–51. Answers will vary.

53.   **55.**

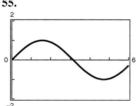

    1                        $\sin x$

57. 1   **59.** 2   **61. and 63.** Answers will vary.

65. $\mu = \tan \theta, W \neq 0$ **67.** Answers will vary.

69. True. Cosine and secant are even functions.

71. False. $\sin^2(x) = (\sin x)^2 \neq \sin(x^2)$

73. $|\sin \theta| = \sqrt{1 - \cos^2 \theta}; \dfrac{7\pi}{4}$

75. $\sqrt{\sin^2 x + \cos^2 x} \neq \sin x + \cos x.$ The left side is 1 for any x, but the right side is not necessarily 1; $\pi/4$.

77. Answers will vary.

79. $4, -9$ **81.** $5, -\frac{3}{2}$ **83.** $\pm\frac{1}{3}$ **85.** $4 \pm 2\sqrt{3}$

87. $\dfrac{4 \pm \sqrt{6}}{2}$ **89.** $\dfrac{-2 \pm 4\sqrt{7}}{9}$ **91.** $\dfrac{-12 \pm \sqrt{78}}{11}$

93. $\theta \approx 2.36$ rad   **95.** Quadrant III   **97.** Quadrant III

Section 2.3   *(page 240)*

1.–5. Answers will vary.   **7.** $-1, 3$ **9.** ± 2

11. $\dfrac{2\pi}{3}, \dfrac{4\pi}{3}$ **13.** $\dfrac{\pi}{6}, \dfrac{11\pi}{6}$ **15.** $\dfrac{\pi}{3}, \dfrac{2\pi}{3}, \dfrac{4\pi}{3}, \dfrac{5\pi}{3}$

17. $\dfrac{\pi}{8}, \dfrac{3\pi}{8}, \dfrac{5\pi}{8}, \dfrac{7\pi}{8}, \dfrac{9\pi}{8}, \dfrac{11\pi}{8}, \dfrac{13\pi}{8}, \dfrac{15\pi}{8}$

19. $\dfrac{\pi}{6}, \dfrac{5\pi}{6}, \dfrac{7\pi}{6}, \dfrac{11\pi}{6}$ **21.** $\dfrac{\pi}{3}, \dfrac{2\pi}{3}, \dfrac{4\pi}{3}, \dfrac{5\pi}{3}$

23. $\dfrac{\pi}{6}, \dfrac{\pi}{3}, \dfrac{2\pi}{3}, \dfrac{5\pi}{6}, \dfrac{7\pi}{6}, \dfrac{4\pi}{3}, \dfrac{5\pi}{3}, \dfrac{11\pi}{6}$ **25.** $0, \dfrac{\pi}{2}, \pi, \dfrac{3\pi}{2}$

27. $0, \dfrac{\pi}{6}, \dfrac{5\pi}{6}, \pi, \dfrac{7\pi}{6}, \dfrac{11\pi}{6}$ **29.** $\dfrac{\pi}{3}, \pi, \dfrac{5\pi}{3}$

31. No solution   **33.** $\dfrac{\pi}{2}$ **35.** $\dfrac{\pi}{2}$ **37.** π

39. $\dfrac{\pi}{6}, \dfrac{5\pi}{6}, \dfrac{7\pi}{6}, \dfrac{11\pi}{6}$ **41.** $2.0344, 5.1760, \dfrac{\pi}{4}, \dfrac{5\pi}{4}$

43. $1.4595, 4.8237$ **45.** $0.8614, 5.4218$

47. $0.7854, 2.3562, 3.6652, 3.9270, 5.4978, 5.7596$

49. $0.5236, 2.6180$ **51.** 4.9172

53. $0, 2.6779, 3.1416, 5.8195$

55. $0.3398, 0.8481, 2.2935, 2.8018$

57. $-1.154, 0.534$ **59.** 1.110

61. (a)

x	0	1	2	3
$f(x)$	Undef.	0.83	-1.36	-2.93

x	4	5	6
$f(x)$	-4.46	-6.34	-13.02

The zero is in the interval $(1, 2)$ because f changes signs in that interval.

(b)

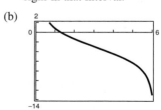

The interval is the same as in part (a).

(c) 1.3065

63. (a)

x	0	1	2	3
$f(x)$	-1	1.39	1.65	-0.70

x	4	5	6
$f(x)$	-1.94	-2.00	-1.48

The zeros are in the intervals $(0, 1)$ and $(2, 3)$ because f changes signs in those intervals.

(b)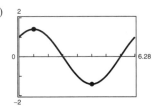

The intervals are the same as in part (a).

(c) 0.4271, 2.7145

65. (a)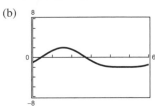

Maximum: $(0.785, 1.41)$

Minimum: $(3.93, -1.41)$

(b) $\dfrac{\pi}{4}, \dfrac{5\pi}{4}$

67. (a)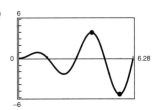

Maximum: $(3.99, 3.96)$

Minimum: $(5.54, -5.52)$

(b) 0, 1.01, 2.46, 3.99, 5.44

69. 1

71. (a) All real numbers x except $x = 0$

(b) y-axis symmetry; Horizontal asymptote: $y = 1$

(c) Oscillates   (d) Infinite number of solutions

(e) Yes, 0.6366

73. 0.04 second, 0.43 second, 0.83 second

75. 122 days $< t <$ 223 days

77. 1.91°   **79.** 2.84 seconds

81. (a) 3.33

(b)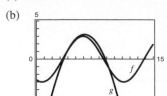

(c) 3.46, 8.81. The zero of $g(x)$ on $[0, 6]$ is 3.46, which is closest to the zero of $f(x)$.

83. False. Some trigonometric equations don't have any solutions in the interval $[0, 2\pi)$.

85. 2.164 rad   **87.** -0.007 rad

89. 24.249   **91.** 24.892

93. 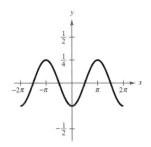   **95.**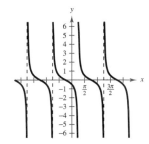

Section 2.4   *(page 248)*

1. (a) 0   (b) $\dfrac{1 + \sqrt{3}}{2}$ **3.** (a) $\dfrac{1}{2}$ (b) $\dfrac{-1 - \sqrt{3}}{2}$

5. (a) $-\dfrac{\sqrt{2}}{2}$ (b) $\dfrac{2 - \sqrt{2}}{2}$

7. (a) $\dfrac{-\sqrt{2} - \sqrt{6}}{4}$ (b) $\dfrac{-\sqrt{2} - \sqrt{3}}{2}$

9. $\sin(75°) = \dfrac{\sqrt{2} + \sqrt{6}}{4}$ **11.** $\sin(105°) = \dfrac{\sqrt{6} + \sqrt{2}}{4}$

$\cos(75°) = \dfrac{\sqrt{6} - \sqrt{2}}{4}$ $\cos(105°) = \dfrac{\sqrt{2} - \sqrt{6}}{4}$

$\tan(75°) = 2 + \sqrt{3}$ $\tan(105°) = -2 - \sqrt{3}$

13. $\sin(195°) = \dfrac{\sqrt{2} - \sqrt{6}}{4}$

$\cos(195°) = \dfrac{-\sqrt{2} - \sqrt{6}}{4}$

$\tan(195°) = 2 - \sqrt{3}$

15. $\sin\left(\dfrac{11\pi}{12}\right) = \dfrac{\sqrt{6} - \sqrt{2}}{4}$ **17.** $\sin\left(-\dfrac{\pi}{12}\right) = \dfrac{\sqrt{2} - \sqrt{6}}{4}$

$\cos\left(\dfrac{11\pi}{12}\right) = \dfrac{-\sqrt{6} - \sqrt{2}}{4}$ $\cos\left(-\dfrac{\pi}{12}\right) = \dfrac{\sqrt{6} + \sqrt{2}}{4}$

$\tan\left(\dfrac{11\pi}{12}\right) = -2 + \sqrt{3}$ $\tan\left(-\dfrac{\pi}{12}\right) = -2 + \sqrt{3}$

19. $\cos 55°$ **21.** $\sin 290°$ **23.** $\tan 239°$

25. $\sin 1.8$ **27.** $\cos \dfrac{12\pi}{35}$

29.

x	0.2	0.4	0.6	0.8
y_1	0.9801	0.9211	0.8253	0.6967
y_2	0.9801	0.9211	0.8253	0.6967

x	1.0	1.2	1.4
y_1	0.5403	0.3624	0.1700
y_2	0.5403	0.3624	0.1700

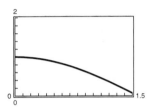

31.

x	0.2	0.4	0.6	0.8
y_1	0.6621	0.7978	0.9017	0.9696
y_2	0.6621	0.7978	0.9017	0.9696

x	1.0	1.2	1.4
y_1	0.9989	0.9883	0.9384
y_2	0.9989	0.9883	0.9384

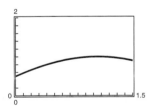

33.

x	0.2	0.4	0.6	0.8
y_1	0.9605	0.8484	0.6812	0.4854
y_2	0.9605	0.8484	0.6812	0.4854

x	1.0	1.2	1.4
y_1	0.2919	0.1313	0.0289
y_2	0.2919	0.1313	0.0289

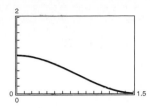

35. $\dfrac{33}{65}$ **37.** $-\dfrac{56}{65}$ **39.** $-\dfrac{3}{5}$ **41.** $\dfrac{44}{125}$

43.–49. Answers will vary.   **51.** 1

53. $\dfrac{2x^2 - \sqrt{1 - x^2}}{\sqrt{4x^2 + 1}}$ **55.** $\dfrac{\pi}{2}$ **57.** $\dfrac{5\pi}{4}, \dfrac{7\pi}{4}$

59. $0, \dfrac{\pi}{3}, \pi, \dfrac{5\pi}{3}$ **61.** 0.7854, 5.4978   **63.** 0, 3.1416

65. Answers will vary.

67. False. $\sin(u \pm v) = \sin u \cos v \pm \cos u \sin v$

69. False. $\sin(75°) = \dfrac{\sqrt{6} + \sqrt{2}}{4}$

71. and 73. Answers will vary.

75. (a) $\sqrt{2} \sin\left(\theta + \dfrac{\pi}{4}\right)$

    (b) $\sqrt{2} \cos\left(\theta - \dfrac{\pi}{4}\right)$

77. (a) $13 \sin(3\theta + 0.3948)$

    (b) $13 \cos(3\theta - 1.1760)$

79. $2 \cos \theta$ **81.** $u + v = \omega$

83.

$h(\theta) = \tan \theta$

85. (a) $(-\infty, 0), (0, \infty)$

(b)

h	0.01	0.02	0.05
$f(h)$	-0.5043	-0.5086	-0.5214
$g(h)$	-0.5043	-0.5086	-0.5214

h	0.1	0.2	0.5
$f(h)$	-0.5424	-0.5830	-0.6915
$g(h)$	-0.5424	-0.5830	-0.6915

(c)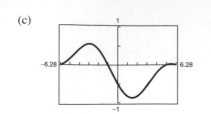

(d) $-\frac{1}{2}$

87. $(38, 0), (0, 19)$ **89.** $(0, 4), (2, 0), (7, 0)$

91. $\dfrac{\pi}{6}$ **93.** $\dfrac{\pi}{2}$

Section 2.5   *(page 258)*

1. $\dfrac{3}{5}$ **3.** $\dfrac{7}{25}$ **5.** $\dfrac{24}{7}$ **7.** $\dfrac{25}{24}$ **9.** $0, \dfrac{\pi}{3}, \pi, \dfrac{5\pi}{3}$

11. $0.1263, 1.4445, 3.2679, 4.5860$ **13.** $\dfrac{\pi}{3}, \pi, \dfrac{5\pi}{3}$

15. $0, \dfrac{\pi}{2}, \pi, \dfrac{3\pi}{2}$ **17.** $4 \sin 2x$ **19.** $5 \cos 2x$

21. $\sin 2u = \dfrac{24}{25}$ **23.** $\sin 2u = \dfrac{4}{5}$
$\cos 2u = \dfrac{7}{25}$ $\cos 2u = \dfrac{3}{5}$
$\tan 2u = \dfrac{24}{7}$ $\tan 2u = \dfrac{4}{3}$

25. $\dfrac{1}{8}(3 + 4 \cos 2x + \cos 4x)$ **27.** $\dfrac{1}{8}(1 - \cos 4x)$

29. $\dfrac{1}{32}(2 + \cos 2x - 2 \cos 4x - \cos 6x)$

31. $\dfrac{5\sqrt{26}}{26}$ **33.** $\dfrac{1}{5}$ **35.** $\sqrt{26}$ **37.** $\dfrac{5}{13}$

39. $\sin 15° = \dfrac{\sqrt{2 - \sqrt{3}}}{2}$ **41.** $\sin 112° 30' = \dfrac{\sqrt{2 + \sqrt{2}}}{2}$

$\cos 15° = \dfrac{\sqrt{2 + \sqrt{3}}}{2}$ $\cos 112° 30' = -\dfrac{\sqrt{2 - \sqrt{2}}}{2}$

$\tan 15° = 2 - \sqrt{3}$ $\tan 112° 30' = -1 - \sqrt{2}$

43. $\sin \dfrac{\pi}{8} = \dfrac{\sqrt{2 - \sqrt{2}}}{2}$ **45.** $\sin \dfrac{3\pi}{8} = \dfrac{\sqrt{2 + \sqrt{2}}}{2}$

$\cos \dfrac{\pi}{8} = \dfrac{\sqrt{2 + \sqrt{2}}}{2}$ $\cos \dfrac{3\pi}{8} = \dfrac{\sqrt{2 - \sqrt{2}}}{2}$

$\tan \dfrac{\pi}{8} = \sqrt{2} - 1$ $\tan \dfrac{3\pi}{8} = \sqrt{2} + 1$

47. $\sin \dfrac{u}{2} = \dfrac{5\sqrt{26}}{26}$ **49.** $\sin \dfrac{u}{2} = \sqrt{\dfrac{89 - 5\sqrt{89}}{178}}$

$\cos \dfrac{u}{2} = \dfrac{\sqrt{26}}{26}$ $\cos \dfrac{u}{2} = -\sqrt{\dfrac{89 + 5\sqrt{89}}{178}}$

$\tan \dfrac{u}{2} = 5$ $\tan \dfrac{u}{2} = \dfrac{5 - \sqrt{89}}{8}$

51. $|\sin 3x|$ **53.** $-|\tan 4x|$ **55.** $\dfrac{\pi}{3}, \dfrac{5\pi}{3}$

57. $\dfrac{\pi}{3}, \pi, \dfrac{5\pi}{3}$ **59.** $3 \sin \dfrac{2\pi}{3}$ **61.** $\dfrac{1}{2}(\sin 8\theta + \sin 2\theta)$

63. $\dfrac{5}{2}(\cos 8\beta + \cos 2\beta)$ **65.** $2 \sin 45° \cos 15°$

67. $2 \cos 3\theta \sin 2\theta$ **69.** $2 \cos \alpha \sin \beta$

71. $-2 \sin \theta \sin \dfrac{\pi}{2} = -2 \sin \theta$

73. $0, \dfrac{\pi}{4}, \dfrac{\pi}{2}, \dfrac{3\pi}{4}, \pi, \dfrac{5\pi}{4}, \dfrac{3\pi}{2}, \dfrac{7\pi}{4}$ **75.** $\dfrac{\pi}{6}, \dfrac{5\pi}{6}$ **77.** $\dfrac{25}{169}$

79. $\dfrac{4}{13}$ **81.–91.** Answers will vary.

93.

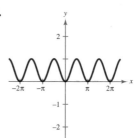

95. (a)   (b) π

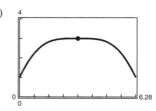

Maximum: $(\pi, 3)$

97. (a)

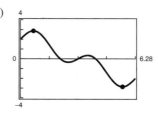

Minimum: $(5.5839, -2.8642)$

Maximum: $(0.6993, 2.8642)$

(b) $0.6993, 2.6078, 3.6754, 5.5839$

99. $2x \sqrt{1 - x^2}$ **101.** $1 - 2x^2$ **103.** $\dfrac{1}{16} v_0 \sin \theta \cos \theta$

105. $\theta = 0.4482$

107. False. $\sin \dfrac{x}{2} = -\sqrt{\dfrac{1 - \cos x}{2}}$ for $\pi \le \dfrac{x}{2} \le 2\pi$.

109. (a)

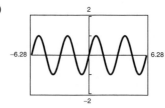

(b) $\sin 2x$

(c) Answers will vary.

111. (a) Complement: $\dfrac{4\pi}{9}$; Supplement: $\dfrac{17\pi}{18}$

(b) Complement: $\dfrac{\pi}{20}$; Supplement: $\dfrac{11\pi}{20}$

113. **115.**

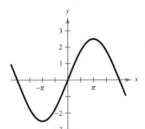 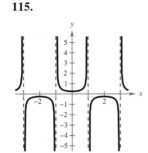

Review Exercises *(page 262)*

1. $\sec x$ **3.** $\cos x$ **5.** $\cot x$ **7.** $\sec x$ **9.** $\sec x$

11. $\tan x = \dfrac{3}{4}$ **13.** $\cos x = \dfrac{\sqrt{2}}{2}$

$\csc x = \dfrac{5}{3}$ $\tan x = -1$

$\sec x = \dfrac{5}{4}$ $\csc x = -\sqrt{2}$

$\cot x = \dfrac{4}{3}$ $\sec x = \sqrt{2}$

   $\cot x = -1$

15. $\sin^2 x$ **17.** $1 + \cot \alpha$ **19.** 1   **21.** $\csc x$

23.–37. Answers will vary.

39. $\dfrac{\pi}{6}, \dfrac{5\pi}{6}$ **41.** $\dfrac{\pi}{3}, \dfrac{2\pi}{3}$ **43.** $\dfrac{\pi}{6}, \dfrac{7\pi}{6}$

45. $\dfrac{\pi}{3}, \dfrac{2\pi}{3}, \dfrac{4\pi}{3}, \dfrac{5\pi}{3}$ **47.** $\dfrac{\pi}{6}, \dfrac{5\pi}{6}, \dfrac{7\pi}{6}, \dfrac{11\pi}{6}$ **49.** $0, \pi$

51. $0, \dfrac{2\pi}{3}, \dfrac{4\pi}{3}$ **53.** $0, \dfrac{\pi}{2}, \pi$ **55.** $\dfrac{\pi}{8}, \dfrac{3\pi}{8}, \dfrac{9\pi}{8}, \dfrac{11\pi}{8}$

57. $0, \dfrac{\pi}{8}, \dfrac{3\pi}{8}, \dfrac{5\pi}{8}, \dfrac{7\pi}{8}, \dfrac{9\pi}{8}, \dfrac{11\pi}{8}, \dfrac{13\pi}{8}, \dfrac{15\pi}{8}$ **59.** $\dfrac{\pi}{2}, \dfrac{3\pi}{2}$

61. $0, \pi$ **63.** 1.2490, 1.8158, 4.3906, 4.9574

65. $\sin 285° = \dfrac{-\sqrt{2} - \sqrt{6}}{4}$ **67.** $\sin \dfrac{5\pi}{12} = \dfrac{\sqrt{6} + \sqrt{2}}{4}$

$\cos 285° = \dfrac{\sqrt{6} - \sqrt{2}}{4}$ $\cos \dfrac{5\pi}{12} = \dfrac{\sqrt{6} - \sqrt{2}}{4}$

$\tan 285° = -2 - \sqrt{3}$ $\tan \dfrac{5\pi}{12} = 2 + \sqrt{3}$

69. $\sin 190°$ **71.** $\tan 35°$ **73.** $\dfrac{-15 - 12\sqrt{7}}{52}$

75. $\dfrac{5\sqrt{7} + 36}{52}$ **77.** $\dfrac{5\sqrt{7} - 36}{52}$

79.–83. Answers will vary.   **85.** $\dfrac{\pi}{6}, \dfrac{11\pi}{6}$

87.–91. Answers will vary.

93. $\sin 2u = \dfrac{20\sqrt{6}}{49}$ **95.** $15°, 75°$

$\cos 2u = -\dfrac{1}{49}$

$\tan 2u = -20\sqrt{6}$

97. $\frac{1}{128}(3 - 4\cos 4x + \cos 8x)$

99. $\frac{1}{8}(3 - 4\cos 4x + \cos 8x)$

101. $\sin 67° 30' = \dfrac{\sqrt{2 + \sqrt{2}}}{2}$

$\cos 67° 30' = \dfrac{\sqrt{2 - \sqrt{2}}}{2}$

$\tan 67° 30' = \sqrt{2} + 1$

103. $\sin \dfrac{11\pi}{12} = \dfrac{\sqrt{2 - \sqrt{3}}}{2}$

$\cos \dfrac{11\pi}{12} = -\dfrac{\sqrt{2 + \sqrt{3}}}{2}$

$\tan \dfrac{11\pi}{12} = -2 + \sqrt{3}$

105. $\tan 3x$ **107.** Answers will vary.

109. $2\cos \dfrac{5\theta}{2} \cos \dfrac{\theta}{2}$ **111.** $-2\sin \dfrac{\pi}{2} \sin \dfrac{\pi}{4}$

113. $\dfrac{1}{2}(\cos \alpha - \cos 5\alpha)$ **115.** $3\sin \dfrac{\pi}{2}$

117. (a) $y = \dfrac{1}{2}\sqrt{10} \sin\!\left(8t - \arctan\dfrac{1}{3}\right)$

(b)   (c) $\dfrac{1}{2}\sqrt{10}$ (d) $\dfrac{4}{\pi}$

119. False. $\sin(x \pm y) = \sin x \cos y \pm \cos x \sin y$

121. True   **123.** No; answers will vary.

125. $y_3 = 1 - y_2$

Chapter Test *(page 266)*

1. $\sin \theta = -\dfrac{6\sqrt{61}}{61}$ **2.** 1 **3.** 1

$\cos \theta = -\dfrac{5\sqrt{61}}{61}$

$\csc \theta = -\dfrac{\sqrt{61}}{6}$

$\sec \theta = -\dfrac{\sqrt{61}}{5}$

$\cot \theta = \dfrac{5}{6}$

4. $\csc \theta \sec \theta$ **5.** $\dfrac{\pi}{2} < \theta \le \pi,\; \dfrac{3\pi}{2} < \theta < 2\pi$

6.

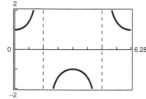

$y_1 = y_2$

7.–12. Answers will vary.

13. $\sin\left(-\dfrac{7\pi}{12}\right) = \dfrac{-\sqrt{6} - \sqrt{2}}{4}$

$\cos\left(-\dfrac{7\pi}{12}\right) = \dfrac{-\sqrt{6} + \sqrt{2}}{4}$

$\tan\left(-\dfrac{7\pi}{12}\right) = 2 + \sqrt{3}$

14. $\dfrac{1}{8}(1 - \cos 4x)$ **15.** $\dfrac{3}{2}(\cos 4\theta - \cos 8\theta)$

16. $2 \cos 4\theta \cos \theta$ **17.** $0, \dfrac{3\pi}{4}, \pi, \dfrac{7\pi}{4}$ **18.** $\dfrac{\pi}{6}, \dfrac{\pi}{2}, \dfrac{5\pi}{6}, \dfrac{3\pi}{2}$

19. $\dfrac{\pi}{6}, \dfrac{5\pi}{6}, \dfrac{7\pi}{6}, \dfrac{11\pi}{6}$ **20.** $\dfrac{\pi}{6}, \dfrac{5\pi}{6}, \dfrac{3\pi}{2}$ **21.** 1.306

22. $\sin 2u = \dfrac{4}{5}$ **23.** Answers will vary.

$\tan 2u = -\dfrac{4}{3}$

24. $-2 - \sqrt{3}$ **25.** 76.52°

Chapter 3

Section 3.1 *(page 274)*

1. $C = 105°,\; b \approx 16.97,\; c \approx 23.18$

3. $C = 110°,\; b \approx 22.44,\; c \approx 24.35$

5. $B \approx 13.6°,\; C \approx 130.4°,\; c \approx 12.96$

7. $B = 10°,\; b \approx 69.46,\; c \approx 136.81$

9. $B = 42°4',\; a \approx 22.05,\; b \approx 14.88$

11. $A \approx 10°11',\; C \approx 154°19',\; c \approx 11.03$

13. $B \approx 18°13',\; C \approx 51°32',\; c \approx 40.05$

15. No solution

17. Two solutions

$B \approx 72.2°,\; C \approx 49.8°,\; c \approx 10.27$

$B \approx 107.8°,\; C \approx 14.2°,\; c \approx 3.3$

19. $B \approx 48.74°,\; C \approx 21.26°,\; c \approx 48.23$

21. (a) $b \le 5,\; b = \dfrac{5}{\sin 36°}$ (b) $5 < b < \dfrac{5}{\sin 36°}$

(c) $b > \dfrac{5}{\sin 36°}$

23. ≈ 28.2 square units **25.** ≈ 1782.3 square units

27. ≈ 1057.1 square units

29. (a)

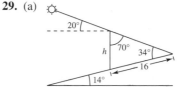

(b) $\dfrac{16}{\sin 70°} = \dfrac{h}{\sin 34°}$ (c) ≈ 9.5 meters

31. S 65° W

33. (a)

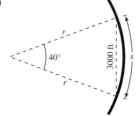

(b) 4385.71 feet

(c) 3061.80 feet

35. Distance from A: 42.3 kilometers

Distance from B: 25.8 kilometers

37. ≈ 4.55 miles

39. (a)

θ	0°	45°	90°	135°	180°
d	0	0.5338	1.6905	2.6552	3

(b) 0.0071 inches

41. $d = \dfrac{2 \sin \theta}{\sin(\phi - \theta)}$

43. False. The triangle can't be solved if only three angles are known.

45. False. For the case SSA there could be two possible triangles, one triangle, or no triangle.

47. $\tan \theta = -\dfrac{12}{5};\; \csc \theta = -\dfrac{13}{12};\; \sec \theta = \dfrac{13}{5};\; \cot \theta = -\dfrac{5}{12}$

49. $\sin \theta = \dfrac{\sqrt{122}}{122}$; $\cos \theta = -\dfrac{11\sqrt{122}}{122}$; $\tan \theta = -\dfrac{1}{11}$;

$\csc \theta = \sqrt{122}$

51. Answers will vary.   **53.** $3(\sin 11\theta + \sin 5\theta)$

55. $\dfrac{3}{2}\left[\sin \dfrac{11\pi}{6} - \sin\left(-\dfrac{9\pi}{6}\right)\right]$

Section 3.2   *(page 281)*

1. $A \approx 26.4°, B \approx 36.3°, C \approx 117.3°$

3. $B \approx 29.4°, C \approx 100.6°, a \approx 23.4$

5. $A \approx 36.9°, B \approx 53.1°, C = 90°$

7. $A \approx 103.5°, B \approx 38.2°, C \approx 38.2°$

9. $b \approx 8.6, A \approx 153.9°, C \approx 17.7°$

	a	b	c	d	θ	ϕ
11.	4	8	11.6	5	30°	150°
13.	10	14	20	13.86	68.2°	111.8°
15.	10	11.58	18	12	67.1°	112.9°

17. ≈ 22.45 square units   **19.** ≈ 15.52 square units

21. 96.82 square units

23.

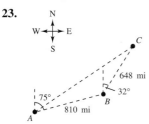

1357.8 miles, S 56° W

25. 483.4 meters   **27.** 43.3 miles

29. (a) N 58.4° W   (b) S 81.5° W   **31.** ≈ 63.7 feet

33. $PQ \approx 9.4, QS \approx 5.0, RS \approx 12.8$

35. (a) $49 = 2.25 + x^2 - 3x \cos \theta$

(b) $x = \frac{1}{2}\left(3\cos\theta + \sqrt{9\cos^2\theta + 187}\right)$

(c)

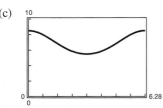

(d) 6 inches

37. ≈ 2.76 feet

39. False. A triangle cannot be formed with sides of length 10 cm, 16 cm, and 5 cm.

41. (a) and (b) Answers will vary.

43. $r \approx 64.5$ feet, track length ≈ 405 feet   **45.** $\dfrac{\pi}{3}, \dfrac{5\pi}{3}$

47. $\dfrac{\pi}{6}, \dfrac{11\pi}{6}$ **49.** $-2 \sin\left(\dfrac{7\pi}{12}\right) \sin\left(\dfrac{\pi}{4}\right)$

Section 3.3   *(page 293)*

1. $\|\mathbf{u}\| = \|\mathbf{v}\| = \sqrt{17}$, $\text{slope}_{\mathbf{u}} = \text{slope}_{\mathbf{v}} = \frac{1}{4}$

$\mathbf{u}$ and $\mathbf{v}$ have the same magnitude and direction so they are equal.

3. $\langle 4, 3 \rangle, \|\mathbf{v}\| = 5$ **5.** $\langle -3, 2 \rangle, \|\mathbf{v}\| = \sqrt{13}$

7. $\langle 0, 5 \rangle, \|\mathbf{v}\| = 5$ **9.** $\langle -\frac{9}{2}, -\frac{5}{2} \rangle, \|\mathbf{v}\| = \frac{1}{2}\sqrt{106}$

11. $\langle 8, 6 \rangle, \|\mathbf{v}\| = 10$

13. $\langle 7.9, -17.9 \rangle, \|\mathbf{v}\| = \sqrt{382.82} \approx 19.6$

15.   **17.**

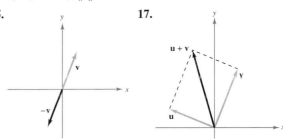

19.

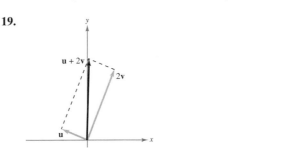

21. (a) $\langle 11, 3 \rangle$ (b) $\langle -3, 1 \rangle$ (c) $\langle -13, 1 \rangle$ (d) $\langle 23, 9 \rangle$

23. (a) $\langle -4, -5 \rangle$ (b) $\langle -6, 1 \rangle$ (c) $\langle -13, 5 \rangle$

     (d) $\langle -19, -11 \rangle$

25. (a) $3\mathbf{i} - 2\mathbf{j}$ (b) $-\mathbf{i} + 4\mathbf{j}$ (c) $-4\mathbf{i} + 11\mathbf{j}$

     (d) $6\mathbf{i} + \mathbf{j}$

27. $\langle 1, 0 \rangle$ **29.** $\left\langle -\dfrac{\sqrt{2}}{2}, \dfrac{\sqrt{2}}{2} \right\rangle$ **31.** $\left\langle -\dfrac{24}{25}, -\dfrac{7}{25} \right\rangle$

33. $\dfrac{4}{5}\mathbf{i} - \dfrac{3}{5}\mathbf{j}$ **35.** $\mathbf{j}$ **37.** $\left\langle \dfrac{5\sqrt{2}}{2}, \dfrac{5\sqrt{2}}{2} \right\rangle$

39. $\frac{21}{5}\mathbf{i} + \frac{28}{5}\mathbf{j}$ **41.** $-8\mathbf{i}$

43. $\mathbf{v} = \left\langle 3, -\frac{3}{2} \right\rangle$

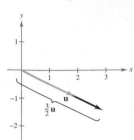

45. $\mathbf{v} = \langle 4, 3 \rangle$

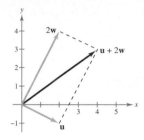

47. $\mathbf{v} = \langle 7, -1 \rangle$

49. $\|\mathbf{v}\| = 5, \theta = 30°$

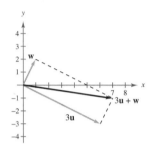

51. $\|\mathbf{v}\| = 6\sqrt{2}, \theta = 315°$ **53.** $\|\mathbf{v}\| = \sqrt{29}, \theta = 111.8°$

55. $\mathbf{v} = \langle 3, 0 \rangle$ **57.** $\mathbf{v} = \left\langle -\dfrac{3\sqrt{6}}{2}, \dfrac{3\sqrt{2}}{2} \right\rangle$

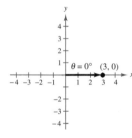

59. $\mathbf{v} = \left\langle \dfrac{\sqrt{10}}{5}, \dfrac{3\sqrt{10}}{5} \right\rangle$

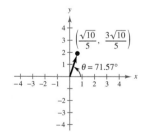

61. $\left\langle \dfrac{5}{2}, \dfrac{5\sqrt{3} + 10}{2} \right\rangle$ **63.** $\left\langle 10\sqrt{2} - 25\sqrt{3}, 10\sqrt{2} + 25 \right\rangle$

65. $90°$ **67.** $63.4°$

69.

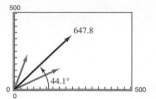

71. $62.7°$ **73.** $8.7°$, 2396.19 newtons

75. Horizontal component: $70 \cos 40° \approx 53.62$ feet per second

Vertical component: $70 \sin 40° \approx 45.00$ feet per second

77. $T_{AC} \approx 3611.1$ pounds, $T_{BC} \approx 2169.5$ pounds

79. (a) 3192.5 pounds

(b) $T = 3000 \sec \theta$; Domain: $0° \le \theta < 90°$

(c)

θ	10°	20°	30°	40°	50°	60°
T	3046.3	3192.5	3464.1	3916.2	4667.2	6000.0

(d)

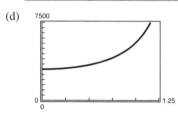

(e) The component in the direction of the motion of the barge decreases.

81. N 26.7° E; 130.35 kilometers per hour

83. (a) $12.1°$, 357.85 newtons

(b) $M = 10\sqrt{660 \cos \theta + 709}$,

$$\alpha = \arctan \frac{15 \sin \theta}{15 \cos \theta + 22}$$

(c)

θ	0°	30°	60°	90°
M	370.0	357.9	322.3	266.3
α	0°	12.1°	23.8°	34.3°

θ	120°	150°	180°
M	194.7	117.2	70.0
α	41.9°	39.8°	0°

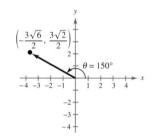

(d)
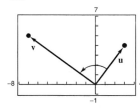

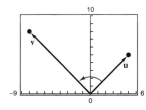

(e) For increasing θ, the two vectors tend to work against each other, resulting in a decrease in the magnitude of the resultant.

85. True   **87.** True. $a = b = 0$

89. (a) $0°$ (b) $180°$

(c) No. The magnitude is equal to the sum when the angle between the vectors is $0°$.

91. Answers will vary.   **93.** $\langle 1, 3 \rangle$ or $\langle -1, -3 \rangle$

95. $7 \cos \theta$ **97.** $10 \csc \theta$

99. $C = 87°, b = 18.16, c = 20.73$

101. $A = 24.15°, B = 30.75°, C = 125.10°$

Section 3.4   *(page 304)*

1. -18 **3.** 11 **5.** 8, scalar   **7.** $\langle 2, -8 \rangle$, vector

9. 4, scalar   **11.** 13 **13.** $5\sqrt{41}$ **15.** 6

17. $90°$ **19.** $70.56°$ **21.** $90°$ **23.** $\dfrac{5\pi}{12}$

25.   **27.**

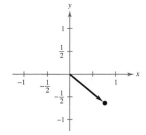   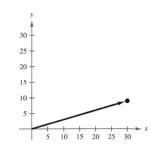

91.33°   90°

29. $26.6°, 63.4°, 90°$ **31.** -20 **33.** Parallel

35. Neither   **37.** Orthogonal   **39.** $\frac{16}{17}\langle 4, 1 \rangle, \frac{13}{17}\langle -1, 4 \rangle$

41. $\frac{45}{229}\langle 2, 15 \rangle, \frac{6}{229}\langle -15, 2 \rangle$ **43.** $\mathbf{u}$ **45.** $\mathbf{0}$

47. $\langle 7, -4 \rangle, \langle -7, 4 \rangle$ **49.** $-\frac{3}{4}\mathbf{i} - \frac{1}{2}\mathbf{j}, \frac{3}{4}\mathbf{i} + \frac{1}{2}\mathbf{j}$ **51.** 32

53. $\$37,289$; Total revenue

55. (a) 6251.3 pounds   (b) $35,453.1$ pounds

57. 735 newton-meters   **59.** 779.4 foot-pounds

61. 725.0 foot-pounds

63. True. The zero vector is orthogonal to every vector.

65. Orthogonal. $\mathbf{u} \cdot \mathbf{v} = 0$

67. (a) $\mathbf{u}$ and $\mathbf{v}$ are parallel.   (b) $\mathbf{u}$ and $\mathbf{v}$ are orthogonal.

69. Answers will vary.   **71.** (a) and (b) Answers will vary.

73. $\dfrac{\pi}{2}, \dfrac{3\pi}{2}$ **75.** $\dfrac{5\pi}{6}, \dfrac{7\pi}{6}$ **77.** 59.43 square units

79. (a) $\langle 7, -1 \rangle$ (b) $\langle 5, -2 \rangle$ (c) $\langle -11, 5 \rangle$

81. (a) $\langle -6, 3 \rangle$ (b) $\langle -6, 12 \rangle$ (c) $\langle 14, -31 \rangle$

83. Buy now.   **85.** 8

Review Exercises   *(page 308)*

1. $C = 110°, b \approx 32.6, c \approx 36.2$ **3.** No solution

5. $A \approx 34.2°, C \approx 30.8°, c \approx 8.2$

7. $B \approx 31.2°, C \approx 133.8°, c \approx 13.9$

$B \approx 148.8°, C \approx 16.2°, c \approx 5.4$

9. $A \approx 40.9°, C \approx 114.1°, c \approx 8.6$

$A \approx 139.1°, C \approx 15.9°, c \approx 2.6$

11. ≈ 9.08 square units   **13.** ≈ 221.3 square units

15. ≈ 31.1 meters   **17.** ≈ 31.0 feet

19. $A \approx 29.7°, B \approx 52.4°, C \approx 97.9°$

21. $c \approx 6.37, A \approx 41.7°, B \approx 93.3°$

23. $A = 35°, C = 35°, b \approx 6.6$

25. $A \approx 39.9°, B \approx 22.4°, C \approx 117.7°$

27. $b \approx 29.1, A \approx 9.9°, C \approx 20.1°$

29. ≈ 4.3 feet, ≈ 12.6 feet   **31.** ≈ 615.1 meters

33. ≈ 9.8 square units   **35.** ≈ 511.7 square units

37. $\|\mathbf{u}\| = \|\mathbf{v}\| = \sqrt{61}$, slope$_\mathbf{u}$ = slope$_\mathbf{v}$ = $\frac{5}{6}$

39. $\langle 7, -5 \rangle$ **41.** $\langle 7, -7 \rangle$ **43.** $\langle -4, 4\sqrt{3} \rangle$

45. $\langle -4, -5 \rangle$ **47.** $\langle 2, -1 \rangle$ **49.** $\langle 3, 33 \rangle$

51. $\left\langle \dfrac{6}{\sqrt{61}}, -\dfrac{5}{\sqrt{61}} \right\rangle$ **53.** $\langle 30, 9 \rangle$

55. $\langle -26, -35 \rangle$

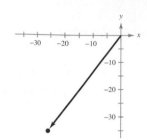

57. $\langle 0, -1 \rangle$ **59.** $\frac{1}{\sqrt{29}}\langle 5, -2 \rangle$ **61.** $\frac{1}{\sqrt{34}}\langle -5, 3 \rangle$

63. $\frac{1}{\sqrt{146}}\langle 5, -11 \rangle$ **65.** $9\mathbf{i} - 8\mathbf{j}$

67. $10\sqrt{2}[(\cos 135°)\mathbf{i} + (\sin 135°)\mathbf{j}]$

69.

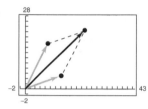

Magnitude: 32.62

Direction: 44.72°

71. 79.9°, 92.2 pounds

73. 422.3 miles per hour, S 49.6° E   **75.** 7

77. $\langle -54, 27 \rangle$ **79.** -36 **81.** $\frac{11\pi}{12}$ **83.** 160.5°

85. Parallel   **87.** Neither

89. (a) $\frac{13}{17}\langle -4, -1 \rangle$ (b) $\langle -\frac{52}{17}, -\frac{13}{17} \rangle + \langle -\frac{16}{17}, \frac{64}{17} \rangle$

91. (a) $-\frac{5}{2}\langle 1, -1 \rangle$ (b) $\langle -\frac{5}{2}, \frac{5}{2} \rangle + \langle \frac{9}{2}, \frac{9}{2} \rangle$

93. 72,000 foot-pounds

95. True   **97.** Direction and magnitude

Chapter Test   *(page 312)*

1. $C = 44°, a \approx 16.0, b \approx 25.9$

2. $A \approx 26.2°, B \approx 136.5°, C \approx 17.3°$

3. $a \approx 9.8, B \approx 46.6°, C \approx 75.5°$

4. $B \approx 48.3°, C \approx 107.1°, c \approx 35.8$

   $B \approx 131.7°, C \approx 23.7°, c \approx 15.1$

5. No solution   **6.** $B \approx 14.7°, C \approx 15.3°, c \approx 4.9$

7. 89.80   **8.** 9.17   **9.** 2.94   **10.** S 59.8° E

11. ≈ 675 feet   **12.** $\mathbf{w} = \langle 12, 13 \rangle, \|\mathbf{w}\| \approx 17.69$

13. $\langle \frac{7}{\sqrt{65}}, \frac{4}{\sqrt{65}} \rangle$

14. (a) $\langle -4, 4 \rangle$ (b) $\langle 6, -16 \rangle$ (c) $\langle 2, -24 \rangle$

15. (a) $\langle 0, 7 \rangle$ (b) $\langle -5, -18 \rangle$ (c) $\langle -11, -20 \rangle$

16. (a) $\langle 13, 17 \rangle$ (b) $\langle -17, -28 \rangle$ (c) $\langle -1, -14 \rangle$

17. (a) $\langle 0, -1 \rangle$ (b) $\langle 5, 9 \rangle$ (c) $\langle 11, 17 \rangle$

18. $\langle \frac{36}{\sqrt{34}}, -\frac{60}{\sqrt{34}} \rangle$ **19.** $\theta \approx 14.9°, \approx 250.2$ pounds

20. $\approx 105.9°$ **21.** No, because $\mathbf{u} \cdot \mathbf{v} = 24$, not 0.

22. $\langle \frac{185}{26}, \frac{37}{26} \rangle, \mathbf{u} = \langle \frac{185}{26}, \frac{37}{26} \rangle + \langle -\frac{29}{26}, \frac{145}{26} \rangle$

Cumulative Test for Chapters 1–3
(page 313)

1. (a)

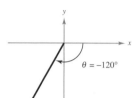

(b) 240°

(c) $-\frac{2\pi}{3}$

(d) 60°

(e) $\sin(-120°) = -\frac{\sqrt{3}}{2}$

   $\cos(-120°) = -\frac{1}{2}$

   $\tan(-120°) = \sqrt{3}$

   $\csc(-120°) = -\frac{2\sqrt{3}}{3}$

   $\sec(-120°) = -2$

   $\cot(-120°) = \frac{\sqrt{3}}{3}$

2. 134.6°   **3.** $\frac{3}{5}$

4.

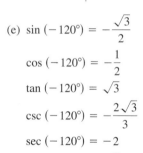

5.

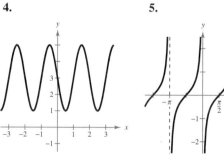

6.

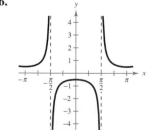

7. $a = -3, b = \pi, c = 0$ **8.** 6.7   **9.** $\frac{3}{4}$

10. $\sqrt{1 - 4x^2}$ **11.** 21.1478 feet   **12.** $2 \tan \theta$

13. and 14. Answers will vary.   **15.** $\frac{\pi}{3}, \frac{\pi}{2}, \frac{3\pi}{2}, \frac{5\pi}{3}$

16. $\frac{\pi}{6}, \frac{5\pi}{6}, \frac{7\pi}{6}, \frac{11\pi}{6}$ **17.** 1.7646, 4.5186

18.

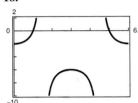

$\frac{\pi}{3}, \frac{5\pi}{3}$

19.

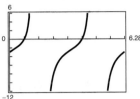

$\frac{\pi}{4}, \frac{5\pi}{4}$

20. $8x^2 - 1$ **21.** $\frac{2x}{x^2 + 1}$ **22.** $\frac{4}{3}$ **23.** $\frac{\sqrt{5}}{5}$

24. $\sin 67° 30' = \dfrac{\sqrt{2 + \sqrt{2}}}{4}$

$\cos 67° 30' = \dfrac{\sqrt{2 - \sqrt{2}}}{2}$

$\tan 67° 30' = \sqrt{2} + 1$

25. $2 \cos 6x \cos 2x$ **26.** $\dfrac{5}{2}\left(\sin \dfrac{\pi}{2} + \sin \pi \right)$

27.–30. Answers will vary.

31. $B = 14.89°$ **32.** $B = 52.48°$
    $C = 119.11°$ $C = 97.52°$
    $c = 17$ $a = 5.04$

33. $C = 101°$ **34.** $A = 26.07°$
    $b = 20.14$ $B = 33.33°$
    $c = 24.13$ $C = 120.60°$

35. 131.71 square inches

36. 85.21 square inches   **37.** $3\mathbf{i} + 5\mathbf{j}$

38. $\left\langle \dfrac{\sqrt{2}}{2}, \dfrac{\sqrt{2}}{2} \right\rangle$ **39.** -5

40. $\left\langle -\frac{1}{13}, -\frac{5}{13} \right\rangle; \left\langle \frac{105}{13}, -\frac{21}{13} \right\rangle$

41. ≈ 4.79 feet

42. 90π radians per minute; 848.23 inches per minute

43. $d = 4 \sin \dfrac{\pi}{4} t$

44. N 32.63° E; 543.88 kilometers per hour

45. $\approx 80.28°$

Chapter 4

Section 4.1 *(page 321)*

1. $a = -9, b = 4$ **3.** $a = 6, b = 5$ **5.** $4 + 5i$

7. 12   **9.** $-1 - 5i$ **11.** -75 **13.** $0.3i$

15. $11 - i$ **17.** $7 - 3\sqrt{2}i$ **19.** $-14 + 20i$

21. $\frac{1}{6} + \frac{7}{6}i$ **23.** $-4.2 + 7.5i$ **25.** $-2\sqrt{3}$ **27.** -10

29. $5 + i$ **31.** $12 + 30i$ **33.** 24   **35.** $-9 + 40i$

37. $\sqrt{-6}\sqrt{-6} = \sqrt{6}i\sqrt{6}i = 6i^2 = -6$ **39.** 25

41. 41   **43.** 484   **45.** 11   **47.** $-6i$

49. $\frac{16}{41} + \frac{20}{41}i$ **51.** $\frac{3}{5} + \frac{4}{5}i$ **53.** $-7 - 6i$

55. $-\frac{9}{1681} + \frac{40}{1681}i$ **57.** $-\frac{1}{2} - \frac{5}{2}i$ **59.** $1 \pm i$

61. $-2 \pm \frac{1}{2}i$ **63.** $-\frac{3}{2}, -\frac{5}{2}$ **65.** $2 \pm \sqrt{2}i$

67. $\dfrac{5}{7} \pm \dfrac{5\sqrt{15}}{7}$ **69.** (a) 1   (b) i (c) -1 (d) $-i$

71. $-4 + 2i$ **73.** i **75.** -8 **77.** $\frac{1}{8}i$

79. (a) 16   (b) 16   (c) 16   (d) 16

81. False. If the complex number is real, the number equals its conjugate.

83. False. The expression equals 1.

85. Answers will vary.   **87.** $x^3 + x^2 + 2x - 6$

89. $4x^2 - 20x + 25$ **91.** $(-4, 0), (2, 0), (0, -8)$

93. $(-1, 0), (1, 0), (0, -1)$ **95.** 88.9 kilometers per hour

Section 4.2 *(page 328)*

1. Three solutions   **3.** Four solutions

5. No real solutions   **7.** Two real solutions

9. Two real solutions   **11.** No real solutions

13. $\pm\sqrt{5}$ **15.** $-5 \pm \sqrt{6}$ **17.** 4, 4   **19.** $-1 \pm 2i$

21. $\frac{1}{2} \pm i$ **23.** $20 \pm 2\sqrt{215}$ **25.** $-3 \pm 2\sqrt{2}i$

27. (a) $4, \pm i$ (b) 4   (c) One real zero, one x-intercept

29. (a) $\pm\sqrt{2}i, \pm\sqrt{2}i$ (b) No real zeros
    (c) No real zeros, no x-intercepts

31. $\pm 5i; (x + 5i)(x - 5i)$

33. $2 \pm \sqrt{3}; (x - 2 - \sqrt{3})(x - 2 + \sqrt{3})$

35. $\pm 3, \pm 3i; (x + 3)(x - 3)(x + 3i)(x - 3i)$

37. $2, 2 \pm i; (x - 2)(x - 2 + i)(x - 2 - i)$

39. $-2, 1 \pm \sqrt{2}i; (x + 2)(x - 1 + \sqrt{2}i)(x - 1 - \sqrt{2}i)$

41. $-\frac{1}{5}, 1 \pm \sqrt{5}i; (5x + 1)(x - 1 + \sqrt{5}i)(x - 1 - \sqrt{5}i)$

43. $2, 2, \pm 2i; (x - 2)^2(x + 2i)(x - 2i)$

45. $\pm i, \pm 3i; (x - i)(x + i)(x - 3i)(x + 3i)$

47. $-\frac{3}{2}, \pm 5i$ **49.** $\pm 2i, 1, -\frac{1}{2}$ **51.** $-3 \pm i, \frac{1}{4}$

53. $2, -3 \pm \sqrt{2}i, 1$ **55.** $f(x) = x^3 - x^2 + 25x - 25$

57. $f(x) = x^3 + 4x^2 - 31x - 174$

59. $f(x) = 3x^4 - 17x^3 + 25x^2 + 23x - 22$

61. No. Setting $h = 64$ and solving the resulting equation yields imaginary roots.

63. False. The most complex zeros it can have is two and the Linear Factorization Theorem guarantees that there are three linear factors, so one zero must be real.

65. r_1, r_2, r_3 **67.** $5 + r_1, 5 + r_2, 5 + r_3$

69. The zeros cannot be determined.

71. (a) $0 < k < 4$ (b) $k = 4$ (c) $k < 0$ (d) $k > 4$

73. (a) $x^2 + b$ (b) $x^2 - 2ax + a^2 + b^2$

75. $-11 + 9i$ **77.** $20 + 40i$ **79.** i

81.

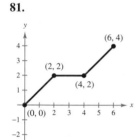

83.

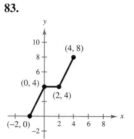

85.

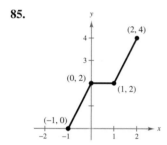

Section 4.3 *(page 335)*

1. 6

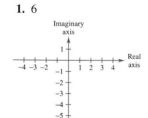

3. 4

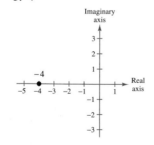

5. $4\sqrt{2}$

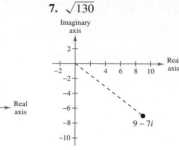

7. $\sqrt{130}$

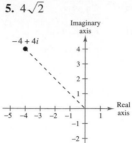

9. $3\left(\cos\dfrac{\pi}{2} + i\sin\dfrac{\pi}{2}\right)$ **11.** $2\sqrt{2}\left(\cos\dfrac{5\pi}{4} + i\sin\dfrac{5\pi}{4}\right)$

13. $5\sqrt{2}\left(\cos\dfrac{7\pi}{4} + i\sin\dfrac{7\pi}{4}\right)$ **15.** $2\left(\cos\dfrac{\pi}{6} + i\sin\dfrac{\pi}{6}\right)$

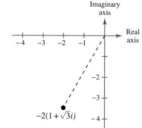

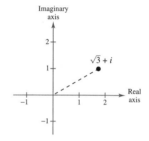

17. $4\left(\cos\dfrac{4\pi}{3} + i\sin\dfrac{4\pi}{3}\right)$ **19.** $8\left(\cos\dfrac{\pi}{2} + i\sin\dfrac{\pi}{2}\right)$

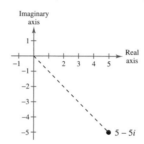

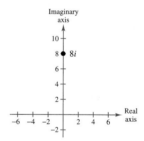

21. $\sqrt{65}(\cos 150.26° + i\sin 150.26°)$

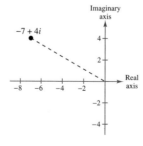

23. $7(\cos 0 + i \sin 0)$

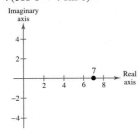

25. $\sqrt{101}(\cos 84.29° + i \sin 84.29°)$

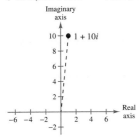

27. $5\sqrt{2}(\cos 188.13° + i \sin 188.13°)$

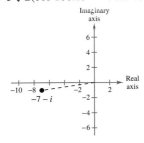

29. $5.385(\cos 21.80° + i \sin 21.80°)$

31. $8.185(\cos 301.22° + i \sin 301.22°)$

33. $-1 + \sqrt{3}i$

35. $\dfrac{3\sqrt{3}}{4} - \dfrac{3}{4}i$

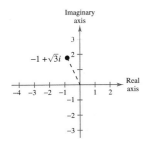

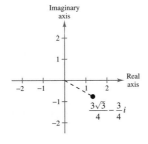

37. $-\dfrac{15\sqrt{2}}{8} + \dfrac{15\sqrt{2}}{8}i$

39. $-4i$

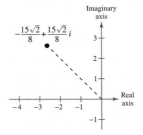

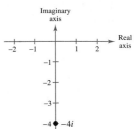

41. $2.8408 + 0.9643i$

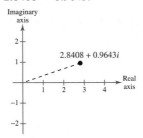

43. $4.6985 + 1.7101i$

45. $4.7693 + 7.6324i$

47. $12\left(\cos\dfrac{\pi}{2} + i \sin\dfrac{\pi}{2}\right)$

49. $\dfrac{10}{9}(\cos 200° + i \sin 200°)$

51. $\dfrac{11}{50}(\cos 130° + i \sin 130°)$

53. $\cos 30° + i \sin 30°$

55. $\dfrac{1}{2}(\cos 80° + i \sin 80°)$

57. $6[\cos(-48°) + i \sin(-48°)]$

59. (a) $2\sqrt{2}\left(\cos\dfrac{\pi}{4} + i \sin\dfrac{\pi}{4}\right)$

$\sqrt{2}\left[\cos\left(-\dfrac{\pi}{4}\right) + i \sin\left(-\dfrac{\pi}{4}\right)\right]$

(b) $4(\cos 0 + i \sin 0) = 4$ (c) 4

61. (a) $2\left[\cos\left(-\dfrac{\pi}{2}\right) + i \sin\left(-\dfrac{\pi}{2}\right)\right]$

$\sqrt{2}\left(\cos\dfrac{\pi}{4} + i \sin\dfrac{\pi}{4}\right)$

(b) $2\sqrt{2}\left[\cos\left(-\dfrac{\pi}{4}\right) + i \sin\left(-\dfrac{\pi}{4}\right)\right] = 2 - 2i$

(c) $2 - 2i$

63. (a) $5(\cos 0 + i \sin 0)$

$\sqrt{13}(\cos 56.31° + i \sin 56.31°)$

(b) $\dfrac{5\sqrt{13}}{13}[\cos(-56.31°) + i \sin(-56.31°)] \approx$

$0.7692 - 1.1538i$

(c) $\dfrac{10}{13} - \dfrac{15}{13}i \approx 0.7692 - 1.1538i$

65.

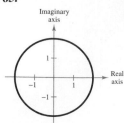

67.

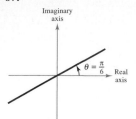

69. True, by the definition of the absolute value of a complex number.

71. Answers will vary. **73.** (a) r^2 (b) $\cos 2\theta + i \sin 2\theta$

75. $B = 68°, b \approx 19.80, c \approx 21.36$

77. $B = 60°, a \approx 65.01, c \approx 130.02$

79. $B = 47°45', a \approx 7.53, b \approx 8.29$

81. $16; 2$ **83.** $\frac{1}{16}; \frac{4}{5}$

Section 4.4 *(page 341)*

1. $-4 - 4i$ **3.** $-32i$ **5.** $-972 + 972i$

7. $-8i$ **9.** $-128\sqrt{3} - 128i$ **11.** $\frac{125}{2} + \frac{125\sqrt{3}}{2}i$

13. $608.02 + 144.69i$ **15.** -1 **17.** $\frac{81}{2} + \frac{81\sqrt{3}}{2}i$

19. $32.3524 - 120.7407i$ **21.** $32i$ **23.** 27

25. (a) $2(\cos 30° + i \sin 30°)$

$2(\cos 150° + i \sin 150°)$

$2(\cos 270° + i \sin 270°)$

(b) and (c) $8i$

27. (a) $\sqrt{5}(\cos 60° + i \sin 60°)$

$\sqrt{5}(\cos 240° + i \sin 240°)$

(b)

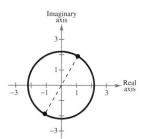

(c) $\dfrac{\sqrt{5}}{2} + \dfrac{\sqrt{15}}{2}i, \ -\dfrac{\sqrt{5}}{2} - \dfrac{\sqrt{15}}{2}i$

29. (a) $2\left(\cos \dfrac{2\pi}{9} + i \sin \dfrac{2\pi}{9}\right)$

$2\left(\cos \dfrac{8\pi}{9} + i \sin \dfrac{8\pi}{9}\right)$

$2\left(\cos \dfrac{14\pi}{9} + i \sin \dfrac{14\pi}{9}\right)$

(b)

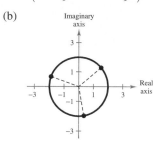

(c) $1.5321 + 1.2856i, -1.8794 + 0.6840i,$

$0.3473 - 1.9696i$

31. (a) $3\left(\cos \dfrac{\pi}{30} + i \sin \dfrac{\pi}{30}\right)$

$3\left(\cos \dfrac{13\pi}{30} + i \sin \dfrac{13\pi}{30}\right)$

$3\left(\cos \dfrac{5\pi}{6} + i \sin \dfrac{5\pi}{6}\right)$

$3\left(\cos \dfrac{37\pi}{30} + i \sin \dfrac{37\pi}{30}\right)$

$3\left(\cos \dfrac{49\pi}{30} + i \sin \dfrac{49\pi}{30}\right)$

(b)

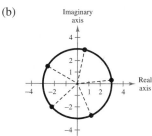

(c) $2.9836 + 0.3136i, 0.6237 + 2.9344i,$

$-2.5981 + 1.5i, -2.2294 - 2.0074i, 1.2202 - 2.7406i$

33. (a) $5\left(\cos\dfrac{3\pi}{4} + i\sin\dfrac{3\pi}{4}\right)$

$5\left(\cos\dfrac{7\pi}{4} + i\sin\dfrac{7\pi}{4}\right)$

(b)

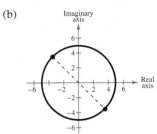

(c) $-\dfrac{5\sqrt{2}}{2} + \dfrac{5\sqrt{2}}{2}i,\ \dfrac{5\sqrt{2}}{2} - \dfrac{5\sqrt{2}}{2}i$

35. (a) $3\left(\cos\dfrac{\pi}{8} + i\sin\dfrac{\pi}{8}\right)$

$3\left(\cos\dfrac{5\pi}{8} + i\sin\dfrac{5\pi}{8}\right)$

$3\left(\cos\dfrac{9\pi}{8} + i\sin\dfrac{9\pi}{8}\right)$

$3\left(\cos\dfrac{13\pi}{8} + i\sin\dfrac{13\pi}{8}\right)$

(b)

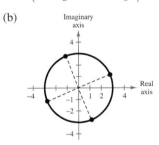

(c) $2.7716 + 1.1481i,\ -1.1481 + 2.7716i,$
$-2.7716 - 1.1481i,\ 1.1481 - 2.7716i$

37. (a) $5\left(\cos\dfrac{4\pi}{9} + i\sin\dfrac{4\pi}{9}\right)$

$5\left(\cos\dfrac{10\pi}{9} + i\sin\dfrac{10\pi}{9}\right)$

$5\left(\cos\dfrac{16\pi}{9} + i\sin\dfrac{16\pi}{9}\right)$

(b)

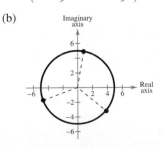

(c) $0.8682 + 4.9240i,\ -4.6985 - 1.7101i,$
$3.8302 - 3.2139i$

39. (a) $2(\cos 0 + i\sin 0)$

$2\left(\cos\dfrac{\pi}{2} + i\sin\dfrac{\pi}{2}\right)$

$2(\cos\pi + i\sin\pi)$

$2\left(\cos\dfrac{3\pi}{2} + i\sin\dfrac{3\pi}{2}\right)$

(b)

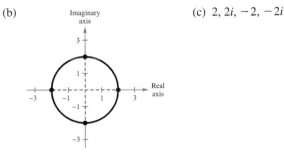

(c) $2, 2i, -2, -2i$

41. (a) $\cos 0 + i\sin 0$

$\cos\dfrac{2\pi}{5} + i\sin\dfrac{2\pi}{5}$

$\cos\dfrac{4\pi}{5} + i\sin\dfrac{4\pi}{5}$

$\cos\dfrac{6\pi}{5} + i\sin\dfrac{6\pi}{5}$

$\cos\dfrac{8\pi}{5} + i\sin\dfrac{8\pi}{5}$

(b)

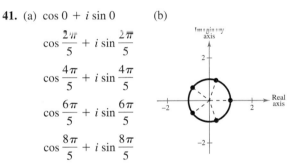

(c) $1, 0.3090 + 0.9511i,\ -0.8090 + 0.5878i,$
$-0.8090 - 0.5878i, 0.3090 - 0.9511i$

43. (a) $5\left(\cos\dfrac{\pi}{3} + i\sin\dfrac{\pi}{3}\right)$

$5(\cos\pi + i\sin\pi)$

$5\left(\cos\dfrac{5\pi}{3} + i\sin\dfrac{5\pi}{3}\right)$

(b)

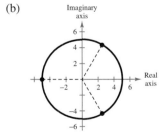

(c) $\dfrac{5}{2} + \dfrac{5\sqrt{3}}{2}i,\ -5,\ \dfrac{5}{2} - \dfrac{5\sqrt{3}}{2}i$

45. (a) $\sqrt[10]{32{,}768}\left(\cos\dfrac{3\pi}{20}+i\sin\dfrac{3\pi}{20}\right)$

$\sqrt[10]{32{,}768}\left(\cos\dfrac{11\pi}{20}+i\sin\dfrac{11\pi}{20}\right)$

$\sqrt[10]{32{,}768}\left(\cos\dfrac{19\pi}{20}+i\sin\dfrac{19\pi}{20}\right)$

$\sqrt[10]{32{,}768}\left(\cos\dfrac{27\pi}{20}+i\sin\dfrac{27\pi}{20}\right)$

$\sqrt[10]{32{,}768}\left(\cos\dfrac{7\pi}{4}+i\sin\dfrac{7\pi}{4}\right)$

(b)

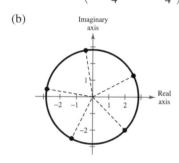

(c) $2.5201+1.2841i,\ -0.4425+2.7936i,$

$-2.7936+0.4425i,\ -1.2841-2.5201i,\ 2-2i$

47. $\cos\dfrac{\pi}{8}+i\sin\dfrac{\pi}{8}$

$\cos\dfrac{5\pi}{8}+i\sin\dfrac{5\pi}{8}$

$\cos\dfrac{9\pi}{8}+i\sin\dfrac{9\pi}{8}$

$\cos\dfrac{13\pi}{8}+i\sin\dfrac{13\pi}{8}$

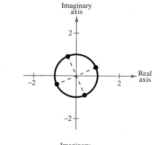

49. $3(\cos 0+i\sin 0)$

$3\left(\cos\dfrac{2\pi}{5}+i\sin\dfrac{2\pi}{5}\right)$

$3\left(\cos\dfrac{4\pi}{5}+i\sin\dfrac{4\pi}{5}\right)$

$3\left(\cos\dfrac{6\pi}{5}+i\sin\dfrac{6\pi}{5}\right)$

$3\left(\cos\dfrac{8\pi}{5}+i\sin\dfrac{8\pi}{5}\right)$

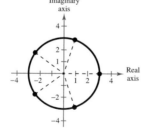

51. $4\left(\cos\dfrac{\pi}{2}+i\sin\dfrac{\pi}{2}\right)$

$4\left(\cos\dfrac{7\pi}{6}+i\sin\dfrac{7\pi}{6}\right)$

$4\left(\cos\dfrac{11\pi}{6}+i\sin\dfrac{11\pi}{6}\right)$

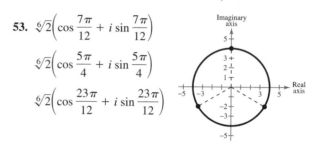

53. $\sqrt[6]{2}\left(\cos\dfrac{7\pi}{12}+i\sin\dfrac{7\pi}{12}\right)$

$\sqrt[6]{2}\left(\cos\dfrac{5\pi}{4}+i\sin\dfrac{5\pi}{4}\right)$

$\sqrt[6]{2}\left(\cos\dfrac{23\pi}{12}+i\sin\dfrac{23\pi}{12}\right)$

55. True   **57.** Answers will vary.   **59.** 12.9410

61. $\left\langle-\dfrac{3\sqrt{58}}{58},\dfrac{7\sqrt{58}}{58}\right\rangle$ **63.** **j**   **65.** $\sqrt{10}+6$

67. -68 **69.** Orthogonal

Review Exercises *(page 344)*

1. $6+5i$ **3.** $2+7i$ **5.** $3+7i$ **7.** $40+65i$

9. $-4-46i$ **11.** -80 **13.** $1-6i$ **15.** $\frac{4}{3}i$

17. $\pm\sqrt{\frac{1}{3}}i$ **19.** $1\pm3i$ **21.** Five solutions

23. Four solutions   **25.** Two real solutions

27. No real solutions   **29.** $0, 2$ **31.** $-4\pm\sqrt{6}$

33. $-\dfrac{1}{2}\pm\dfrac{\sqrt{5}}{2}i$ **35.** $1,\dfrac{3}{4}$ **37.** $-1,\dfrac{3}{2},3,\dfrac{2}{3}$

39. Yes. A price of \$95.41 or \$119.59 per unit would yield a profit of 9 million dollars.

41. $-2,-\frac{3}{5},\frac{1}{2};\ (x+2)(5x+3)(2x-1)$

43. $4, 2\pm3i;\ (x-4)(x-2-3i)(x-2+3i)$

45. $\frac{4}{5},\pm2i;\ (5x-4)(x-2i)(x+2i)$

47. $2\pm\sqrt{3}i, 1\pm\sqrt{2}i;$

$\left(x-2-\sqrt{3}i\right)\left(x-2+\sqrt{3}i\right)\left(x-1-\sqrt{2}i\right)\left(x-1+\sqrt{2}i\right)$

49. $f(x)=x^4-6x^3+5x^2+24x-36$

51. $f(x)=x^3-7x^2+9x+5$

53. $f(x)=x^4-x^3-3x^2+17x-30$

55. $f(x)=x^4+20x^2+64$

57.

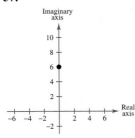

6

59.

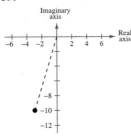

$\sqrt{109}$

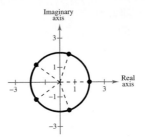

61. $3(\cos 19.5° + i \sin 19.5°)$

63. $6(\cos 150° + i \sin 150°)$

65. $7.62(\cos 156.8° + i \sin 156.8°)$

67. (a) $z_1 = 3\sqrt{2}\left(\cos \dfrac{5\pi}{4} + i \sin \dfrac{5\pi}{4}\right)$

$z_2 = 4\left(\cos \dfrac{\pi}{6} + i \sin \dfrac{\pi}{6}\right)$

(b) $z_1 z_2 = 12\sqrt{2}\left(\cos \dfrac{17\pi}{12} + i \sin \dfrac{17\pi}{12}\right)$

$\dfrac{z_1}{z_2} = \dfrac{3\sqrt{2}}{4}\left(\cos \dfrac{13\pi}{12} + i \sin \dfrac{13\pi}{12}\right)$

69. $6\left(\cos\dfrac{5\pi}{6} + i \sin\dfrac{5\pi}{6}\right)$ **71.** $\dfrac{1}{3}(\cos 135° + i \sin 135°)$

73. $-16 - 16\sqrt{3}i$ **75.** 16

77. $4(\cos 0 + i \sin 0) = 4$

$4\left(\cos \dfrac{\pi}{2} + i \sin \dfrac{\pi}{2}\right) = 4i$

$4(\cos \pi + i \sin \pi) = -4$

$4\left(\cos \dfrac{3\pi}{2} + i \sin \dfrac{3\pi}{2}\right) = -4i$

79. $2(\cos 0 + i \sin 0) = 2$

$2\left(\cos \dfrac{2\pi}{5} + i \sin \dfrac{2\pi}{5}\right) = 0.6180 + 1.9021i$

$2\left(\cos \dfrac{4\pi}{5} + i \sin \dfrac{4\pi}{5}\right) = -1.6180 + 1.1756i$

$2\left(\cos \dfrac{6\pi}{5} + i \sin \dfrac{6\pi}{5}\right) = -1.6180 - 1.1756i$

$2\left(\cos \dfrac{8\pi}{5} + i \sin \dfrac{8\pi}{5}\right) = 0.6180 - 1.9021i$

81. $\cos 0 + i \sin 0 = 1$

$\cos \dfrac{\pi}{2} + i \sin \dfrac{\pi}{2} = i$

$\cos \dfrac{2\pi}{3} + i \sin \dfrac{2\pi}{3} = -\dfrac{1}{2} + \dfrac{\sqrt{3}}{2}i$

$\cos \dfrac{4\pi}{3} + i \sin \dfrac{4\pi}{3} = -\dfrac{1}{2} - \dfrac{\sqrt{3}}{2}i$

$\cos \dfrac{3\pi}{2} + i \sin \dfrac{3\pi}{2} = -i$

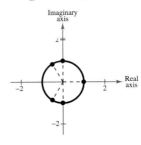

83. False.

$\sqrt{-18}\sqrt{-2} = 3\sqrt{2}i\sqrt{2}i$ and $\sqrt{(-18)(-2)} = \sqrt{36}$

$\qquad\qquad = 3\sqrt{4}i^2 \qquad\qquad\qquad = 6$

$\qquad\qquad = 6i^2$

$\qquad\qquad = -6$

85. False. A fourth-degree polynomial can have at most four zeros, and complex zeros occur in conjugate pairs.

87. $z_1 z_2 = -4$

$\dfrac{z_1}{z_2} = -\dfrac{1}{4}z_1{}^2$

Chapter Test *(page 348)*

1. $-3 + 9i$ **2.** $-3 + 5i$ **3.** $-32 + 24i$

4. 7   **5.** $2 - i$ **6.** $\dfrac{1}{2} \pm \dfrac{\sqrt{5}}{2}i$

7. Five solutions   **8.** Four solutions

9. $6, \pm\sqrt{5}i$ **10.** $\pm\sqrt{6}, \pm 2i$

11. $\pm 2, \pm\sqrt{2}i;\ (x + 2)(x - 2)(x + \sqrt{2}i)(x - \sqrt{2}i)$

12. $\dfrac{3}{2}, 2 \pm i;\ (2v - 3)(v - 2 - i)(v - 2 + i)$

13. $f(x) = x^4 - 9x^3 + 28x^2 - 30x$

14. $f(x) = x^4 - 8x^3 + 28x^2 - 60x + 63$

15. No. If $a + bi$ is a zero, its conjugate $a - bi$ is also a zero.

16. $z = 2\sqrt{2}\left(\cos\dfrac{3\pi}{4} + i\sin\dfrac{3\pi}{4}\right)$

17. $-50 - 50\sqrt{3}i$ **18.** $12\sqrt{3} - 12i$

19. $-\dfrac{6561}{2} + \dfrac{6561\sqrt{3}}{2}i$ **20.** $5832i$

21. $4\left(\cos\dfrac{\pi}{12} + i\sin\dfrac{\pi}{12}\right) \approx 3.8637 + 1.0353i$

$4\left(\cos\dfrac{7\pi}{12} + i\sin\dfrac{7\pi}{12}\right) \approx 1.0353 + 3.8637i$

$4\left(\cos\dfrac{13\pi}{12} + i\sin\dfrac{13\pi}{12}\right) \approx -3.8637 - 1.0353i$

$4\left(\cos\dfrac{19\pi}{12} + i\sin\dfrac{19\pi}{12}\right) \approx 1.0353 - 3.8637i$

22. $5\left(\cos\dfrac{\pi}{8} + i\sin\dfrac{\pi}{8}\right)$

$5\left(\cos\dfrac{5\pi}{8} + i\sin\dfrac{5\pi}{8}\right)$

$5\left(\cos\dfrac{9\pi}{8} + i\sin\dfrac{9\pi}{8}\right)$

$5\left(\cos\dfrac{13\pi}{8} + i\sin\dfrac{13\pi}{8}\right)$

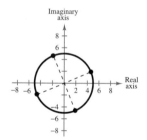

Chapter 5

Section 5.1   *(page 359)*

1. 4112.033   **3.** 77,494.076   **5.** 19.568   **7.** 1.649

9. 9897.129   **11.** $f(x) = h(x)$

13. $f(x) = g(x) = h(x)$ **15.** (c)   **17.** (e)   **19.** (g)

21. (a)   **23.** Right shift of 5 units

25. Left shift of 4 units and reflection in x-axis

27.

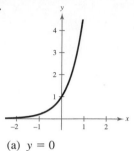

(a) $y = 0$

(b) $(0, 1)$

(c) Increasing

29.

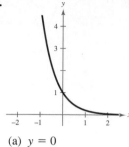

(a) $y = 0$

(b) $(0, 1)$

(c) Decreasing

31.

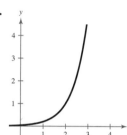

(a) $y = 0$

(b) $\left(0, \dfrac{1}{25}\right)$

(c) Increasing

33.

(a) $y = -3$

(b) $(0, -2)$, $(-0.683, 0)$

(c) Decreasing

35.

x	-1	0	1	2	3
$f(x)$	0.4	1	2.5	6.3	15.6

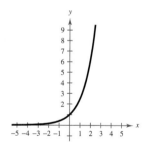

37.

x	-1	0	1	2
$f(x)$	0.2	1	6	36

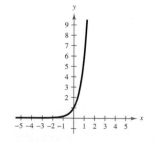

39.

x	-3	-2	0	1
$f(x)$	$\frac{1}{3}$	1	9	27

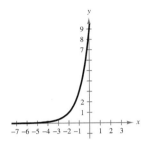

41.

x	-7	-6	-5	-4	-3
$f(x)$	0.1	0.4	1.1	3	8

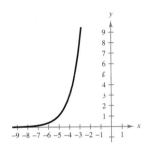

43.

x	2	3	4	5	6	7
$f(x)$	2.0	2.1	2.4	3	4.7	9.3

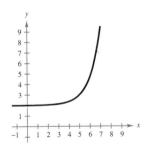

45.

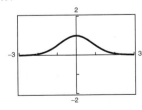

$y = 0$

47.

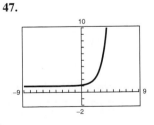

$y = 1$

49.

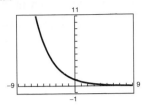

$y = 0$

51.

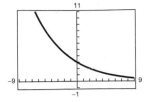

$s(t) = 0$

53. (a)

x	-1	-0.5	0	0.5	1
$f(x)$	0.3333	0.5774	1	1.7321	3
$g(x)$	0.25	0.5	1	2	4

$x < 0$

(b)

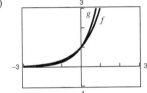

(i) $4^x < 3^x$ when $x < 0$ (ii) $4^x > 3^x$ when $x > 0$

55. (a)

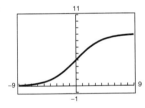

(b)

x	-30	-20	-10	0
$f(x)$	0.0000024	0.00036	0.054	4

x	10	20	30
$f(x)$	7.95	7.9996	7.999998

$y = 0, y = 8$

57. (a)

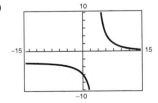

(b)

x	−20	−10	0	3	3.4	3.46
f(x)	−3.03	−3.22	−6	−34	−230	−2617

x	3.47	4	5	7	10	15	25
f(x)	3516	27	8	2.9	1.1	0.3	0.04

$y = -3, y = 0, x \approx 3.46$

59. (a)

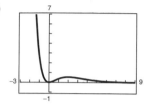

(b) Decreasing on $(-\infty, 0), (2, \infty)$

Increasing on $(0, 2)$

(c) Relative minimum at $(0, 0)$

Relative maximum at $(2, 0.541)$

61. (a)

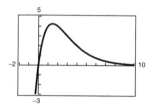

(b) Increasing on $(-\infty, 1.443)$

Decreasing on $(1.443, \infty)$

(c) Relative maximum at $(1.443, 4.246)$

63.

n	1	2	4	12	365	Continuous
A	5397.3	5477.8	5520.1	5549.1	5563.4	5563.85

65.

n	1	2	4	12	365	Continuous
A	11,652	12,003	12,189	12,317	12,380	12,382.58

67.

t	1	10	20	30	40	50
A	12,999	26,706	59,436	132,278	294,390	655,178

69.

t	1	10	20	30	40	50
A	12,804	22,946	43,877	83,902	160,435	306,782

71. (a)

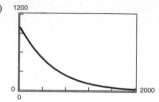

(b) $421.12

(c) $350.13

(d)

x	100	200	300	400
P	849.53	717.64	603.25	504.94

x	500	600	700
P	421.12	350.13	290.35

73. (a)

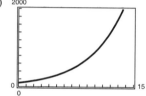

(b) $P(0) = 100; P(5) \approx 300; P(10) \approx 900$

75. (a) 25 units (b) 16.30 units

(c)

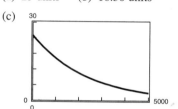

(d) Never. The graph has a horizontal asymptote at $Q = 0$.

77. (a) and (b)

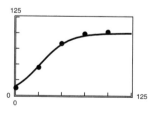

The model fits the data well.

(c)

x	0	25	50	75	100
y	15	47	82	96	99

(d) 64.7% (e) 37.4

79. (a) 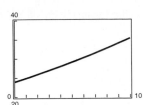 (b) $35.45

81. True

83.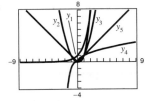

(a) $y = e^x$

(b) The exponential function increases at a faster rate.

(c) It usually implies rapid growth.

85. $1 < \sqrt{2} < 2$, so $2^1 < 2^{\sqrt{2}} < 2^2$

87.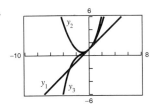

Answers will vary.

89. $f^{-1}(x) = -\frac{3}{2}\left(x - \frac{5}{2}\right)$ **91.** No inverse function

93. $f(x) = x^4 + 17x^2 + 16$

95. $f(x) = x^5 - 4x^4 + 6x^3 - 4x^2$

Section 5.2   *(page 370)*

1. $4^3 = 64$ **3.** $7^{-2} = \frac{1}{49}$ **5.** $32^{2/5} = 4$ **7.** $e^0 = 1$

9. $\log_5 125 = 3$ **11.** $\log_{81} 3 = \frac{1}{4}$ **13.** $\log_6 \frac{1}{36} = -2$

15. $\ln 20.0855\ldots = 3$ **17.** $\ln 13.463\ldots = 2.6$

19. 4   **21.** $-\frac{1}{2}$ **23.** -2 **25.** 9   **27.** 8   **29.** 2

31. 2.538   **33.** -0.097 **35.** 1.746   **37.** 1.869

39. 7.022   **41.** 22.276

43. 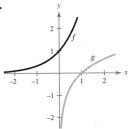   **45.**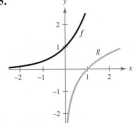

Reflections in the line $y = x$ Reflections in the line $y = x$

$g = f^{-1}$ $g = f^{-1}$

47. (c)   **49.** (d)   **51.** (b)

53. Domain: $(0, \infty)$

Vertical asymptote: $x = 0$

Intercept: $(1, 0)$

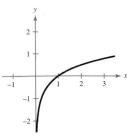

55. Domain: $(3, \infty)$

Vertical asymptote: $x = 3$

Intercept: $(4, 0)$

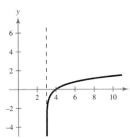

57. Domain: $(0, \infty)$

Vertical asymptote: $x = 0$

Intercept: $(9, 0)$

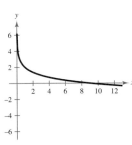

59. Domain: $(3, \infty)$

Vertical asymptote: $x = 3$

Intercept:
$(3 + 6^{-6}, 0) \approx (3, 0)$

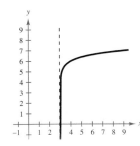

61. Domain: $(0, \infty)$

Vertical asymptote: $x = 0$

Intercept: $(5, 0)$

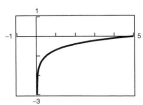

63. Domain: $(2, \infty)$

Vertical asymptote: $x = 2$

Intercept: $(3, 0)$

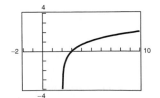

65. Domain: $(-\infty, 0)$

Vertical asymptote: $x = 0$

Intercept: $(-1, 0)$

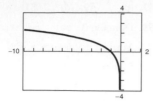

67. (a) 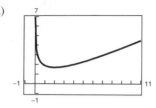   (b) Domain: $(0, \infty)$

(c) Decreasing: $(0, 2)$; Increasing: $(2, \infty)$

(d) Relative minimum: $(2, 1.693)$

69. (a) 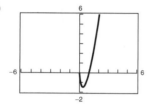   (b) Domain: $(0, \infty)$

(c) Decreasing: $(0, 0.37)$; Increasing: $(0.37, \infty)$

(d) Relative minimum: $(0.37, -1.47)$

71. 23.68 years   **73.** (a) 80   (b) 68.1   (c) 62.3

75. (a)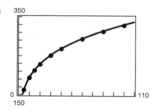

The model fits the data well.

(b) 67.3 pounds per square inch   (c) 306.5°F

77.

r	0.005	0.010	0.015
t	138.6	69.3	46.2

r	0.020	0.025	0.030
t	34.7	27.7	23.1

As the growth rate of the population increases, the time for the population to double decreases.

79. (a) 120 decibels   (b) 100 decibels   (c) No

81. (a) 15 cubic feet per minute   (b) 382 cubic feet

(c) 382 square feet

83. 10 years

85. Total amount: $199,108.80; Interest: $49,108.80

87. True   **89.**

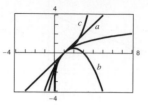

91. (a)

x	1	5	10	10^2	10^4	10^6
$f(x)$	0	0.322	0.230	0.046	0.00092	0.0000138

(b) 0

(c)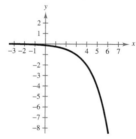

93. 33.115   **95.** 0.002

97.   **99.**

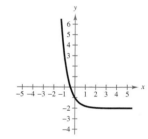

Section 5.3   *(page 378)*

1.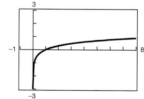

$f = g$

3. 1.771   **5.** -2 **7.** -0.102 **9.** 2.691

11. (a) $\dfrac{\log_{10} x}{\log_{10} 5}$ (b) $\dfrac{\ln x}{\ln 5}$ **13.** (a) $\dfrac{\log_{10} x}{\log_{10} \frac{1}{5}}$ (b) $\dfrac{\ln x}{\ln \frac{1}{5}}$

15. (a) $\dfrac{\log_{10} \frac{3}{10}}{\log_{10} x}$ (b) $\dfrac{\ln \frac{3}{10}}{\ln x}$

17. (a) $\dfrac{\log_{10} x}{\log_{10} 2.6}$ (b) $\dfrac{\ln x}{\ln 2.6}$

19. $\dfrac{\log_{10} x}{\log_{10} 2}$

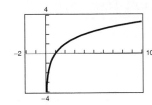

21. $\dfrac{\log_{10} x}{\log_{10} \frac{1}{2}}$

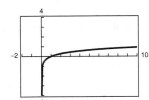

23. $\dfrac{\ln x}{\ln 11.8}$

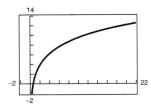

25. $\dfrac{\frac{1}{2}\ln x}{\ln 3}$

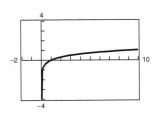

27. $\log_{10} 5 + \log_{10} x$ **29.** $\log_{10} 5 - \log_{10} x$

31. $4\log_8 x$ **33.** $\frac{1}{2}\ln z$ **35.** $\ln x + \ln y + \ln z$

37. $\frac{1}{2}\ln(a-1)$ **39.** $\ln z + 2\ln(z-1)$

41. $\frac{1}{3}\ln x - \frac{1}{3}\ln y$ **43.** $4\ln x + \frac{1}{2}\ln y - 5\ln z$

45. $2\log_b x - 2\log_b y - 3\log_b z$

47.

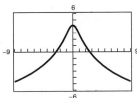

$y_1 = y_2$ for positive values of x.

49. $\ln 4x$ **51.** $\log_4 \dfrac{z}{y}$ **53.** $\log_2(x+3)^2$

55. $\log_3 \sqrt[3]{7x}$ **57.** $\ln \dfrac{x}{(x+1)^3}$ **59.** $\ln \dfrac{x-2}{x+2}$

61. $\ln \dfrac{x}{(x^2-4)^2}$ **63.** $\ln \sqrt[3]{\dfrac{x(x+3)^2}{x^2-1}}$

65. $\ln \dfrac{\sqrt[3]{y(y+4)^2}}{y-1}$ **67.** $\ln \dfrac{9}{\sqrt{x^2+1}}$

69.

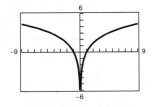

$y_1 = y_2$

71.

No. The domains differ.

73. 2 **75.** 6.8 **77.** -4 is not in the domain of $\log_2 x$.

79. 2 **81.** -4 **83.** 0 is not in the domain of $\log_{10} x$.

85. 8.5 **87.** $\frac{3}{2}$ **89.** $\frac{1}{2}(1 + \log_7 10)$

91. $-3 - \log_5 2$ **93.** $6 + \ln 5$

95. (a) $120 + 10\log_{10} I$

(b) and (c)

I	10^{-4}	10^{-6}	10^{-8}	10^{-10}	10^{-12}	10^{-14}
B	80	60	40	20	0	-20

97. (a)

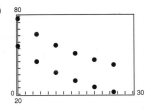

(b) $T = 54.438(0.964)^x + 21$

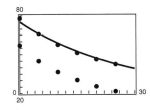

(c) $\ln(T-21) = -0.037t + 3.997$

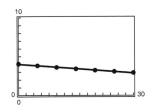

$T = e^{(-0.037t + 3.997)} + 21$

(d)

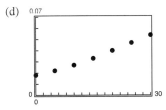

$T = \dfrac{1}{0.0012t + 0.0162} + 21$

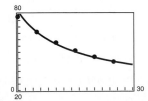

99. False. $\ln 1 = 0$ **101.** False. $f(x) - f(2) = \ln \dfrac{x}{2}$

103. False. $u = v^2$

105.

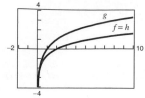

$f(x) = h(x)$

The graphs are identical because for each positive value of x, Property 2 of logarithms holds.

107. Answers will vary.   **109.** $\dfrac{3x^4}{2y^3}$

111. $1,\ x \neq 0,\ y \neq 0$ **113.** $3 \pm \sqrt{7}$ **115.** $\pm 4, \pm \sqrt{3}$

117. $\pm 2, 6$ **119.** 0.052 **121.** $15{,}235.494$

123. 2.342 **125.** 0.697

Section 5.4   *(page 388)*

1. (a) Yes   (b) No   **3.** (a) No   (b) Yes   (c) Yes

5. (a) No   (b) No   (c) Yes

7. (a) Yes   (b) Yes   (c) No

9.

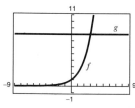

$(3, 8)$

11.

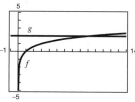

$(9, 2)$

13.

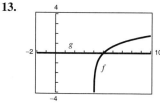

$(5, 0)$

15. 2 **17.** 4 **19.** $\frac{2}{3}$ **21.** -3 **23.** 4 **25.** 5

27. $\ln 4 \approx 1.386$ **29.** $e^{-7} \approx 0.00091$ **31.** 5

33. 0.1 **35.** 1 **37.** x^2 **39.** $5x + 2$ **41.** x^2

43. 2.756 **45.** $\ln 10 \approx 2.303$ **47.** 2 **49.** -6.142

51. 0.511 **53.** 0 **55.** $\ln 5 \approx 1.609$

57. $2 \ln 108 \approx 9.364$ **59.** 6.960

61.

x	0.6	0.7	0.8	0.9	1.0
$f(x)$	6.05	8.17	11.02	14.88	20.09

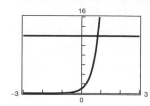

0.828

63.

x	5	6	7	8	9
$f(x)$	1756	1598	1338	908	200

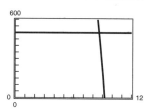

8.635

65. 1.881 **67.** 0.051 **69.** 6.146 **71.** 21.330

73. 3.656

75.

77.

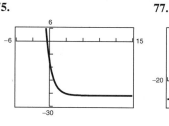

-0.427 12.207

79. 0.050 **81.** 2.042 **83.** 4453.242 **85.** 103

87. 17.945 **89.** 5.389 **91.** $1.718, -3.718$ **93.** 2

95. No real solution   **97.** 180.384

99.

x	2	3	4	5	6
$f(x)$	1.39	1.79	2.08	2.30	2.48

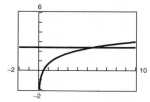

5.512

101.

x	12	13	14	15	16
$f(x)$	9.79	10.22	10.63	11.00	11.36

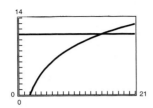

14.988

103. 14   **105.** 5.294   **107.** 3.729   **109.** 5.275

111.

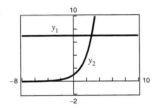

(2.807, 7)

113.

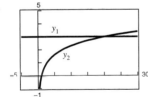

(−3.466, 8)

115.

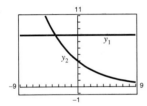

(20.086, 3)

117. (a) 8.2 years   (b) 12.9 years

119. (a)

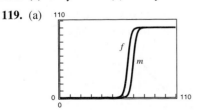

(b) $y = 100$ and $y = 0$

(c) Males: 69.71 inches; Females: 64.51 inches

121. (a) 1426 units   (b) 1498 units

123. (a)

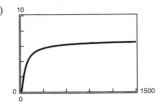

(b) $y = 6.7$. Yield will approach 6.7 million cubic feet per acre.

(c) 29.3 years

125. (a)

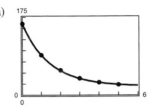

(b) $y = 20$. Room temperature   (c) 0.81 hour

127. (a) $y = 15.17x - 46.15$

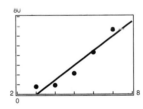

1999

(b) $\ln y = 2.706 \ln x - 1.175$

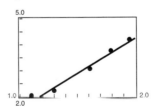

1998

(c) $y = e^{2.706 \ln x - 1.175}$

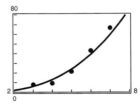

This model better fits the original data and will better predict future shipping levels.

129. False.

For example, $\ln(2x - 1) + \ln(x + 2) = \ln(x^2 - x + 1)$ has 2 extraneous solutions, $x = -1$ and $x = -3$.

131. No. It is dependent on the interest rate.

133.   **135.**

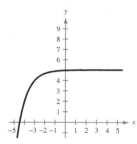 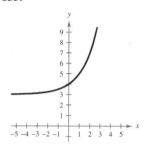

137. 2.814   **139.** 1.623

Section 5.5 *(page 400)*

1. (c)   **3.** (b)   **5.** (d)   **7.** Logarithmic model

9. Gaussian model   **11.** Exponential model

13. Gaussian model

15.   **17.**

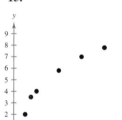 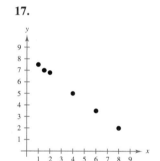

Logarithmic model   Linear model

Initial Investment	Annual % Rate	Time to Double	Amount After 10 Years
19. $1000	12%	5.78 yr	$3320.12
21. $750	8.94%	7.75 yr	$1833.67
23. $500	9.5%	7.30 yr	$1292.85
25. $6376.28	4.5%	15.4 yr	$10,000.00

27. (a)

r	2%	4%	6%	8%	10%	12%
t	54.93	27.47	18.31	13.73	10.99	9.16

(b)

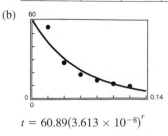

$$t = 60.89(3.613 \times 10^{-8})^r$$

29.

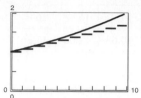

Continuous compounding

Isotope	Half-Life (years)	Initial Quantity	Amount After 1000 Years
31. ^{226}Ra	1620	10 g	6.52 g
33. ^{14}C	5730	3 g	2.66 g

35. 2023   **37.** $k = 0.0112$; 2796   **39.** 3.15 hours

41. 95.8%

43. (a) $V = -4500t + 22{,}000$ (b) $V = 22{,}000e^{-0.263t}$

(c)

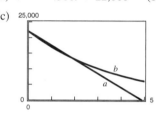

Exponential

(d) 1 year. Straight-line: $17,500; Exponential: $16,912   3 years. Straight-line: $8500; Exponential: $9995

(e) Value decreases $4500 per year.

45. (a) $S(t) = 100(1 - e^{-0.1625t})$

(b)   (c) 55,625

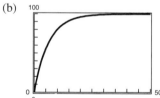

47. (a) $N = 30(1 - e^{-0.050t})$ (b) 36 days

(c) No. It is not a linear function.

49. (a) 7.6   (b) 7.1

51. (a) 20 decibels   (b) 70 decibels   (c) 120 decibels

53. 95%   **55.** 4.64   **57.** 10,000,000 times

59. (a)

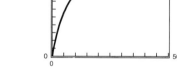

(b) Interest. $t \approx 20.7$ years

(c)

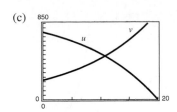

Interest. $t \approx 10.7$ years
Less interest is paid over the life of the loan.

61. $y = e^{0.768x}$ **63.** $y = \frac{1}{2}e^{0.576x}$

65. (a) $t_3 = 0.2729s - 6.0143$; $t_4 = 1.5385e^{0.0291s}$

(b)

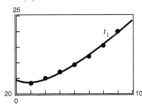 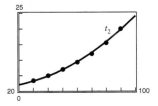

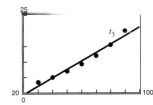 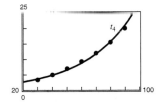

(c)

s	30	40	50	60	70	80	90
t_1	3.6	4.7	6.7	9.4	12.5	15.9	19.6
t_2	3.3	4.9	7.0	9.5	12.5	15.9	19.9
t_3	2.2	4.9	7.6	10.4	13.1	15.8	18.5
t_4	3.7	4.9	6.6	8.8	11.8	15.8	21.1

(d) Model: t_1; Sum ≈ 1.9

    Model: t_2; Sum ≈ 1.1

    Model: t_3; Sum ≈ 5.6

    Model: t_4; Sum ≈ 2.6

    Quadratic model fits best.

67. 7:30 A.M.

69. (a) $y = 0.082x + 4.45$

(b) $y = 4.536(1.015)^x$

(c) The linear model fits the data better; Answers will vary.

(d) Linear: 6.50 billion; Exponential: 6.58 billion

71. (a) $y = 298.794(1.085)^x$

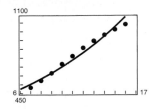

(b) $y = -837.7 + 673.619 \ln x$

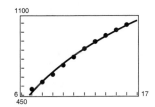

(c) The logarithmic model is better. It will continue to be better if growth of healthcare costs is slowed.

73. (a) $y_1 = -1.81x^3 + 14.58x^2 + 16.39x + 10.00$

    $y_2 = 23.07 + 121.00 \ln x$

    $y_3 = 38.38(1.4227)^x$

(b)

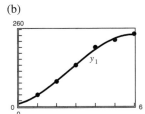

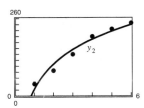

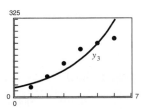

Cubic model

(c)

x	y	$y - y_1$	$(y - y_1)^2$	$y - y_2$	$(y - y_2)^2$
1	40	0.84	0.71	16.93	286.62
2	85	-1.62	2.62	-22.00	483.84
3	140	-1.52	2.31	-16.09	258.89
4	200	7.00	49.00	9.08	82.40
5	225	-5.20	27.04	7.06	49.83
6	245	2.74	7.51	4.98	24.84

x	$y - y_3$	$(y - y_3)^2$
1	-14.60	213.25
2	7.32	53.52
3	29.48	869.01
4	42.76	1828.56
5	1.30	1.68
6	-73.26	5367.34

(d) y_1: 89.19; y_2: 1186.42; y_3: 8333.36

Cubic model

(e) Sum of the squares of the errors

75. False. For example, $y = \dfrac{3}{1 + 2e^{-0.5x}}$ does not have an x-intercept.

77. True   **79.** (b); $(0, -3)$, $\left(\frac{9}{4}, 0\right)$ **81.** (f); $(0, 25)$, $\left(\frac{100}{9}, 0\right)$

83. (d); $(0, 3)$

85.

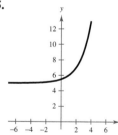

87.

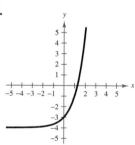

Review Exercises   *(page 409)*

1. 10.325   **3.** 0.201   **5.** (e)   **7.** (b)   **9.** (d)

11.

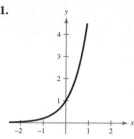

13.

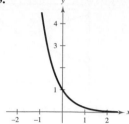

15.

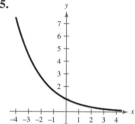

17.

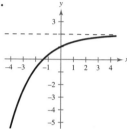

19.

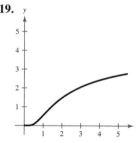

21.

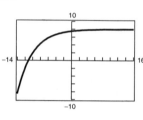

$y = 8$

23.

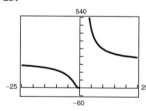

$y = 200, x = 0$

25.

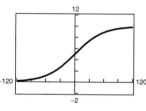

$y = 0, y = 10$

27. 0.069   **29.** 369.355

31.

t	1	10	20
P	\$184,623.27	\$89,865.79	\$40,379.30

t	30	40	50
P	\$18,143.59	\$8152.44	\$3663.13

33. (a)

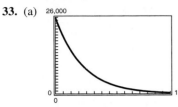

(b) $14,625

(c) When it is first sold; Yes; Answers will vary.

35. $\log_4 64 = 3$ **37.** $\log_{25} 125 = \frac{3}{2}$ **39.** -0.34

41. 3   **43.** $-\frac{1}{2}$

45.

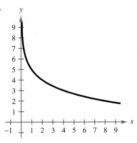

47.

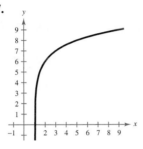

49.

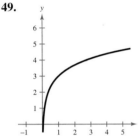

51.

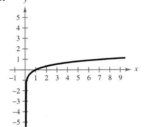

53.

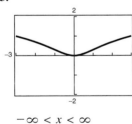

$-\infty < x < \infty$

55.

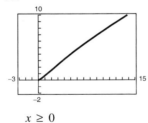

$x \geq 0$

57. 7   **59.** -18

61. (a) $0 \leq h < 18,000$

(b)

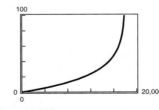

$h = 18,000$

(c) The time required to further increase its altitude increases.

(d) 5.46 minutes

63. 1.585   **65.** 2.132   **67.** $\ln 4 + \ln 5$

69. $\ln 5 - 3 \ln 4$ **71.** 1.6542   **73.** 0.2823

75. $1 + 2 \log_5 x$ **77.** $\log_{10} 5 + \frac{1}{2} \log_{10} y - 2 \log_{10} x$

79. $\ln(x^2 + 1) + \ln(x - 1)$ **81.** $\log_2 5x$

83. $\ln \dfrac{\sqrt{|2x - 1|}}{(x + 1)^2}$ **85.** $\ln \dfrac{3 \sqrt[3]{4 - x^2}}{x}$

87. (a)

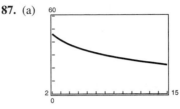

(b)

h	4	6	8	10	12	14
s	38	33	30	27	25	23

(c) The decrease in productivity starts to level off.

89. 3   **91.** -3 **93.** 2401   **95.** $\ln 12 \approx 2.485$

97. $-\dfrac{\ln 44}{5} \approx -0.757$ **99.** $\ln 22 \approx 3.091$

101. $\log_5 17 \approx 1.760$ **103.** $\ln 5 \approx 1.609, \ln 2 \approx 0.693$

105. $\frac{1}{3}e^{8.2} \approx 1213.650$ **107.** $\frac{1}{4}e^{15/2} \approx 452.011$

109. $3e^2 \approx 22.167$ **111.** $e^4 - 1 \approx 53.598$

113. No solution   **115.** $\frac{9}{10}$ **117.** ≈ 15.2 years

119. (e)   **121.** (f)   **123.** (a)   **125.** 2025

127. (a) 5.78%   (b) $10,595.03   (c) 5.95%

129. (a) 7.7 weeks   (b) 13.3 weeks

131. $y = 2e^{0.1014x}$ **133.** $y = \frac{1}{2}e^{0.4605x}$

135. $y = 234.684(0.8746)^x$

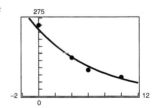

137. True   **139.** False. $\ln(xy) = \ln x + \ln y$ **141.** True

Chapter Test   *(page 414)*

1.

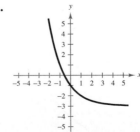

2.

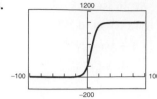

$y = 0$, $y = 1000$

3. \$40,386.38   **4.** $4^3 = 64$

5.

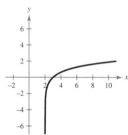

6. $\ln 6 + 2 \ln x - \frac{1}{2}\ln(x^2 + 1)$ **7.** 2   **8.** -3

9. 14.955   **10.** 18.447   **11.** -3.975

12. \$10,204   **13.** (a) 0.154   (b) 0.487   (c) 0.811

14. (a) 300   (b) 570   (c) At the end of the 8th year

15. (c); it passes through the point $(0, 0)$. Symmetric to the y-axis, and $y = 6$ is a horizontal asymptote.

16. $y = 6.775(1.361)^x$

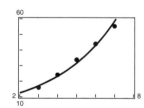

Chapter 6

Section 6.1 *(page 421)*

1. (e)   **3.** (d)   **5.** (a)

7. Vertex: $(0, 0)$
  Focus: $\left(0, \frac{1}{2}\right)$
  Directrix: $y = -\frac{1}{2}$

9. Vertex: $(0, 0)$
  Focus: $\left(-\frac{3}{2}, 0\right)$
  Directrix: $x = \frac{3}{2}$

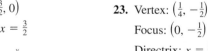

11. Vertex: $(0, 0)$
  Focus: $\left(0, -\frac{3}{2}\right)$
  Directrix: $y = \frac{3}{2}$

13. Vertex: $(1, -2)$
  Focus: $(1, -4)$
  Directrix: $y = 0$

15. Vertex: $\left(-\frac{3}{2}, 2\right)$
  Focus: $\left(-\frac{3}{2}, 3\right)$
  Directrix: $y = 1$

17. Vertex: $(1, 1)$
  Focus: $(1, 2)$
  Directrix: $y = 0$

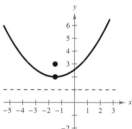

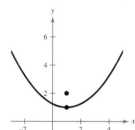

19. Vertex: $(-2, -3)$
  Focus: $(-4, -3)$
  Directrix: $x = 0$

21. Vertex: $(-2, 1)$
  Focus: $\left(-2, -\frac{1}{2}\right)$
  Directrix: $y = \frac{5}{2}$

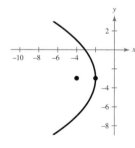

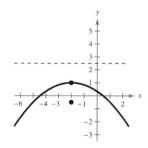

23. Vertex: $\left(\frac{1}{4}, -\frac{1}{2}\right)$
  Focus: $\left(0, -\frac{1}{2}\right)$
  Directrix: $x = \frac{1}{2}$

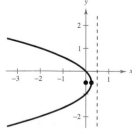

25. $y = -\sqrt{-6x}$ **27.** $y = \frac{2}{3}x^2$ **29.** $x^2 = -6y$

31. $y^2 = -8x$ **33.** $x^2 = 4y$ **35.** $y^2 = -8x$

37. $y^2 = 9x$ **39.** $(x - 3)^2 = -(y - 1)$

41. $y^2 = 2(x + 2)$ **43.** $(y - 2)^2 = -8(x - 5)$

45. $x^2 = 8(y - 4)$ **47.** $(y - 2)^2 = 8x$

49.

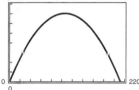

(2, 4)

51. $4x - y - 8 = 0;\ (2, 0)$ **53.** $4x - y + 2 = 0;\ \left(-\frac{1}{2}, 0\right)$

55.

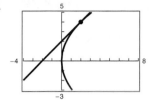

About 106 sales

57. $y = \frac{1}{14}x^2$ **59.** (a) $x^2 = -640y$ (b) 8 feet

61. (a) $17,500\sqrt{2}$ miles per hour

(b) $x^2 = -16,400(y - 4100)$

63. (a) $y = -\frac{1}{64}x^2 + 75$ (b) 69.3 feet

65. False. If the graph intersected the directrix, there would exist points nearer the directrix then the focus.

67. x-intercept: $(7, 0)$ **69.** x-intercepts: $\left(5 \pm \sqrt{5}, 0\right)$

y-intercept: $(0, 49)$ y-intercept: $(0, 20)$

71. x-intercepts: $\left(\dfrac{7 \pm \sqrt{53}}{2}, 0\right)$

y-intercept: $(0, -1)$

73. $f(x) = 3\left(x + \frac{1}{3}\right)^2 - \frac{49}{3}$ **75.** $f(x) = 5\left(x + \frac{17}{5}\right)^2 - \frac{324}{5}$

Vertex: $\left(-\frac{1}{3}, -\frac{49}{3}\right)$ Vertex: $\left(-\frac{17}{5}, -\frac{324}{5}\right)$

77. $f(x) = 6\left(x - \frac{1}{12}\right)^2 - \frac{289}{24}$ Vertex: $\left(\frac{1}{12}, -\frac{289}{24}\right)$

Section 6.2 *(page 430)*

1. (b)   **3.** (d)   **5.** (a)

7. Center: $(0, 0)$

Vertices: $(\pm 5, 0)$

Foci: $(\pm 3, 0)$

Eccentricity: $\frac{3}{5}$

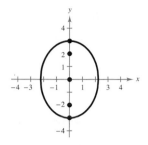

9. Center: $(0, 0)$

Vertices: $(0, \pm 3)$

Foci: $(0, \pm 2)$

Eccentricity: $\frac{2}{3}$

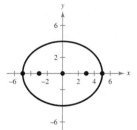

11. Center: $(-3, 5)$

Vertices: $(-3, 0), (-3, 10)$

Foci: $(-3, 8), (-3, 2)$

Eccentricity: $\frac{3}{5}$

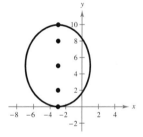

13. Center: $(-5, 1)$

Vertices: $\left(-\frac{7}{2}, 1\right), \left(-\frac{13}{2}, 1\right)$

Foci: $\left(-5 \pm \dfrac{\sqrt{5}}{2}, 1\right)$

Eccentricity: $\dfrac{\sqrt{5}}{3}$

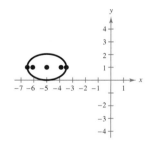

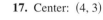

15. Center: $(-2, 3)$

Vertices:

$(-2, 6), (-2, 0)$

Foci: $\left(-2, 3 \pm \sqrt{5}\right)$

Eccentricity: $\dfrac{\sqrt{5}}{3}$

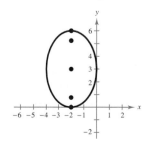

17. Center: $(4, 3)$

Vertices:

$(-6, 3), (14, 3)$

Foci: $\left(4 \pm 4\sqrt{5}, 3\right)$

Eccentricity: $\dfrac{2\sqrt{5}}{5}$

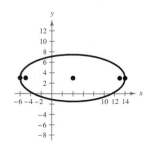

19. Center: $\left(-\frac{3}{2}, \frac{5}{2}\right)$

Vertices: $\left(-\frac{3}{2}, \frac{5 \pm 4\sqrt{3}}{2}\right)$

Foci: $\left(-\frac{3}{2}, \frac{5}{2} \pm 2\sqrt{2}\right)$

Eccentricity: $\frac{\sqrt{6}}{3}$

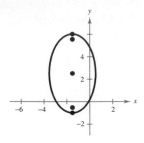

21. Center: $(1, -1)$

Vertices: $\left(\frac{9}{4}, -1\right), \left(-\frac{1}{4}, -1\right)$

Foci: $\left(\frac{7}{4}, -1\right), \left(\frac{1}{4}, -1\right)$

Eccentricity: $\frac{3}{5}$

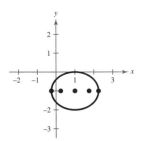

23. Center: $(0, 0)$

Vertices: $\left(0, \pm\sqrt{5}\right)$

Foci: $\left(0, \pm\sqrt{2}\right)$

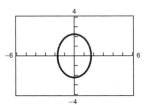

25. Center: $\left(\frac{1}{2}, -1\right)$

Vertices: $\left(\frac{1}{2} \pm \sqrt{5}, -1\right)$

Foci: $\left(\frac{1}{2} \pm \sqrt{2}, -1\right)$

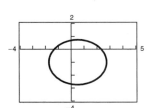

27. $\dfrac{x^2}{4} + \dfrac{y^2}{16} = 1$ **29.** $\dfrac{x^2}{36} + \dfrac{y^2}{32} = 1$

31. $\dfrac{x^2}{36} + \dfrac{y^2}{11} = 1$ **33.** $\dfrac{21x^2}{400} + \dfrac{y^2}{25} = 1$

35. $\dfrac{(x-2)^2}{1} + \dfrac{(y-3)^2}{9} = 1$

37. $\dfrac{(x+2)^2}{16} + \dfrac{(y-3)^2}{9} = 1$

39. $\dfrac{(x-2)^2}{4} + \dfrac{(y-4)^2}{1} = 1$ **41.** $\dfrac{x^2}{48} + \dfrac{(y-4)^2}{64} = 1$

43. $\dfrac{(x-3)^2}{9} + \dfrac{(y-5)^2}{16} = 1$ **45.** $\dfrac{x^2}{16} + \dfrac{(y-4)^2}{12} = 1$

47. $\dfrac{x^2}{25} + \dfrac{y^2}{16} = 1$ **49.** $\left(\pm\sqrt{5}, 0\right)$; 6 feet   **51.** 40

53. $\dfrac{x^2}{4.88} + \dfrac{y^2}{1.39} = 1$ **55.** Answers will vary.

57.

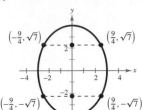

59.

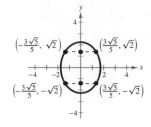

61. True   **63.** False

65. (a) $A = \pi a(20 - a)$

(b) $\dfrac{x^2}{196} + \dfrac{y^2}{36} = 1$

(c)

a	8	9	10	11	12	13
A	301.6	311.0	314.2	311.0	301.6	285.9

$a = 10$. Circle

(d)

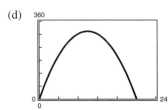

$a = 10$

67. $\frac{3}{2}, \pm 5i$

69.

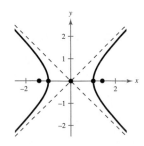

$\pm 3, -1, \frac{1}{2}$

Section 6.3   *(page 440)*

1. (b)   **3.** (a)

5. Center: $(0, 0)$

Vertices: $(\pm 1, 0)$

Foci: $\left(\pm\sqrt{2}, 0\right)$

Asymptotes: $y = \pm x$

7. Center: $(0, 0)$

Vertices: $(0, \pm 1)$

Foci: $\left(0, \pm \sqrt{5}\right)$

Asymptotes: $y = \pm \frac{1}{2}x$

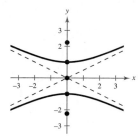

9. Center: $(0, 0)$

Vertices: $(0, \pm 5)$

Foci: $\left(0, \pm \sqrt{106}\right)$

Asymptotes: $y = \pm \frac{5}{9}x$

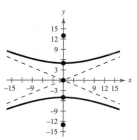

11. Center: $(1, -2)$

Vertices: $(3, -2), (-1, -2)$

Foci: $\left(1 \pm \sqrt{5}, -2\right)$

Asymptotes:

$y = -2 \pm \frac{1}{2}(x - 1)$

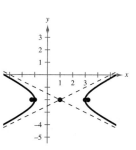

13. Center: $(2, -6)$

Vertices: $(2, -5), (2, -7)$

Foci: $\left(2, -6 \pm \sqrt{2}\right)$

Asymptotes:

$y = -6 \pm (x - 2)$

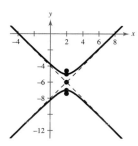

15. Center: $(2, -3)$

Vertices: $(3, -3), (1, -3)$

Foci: $\left(2 \pm \sqrt{10}, -3\right)$

Asymptotes:

$y = -3 \pm 3(x - 2)$

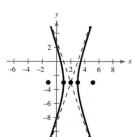

17. The graph of this equation is two lines intersecting at $(-1, -3)$.

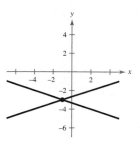

19. Center: $(0, 0)$

Vertices: $\left(\pm \sqrt{3}, 0\right)$

Foci: $\left(\pm \sqrt{5}, 0\right)$

21. Center: $(1, -3)$

Vertices: $\left(1, -3 \pm \sqrt{2}\right)$

Foci: $\left(1, -3 \pm 2\sqrt{5}\right)$

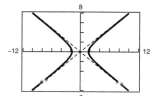

23. $\dfrac{y^2}{4} - \dfrac{x^2}{12} = 1$ **25.** $\dfrac{x^2}{1} - \dfrac{y^2}{25} = 1$

27. $\dfrac{17y^2}{1024} - \dfrac{17x^2}{64} = 1$ **29.** $\dfrac{(x - 4)^2}{4} - \dfrac{y^2}{12} = 1$

31. $\dfrac{(y - 5)^2}{16} - \dfrac{(x - 4)^2}{9} = 1$ **33.** $\dfrac{y^2}{9} - \dfrac{4(x - 2)^2}{9} = 1$

35. $\dfrac{(y - 2)^2}{4} - \dfrac{x^2}{4} = 1$ **37.** $\dfrac{(x - 2)^2}{1} - \dfrac{(y - 2)^2}{1} = 1$

39. $\dfrac{(x - 3)^2}{9} - \dfrac{(y - 2)^2}{4} = 1$ **41.** $(3300, -2750)$

43. $\left(12\left(\sqrt{5} - 1\right), 0\right) \approx (14.83, 0)$ **45.** Ellipse

47. Parabola   **49.** Hyperbola   **51.** Circle

53. False. $b = 0$ yields an undefined term. For the trivial solution of two intersecting lines to occur, the standard form of the equation of the hyperbola would be equal to zero.

$\dfrac{(x - h)^2}{a^2} - \dfrac{(y - k)^2}{b^2} = 0$ or $\dfrac{(y - k)^2}{a^2} - \dfrac{(x - h)^2}{b^2} = 0$.

55. Answers will vary.   **57.** $0, \pm i$ **59.** $\frac{5}{4}$

61. (a) -24 (b) 144 (c) $x^2 - 18x + 56$

63. (a) -3 (b) $-\frac{87}{16}$ (c) $139 - 2\sqrt{3}$

Section 6.4 *(page 449)*

1. $(4, 0)$ **3.** $\left(\dfrac{6 + \sqrt{3}}{2}, \dfrac{6\sqrt{3} - 1}{2}\right)$

5. $\dfrac{(y')^2}{2} - \dfrac{(x')^2}{2} = 1$

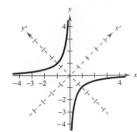

7. $\dfrac{(x')^2}{1/3} - \dfrac{(y')^2}{1/5} = 1$

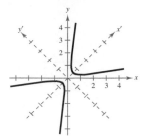

9. $\dfrac{(x' - 3\sqrt{2})^2}{16} - \dfrac{(y' - \sqrt{2})^2}{16} = 1$

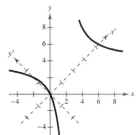

11. $\dfrac{(x')^2}{6} + \dfrac{(y')^2}{3/2} = 1$

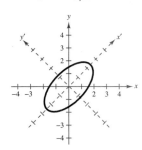

13. $x' = -(y')^2$

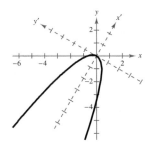

15. $y' = \frac{1}{6}(x')^2 - \frac{1}{3}x'$

17.

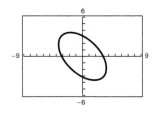

$\theta = 45°$

19.

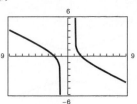

$\theta = 26.57°$

21.

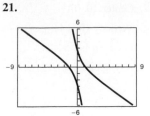

$\theta = 31.72°$

23. (e)   **25.** (b)   **27.** (d)

29. (a) Parabola

(b) $y = \dfrac{24x + 40 \pm \sqrt{3000x + 1600}}{18}$

(c)

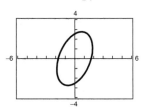

31. (a) Ellipse

(b) $y = \dfrac{8x \pm \sqrt{-356x^2 + 1260}}{14}$

(c)

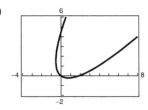

33. (a) Hyperbola

(b) $y = \dfrac{6x \pm \sqrt{56x^2 + 80x - 440}}{-10}$

(c)

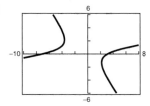

35. (a) Parabola

(b) $y = \dfrac{-4x + 1 \pm \sqrt{72x + 49}}{8}$

(c)

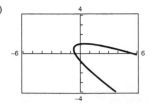

37.

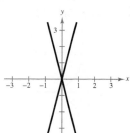

39.

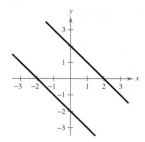

(b)

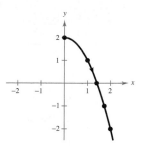

41. $(2, 2), (2, 4)$ **43.** $(-8, 12)$ **45.** $(0, 8), (12, 8)$

47. $(0, 4)$ **49.** $\left(1, \sqrt{3}\right), \left(1, -\sqrt{3}\right)$ **51.** No solution

53. $(-3, 0), \left(0, \frac{3}{2}\right)$

55. True. The discriminant will be greater than zero.

57. Answers will vary.

(c)

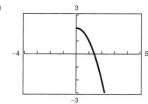

59.

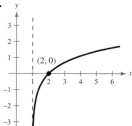

Asymptote at $y = 0$, y-intercept at $y = \frac{1}{2}$

(d) $y = 2 - x^2$

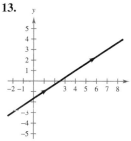

61.

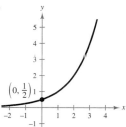

Asymptote at $x = 1$, x-intercept at $x = 2$

The graph is an entire parabola rather than just the right half.

11.

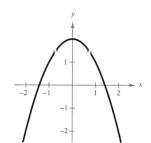

$y = -4x$

13.

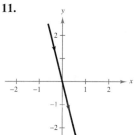

$y = \frac{2}{3}x - \frac{5}{3}$

63. $x = -1.752$ **65.** $x = 0.002$

Section 6.5   *(page 456)*

1. (c)   **3.** (b)   **5.** (a)   **7.** (f)

9. (a)

t	0	1	2	3	4
x	0	1	1.414	1.732	2
y	2	1	0	-1	-2

15.

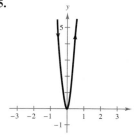

$y = 16x^2$

17.

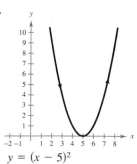

$y = (x - 5)^2$

19.

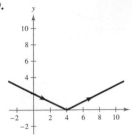

$$y = \tfrac{1}{2}|x - 4|$$

21.

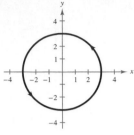

$$x^2 + y^2 = 9$$

23.

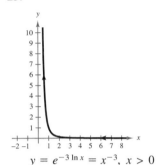

$$y = e^{-3 \ln x} = x^{-3}, \ x > 0$$

25.

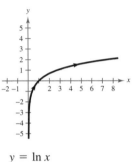

$$y = \ln x$$

27.

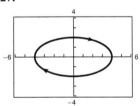

29.

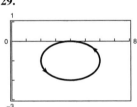

31.

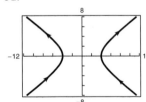

33.

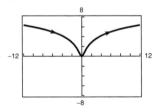

35. Each curve represents a portion of the line $y = 2x + 1$.

	Domain	*Orientation*
(a)	$(-\infty, \infty)$	Left to right
(b)	$[-1, 1]$	Depends on θ
(c)	$(0, \infty)$	Right to left
(d)	$(0, \infty)$	Left to right

37. $y - y_1 = \dfrac{y_2 - y_1}{x_2 - x_1}(x - x_1)$

39. $\dfrac{(x - h)^2}{a^2} + \dfrac{(y - k)^2}{b^2} = 1$

41. $x = 5t$
   $y = -2t$

43. $x = 2 + 4 \cos \theta$
   $y = 1 + 4 \sin \theta$

45. $x = 5 \cos \theta$
   $y = 3 \sin \theta$

47. $x = t, \ y = 3t - 2$
   $x = t^3, \ y = 3t^3 - 2$

49.

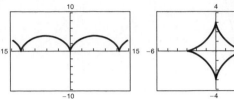

51.

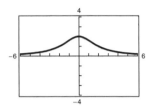

53.

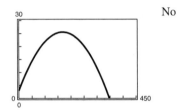

55. (b)   **57.** (d)

59. (a) $x = (146.67 \cos \theta)t$
   $y = 3 + (146.67 \sin \theta)t - 16t^2$

   (b) $x = 141.7t$
   $y = 3 + 38.0t - 16t^2$

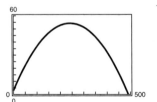

No

   (c) $x = 135.0t$
   $y = 3 + 57.3t - 16t^2$

Yes

   (d) About $19.4°$

61. True   **63.** Only the direction of the curve would change.

65. $x = 3 \pm \sqrt{5}$ **67.** $x = \pm\sqrt{9 + \sqrt{63}}, \pm\sqrt{9 - \sqrt{63}}$

69.

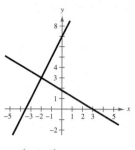

$(-2, 3)$

71.

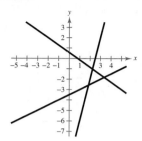

No solution

Section 6.6 *(page 463)*

1. $(0, 4)$ **3.** $\left(\dfrac{\sqrt{2}}{2}, \dfrac{\sqrt{2}}{2}\right)$

5.

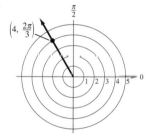

$\left(-4, -\dfrac{\pi}{3}\right), \left(-4, \dfrac{5\pi}{3}\right), \left(4, -\dfrac{4\pi}{3}\right)$

7.

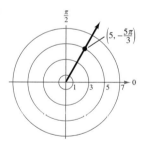

$\left(5, \dfrac{\pi}{3}\right), \left(-5, \dfrac{4\pi}{3}\right), \left(-5, -\dfrac{2\pi}{3}\right)$

9.

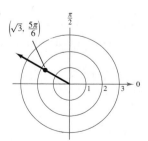

$\left(-\sqrt{3}, \dfrac{11\pi}{6}\right), \left(\sqrt{3}, -\dfrac{7\pi}{6}\right), \left(-\sqrt{3}, -\dfrac{\pi}{6}\right)$

11.

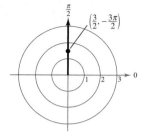

$\left(\dfrac{3}{2}, \dfrac{\pi}{2}\right), \left(-\dfrac{3}{2}, \dfrac{3\pi}{2}\right), \left(-\dfrac{3}{2}, -\dfrac{\pi}{2}\right)$

13.

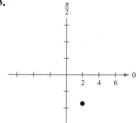

$(2, -2\sqrt{3})$

15.

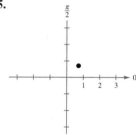

$\left(\dfrac{\sqrt{2}}{2}, \dfrac{\sqrt{2}}{2}\right)$

17.

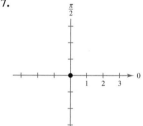

$(0, 0)$

19.

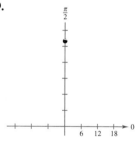

$(0, 32)$

21.

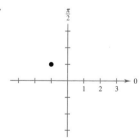

$(-1.004, 0.996)$

23. $\left(-\sqrt{2}, \sqrt{2}\right)$ **25.** $(-1.204, -4.336)$

27.

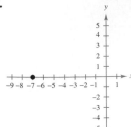

$(7, \pi), (-7, 0)$

29.

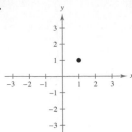

$\left(\sqrt{2}, \dfrac{\pi}{4}\right), \left(-\sqrt{2}, \dfrac{5\pi}{4}\right)$

31.

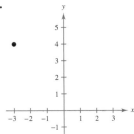

$(5, 2.214), (-5, 5.356)$

33.

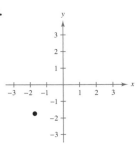

$\left(\sqrt{6}, \dfrac{5\pi}{4}\right), \left(-\sqrt{6}, \dfrac{\pi}{4}\right)$

35.

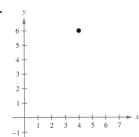

$\left(2\sqrt{13}, 0.983\right), \left(-2\sqrt{13}, 4.124\right)$

37. $\left(\sqrt{13}, -0.588\right)$ **39.** $\left(\sqrt{7}, 0.857\right)$ **41.** $\left(\dfrac{17}{6}, 0.490\right)$

43. $\left(5, -\dfrac{\pi}{2}\right)$ **45.** (a) $r = 7$ (b) $r = a$

47. (a) $r = 2a \cos \theta$ (b) $r = 2a \sin \theta$

49. (a) $r = 12 \sec \theta$ (b) $r = a \sec \theta$

51. (a) $r^2 = 4 \csc \theta \sec \theta = 8 \csc 2\theta$

   (b) $r^2 = \frac{1}{2} \csc \theta \sec \theta = \csc 2\theta$

53. (a) $r = \tan^2 \theta \sec \theta$ (b) $r = \cot^2 \theta \csc \theta$

55. $x^2 + y^2 - 4y = 0$ **57.** $\sqrt{3}x - 3y = 0$

59. $x^2 + y^2 = 16$ **61.** $y = -3$ **63.** $(x^2 + y^2)^3 = x^2$

65. $(x^2 + y^2)^2 = 6x^2y - 2y^3$ **67.** $y^2 = 2x + 1$

69. $4x^2 - 5y^2 - 36y - 36 = 0$

71. The graph is a circle centered at the origin with a radius of 3; $x^2 + y^2 = 9$.

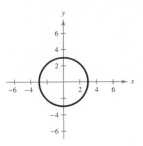

73. The graph consists of all points on the line that make an angle of $\pi/4$ with the positive x-axis; $x - y = 0$.

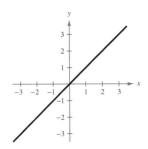

75. The graph is not evident by simple inspection; $x - 3 = 0$.

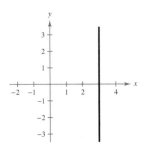

77. True. Because r is a directed distance, (r, θ) can be represented by $(-r, \theta \pm (2n + 1)\pi)$, so $|r| = |-r|$.

79. (a) Answers will vary.

   (b) The points lie on a line.

   $$d = \sqrt{r_1^2 + r_2^2 - 2r_1r_2} = |r_1 - r_2|$$

   (c) $d = \sqrt{r_1^2 + r_2^2}$ (Pythagorean Theorem)

   (d) Answers will vary. The distance formulas should give the same results.

81. $\ln 19 \approx 2.944$ **83.** $\log 84 \approx 1.924$

85. $e^4 \approx 54.598$ **87.** $\dfrac{\pi}{6} + n\pi$

89. $\dfrac{\pi}{3} + n\pi, \dfrac{2\pi}{3} + n\pi$ **91.** $\dfrac{\pi}{2} + n\pi$

Section 6.7   *(page 472)*

1. Rose curve   **3.** Circle   **5.** Rose curve

7. Polar axis   **9.** $\theta = \dfrac{\pi}{2}$ **11.** $\theta = \dfrac{\pi}{2}$

13. $\theta = \dfrac{\pi}{2}$, polar axis, pole   **15.** Pole

17. Maximum: $|r| = 20$ when $\theta = \dfrac{3\pi}{2}$

   Zero: $r = 0$ when $\theta = \dfrac{\pi}{2}$

19. Maximum: $|r| = 4$ when $\theta = 0, \dfrac{\pi}{3}, \dfrac{2\pi}{3}, \pi$

   Zeros: $r = 0$ when $\theta = \dfrac{\pi}{6}, \dfrac{\pi}{2}, \dfrac{5\pi}{6}$

21.

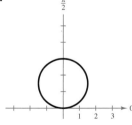

23.

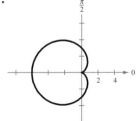

25.

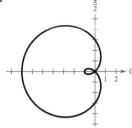

27.

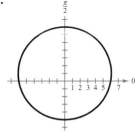

29.

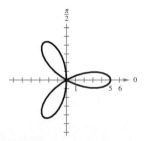

31.

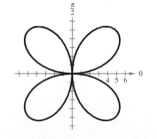

33.

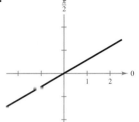

35.

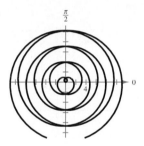

37.

39.

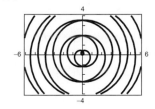

$-10\pi \le \theta < 10\pi$

41.

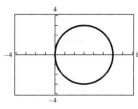

$0 \le \theta < \pi$

43.

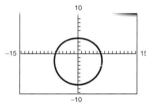

$0 \le \theta < 2\pi$

45.

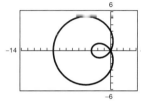

$0 \le \theta < 2\pi$

47.

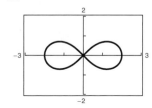

$0 \le \theta < \dfrac{\pi}{2}$

49.

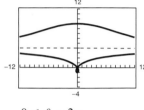

$-2\pi \le \theta < 2\pi$

51.

$0 \le \theta < \pi$

53.

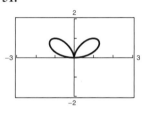

$0 \le \theta < 2\pi$

55.

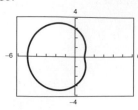

$0 \leq \theta < 2\pi$

57.

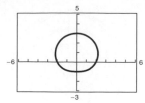

$0 \leq \theta < 2\pi$

59.

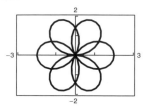

$0 \leq \theta < 4\pi$

61.

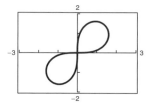

$0 \leq \theta < \dfrac{\pi}{2}$

63.

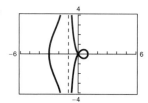

65.

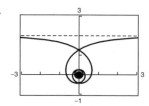

67. False. When the coordinates are substituted into the equation you get $6 = 4$, which is a false statement.

69. False. The rose curve will have 5 petals.

71. (a)

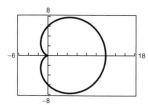

(b)

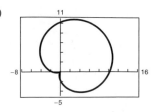

(c)

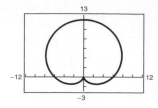

$r = 6(1 + \sin \theta)$

ϕ rotates the graph around the pole.

73. (a), (b), and (c) Answers will vary.

75. (a) $r = 4 \sin\left(\theta - \dfrac{\pi}{6}\right) \cos\left(\theta - \dfrac{\pi}{6}\right)$

(b) $r = -4 \sin \theta \cos \theta$

(c) $r = 4 \sin\left(\theta - \dfrac{2\pi}{3}\right) \cos\left(\theta - \dfrac{2\pi}{3}\right)$

(d) $r = 4 \sin \theta \cos \theta$

77.

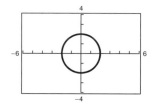

$k = 0$: circle

$k = 1$: convex limaçon

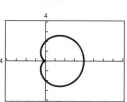

$k = 2$: cardioid

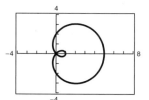

$k = 3$: limaçon with inner loop

79. $\dfrac{\sqrt{2}}{10}$ **81.** $\dfrac{7\sqrt{2}}{10}$

83. $\sin 2u = -\dfrac{24}{25}$

$\cos 2u = -\dfrac{7}{25}$

$\tan 2u = \dfrac{24}{7}$

Section 6.8   *(page 478)*

1.
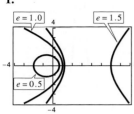

3.
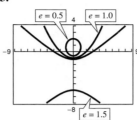

5. (b)   **7.** (d)   **9.** Parabola   **11.** Ellipse

13. Ellipse   **15.** Hyperbola   **17.** Hyperbola

19.
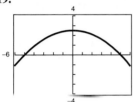

Parabola

21.
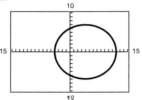

Ellipse

23.
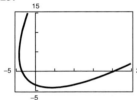

25.
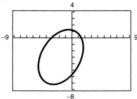

27. $r = \dfrac{1}{1 - \cos \theta}$ **29.** $r = \dfrac{1}{2 + \sin \theta}$

31. $r = \dfrac{2}{1 + 2 \cos \theta}$ **33.** $r = \dfrac{2}{1 - \sin \theta}$

35. $r = \dfrac{16}{5 + 3 \cos \theta}$ **37.** $r = \dfrac{9}{4 - 5 \sin \theta}$

39. Answers will vary.

41. $r = \dfrac{9.2931 \times 10^7}{1 - 0.0167 \cos \theta}$

Perihelion: 9.1405×10^7 miles

Aphelion: 9.4509×10^7 miles

43. $r = \dfrac{7977.2}{1 - 0.937 \cos \theta}$

11,008 miles

45. False. The equation can be rewritten as

$$r = \frac{-4/3}{1 + \sin \theta}.$$

Because ep is negative, you know that e is positive and p is negative and that p represents the distance between the pole and the directrix, so the directrix has to be below the pole.

47. Answers will vary.   **49.** $r^2 = \dfrac{24{,}336}{169 - 25 \cos^2 \theta}$

51. $-\frac{1}{2}$ **53.** $-\frac{1}{2}$

55.
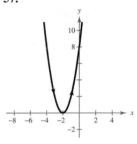

$y = \dfrac{-x + 5}{2}$

57.
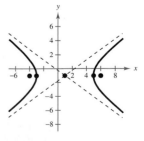

$y = 2(x + 2)^2$

Review Exercises   *(page 481)*

1. Hyperbola   **3.** $(x - 4)^2 = -8(y - 2)$

5. $(y - 2)^2 = 12x$ **7.** $2x + y - 2 = 0$; $(1, 0)$

9. About 19.6 meters   **11.** $\dfrac{(x - 2)^2}{25} + \dfrac{y^2}{21} = 1$

13. $\dfrac{(x - 2)^2}{4} + (y - 1)^2 = 1$ **15.** 3 feet atop the pillars

17. Center: $(1, -4)$ **19.** Center: $(-2, 1)$

Vertices: Vertices:

$(1, 0), (1, -8)$ $(-2, 11), (-2, -9)$

Foci: $\left(1, -4 \pm \sqrt{7}\right)$ Foci: $\left(-2, 1 \pm \sqrt{19}\right)$

Eccentricity: $\dfrac{\sqrt{7}}{4}$ Eccentricity: $\dfrac{\sqrt{19}}{10}$

21. $\dfrac{(x + 2)^2}{64} - \dfrac{(y - 3)^2}{36} = 1$ **23.** $\dfrac{5(x - 4)^2}{16} - \dfrac{5y^2}{64} = 1$

25. Center: $(1, -1)$

Vertices: $(5, -1), (-3, -1)$

Foci: $(6, -1), (-4, -1)$

Asymptotes:

$y = -1 \pm \frac{3}{4}(x - 1)$

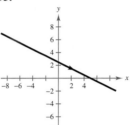

27. Center: $(3, -5)$

Vertices: $(7, -5), (-1, -5)$

Foci: $\left(3 \pm 2\sqrt{5}, -5\right)$

Asymptotes:

$y = -5 \pm \frac{1}{2}(x - 3)$

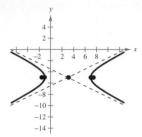

29. 72 miles   **31.** Ellipse

33. $\dfrac{(x')^2}{8} - \dfrac{(y')^2}{8} = 1$ **35.** $\dfrac{(x')^2}{3} + \dfrac{(y')^2}{2} = 1$

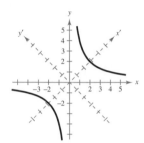   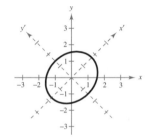

37. (a) Parabola

(b) $y = \dfrac{8x - 5 \pm \sqrt{(8x - 5)^2 - 4(16x^2 - 10x)}}{2}$

(c)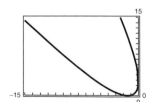

39. (a) Parabola

(b) $y =$

$\dfrac{-\left(2x - 2\sqrt{2}\right) \pm \sqrt{\left(2x - 2\sqrt{2}\right)^2 - 4\left(x^2 + 2\sqrt{2}x + 2\right)}}{2}$

(c)

41. $(-10, 12)$ **43.** $x = 3, y = 0$ **45.** $x = \dfrac{3\sqrt{3}}{2}, y = \dfrac{1}{2}$

47.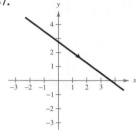

$y = -\dfrac{3}{4}x + \dfrac{11}{4}$

49.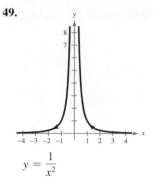

$y = \dfrac{1}{x^2}$

51.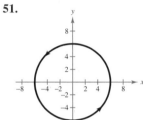

$x^2 + y^2 = 36$

53.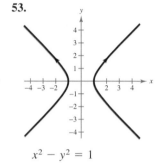

$x^2 - y^2 = 1$

55. $x = -3 + 4\cos\theta$

$y = 4 + 3\sin\theta$

57.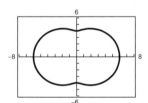

59.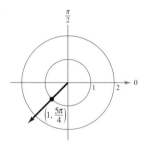

$\left(1, -\dfrac{3\pi}{4}\right), \left(-1, \dfrac{\pi}{4}\right), \left(-1, -\dfrac{7\pi}{4}\right)$

61.

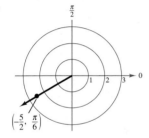

$$\left(-\frac{5}{2}, -\frac{11\pi}{6}\right), \left(\frac{5}{2}, \frac{7\pi}{6}\right), \left(\frac{5}{2}, -\frac{5\pi}{6}\right)$$

63.

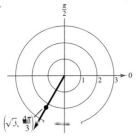

$$\left(\sqrt{5}, -\frac{2\pi}{3}\right), \left(-\sqrt{5}, \frac{\pi}{3}\right), \left(-\sqrt{5}, -\frac{5\pi}{3}\right)$$

65. **67.**

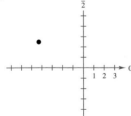

$$\left(-\frac{5\sqrt{3}}{2}, \frac{5}{2}\right) \qquad (0, -12)$$

69. $\left(-9, \frac{\pi}{2}\right), \left(9, \frac{3\pi}{2}\right)$ **71.** $\left(-5\sqrt{2}, \frac{3\pi}{4}\right), \left(5\sqrt{2}, \frac{7\pi}{4}\right)$

73. $x^2 + y^2 = 25$ **75.** $x^2 + y^2 = 3x$

77. $(x^2 + y^2)^2 - x^2 + y^2 = 0$

79. $4x^2 + 3y^2 - 4y - 4 = 0$ **81.** $r = 3$

83. $r = 6 \csc \theta$ **85.** $r = 4 \cos \theta$ **87.** $r^2 = 5 \sec \theta \csc \theta$

89. **91.**

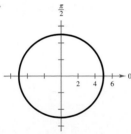 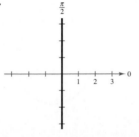

93. **95.**

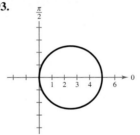 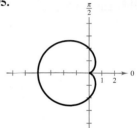

97. **99.**

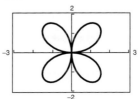 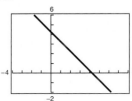

Rose curve with Straight line, $y = 4 - x$
four petals

101.

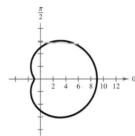

Symmetry: Polar axis

Maximum: $|r| = 9$ when $\theta = 0, 2\pi$

Zeros of r: None

103.

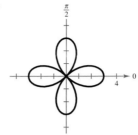

Symmetry: Pole, polar axis, and the line $\theta = \dfrac{\pi}{2}$

Maximum: $|r| = 3$ when $\theta = 0, \dfrac{\pi}{2}, \pi, \dfrac{3\pi}{2}, 2\pi$

Zeros of r: when $\theta = \dfrac{\pi}{4}, \dfrac{3\pi}{4}, \dfrac{5\pi}{4}, \dfrac{7\pi}{4}$

105.

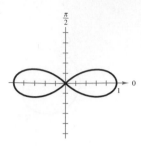

Symmetry: Pole, polar axis, and the line $\theta = \dfrac{\pi}{2}$

Maximum: $|r| = 1$ when $\theta = 0, \pi, 2\pi$

Zeros of r: when $\theta = \dfrac{\pi}{4}, \dfrac{3\pi}{4}, \dfrac{5\pi}{4}, \dfrac{7\pi}{4}$

107. Limaçon

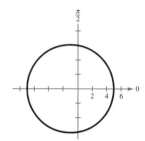

109. Limaçon

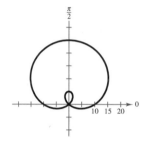

111. Rose curve

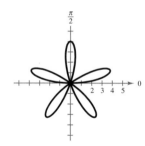

113. Lemniscate

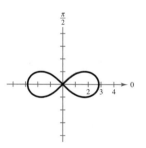

115. Parabola

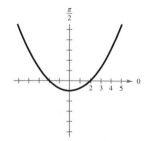

117. Ellipse

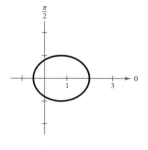

119. Ellipse

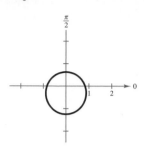

121. $r = 16 \sin \theta$ **123.** $r = \dfrac{5}{3 - 2 \cos \theta}$

125. $r = \dfrac{1.512}{1 - 0.092 \cos \theta}$

Perihelion: 1.3847 astronomical units

Aphelion: 1.6653 astronomical units

127. False. The following are two sets of parametric equations for the line.

$x = t, \quad y = 3 - 2t$

$x = 3t, \quad y = 3 - 6t$

129. (a) Vertical translation

(b) Horizontal translation

(c) Reflection in the y-axis

(d) Parabola opens more slowly.

131. 5; The ellipse becomes more circular and approaches a circle of radius 5.

Chapter Test   *(page 486)*

1.

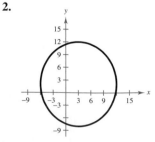

Vertex: $(0, 0)$

Focus: $(2, 0)$

2.

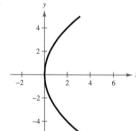

Vertices: $(3, 12), (3, -8)$

Foci: $\left(3, 2 \pm \sqrt{19}\right)$

3.

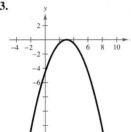

Vertex: $(3, 0)$

Focus: $\left(3, -\frac{1}{2}\right)$

5. $(y + 2)^2 = 8(x - 6)$

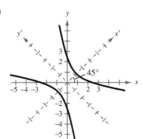

6. $\dfrac{(x + 6)^2}{16} + \dfrac{(y - 3)^2}{49} = 1$ **7.** $\dfrac{y^2}{9} - \dfrac{x^2}{4} = 1$

8. (a) $45°$

(b)

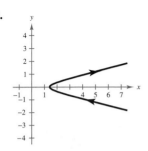

10.

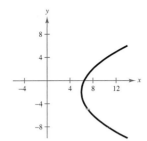

$y = \dfrac{\pm\sqrt{x^2 - 2}}{4}, \; x \ge \sqrt{2}$

11. $x = t + 4, \; y = \dfrac{t}{4} - 4$

$x = 2t, \; y = \dfrac{1}{2}t - 5$

(Solutions are not unique.)

4.

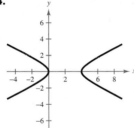

Vertices: $(0, 0), (4, 0)$

Foci: $\left(2 \pm \sqrt{5}, 0\right)$

12. $\left(-7, 7\sqrt{3}\right)$

13. $\left(2\sqrt{2}, \dfrac{5\pi}{4}\right), \left(-2\sqrt{2}, \dfrac{\pi}{4}\right), \left(2\sqrt{2}, -\dfrac{3\pi}{4}\right)$

14. $r = 12 \sin \theta$

15.   **16.**

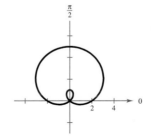

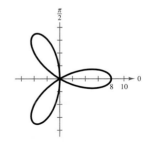

Limaçon Rose curve

17.

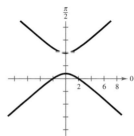

Hyperbola

18. $r = \dfrac{1}{1 + \frac{1}{4} \sin \theta}$

19. $r = \dfrac{34{,}478{,}231}{1 + 0.2056 \sin \theta}$

Perihelion: 28,598,400 miles

Aphelion: 43,401,600 miles

Cumulative Test for Chapters 4–6
(page 487)

1. $3 - 5i$ **2.** $-2 - 3i$ **3.** $5 - 12i$ **4.** 4

5. $\dfrac{8}{5} + \dfrac{4}{5}i$ **6.** $-2, \pm 2i$ **7.** $-7, 0, 0, 3$

8. $f(x) = x^4 + 3x^3 - 11x^2 + 9x + 70$

9. $2\sqrt{2}\left(\cos\dfrac{3\pi}{4} + i \sin\dfrac{3\pi}{4}\right)$ **10.** $-12\sqrt{3} + 12i$

11. $-8 + 8\sqrt{3}i$

12. $\cos 0 + i \sin 0 = 1$

$\cos\dfrac{2\pi}{3} + i \sin\dfrac{2\pi}{3} = -\dfrac{1}{2} + \dfrac{\sqrt{3}}{2}i$

$\cos\dfrac{4\pi}{3} + i \sin\dfrac{4\pi}{3} = -\dfrac{1}{2} - \dfrac{\sqrt{3}}{2}i$

13. $3\left(\cos\dfrac{\pi}{8} + i\sin\dfrac{\pi}{8}\right)$

$3\left(\cos\dfrac{5\pi}{8} + i\sin\dfrac{5\pi}{8}\right)$

$3\left(\cos\dfrac{9\pi}{8} + i\sin\dfrac{9\pi}{8}\right)$

$3\left(\cos\dfrac{13\pi}{8} + i\sin\dfrac{13\pi}{8}\right)$

14. Reflect f in x-axis and y-axis, shift f 3 units to the right.

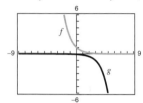

15. Reflect f in x-axis and shift f 4 units upward.

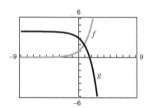

16. 1.991   **17.** -0.067 **18.** 1.717   **19.** 0.281

20. 0.302   **21.** -1.733 **22.** -4.087

23. $\ln(x - 4) + \ln(x + 4) - 4\ln x$

24. $\ln\dfrac{x^2}{\sqrt{x + 5}},\ x > 0$

25. Hyperbola

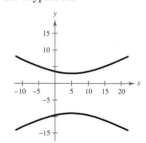

26. Ellipse

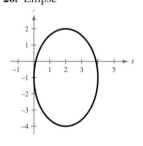

27. Hyperbola

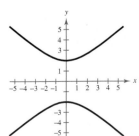

28. Circle

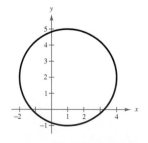

29. $(x - 2)^2 = -\frac{4}{3}(y - 3)$

30. $\dfrac{(x - 1)^2}{25} + \dfrac{(y - 4)^2}{4} = 1$ **31.** $\dfrac{(y + 4)^2}{4} - \dfrac{x^2}{16/3} = 1$

32. $\dfrac{(y - 2)^2}{4/5} - \dfrac{x^2}{16/5} = 1$

33.

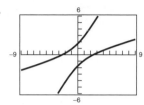

$\theta \approx 38°$

34. $(0, 6), (12, 6)$

35.

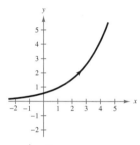

$y = \frac{1}{2}e^{x/2}$

36.

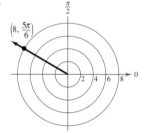

$\left(8, -\dfrac{7\pi}{6}\right), \left(-8, -\dfrac{\pi}{6}\right), \left(-8, \dfrac{11\pi}{6}\right)$

37.

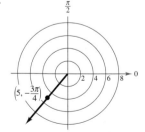

$\left(5, \dfrac{5\pi}{4}\right), \left(-5, -\dfrac{7\pi}{4}\right), \left(-5, \dfrac{\pi}{4}\right)$

38.

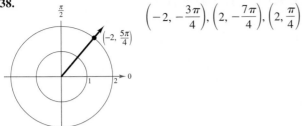

$$\left(-2, -\frac{3\pi}{4}\right), \left(2, -\frac{7\pi}{4}\right), \left(2, \frac{\pi}{4}\right)$$

39.

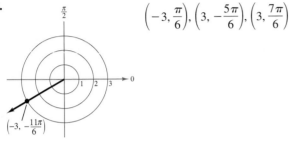

$$\left(-3, \frac{\pi}{6}\right), \left(3, -\frac{5\pi}{6}\right), \left(3, \frac{7\pi}{6}\right)$$

40. $r = \dfrac{5}{8\cos\theta + 3\sin\theta}$ **41.** $9x^2 - 16y^2 + 20x + 4 = 0$

42.

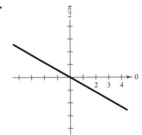

Line

43.

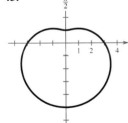

Dimpled limaçon

44.

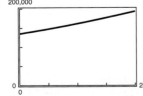

Limaçon

45. $r = \dfrac{14}{4 - 3\cos\theta}$ **46.** \$108.63

47.

about 2009; Answers will vary.

Appendix B

Section B.1   *(page A16)*

1. (a) \$1.109   (b) [\$0.999, \$1.189]

3. Quiz 1:

```
            ×
            ×       × × ×                    15
          × × ×     × × × ×
          × × ×     × × × ×
  ×   ×   × × ×     × × × × ×     ×
  +---+---+---+---+---+---+---+---+---+
  10  12  14  16  18  20  22  24
```

Quiz 2:

```
                      × ×
                      × ×
                      × ×
                    × × ×
                  × × × ×
                  × × × ×
      ×           × × × ×
  ×   ×       ×   × × × × ×
  +---+---+---+---+---+---+---+
  14  16  18  20  22  24  26
```

22 and 23

5.

Stems	Leaves
7	0 5 5 5 7 7 8 8 8
8	1 1 1 1 2 3 4 5 5 5 5 7 8 9 9 9
9	0 2 8
10	0 0

7.

Stems	Leaves
0	89 66 67 65 80 98 62 93
1	09 01 46 24 96 90
2	92 55 40 61
3	68 35
4	12 96 80 38
5	81 18 50 70 66
6	44 00 34 01
7	61 66 00
8	11 57 41 90
9	
10	
11	60 59 33
12	92
13	19 17 37
27	22
31	32
46	80
65	14

9. (a)

Interval	Tally
90–109	ℍℍ
110–129	ℍℍ ℍℍ I
130–149	ℍℍ IIII
150–169	ℍℍ I
170–189	I
190–209	III

(b)

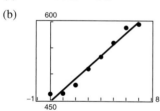

Amount contributed (in dollars)

11. (a)

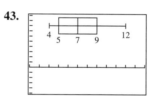

(b) Answers will vary.   (c) $y = 4x + 134$

(d) 2000: 214; 2002: 222; 2005: 234

13. (a) $P = 21.91t + 447$

(b)

The model fits the data well.

(c)

Year	1990	1991	1992	1993
P (actual)	461.4	462.4	478.6	509.5
P (model)	447	468.9	490.8	512.7

Year	1994	1995	1996	1997
P (actual)	532.9	559.9	589.1	595.6
P (model)	534.6	556.6	578.5	600.4

Section B.2   *(page A24)*

1. Mean: 8.86; median: 8; mode: 7

3. Mean: 10.29; median: 8; mode: 7

5. Mean: 9; median: 8; mode: 7

7. The mean is sensitive to extreme values.

9. Mean: $67.14; median: $65.35

11. Mean: 3.07; median: 3; mode: 3

13. One possibility: {4, 4, 10}

15. The median gives the most representative description.

17. $\bar{x} = 6$, $v = 10$, $\sigma = 3.16$

19. $\bar{x} = 2$, $v = \frac{4}{3}$, $\sigma = 1.15$ **21.** $\bar{x} = 4$, $v = 4$, $\sigma = 2$

23. $\bar{x} = 47$, $v = 226$, $\sigma = 15.03$ **25.** 3.42

27. 101.55   **29.** 1.65

31. (a) $\bar{x} = 12$; $\sigma = 2.83$ (b) $\bar{x} = 20$; $\sigma = 2.83$

(c) $\bar{x} = 12$; $\sigma = 1.41$ (d) $\bar{x} = 9$; $\sigma = 1.41$

33. $\bar{x} = 12$ and $|x_i - 12| = 8$ for all x_i

35. It will increase the mean by 5, but the standard deviation will not change.

37. First histogram

39. 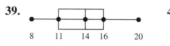   **41.**

43. 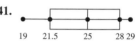   **45.**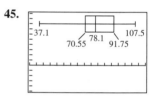

Appendix C   *(page A29)*

1. 3   **3.** 7   **5.** 4   **7.** 20   **9.** 4   **11.** 3

13. $\frac{4}{3}$ **15.** 20   **17.** No solution   **19.** $x < 2$

21. $x < 9$ **23.** $x \leq -14$ **25.** $x > 10$

27. $x < 4$ **29.** $x < 3$ **31.** $x \geq 2$ **33.** $x \leq -5$

35. $x < 6$

Index of Applications

Index

COMMON FORMULAS

Temperature

$$F = \frac{9}{5}C + 32$$

F = degrees Fahrenheit

C = degrees Celsius

Distance

$$d = rt$$

d = distance traveled

t = time

r = rate

Simple Interest

$$I = Prt$$

I = interest

P = principal

r = annual interest rate

t = time in years

Compound Interest

$$A = P\left(1 + \frac{r}{n}\right)^{nt}$$

A = balance

P = principal

r = annual interest rate

n = compoundings per year

t = time in years

Coordinate Plane: Midpoint Formula

$$\left(\frac{x_1 + x_2}{2}, \frac{y_1 + y_2}{2}\right)$$

midpoint of line segment joining (x_1, y_1) and (x_2, y_2)

Coordinate Plane: Distance Formula

$$d = \sqrt{(x_2 - x_1)^2 + (y_2 - y_1)^2}$$

d = distance between points (x_1, y_1) and (x_2, y_2)

Quadratic Formula

If $p(x) = ax^2 + bx + c, a \neq 0$ and $b^2 - 4ac \geq 0$, then the real zeros of p are

$$x = \frac{-b \pm \sqrt{b^2 - 4ac}}{2a}.$$

CONVERSIONS

Length and Area

1 foot = 12 inches
1 mile = 5280 feet
1 kilometer = 1000 meters
1 kilometer $\approx$ 0.621 mile
1 meter $\approx$ 3.281 feet
1 foot $\approx$ 0.305 meter

1 yard = 3 feet
1 mile = 1760 yards
1 meter = 100 centimeters
1 mile $\approx$ 1.609 kilometers
1 meter $\approx$ 39.370 inches
1 foot $\approx$ 30.480 centimeters

1 meter = 1000 millimeters
1 centimeter $\approx$ 0.394 inch
1 inch $\approx$ 2.540 centimeters
1 acre = 4840 square yards
1 square mile = 640 acres

Volume

1 gallon = 4 quarts
1 gallon = 231 cubic inches
1 liter = 1000 milliliters
1 liter $\approx$ 1.057 quarts
1 gallon $\approx$ 3.785 liters

1 quart = 2 pints
1 gallon $\approx$ 0.134 cubic foot
1 liter = 100 centiliters
1 liter $\approx$ 0.264 gallon
1 quart $\approx$ 0.946 liter

1 pint = 16 fluid ounces
1 cubic foot $\approx$ 7.48 gallons

Weight and Mass on Earth

1 ton = 2000 pounds
1 kilogram $\approx$ 2.205 pounds

1 pound = 16 ounces
1 pound $\approx$ 0.454 kilogram

1 kilogram = 1000 grams
1 gram $\approx$ 0.035 ounce

FORMULAS FROM GEOMETRY

Triangle

$h = a \sin \theta$

$\text{Area} = \dfrac{1}{2}bh$

Laws of Cosines:

$c^2 = a^2 + b^2 - 2ab \cos \theta$

Right Triangle

Pythagorean Theorem:

$c^2 = a^2 + b^2$

Equilateral Triangle

$h = \dfrac{\sqrt{3}s}{2}$

$\text{Area} = \dfrac{\sqrt{3}s^2}{4}$

Parallelogram

$\text{Area} = bh$

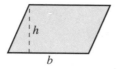

Trapezoid

$\text{Area} = \dfrac{h}{2}(a + b)$

Circle

$\text{Area} = \pi r^2$

$\text{Circumference} = 2\pi r$

Sector of Circle

$\text{Area} = \dfrac{\theta r^2}{2}$

$s = r\theta$

(θ in radians)

Circular Ring

$\text{Area} = \pi(R^2 - r^2)$

$\qquad = 2\pi pw$

(p = average radius,

w = width of ring)

Ellipse

$\text{Area} = \pi ab$

$\text{Circumference} \approx 2\pi \sqrt{\dfrac{a^2 + b^2}{2}}$

Cone

(A = area of base)

$\text{Volume} = \dfrac{Ah}{3}$

Right Circular Cone

$\text{Volume} = \dfrac{\pi r^2 h}{3}$

$\text{Lateral Surface Area} = \pi r \sqrt{r^2 + h^2}$

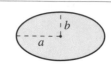

Frustum of Right Circular Cone

$\text{Volume} = \dfrac{\pi(r^2 + rR + R^2)h}{3}$

$\text{Lateral Surface Area} = \pi s(R + r)$

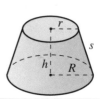

Right Circular Cylinder

$\text{Volume} = \pi r^2 h$

$\text{Lateral Surface Area} = 2\pi rh$

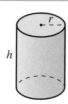

Sphere

$\text{Volume} = \dfrac{4}{3}\pi r^3$

$\text{Surface Area} = 4\pi r^2$